Introduction to Probability and Statistics

Seymour Lipschutz, Ph.D.
Professor of Mathematics
Temple University

John J. Schiller, Ph.D.
Associate Professor of Mathematics
Temple University

Schaum's Outline Series

New York Chicago San Francisco Lisbon London Madrid
Mexico City Milan New Delhi San Juan Seoul
Singapore Sydney Toronto

W

The **McGraw-Hill** Companies

SEYMOUR LIPSCHUTZ, who is presently on the mathematics faculty of Temple University, formerly taught at the Polytechnic Institute of Brooklyn and was visiting professor in the Computer Science Department of Brooklyn College. He received his Ph.D. in 1960 at the Courant Institute of Mathematical Sciences of New York University. Some of his other books in the Schaum's Outline Series are *Beginning Linear Algebra, Discrete Mathematics,* and *Linear Algebra*.

JOHN J. SCHILLER is an associate professor of Mathematics at Temple University. He received his Ph.D. at the University of Pennsylvania and has published research papers in the areas of Riemann surfaces, discrete mathematics, and mathematical biology. He has also coauthored texts in finite mathematics, precalculus, and calculus.

2 3 4 5 6 7 8 9 10 CUS/CUS 1 9 8 7 6 5 4 3 2

ISBN 978-0-07-176249-6
MHID 0-07-176249-3

Preface

Probability and statistics appear explicitly or implicitly in many disciplines, including computer and information science, physics, chemistry, geology, biology, medicine, psychology, sociology, political science, education, economics, business, operations research, and all branches of engineering.

The purpose of this book is to present an introduction to principles and methods of probability and statistics which would be useful to all individuals regardless of their fields of specialization. It is designed for use as a supplement to all current standard texts, or as a textbook in a beginning course in probability and statistics with high school algebra as the only prerequisite.

The material is divided into two parts, since the logical development is not disturbed by the division while the usefulness as a text and reference book is increased.

Part I covers descriptive statistics and elements of probability. The first chapter treats descriptive statistics which motivates various concepts appearing in the chapters on probability, and the second chapter covers sets and counting which are needed for a modern treatment of probability. Part I also includes a chapter on random variables where we define expectation, variance, and standard deviation of random variables, and where we discuss and prove Chebyshev's inequality and the law of large numbers. This is followed by a separate chapter on the binomial and normal distributions, where the central limit theorem is discussed in the context of the normal approximation to the binomial distribution.

Part II treats inferential statistics. It begins with a chapter on sampling distributions for sampling with and without replacement and for small and large samples. Then there are chapters on estimation (confidence intervals) and hypothesis testing for a single population, and then a separate chapter covering these topics for two populations. Lastly, there is a chapter on chi-square tests and analysis of variance.

Each chapter begins with clear statements of pertinent definitions, principles, and theorems together with illustrative and other descriptive material. This is followed by graded sets of solved and supplementary problems. The solved problems serve to illustrate and amplify the material, and provide the repetition of basic principles so vital to effective learning. The supplementary problems serve as a complete review of the material in the chapter.

We wish to thank many friends and colleagues for invaluable suggestions and critical review of the manuscript. We also wish to express our gratitude to the staff of McGraw-Hill, particularly to Barbara Gilson and Mary Loebig Giles, for their excellent cooperation.

SEYMOUR LIPSCHUTZ
JOHN J. SCHILLER

Temple University

Contents

PART I *Descriptive Statistics and Probability*

PART II *Inferential Statistics*

PART I: **Descriptive Statistics and Probability**

Chapter 1

Preliminary: Descriptive Statistics

1.1 INTRODUCTION

Statistics, on the one hand, means lists of numerical values; for example, the salaries of the employees of a company, or the SAT scores of the incoming students of a university. Statistics as a science, on the other hand, is the branch of mathematics which organizes, analyzes, and interprets such raw data. Statistical methods are applicable to any area of human endeavor where numerical data are collected for some type of decision-making process.

This preliminary chapter simply covers topics related to gathering and describing data called *Descriptive Statistics.* It will be used in both the first part of the text, which mainly treats Probability Theory, and the second part of the text, which mainly treats Inferential Statistics.

Real Line R

The notation **R** will be used to denote the set of real numbers, which are the numbers we use for numerical data. We assume the reader is familiar with the graphical representation of **R** as points on a straight line, as pictured in Fig. 1-1. We refer to such a line as the *real line* or the *real line* **R**.

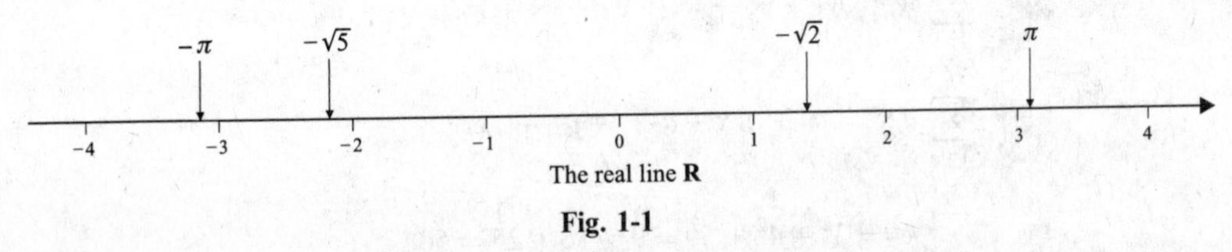

The real line **R**

Fig. 1-1

Frequently we will deal with sets of numbers called *intervals.* Specifically, for any real numbers a and b, with $a < b$, we denote and define *intervals from a to b* as follows:

$$(a, b) = \{x : a < x < b\}, \text{ open interval}$$

$$[a, b] = \{x : a \leq x \leq b\}, \text{ closed interval}$$

$$[a, b) = \{x : a \leq x < b\}, \text{ closed-open interval}$$

$$(a, b] = \{x : a < b \leq b\}, \text{ open-closed interval}$$

That is, each interval consists of all the points between a and b; the term "closed" and a bracket are used to indicate that the endpoint belongs to the interval and the term "open" and a parenthesis are used to indicate that an endpoint does not belong to the interval.

Subscript Notation, Summation Symbol

Consider a list of numerical data, say the weights of eight students. They may all be denoted by:

$$w_1, \quad w_2, \quad w_3, \quad w_4, \quad w_5, \quad w_6, \quad w_7, \quad w_8$$

The numbers $1, 2, \ldots, 8$ written below the ws are called *subscripts.* An arbitrary element in the list will be denoted by w_j. The subscript j is called an *index* because it gives the position of the element in the list. (The letters i and k are also frequently used as index symbols.)

The sum of the eight weights of the students may be expressed in the form

$$w_1 + w_2 + w_3 + w_4 + w_5 + w_6 + w_7 + w_8$$

Clearly, this expression for the sum would be very long and awkward to use if there were many more numbers in the list. Mathematics has developed a shorthand for such sums which is independent of the number of items in the list.

Summation notation uses the *summation symbol* $\sum$ (the Greek letter sigma). Specifically, given a list $x_1, x_2, \ldots, x_n$ of n numbers, its sum may be denoted by

$$\sum_{j=1}^{n} x_j \quad \text{or} \quad \textstyle\sum_{j=1}^{n} x_j$$

which is read:

The sum of the *x*-sub-*j*s as *j* goes from *1* to *n*. If the number n of items is understood we may simply write

$$\sum x_j$$

More generally, suppose $f(k)$ is an algebraic expression involving the variable k, and n_1 and n_2 are integers for which $n_1 \leq n_2$. Then we define

$$\sum_{k=n_1}^{n_2} f(k) = f(n_1) + f(n_1 + 1) + f(n_1 + 2) + \cdots + f(n_2)$$

Thus we have, for example,

$$\sum_{j=1}^{8} w_j = w_1 + w_2 + w_3 + w_4 + w_5 + w_6 + w_7 + w_8$$

$$\sum_{i=0}^{n} a_i x^i = a_0 + a_1 x + a_2 x^2 + \cdots + a_n x^n$$

$$\sum_{k=3}^{5} k^2 = 3^2 + 4^2 + 5^2 = 9 + 16 + 25 = 50$$

$$\sum a_k b_k = a_1 b_1 + a_2 b_2 + \cdots + a_n b_n$$

$$\sum (x_j - \bar{x})^2 = (x_1 - \bar{x})^2 + (x_2 - \bar{x})^2 + \cdots + (x_n - \bar{x})^2$$

(We assume the index goes from 1 to n in the last two sums.)

1.2 FREQUENCY TABLES, HISTOGRAMS

One of the first things one usually does with a large list of numerical data is to form some type of *frequency table*, where the table shows the number of times an individual item occurs or the number of items that fall within a given interval. These frequency distributions may be pictured using *histograms*. We illustrate this technique with two examples.

EXAMPLE 1.1 An apartment house has 45 apartments, with the following number of tenants:

2 1 3 5 2 2 2 1 4 2 6 2 4 3 1

2 4 3 1 4 4 2 4 4 2 2 3 1 4 2

3 1 5 2 4 1 3 2 4 4 2 5 1 3 4

Observe that the only numbers which appear in the list are 1, 2, 3, 4, 5, and 6. The frequency distribution of these numbers appears in Fig. 1-2. Specifically, column 1 lists the given numbers and column 2 gives the frequency of each number. (These frequencies can be obtained by some sort of "tally count" as in Problem 1.2.) Figure 1-2 also gives the cumulative frequency distribution. Specifically, column 3 gives the cumulative frequency of each number, which is the number of tenant numbers not exceeding the given number. The cumulative frequency is obtained by simply adding up the frequencies until the given frequency. Clearly, the last cumulative frequency number 45 is the same as the sum of all frequencies, that is, the number of apartments.

The frequency distribution in Fig. 1-2 may be pictured by a histogram shown in Fig. 1-3. A *histogram* is simply a bar graph where the height of the bar gives the number of times the given number appears in the list. Similarly, the cumulative frequency distribution could be presented as a histogram, the heights of the bars would be 8, 22, 29, ..., 45.

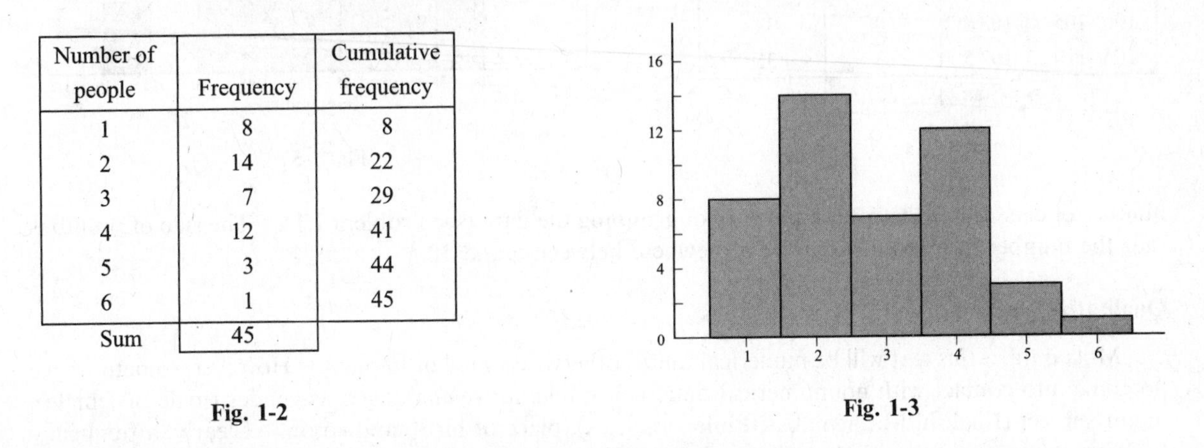

Number of people	Frequency	Cumulative frequency
1	8	8
2	14	22
3	7	29
4	12	41
5	3	44
6	1	45
Sum	45	

Fig. 1-2

Fig. 1-3

EXAMPLE 1.2 Suppose the 6:00 P.M. temperatures (in degrees Fahrenheit) for a 35-day period are as follows:

72 78 86 93 106 107 98 82 81 77 87 82

91 95 92 83 76 78 73 81 86 92 93 84

107 99 94 86 81 77 73 76 80 88 91

Rather than find the frequency distribution of each individual data item, it is more useful to construct a frequency table which counts the number of times the observed temperature falls in a given class, i.e. an interval with certain limits. This is done in Fig. 1-4.

The numbers 70, 75, 80, ... are called the *class boundaries* or *class limits*. If a data item falls on a class boundary, it is usually assigned to the higher class; for example, the number 95° was placed in the 95–100 class. Sometimes a frequency table also lists each *class value*, i.e. the midpoint of the class interval which serves as an approximation to the values in the interval.

Figure 1-5 shows the histogram which corresponds to the frequency distribution in Fig. 1-4. It also shows the *frequency polygon*, which is a line graph obtained by connecting the midpoints of the tops of the rectangles in the histogram. Observe that the line graph is extended to the class value 67.5 on the left and to 112.5 on the right. In such a case, the sum of the areas of the rectangles equals the area bounded by the frequency polygon and the x-axis.

Interval Notation, Number of Classes

The entries forming a class can be denoted using interval notation. Since a bracket indicates that a class boundary belongs to an interval, but a parenthesis means that it does not, the classes in Fig. 1-3 can be denoted by

$$[70, 75),\ [75, 80), \ldots,\ [105, 110)$$

respectively. Also, there is no fixed rule for the number of classes that should be formed for data. The fewer the number of classes, the less specific is the information displayed by the histogram, but a larger

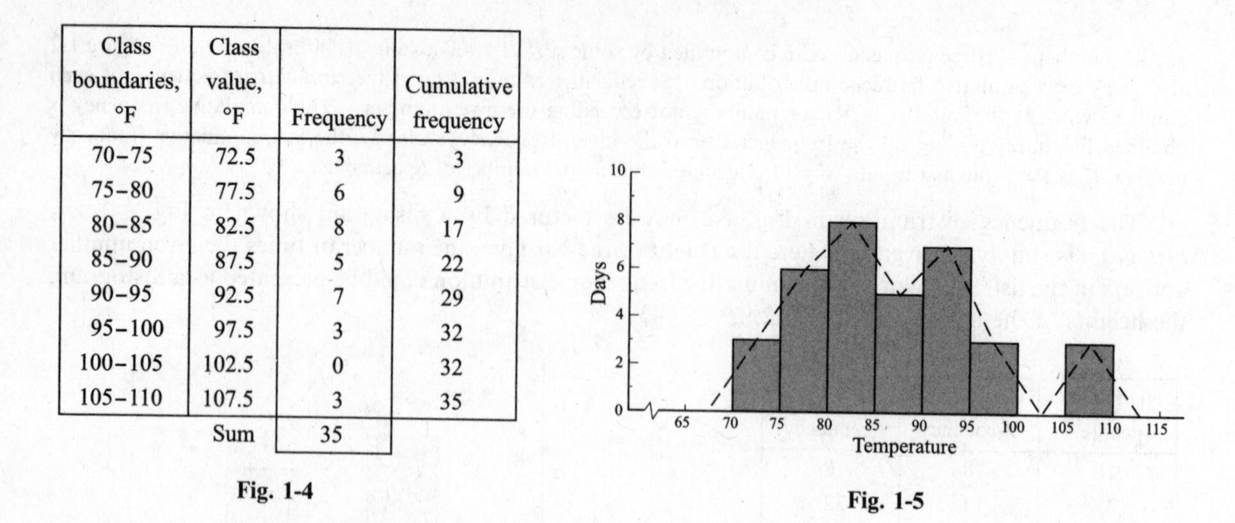

Class boundaries, °F	Class value, °F	Frequency	Cumulative frequency
70–75	72.5	3	3
75–80	77.5	6	9
80–85	82.5	8	17
85–90	87.5	5	22
90–95	92.5	7	29
95–100	97.5	3	32
100–105	102.5	0	32
105–110	107.5	3	35
	Sum	35	

Fig. 1-4

Fig. 1-5

number of classes may defeat the purpose of grouping the data (see Problem 1.1). The rule of thumb is that the number of classes should lie somewhere between 5 and 10.

Qualitative Data, Bar and Circular Graphs

Most data in this text will be numerical unless otherwise stated or implied. However, sometimes we do come into contact with nonnumerical data, called *qualitative data*, such as gender (male or female), major subject (English, Mathematics, Philosophy, ...), place of birth, and so on. Clearly, a frequency table can be formed for such data (but a cumulative frequency table would have no meaning). Instead of a histogram, such data may be pictured as (*a*) a *bar graph* and/or (*b*) a *circular graph* (also called a *pie graph* or *pie chart*).

EXAMPLE 1.3 Suppose the students at a small Community College in Philadelphia are partitioned into five groups according to their home address: (1) Philadelphia, (2) suburbs of Philadelphia, (3) Pennsylvania (outside Philadelphia and its suburbs), (4) New Jersey, and (5) elsewhere; and suppose the following is the frequency distribution for the college during some semester:

	Philadelphia	Suburbs	PA	NJ	Elsewhere	Sum
Number of students:	225	100	60	75	40	500

Draw (*a*) the bar graph, and (*b*) the circular graph of the data.

(*a*) Figure 1-6 shows a bar graph for the data. The length of each bar is proportional to the number of students living in the area. The bar graph is not a horizontal histogram. Specifically, the order of the data can be interchanged in the bar graph, e.g. putting New Jersey before Pennsylvania, without essentially changing the graph. This cannot be done with a histogram, since the data is numerical and has a given order. (A histogram may be viewed as a special kind of bar graph.)

(*b*) Figure 1-7 shows a circular graph for the data. If d_j is the number of degrees in a "slice" (sector) corresponding to a group with n_j items out of SUM items, then

$$d_j = (n_j/\text{SUM})\,(360)$$

For example, Philadelphia is assigned a slice with

$$[(225)/(500)](360) = 162 \text{ degrees}$$

Clearly, the sum of the degrees assigned to the data must equal 360 degrees.

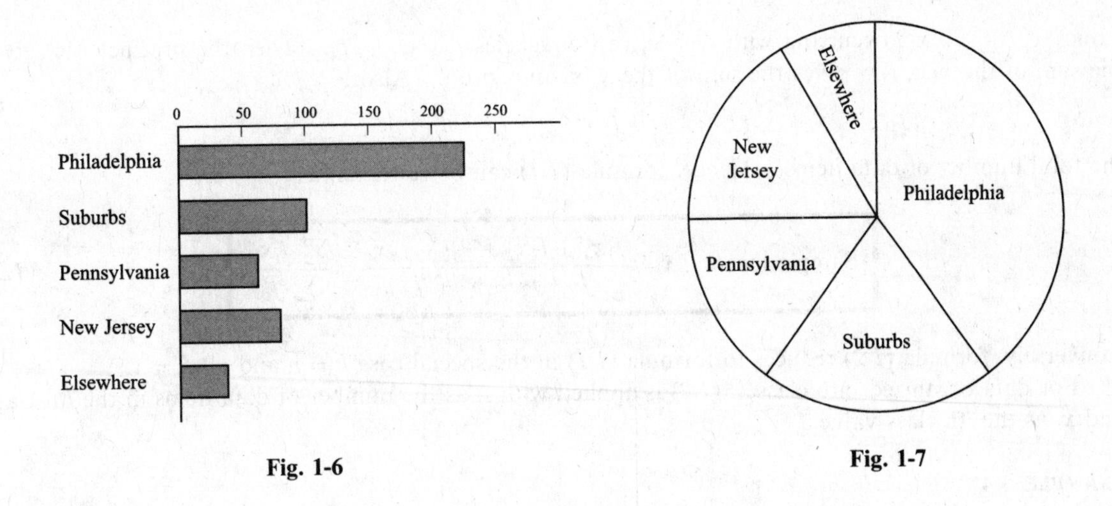

Fig. 1-6

Fig. 1-7

1.3 MEASURES OF CENTRAL TENDENCY: MEAN AND MEDIAN

There are various ways of giving an overview of data. One way is by the graphical descriptions discussed above. There are also numerical descriptions of data. Numbers such as the mean and median give, in some sense, the central or middle values of the data. Other numbers, such as variance and standard deviation, measure the dispersion or spread of the data about the mean. The central tendency of data is discussed in this section and dispersion in the following section.

The data we discuss will come either from a random sample of a larger population or from the larger population itself. We distinguish these two cases using different notation as follows:

$n =$ number of items in the sample,	$N =$ number of elements in the population
$\bar{x}$ (read: x-bar) $=$ sample mean,	μ (read: mu) $=$ population mean
$s^2 =$ sample variance,	$\sigma^2 =$ population variance

Note: Greek letters are used with the population and are called *parameters*. Latin letters are used with the samples and are called *statistics*.

Mean

Suppose a sample consists of the eight numbers:

$$7, \quad 11, \quad 11, \quad 8, \quad 12, \quad 7, \quad 6, \quad 6$$

The sample mean $\bar{x}$ is defined to be the sum of the values divided by the number of values; that is,

$$\bar{x} = \frac{7 + 11 + 11 + 8 + 12 + 7 + 6 + 6}{8} = \frac{68}{8} = 8.5$$

Generally speaking, suppose $x_1, x_2, \ldots, x_n$ are n numerical values of some sample. Then:

$$\text{Sample mean:} \quad \bar{x} = \frac{x_1 + x_2 + \cdots + x_n}{n} = \frac{\sum x_i}{n} \tag{1.1}$$

Now suppose that the data are organized into a frequency table; let there be k *distinct* numerical values $x_1, x_2, \ldots, x_k$, occurring with respective frequencies $f_1, f_2, \ldots, f_k$. Then the product f_1x_1 gives the sum of the x_1's, f_2x_2 gives the sum of the x_2's, and so on. Also,

$$f_1 + f_2 + \cdots + f_k = n$$

the total number of data items. Hence, formula (*1.1*) can be rewritten as

$$\text{Sample mean:} \quad \bar{x} = \frac{f_1x_1 + f_2x_2 + \cdots + f_kx_k}{f_1 + f_2 + \cdots + f_k} = \frac{\sum f_ix_i}{\sum f_i} \tag{1.2}$$

Conversely, formula (*1.2*) reduces to formula (*1.1*) in the special case $k = n$ and all $f_i = 1$.

For data organized into classes, (*1.2*) is applied with f_i as the number of data items in the ith class and x_i as the ith class value.

EXAMPLE 1.4

(*a*) Consider the data of Example 1.1, of which the frequency distribution is given in Fig. 1-2. The mean is

$$\bar{x} = \frac{8(1) + 14(2) + 7(3) + 12(4) + 3(5) + 1(6)}{45} = \frac{126}{45} = 2.8$$

In other words, there is an average of 2.8 people living in an apartment.

(*b*) Consider the data of Example 1.2, of which the frequency distribution is given in Fig. 1-4. Using the class values as approximations to the original values, we obtain

$$\bar{x} = \frac{3(72.5) + 6(77.5) + 8(82.5) + 5(87.5) + 7(92.5) + 3(97.5) + 0(102.5) + 3(107.5)}{35} = \frac{3042.5}{35} \approx 86.9$$

i.e. the mean 6:00 P.M. temperature is approximately 86.9 °F.

Remark: The formula for the population mean μ is the same as the formula for the sample mean $\bar{x}$. That is, suppose $x_1, x_2, \ldots, x_N$ are the N numerical values of the entire population. Then:

$$\text{Population mean:} \quad \mu = \frac{x_1 + x_2 + \cdots + x_N}{N} = \frac{\sum x_i}{N}$$

The reader may wonder why we give separate formulas for the sample mean $\bar{x}$ and population mean μ, since the formulas are the same. The reason is that the formulas will not be the same when we discuss the sample variance s^2 and population variance σ^2 in Section 1.4.

Median

Consider a list $x_1, x_2, \ldots, x_n$ of n data values which are sorted in increasing order. The *median* of the data, denoted by

$$\tilde{x} \text{ (read: x-tilde)}$$

is defined to be the "middle value". That is,

$$\text{Median:} \quad \tilde{x} = \begin{cases} [(n+1)/2]\text{th term} & \text{when } n \text{ is odd,} \\ \dfrac{(n/2)\text{th term} + [(n/2)+1]\text{th term}}{2} & \text{when } n \text{ is even.} \end{cases}$$

Note that $\tilde{x}$ is the average of the $(n/2)$th and $[(n/2)+1]$th terms when n is even.

Suppose, for example, the following two lists of sorted numbers are given:

$$\text{List A:} \quad 11, 11, 16, 17, 25$$

$$\text{List B:} \quad 1, 4, 8, 8, 10, 16, 16, 19$$

List A has five terms; its median $\tilde{x} = 16$, the middle or third term. List B has eight terms; its median $\tilde{x} = 9$, the average of the fourth term (8) and the fifth term (10).

One property of the median $\tilde{x}$ is that there are just as many numbers less than $\tilde{x}$ as there are greater than $\tilde{x}$.

The cumulative frequency distribution can be used to find the median of an arbitrary set of data.

EXAMPLE 1.5

(*a*) Consider the data in Fig. 1-2, which gives the number of tenants in 45 apartments. Here $n = 45$. The cumulative frequency column tells us that the median $\tilde{x} = 3$, the 23rd value.

(*b*) Consider the data in Fig. 1-4 which gives the 6:00 P.M. temperatures for a 35-day period. The median $\tilde{x} = 87.5$, the approximate 18th value.

Comparison of Mean and Median

Although the mean and median each locate, in some sense, the center of the data, the mean is sensitive to the *magnitude* of the values on either side of it, whereas the median is sensitive only to the *number* of values on either side of it.

EXAMPLE 1.6 The owner of a small company has 15 employees. Five employees earn $25,000 per year, seven earn $30,000, three earn $40,000, and the owner's annual salary is $153,000. (*a*) Find the mean and median salaries of all 16 persons in the company. (*b*) Find the mean and median salaries if the owner's salary is increased by $80,000.

(*a*) The mean salary is

$$\bar{x} = \frac{5 \cdot 25{,}000 + 7 \cdot 30{,}000 + 3 \cdot 40{,}000 + 153{,}000}{16}$$

$$= \frac{608{,}000}{16} = \$38{,}000$$

Since there are 16 persons, the median is the average of the eighth $(\frac{16}{2})$ and ninth $(\frac{16}{2} + 1)$ salaries when the salaries are arranged in increasing order from left to right. The eighth and ninth salaries are each $30,000. Therefore, the median is

$$\tilde{x} = \$30{,}000$$

(*b*) The new mean salary is

$$\bar{x} = \frac{608{,}000 + 80{,}000}{16} = \frac{688{,}000}{16} = \$43{,}000$$

The median is still $30,000, the average of the eighth and ninth salaries, which did not change. Hence, the mean moves in the direction of the increased salary, but the median does not change.

1.4 MEASURES OF DISPERSION: VARIANCE AND STANDARD DEVIATION

Consider the following two samples of numerical values:

$$\text{List A:} \quad 12, 10, 9, 9, 10$$

$$\text{List B:} \quad 5, 10, 16, 15, 4$$

For both A and B, the sample mean is $\bar{x} = 10$. However, observe that the values in A are clustered more closely about the mean than the values in B. To distinguish between A and B in this regard, we define a measure of the dispersion or spread of the values about the mean, called the *sample variance*, and its square root, called the *sample standard deviation*.

Let $\bar{x}$ be the sample mean of the n values $x_1, x_2, \ldots, x_n$. The difference $x_i - \bar{x}$ is called the *deviation* of the data value about the mean $\bar{x}$; it is positive or negative according as x_i is greater or less than $\bar{x}$. The sample variance s^2 is defined as follows:

$$\text{Sample variance:} \quad s^2 = \frac{(x_1 - \bar{x})^2 + (x_2 - \bar{x})^2 + \cdots + (x_n - \bar{x})^2}{n-1} = \frac{\sum (x_i - \bar{x})^2}{n-1} \tag{1.3}$$

The sample standard deviation s is the nonnegative square root of the sample variance; that is:

$$\text{Sample standard deviation:} \quad s = \sqrt{s^2} \tag{1.4}$$

Since each squared deviation is nonnegative, so is s^2. Moreover s^2 is zero precisely when each data value x_i is equal to $\bar{x}$. The more spread out the data values are, the larger the sample variance and standard deviation will be.

EXAMPLE 1.7 Consider the lists A and B above.

(*a*) In list A, whose sample mean is $x = 10$, the deviations of the five data are as follows:

$$12 - 10 = 2, \quad 10 - 10 = 0, \quad 9 - 10 = -1, \quad 9 - 10 = -1, \quad 10 - 10 = 0$$

The squares of the deviations are then

$$2^2 = 4, \quad 0^2 = 0, \quad (-1)^2 = 1, \quad (-1)^2 = 1, \quad 0^2 = 0$$

Also $n - 1 = 5 - 1 = 4$. Thus the sample variance s^2 and standard deviation s are as follows:

$$s^2 = \frac{4 + 0 + 1 + 1 + 0}{4} = \frac{6}{4} = 1.5$$

and

$$s = \sqrt{1.5} \approx 1.22$$

(*b*) In list B, we obtain the following:

$$s^2 = \frac{(5 - 10)^2 + (10 - 10)^2 + (16 - 10)^2 + (15 - 10)^2 + (4 - 10)^2}{5 - 1}$$

$$= \frac{25 + 0 + 16 + 25 + 36}{4} = \frac{122}{4} = 30.5$$

and

$$s = \sqrt{30.5} \approx 5.52$$

Note that B, which exhibits more dispersion than A, has a much larger variance and standard deviation than A.

The following is another formula for the sample variance; that is, it is equivalent to (*1.3*):

$$\text{Sample variance:} \quad s^2 = \frac{\sum x_i^2 - (\sum x_i)^2/n}{n-1} \tag{1.5}$$

Although formula (*1.5*) may look more complicated than formula (*1.3*), it is actually more convenient to use than formula (*1.3*), especially when the data are given in tabular form. In particular, this formula can be used without calculating the sample mean $\bar{x}$.

EXAMPLE 1.8 Consider the following values:

$$3, \quad 5, \quad 8, \quad 9, \quad 10, \quad 12, \quad 13, \quad 15, \quad 20$$

Find: (*a*) the sample mean x and (*b*) the sample variance s^2.

First construct the following table:

										Sum
x_i	3	5	8	9	10	12	13	15	20	95
x_i^2	9	25	64	81	100	144	169	225	400	1217

(*a*) By formula (*1.1*), where $n = 9$,

$$\bar{x} = \left(\sum x_i\right)/n = 95/9 \approx 10.56$$

(*b*) Here we use formula (*1.5*) with $n = 9$ and $n - 1 = 8$:

$$s^2 = \frac{1217 - (95)^2/9}{8} \approx \frac{1217 - 1002.7778}{8} \approx 26.78$$

Note that if we used formula (*1.3*) we would need to subtract $x = 10.56$ from each x_i before squaring.

Remark: The formula for the population variance σ^2 is not the same as the formula for the sample variance s^2 in that, when computing σ^2, we divide by N and not $N - 1$. That is, suppose $x_1, x_2, \ldots, x_N$ are the N numerical values of the entire population and suppose μ is the population mean. Then:

$$\text{Population variance:} \quad \sigma^2 = \frac{(x_1 - \mu)^2 + (x_2 - \mu)^2 + \cdots + (x_N - \mu)^2}{N} = \frac{\sum(x_i - \mu)^2}{N}$$

$$\text{Population standard deviation:} \quad \sigma = \sqrt{\sigma^2}$$

Some texts do define s^2 using n rather than $n - 1$. The reason that $n - 1$ is usually used for the sample variance s^2 is that one wants to use s^2 as an estimate of the population variance σ^2. One can prove that using n rather than $n - 1$ for s^2 tends to underestimate σ^2.

Sample Variance with a Frequency Distribution

For n data items organized into a frequency distribution consisting of k distinct values $x_1, x_2, \ldots, x_k$ with respective frequencies $f_1, f_2, \ldots, f_k$, the product $f_i(x_i - \bar{x})^2$ gives the sum of the squares of the deviations of each x_i from $\bar{x}$. Also, $f_1 + f_2 + \cdots + f_k = n$. Hence we can rewrite formulas (*1.3*) and (*1.5*) as follows:

$$\text{Sample variance:} \quad s^2 = \frac{f_1(x_1 - \bar{x})^2 + f_2(x_2 - \bar{x}^2) + \cdots + f_k(x_k - \bar{x})^2}{(f_1 + f_2 + \cdots + f_k) - 1} = \frac{\sum f_i(x_i - \bar{x})^2}{(\sum f_i) - 1} \tag{1.6}$$

and

$$\text{Sample variance:} \quad s^2 = \frac{\sum f_i x_i^2 - (\sum f_i x_i)^2 / \sum f_i}{(\sum f_i) - 1} \tag{1.7}$$

If the data are organized into classes, we use the *i*th class value for x_i in the above formulas (*1.6*) and (*1.7*).

EXAMPLE 1.9 Consider the data in Example 1.1, which gives the number of tenants in 45 apartments. By Example 1.4(*a*), the sample mean is $x = 2.8$. Find the sample variance s^2 and the sample standard deviation s.

First extend the frequency distribution table of the data in Fig. 1-2 to obtain the table in Fig. 1-8. We then obtain, using formula (*1.7*),

$$s^2 = \frac{430 - (126)^2/45}{44} \approx 1.75 \qquad \text{and} \qquad s \approx 1.32$$

Note $n = 45$ and $n - 1 = 44$.

Number of people, x_i	Frequency, f_i	$f_i x_i$	x_i^2	$f_i x_i^2$
1	8	8	1	8
2	14	28	4	56
3	7	21	9	63
4	12	48	16	192
5	3	15	25	75
6	1	6	36	36
Sums	45	126		430

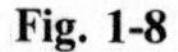
Fig. 1-8

EXAMPLE 1.10 Three hundred incoming students take a mathematics exam consisting of 75 multiple-choice questions. Suppose the following is the distribution of the scores on the exam:

Test scores	5–15	15–25	25–35	35–45	45–55	55–65	65–75
Number of students	2	0	8	36	110	78	66

Find the sample mean $\bar{x}$, variance s^2, and standard deviation s.

First enter the data in a table as in Fig. 1-9. Then, by formulas (*1.2*) and (*1.7*),

$$\bar{x} = \frac{16{,}500}{300} = 55, \quad s^2 = \frac{944{,}200 - (16{,}500)^2/300}{299} \approx 122.74 \qquad \text{and} \qquad s \approx 11.08$$

Class limits	Class value, x_i	Frequency, f_i	$f_i x_i$	x_i^2	$f_i x_i^2$
5–15	10	2	20	100	200
15–25	20	0	0	400	0
25–35	30	8	240	400	7,200
35–45	40	36	1,440	1600	57,600
45–55	50	110	5,500	2500	275,000
55–65	60	78	4,680	3600	280,800
65–75	70	66	4,620	4900	323,400
	Sums	300	16,500		944,200

Fig. 1-9

1.5 MEASURES OF POSITION: QUARTILES AND PERCENTILES

The preceding two sections discussed numerical measures of central tendency and of dispersion for a sample of data values. Now we consider numerical measures of position within the values when they are arranged in increasing order.

Quartiles

The median $\tilde{x}$ of n data values arranged in increasing order has been defined as a number for which at most half the values are less than $\tilde{x}$ and at most half are greater than $\tilde{x}$. Here, "half" means $n/2$ if n is even and $(n-1)/2$ if n is odd. The *first quartile*, Q_1, is defined as the median of the first half of the values, and the *third quartile*, Q_3, is the median of the second half. Hence about one-quarter of the data values are less than Q_1 and three-quarters are greater than Q_1. Similarly, about three-quarters are less than Q_3, and one-quarter are greater than Q_3. The *second quartile*, Q_2, is defined to be the median $\tilde{x}$.

EXAMPLE 1.11 Consider the following ten numerical values:

$$2 \quad 5 \quad 3 \quad 4 \quad 7 \quad 0 \quad 11 \quad 2 \quad 3 \quad 8$$

Find Q_1, Q_2, and Q_3 for the data.

First arrange the values in increasing order:

$$0 \quad 2 \quad 2 \quad 3 \quad 3 \quad 4 \quad 5 \quad 7 \quad 8 \quad 11$$

Since $n = 10$, the median $\tilde{x} = Q_2$ is the average of the fifth and sixth values:

$$Q_2 = \frac{3+4}{2} = 3.5$$

Q_1 is the median of the first five values, which are 0, 2, 2, 3, 3, and Q_3 is the median of the last five values, which are 4, 5, 7, 8, 11; hence

$$Q_1 = 2 \quad \text{and} \quad Q_3 = 7$$

Percentiles

Suppose n data values are arranged in increasing order. The *kth percentile*, denoted by P_k, is a number for which at most k percent of the values are less than P_k and at most $(100 - k)$ percent are greater than P_k. Specifically, P_k is defined as follows.

First compute $kn/100$ and break it into its integer part I and its decimal part D; that is, set

$$\frac{kn}{100} = I + D$$

Then:

$$P_k = \begin{cases} (I+1)\text{th value} & \text{when } D \neq 0 \\ \dfrac{I\text{th value} + (I+1)\text{th value}}{2} & \text{when } D = 0 \end{cases}$$

EXAMPLE 1.12 Suppose 50 data values are arranged in increasing order. Find (*a*) P_{35} and (*b*) P_{30}.

(*a*) Given $n = 50$, $k = 35$. Thus

$$\frac{kn}{100} = \frac{35(50)}{100} = 17.5 = 17 + 0.5$$

Here, $I = 17$ and $D = 0.5$. Since $D \neq 0$ and $I + 1 = 18$,

$$P_{35} = 18\text{th value}$$

(*b*) Given $n = 50$, $k = 30$. Thus

$$\frac{kn}{100} = \frac{30(50)}{100} = 15 = 15 + 0$$

Here, $I = 15$ and $D = 0$. Since $D = 0$,

$$P_{30} = \frac{\text{15th value} + \text{16th value}}{2}$$

EXAMPLE 1.13 Consider the following 50 values, listed in order, column by column:

10	20	35	44	55	64	75	81	87	99
11	22	36	48	56	68	76	82	89	101
13	23	38	49	57	69	76	83	90	102
15	23	41	50	60	70	78	83	94	105
18	30	44	50	63	73	80	85	96	107

Find P_{35} and P_{30}.

According to Example 1.12,

$$P_{35} = \text{18th value} = 49$$

$$P_{30} = \frac{\text{15th value} + \text{16th value}}{2} = \frac{44 + 44}{2} = 44$$

EXAMPLE 1.14 Consider the 50 data values in Example 1.13. Find (*a*) P_{25}, (*b*) P_{50}, (*c*) P_{75}. Compare these values with Q_1, Q_2, and Q_3, respectively.

(*a*) Given $n = 50$, $k = 25$. Thus

$$\frac{kn}{100} = \frac{25(50)}{100} = 12.5 = 12 + 0.5$$

Note $D = 0.5 \neq 0$, and $I + 1 = 13$. Hence

$$P_{25} = \text{13th value} = 38$$

Q_1 is the median of the first 25 values, which is the 13th value or 38. Hence $Q_1 = P_{25}$.

(*b*) Given $n = 50$, $k = 50$. Thus

$$\frac{kn}{100} = \frac{50(50)}{100} = 25 = 25 + 0$$

Since $D = 0$,

$$P_{50} = \frac{\text{25th value} + \text{26th value}}{2} = \frac{63 + 64}{2} = 63.5$$

Q_2 is the median of the 50 values, which is the same as P_{50}. That is, $Q_2 = P_{50}$.

(*c*) Given $n = 50$, $k = 75$. Thus

$$\frac{kn}{100} = \frac{75(50)}{100} = 37.5 = 37 + 0.5$$

Note $D = 0.5 \neq 0$, and $I + 1 = 38$. Hence

$$P_{75} = \text{38th value} = 83$$

Q_3 is the median of the last 25 values, which is the 13th of these values, or 83. Hence $Q_3 = P_{75}$.

Remark: The results in Example 1.14 are true for any set of numerical values; that is:

$$Q_1 = P_{25}, \qquad Q_2 = P_{50}, \qquad Q_3 = P_{75}$$

In other words, the percentiles form a generalization of the quartiles.

Five-Number Summary

The *5-number summary* of a collection of numerical data consists of the lowest value L, the quartiles Q_1, Q_2, Q_3, and the highest value H. Thus the 5-number summary of the 50 values in Example 1.13 follows:

$$L = 10, \qquad Q_1 = 38, \qquad Q_2 = 63.5, \qquad Q_3 = 83, \qquad H = 107$$

(The quartiles were obtained in Example 1.14.) We note that each of the four intervals

$$[L, Q_1], \quad [Q_1, Q_2], \quad [Q_2, Q_3], \quad [Q_3, L]$$

will contain about 25% (one-quarter) of the data items.

1.6 MEASURES OF COMPARISON: STANDARD UNITS AND COEFFICIENT OF VARIATION

Sometimes we want to compare data which come from different samples or populations. This is sometimes done using standard units and/or the coefficient of variation.

Standard Units

Suppose x is a value coming from a sample (or population) with mean $\bar{x}$ (or μ) and standard deviation s (or σ). Then the value of x in *standard units*, denoted by z, is defined as follows:

$$\text{Standard units:} \quad z = \frac{x - \bar{x}}{s} \quad \text{or} \quad z = \frac{x - \mu}{\sigma}$$

Standard units tell the number of standard deviations a given value lies above or below the mean of its sample (or population). It can also be used to compare values from different samples (or populations).

EXAMPLE 1.15 Student A got a score of 85 in a test whose scores had mean 79 and standard deviation 8. Student B got a score of 74 in a test whose scores had a mean of 70 and standard deviation 5. Which student got a "higher score"?

The standard scores for students A and B are, respectively,

$$z_A = \frac{85 - 79}{8} = \frac{6}{8} = 0.75 \quad \text{and} \quad z_B = \frac{74 - 70}{5} = \frac{4}{5} = 0.8$$

Thus student B did better than student A, even though his actual score, 74, was less than 85.

Coefficient of Variation

One major disadvantage of the standard deviation as a measure of variation or dispersion is that it depends on the units of measurements and on the sample (or population).

Clearly, a variation of 2 pounds when measuring a weight of 40 pounds represents a different effect than the same variation of 2 pounds when measuring a weight of 160 pounds. This effect is measured by the relative variation, defined by:

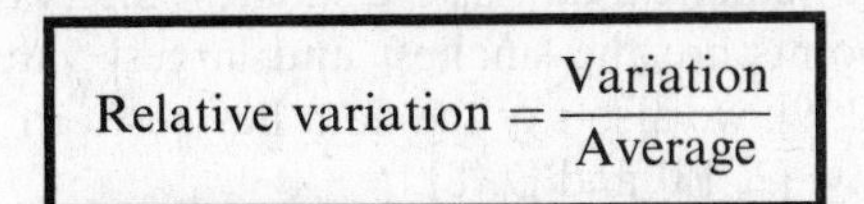

$$\text{Relative variation} = \frac{\text{Variation}}{\text{Average}}$$

Thus, for the above data, there is a relative variation of $2/40 = 0.05 = 5$ percent in the first case, but a relative variation of $2/160 = 0.0125 = 1.25$ percent in the second case.

Now suppose the variation is the standard deviation s (or σ) and the average is the mean $\bar{x}$ (or μ); then the relative variation is called the *coefficient of variation*, denoted by V and usually expressed as a percentage. That is:

$$\text{Coefficient of variation:} \quad V = \frac{s}{\bar{x}}(100 \text{ percent}) \quad \text{or} \quad V = \frac{\sigma}{\mu}(100 \text{ percent})$$

EXAMPLE 1.16 Suppose measurements of an item with a metric micrometer A had a mean of 3.25 mm and a standard deviation of 0.01 mm, and suppose measurements of another item with an English micrometer B had a mean of 0.80 in and a standard deviation of 0.002 in. Which micrometer is relatively "more" precise?

Calculating the two coefficients of variation yields

$$V_A = \frac{0.01}{3.25}(100 \text{ percent}) \approx 0.31 \text{ percent} \quad \text{and} \quad V_B = \frac{0.002}{0.80}(100 \text{ percent}) = 0.25 \text{ percent}$$

Thus micrometer B is more precise.

1.7 ADDITIONAL DESCRIPTIONS OF DATA

There are additional descriptions of data, besides the mean, median, variance, and standard deviation. Some of them will be discussed in this section.

Mode

The *mode* of a list of numerical data is the value which occurs most often and more than once. The mode may not exist (e.g. every value may occur only once), and if it does exist it may not be unique. Geometrically, the mode is the highest point in the histogram or the frequency polygon.

Consider, for example, the following three lists:

List A: 2, 3, 3, 5, 7, 7, 7, 8, 9
List B: 2, 3, 5, 7, 8, 9, 11, 13
List C: 2, 3, 3, 3, 5, 7, 7, 7, 8

List A has the unique mode 7; it is said to be *unimodal*. List B has no mode. List C has two modes, 3 and 7; it is said to be *bimodal*.

The mode also applies to nonnumerical (qualitative) data. For example, the mode of the data in Example 1.3 is Philadelphia. Geometrically, it is the item with the longest bar in the bar graph or the largest sector in the circular graph.

The mode of grouped data is usually the class value of the class with the greatest frequency. For example, the mode of the grouped temperature data in Example 1.2 is the class value 82.5 °F.

Range Interval, Range, and Midrange

The *range interval* of a set of numerical data is the smallest interval containing the data or, in other words, the interval whose endpoints are the smallest and largest values. Thus the range interval of the above list A is the interval [2, 9]; we also say the data lies between 2 and 9. The range intervals of the lists B and C are, respectively, [2, 13] and [2, 8].

The *range* of a set of numerical data is the difference between the largest and smallest values or, in other words, the length of the range interval. Thus the ranges of the above lists A, B, and C are, respectively, 7, 11, and 6.

The *midrange* of a set of numerical data is the average of the smallest and largest values or, in other words, the midpoint of the range interval. Thus the midranges of the above lists A, B, and C are, respectively, 5.5, 7.5, and 5.

Weighted Mean

Sometimes numerical data $x_1, x_2, \ldots, x_n$ are assigned respective weights $w_1, w_2, \ldots, w_n$. For example, each weight may be the frequency that an item occurs, or the probability that the item occurs, or some measure of the "importance" of the item. The *weighted mean*, denoted by $\bar{x}_w$, is defined as follows:

$$\text{Weighted mean:}\quad \bar{x}_w = \frac{x_1w_1 + x_2w_2 + \cdots + x_nw_n}{w_1 + w_2 + \cdots + w_n} = \frac{\sum x_iw_i}{\sum w_i}$$

Observe that the weighted mean formula is the same as the sample mean formula (*1.2*) when the weights represent frequencies.

Grand Mean

Suppose we want to find the overall mean of a collection of data where the data has been partitioned into t sets, where:

$n_1, n_2, \ldots, n_t$ are the numbers of elements in the sets

$\bar{x}_1, \bar{x}_2, \ldots, \bar{x}_t$ are the means of the corresponding sets

Then the *grand mean* of the total collection of data, denoted by $\bar{\bar{x}}$ (read: x-double bar), is defined as follows:

$$\text{Grand mean:}\quad \bar{\bar{x}} = \frac{n_1\bar{x}_1 + n_2\bar{x}_2 + \cdots + n_t\bar{x}_t}{n_1 + n_2 + \cdots + n_t} = \frac{\sum n_i\bar{x}_i}{\sum n_i}$$

One may view the grand mean as a special case of the weighted mean.

EXAMPLE 1.17 A philosophy class contains 10 freshmen, 20 sophomores, 15 juniors, and 5 seniors. The class is given an exam where the freshmen average 75, the sophomores 78, the juniors 80, and the seniors 82. Find the mean grade for the class.

Use the grand mean formula with

$$n_1 = 10,\quad n_2 = 20,\quad n_3 = 15,\quad n_4 = 5,\quad x_1 = 75,\quad x_2 = 78,\quad x_3 = 80,\quad x_4 = 82$$

This yields

$$\bar{\bar{x}} = \frac{10(75) + 20(78) + 15(80) + 5(82)}{10 + 20 + 15 + 5} = \frac{3920}{50} = 78.4$$

That is, 78.4 is the grand mean grade for the class.

1.8 BIVARIATE DATA, SCATTERPLOTS

Quite often in statistics it is desired to determine the relationship, if any, between two variables such as age and weight, weight and height, years of education and salary, or amount of daily exercise and cholesterol level. Letting x and y denote the two variables. The data will consist of a list of pairs of numerical values:

$$(x_1, y_1),\quad (x_2, y_2),\quad (x_3, y_3), \ldots, (x_n, y_n)$$

where the first values correspond to the variable x and the second values correspond to y.

As with a single variable, we can describe such *bivariate data* both graphically and numerically. Our primary concern is to determine whether there is a mathematical relationship, such as a linear relationship, between the data.

It should be kept in mind that a statistical relationship between two variables does not necessarily imply that there is a *causal* relationship between them. For example, a strong relationship between weight and height does not imply that one variable causes the other. Specifically, eating more does usually increase the weight of a person, but it does not usually mean that there will be an increase in the height of the person.

This section will give geometrical descriptions of bivariate data. The next section will discuss numerical descriptions of such data.

Cartesian Plane $\mathbf{R}^2$

The notation $\mathbf{R}^2$ is used to denote the collection of all ordered pairs (a, b) of real numbers. (By definition, $(a, b) = (c, d)$ if and only if $a = c$ and $b = d$.) Just as we can identify $\mathbf{R}$ with points on a line as in Fig. 1-1, so can we identify $\mathbf{R}^2$ with points in the plane. This identification, discussed below, is called the *cartesian plane* (named after the French mathematician René Descartes (1596–1650)), the *coordinate plane*, or simply the *plane* $\mathbf{R}^2$.

Two perpendicular lines L_1 and L_2 are chosen in the plane; the first line L_1 is pictured horizontally and the second line L_2 is pictured vertically. The point of intersection of the lines is called the *origin* and is denoted by 0. These lines, called *axes*, are now viewed as number lines, each with zero at the common origin and with the positive direction to the right on L_1 and upward on L_2. Also, L_1 is usually called the *x-axis* and L_2 the *y-axis* (see Fig. 1-10). Normally, we choose the same unit length on both axes, but this is not an absolute requirement.

Now each point P in the plane corresponds to a pair of real numbers (a, b), called the *coordinates of* P, as pictured in Fig. 1-10; that is, where the vertical line through P intersects the x-axis at a and where the horizontal line through P intersects the y-axis at b. (We will frequently write $P(a, b)$ when we want to indicate a point P and its coordinates a and b.) Note that this correspondence is one-to-one, i.e. each point P corresponds to a unique ordered pair (a, b), and vice versa. Thus, in this context, the terms point and ordered pair of real numbers are used interchangeably.

The two axes partition the plane into four regions, called *quadrants*, which are usually numbered using the Roman numerals I, II, III, and IV, as pictured in Fig. 1-11. That is:

Quadrant I: Both coordinates are positive, $(+, +)$.
Quadrant II: First coordinate negative, second positive $(-, +)$.
Quadrant III: Both coordinates are negative, $(-, -)$.
Quadrant IV: First coordinate positive, second negative, $(+, -)$.

Thus the quadrants are numbered counterclockwise from the upper-right-hand position.

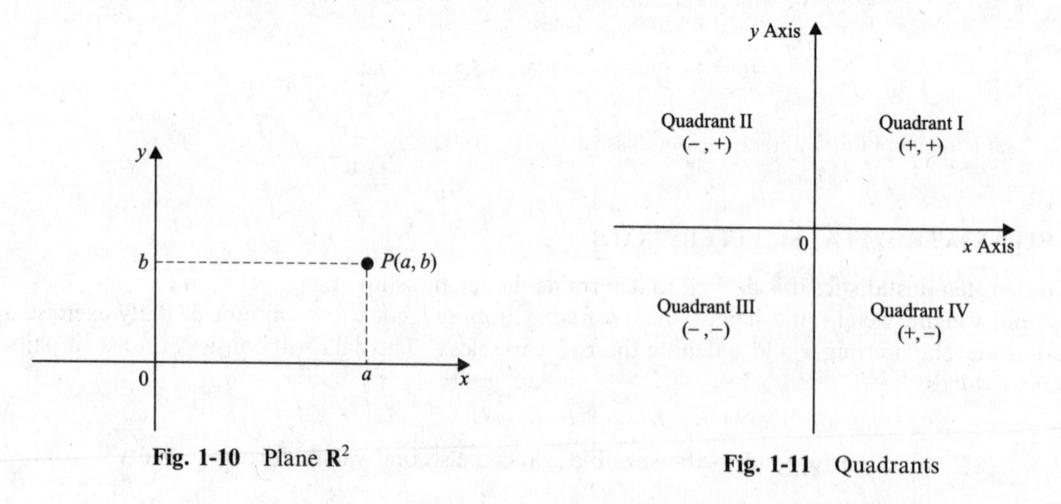

Fig. 1-10 Plane $\mathbf{R}^2$

Fig. 1-11 Quadrants

EXAMPLE 1.18 Locate and find the quadrant containing each of the following points in the plane $\mathbf{R}^2$:

$$A(2,-5), \quad B(-5,2), \quad C(-3,-7), \quad D(4,4), \quad E(0,6), \quad F(-7,0)$$

To locate the point $P(x,y)$, start at the origin, go x directed units along the x-axis and then y directed units parallel to the y-axis. The final point is $P(x,y)$. Figure 1-12 shows the given points in the plane. Thus $D(4,4)$ is in quadrant I, $B(-5,2)$ in quadrant II, $C(-3,-7)$ in quadrant III, and $A(-2,5)$ in quadrant IV. The points $E(0,6)$ and $F(-7,0)$ lie on the axes, so they do not belong to any quadrant.

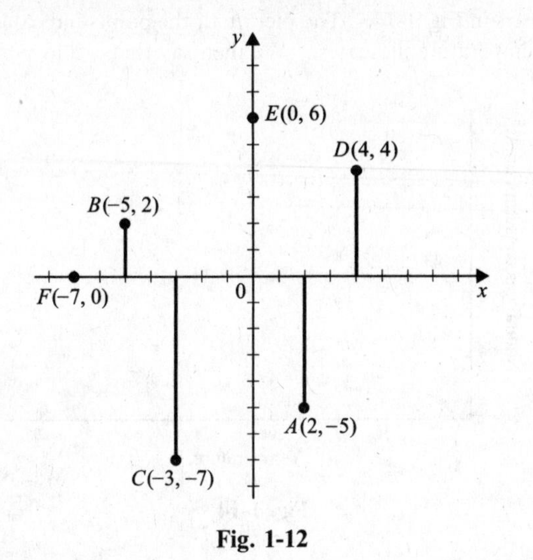

Fig. 1-12

Scatterplots

Consider a list of pairs of numerical values representing variables x and y. The *scatterplot* of the data is simply a picture of the pairs of values as points in a coordinate plane $\mathbf{R}^2$. The picture sometimes indicates a relationship between the points, as illustrated in the following examples.

EXAMPLE 1.19 Consider the following data, where x denotes the respective number of branches that 10 different banks have in some metropolitan area and y denotes the corresponding share of the total deposits held by the banks:

x	198	186	116	89	120	109	28	58	34	31
y	22.7	16.6	15.9	12.5	10.2	6.8	6.8	4.0	2.7	2.8

The scatterplot of the data appears in Fig. 1-13. The picture of the points indicates, roughly speaking, that the market share increases as the number of branches increases. We then say that x and y have a *positive correlation*.

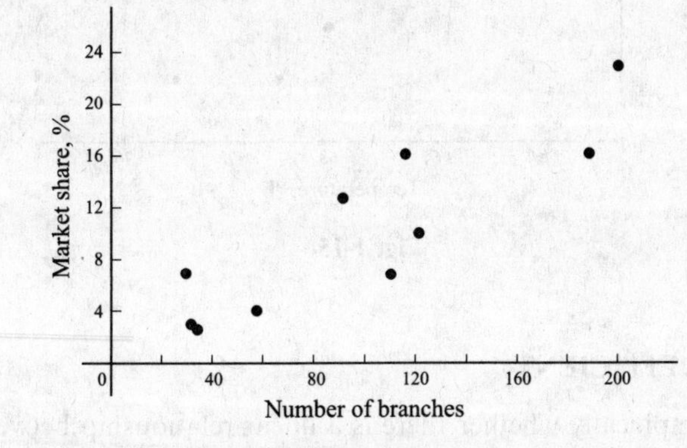

Fig. 1-13

EXAMPLE 1.20 Consider the following data, where x denotes the average daily temperature in degrees Fahrenheit and y denotes the corresponding daily natural gas consumption in cubic feet:

x, °F	50	45	40	38	32	40	55
y, ft³	2.5	5.0	6.2	7.4	8.3	4.7	1.8

The scatterplot of the data appears in Fig. 1-14. The picture of the points indicates, roughly speaking, that the gas consumption decreases as the temperature increases. We then say that x and y have a *negative correlation.*

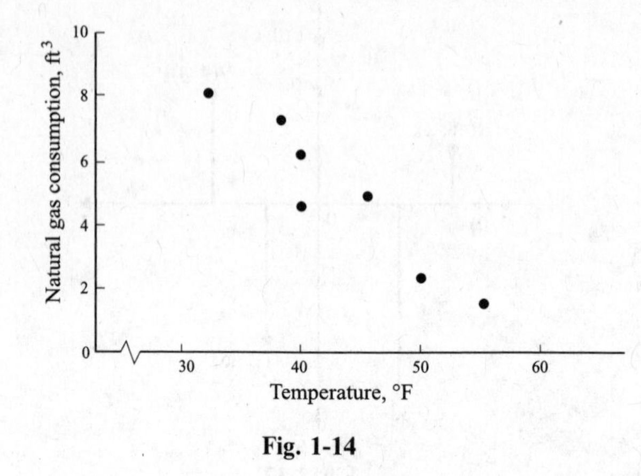

Fig. 1-14

EXAMPLE 1.21 Consider the following data, where x denotes the average daily temperature in degrees Fahrenheit over a 10-day period and y denotes the corresponding daily stock index average (in 1998):

x	63	72	76	70	71	65	70	74	68	61
y	8385	8330	8325	8320	8330	8325	8280	8280	8300	8265

The scatterplot of the data appears in Fig. 1-15. The picture of the points indicate that there is no apparent relationship between x and y.

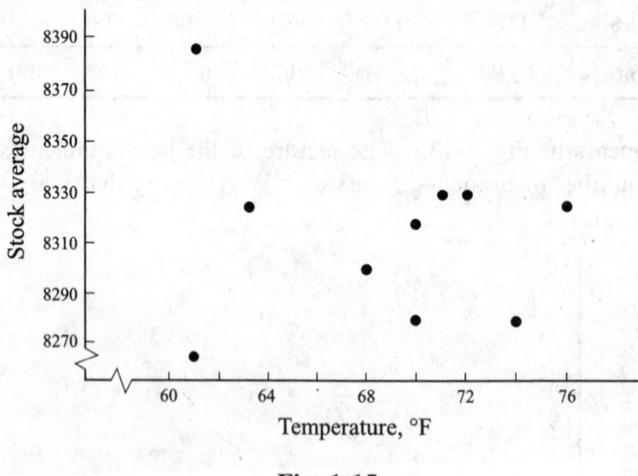

Fig. 1-15

1.9 CORRELATION COEFFICIENT

Scatterplots indicate graphically whether there is a linear relationship between two variables x and y. A numeric indicator of such a linear relationship is the *sample correlation coefficient* r of x and y, which is defined as follows:

$$\text{Sample correlation coefficient:} \quad r = \frac{\sum(x_i - \bar{x})(y_i - \bar{y})}{\sqrt{\sum(x_i - \bar{x})^2 \sum(y_i - \bar{y})^2}} \tag{1.8}$$

We assume that the denominator in formula (*1.8*) is not zero. It can be shown that the correlation coefficient r has the following properties:

(1) $-1 \le r \le 1$
(2) $r > 0$ if y tends to increase as x increases and $r < 0$ if y tends to decrease as x increases
(3) The stronger the linear relationship between x and y, the closer r is to -1 or 1; the weaker the linear relationship between x and y, the closer r is to 0

An alternative formula for computing r is given below; we then illustrate the above properties of r with examples.

Formula (*1.8*) can be written in the more compact form as

$$r = \frac{s_{xy}}{s_x s_y} \tag{1.9}$$

where s_x and s_y are the sample standard deviations of x and y, respectively [see formulas (*1.3*) and (*1.4*)], and where s_{xy}, called the *sample covariance* of x and y, is defined by

$$s_{xy} = \frac{\sum(x_i - \bar{x})(y_i - \bar{y})}{n-1} \tag{1.10}$$

An alternative formula for computing r follows:

$$r = \frac{\sum x_i y_i - (\sum x_i)(\sum y_i)/n}{\sqrt{\sum x_i^2 - (\sum x_i)^2/n}\sqrt{\sum y_i^2 - (\sum y_i)^2/n}} \tag{1.11}$$

This formula is very convenient to use after forming a table with the values of $x_i, y_i, x_i^2, y_i^2, x_i y_i$, and their sums, as illustrated below.

EXAMPLE 1.22 Find the correlation coefficient r for the data in (*a*) Example 1.19, (*b*) Example 1.20, (*c*) Example 1.21.

(*a*) Construct the table in Fig. 1-16. Then use formula (*1.11*) and that the number of points is $n = 10$ to obtain:

$$r = \frac{13{,}105.3 - (969)(101)/10}{\sqrt{127{,}723 - (969)^2/10}\sqrt{1427.56 - (101)^2/10}} \approx 0.8938$$

Here r is close to 1, which is expected since the scatterplot Fig. 1-13 indicates a strong positive linear relationship between x and y.

(*b*) Construct the table in Fig. 1-17. By formula (*1.11*), with $n = 7$,

$$r = \frac{1431.8 - (300)(35.9)/7}{\sqrt{13.218 - (300)^2/7}\sqrt{218.67 - (35.9)^2/7}} \approx -0.9562$$

Here r is close to -1, and the scatterplot Fig. 1-14 does indicate a strong negative linear relationship between x and y.

	x_i	y_i	x_i^2	y_i^2	x_iy_i
	198	22.7	39,204	515.29	4494.6
	186	16.6	34,596	275.56	3087.6
	116	15.9	13,456	252.81	1844.4
	89	12.5	7,921	156.25	1112.5
	120	10.2	14,400	104.04	1224.0
	109	6.8	11,881	46.24	741.2
	28	6.8	784	46.24	190.4
	58	4.0	3,364	16.00	232.0
	34	2.7	1,156	7.29	91.8
	31	2.8	961	7.84	86.8
Sums	969	101.0	127,723	1427.56	13,105.3

Fig. 1-16

	x_i	y_i	x_i^2	y_i^2	x_iy_i
	50	2.5	2,500	6.25	125.0
	45	5.0	2,025	25.00	225.0
	40	6.2	1,600	38.44	248.0
	38	7.4	1,444	54.76	281.2
	32	8.3	1,024	68.89	265.6
	40	4.7	1,600	22.09	188.0
	55	1.8	3,025	3.24	99.0
Sums	300	35.9	13,218	218.67	1431.8

Fig. 1-17

(c) Construct the table in Fig. 1-18. By formula (*1.11*), with $n = 10$,

$$r = \frac{2{,}286{,}555 - (690)(33{,}140)/10}{\sqrt{47{,}816 - (690)^2/10}\sqrt{109{,}836{,}700 - (33{,}140)^2/10}} \approx -0.0706$$

Here r is close to 0, which is expected since the scatterplot Fig. 1-15 indicates no linear relationship between x and y.

	x_i	y_i	x_i^2	y_i^2	x_iy_i
	63	3,385	3,969	11,458,225	213,255
	72	3,330	5,184	11,088,900	239,760
	76	3,325	5,776	11,055,625	252,700
	70	3,320	4,900	11,022,400	232,400
	71	3,330	5,041	11,088,900	236,430
	65	3,325	4,225	11,055,625	216,125
	70	3,280	4,900	10,758,400	229,600
	74	3,280	5,476	10,758,400	242,720
	68	3,300	4,624	10,890,000	224,400
	61	3,265	3,721	10,660,225	199,165
Sums	690	33,140	47,816	109,836,700	2,286,555

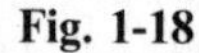

Fig. 1-18

1.10 METHODS OF LEAST SQUARES, REGRESSION LINE, CURVE FITTING

Suppose a scatterplot of the data points (x_i, y_i) indicates a linear relationship between variables x and y or, alternatively, suppose the correlation coefficient r of x and y is close to 1 or -1. Then the next step is to find a line L that, in some sense, fits the data. The line L we choose is called the *least-squares line*. We discuss this line in this section, and then we discuss more general types of curve fitting.

Least-Squares Line

Consider a given set of data points $P_i(x_i, y_i)$ and any (nonvertical) linear equation L. Let y_i^* denote the y value of the point on L corresponding to x_i. Furthermore, let $d_i = y_i - y_i^*$, the difference between the actual value of y and the value of y on the curve or, in other words, the vertical (directed) distance between the point P_i and the line L as shown in Fig. 1-19. The sum

$$\sum d_i^2 = d_1^2 + d_2^2 + \cdots + d_n^2$$

is called the *squares error* between the line L and the data points.

The *least-squares line* or the *line of best fit* or the *regression line* of y on x is, by definition, the line L whose squares error is as small as possible. It can be shown that such a line L exists and is unique. Let a denote the y-intercept of the line L and let b denote its slope; that is, let

$$y = a + bx \tag{1.12}$$

be the equation of L. Then a and b can be obtained from the following two equations in the two unknowns a and b, where n is the number of points:

$$\begin{aligned} na + (\textstyle\sum x_i)b &= \textstyle\sum y_i \\ (\textstyle\sum x_i)a + (\textstyle\sum x_i^2)b &= \textstyle\sum x_i y_i \end{aligned} \tag{1.13}$$

In particular, the slope b and y-intercept a can also be obtained from the following:

$$\boxed{b = \frac{rs_y}{s_x} \quad \text{and} \quad a = \bar{y} - b\bar{x}} \tag{1.14}$$

The second equation in (*1.14*) tells us that $(\bar{x}, \bar{y})$ lies on the regression line L, since

$$\bar{y} = (\bar{y} - b\bar{x}) + b\bar{x} = a + b\bar{x}$$

The first equation in (*1.14*) then tells us that the point $(\bar{x} + s_x, \bar{y} + rs_y)$ is also on L, as in Fig. 1-20.

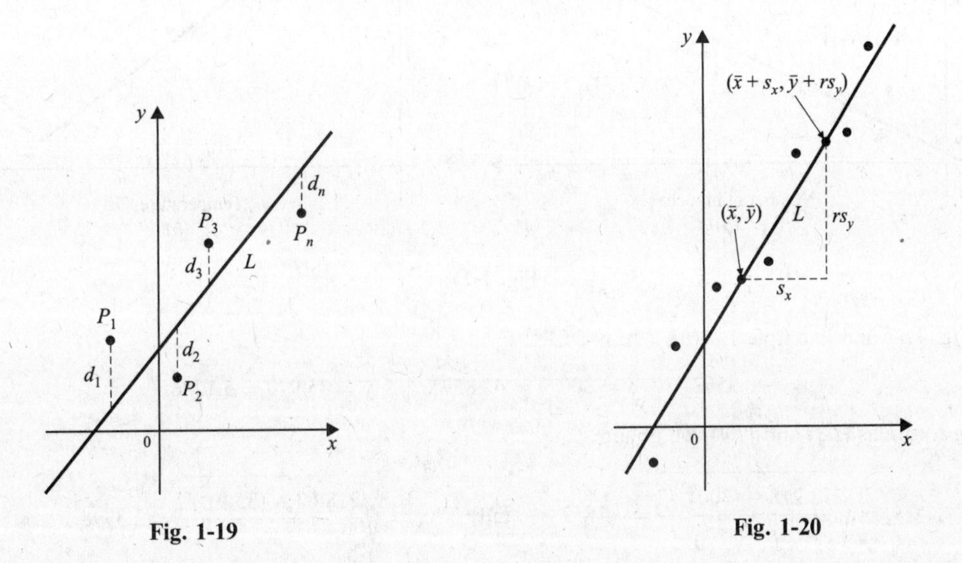

Fig. 1-19 **Fig. 1-20**

Remark: Recall that the above line L which minimizes the squares of the vertical distances from the given points P_i to L is called the *regression line* of y on x; it is usually used when one views y as a function of x. There also exists a line L' which minimizes the squares of the horizontal distances of the points P_i from L'; it is called the *regression line* of x on y. Given any two variables, the data usually indicates that one of them depends upon the other; we then let x denote the independent variable and let y denote the dependent variable. For example, suppose the variables are age and height. We normally assume height is a function of age, so we would let x denote age and y denote height. Accordingly, our least-squares lines will be regression lines of y on x, unless otherwise stated.

EXAMPLE 1.23 Find the line of best fit for the scatterplots in (*a*) Fig. 1-13, (*b*) Fig. 1-14.

(*a*) By Fig. 1-16 and Example 1.22(*a*),

$$r = 0.8938, \qquad \bar{x} = 969/10 = 96.9, \qquad \bar{y} = 101.0/10 = 10.1$$

Using formulas (*1.3*) and (*1.4*), we obtain

$$s_x = \sqrt{\frac{127{,}723 - (969)^2/10}{9}} = 61.3070 \quad \text{and} \quad s_y = \sqrt{\frac{1427.56 - (101)^2/10}{9}} = 6.7285$$

Substituting these values in (*1.14*), we get

$$b = \frac{(0.8938)(6.7285)}{61.3070} = 0.0981 \quad \text{and} \quad a = 10.1 - (0.0981)(96.9) = 0.5941$$

Thus the line L of best fit is

$$y = 0.5941 + 0.0981x$$

To graph L, we need only plot two points on L, and then draw the line through these points. Here we plot

$$(0, a) = (0, 0.5941) \quad \text{and} \quad (\bar{x}, \bar{y}) = (96.9, 10.1)$$

(approximately), and then draw L, as shown in Fig. 1-21(*a*).

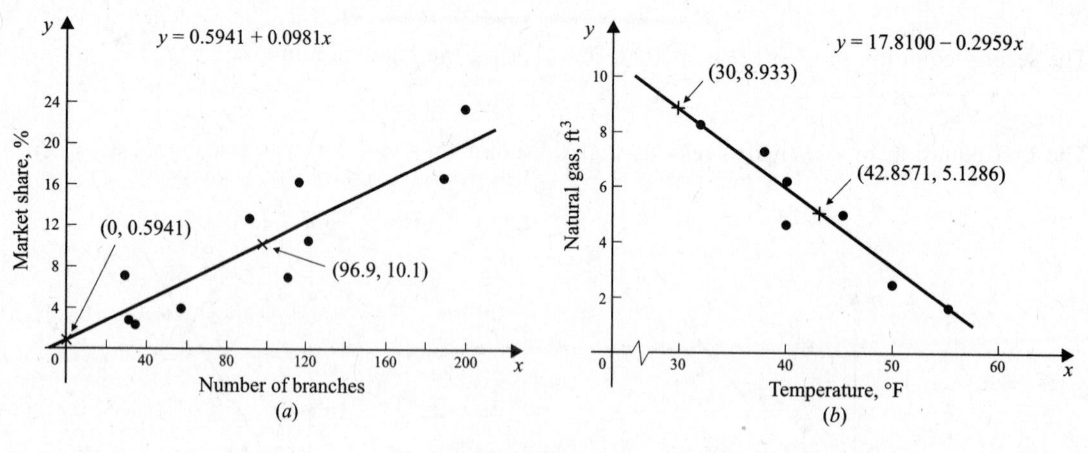

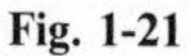

Fig. 1-21

(*b*) By Fig. 1-17 and Example 1.22(*b*),

$$r = -0.9562, \qquad \bar{x} = 300/7 = 42.8571, \qquad \bar{y} = 35.9/7 = 5.1286$$

Using formulas (*1.3*) and (*1.4*), we obtain

$$s_x = \sqrt{\frac{13{,}218 - (300)^2/7}{6}} = 7.7552 \quad \text{and} \quad s_y = \sqrt{\frac{218.67 - (35.9)^2/7}{6}} = 2.3998$$

Substituting these values in (*1.14*), we get

$$b = \frac{(-0.9562)(2.3998)}{7.7552} = -0.2959 \quad \text{and} \quad a = 5.1286 - (-0.2959)(42.8571) = 17.8100$$

Thus the line L of best fit is

$$y = 17.8100 - 0.2959x$$

The graph of L, obtained by plotting (30, 8.933) and (42.8571, 5.1286) (approximately) and drawing the line through these points, is shown in Fig. 1.21(*b*).

Curve Fitting

Sometimes the scatterplot does not indicate a linear relationship between the variables x and y, but one may visualize some other standard (well-known) curve $y = f(x)$ which may approximate the data, called an *approximate curve*. Several such standard curves, where letters other than x and y denote constants, follow:

(1) Parabolic curve: $y = a_0 + a_1x + a_2x^2$

(2) Polynomial curve: $y = a_0 + a_1x + a_2x^2 + \cdots + a_nx^n$

(3) Hyperbolic curve: $y = \dfrac{1}{a + bx}$ or $\dfrac{1}{y} = a + bx$

(4) Exponential curve: $y = ab^x$ or $\log y = a_0 + a_1x$

(5) Geometric curve: $y = ax^b$ or $\log y = \log a + b \log x$

Pictures of some of these standard curves appear in Fig. 1-22.

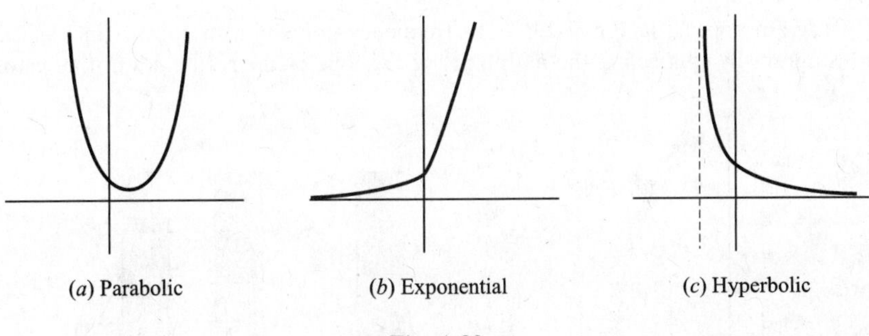

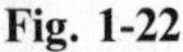

Fig. 1-22

It is generally not easy to decide which curve to use for a given set of data points. On the other hand, it is usually easier to determine a linear relationship by looking at the scatterplot or by using the correlation coefficient. Thus it is standard procedure to find the scatterplot of transformed data. Specifically:

(*a*) If log y versus x indicates a linear relationship, use the exponential curve (4).

(*b*) If $1/y$ versus x indicates a linear relationship, use the hyperbolic curve (3).

(*c*) If log y versus log x indicates a linear relationship, use the power curve (5).

Once one decides upon the kind of curve that is to be used, then the particular curve that one uses is the one that minimizes the squares error. We state this formally:

Definition: Consider a collection of curves and a given set of data points. The *best-fitting* or *least-squares* curve C in the collection is the curve which minimizes the sum

$$\sum d_i^2 = d_1^2 + d_2^2 + \cdots + d_n^2$$

(where d_i denotes the vertical distance from a data point $P_i(x_i, y_i)$ to the curve C).

Just as there are formulas to compute the constants a and b in the regression line L for a set of data points, so there are formulas to compute the constants in the best-fitting curve C in any of the above types (collections) of curves. Further discussion of curve fitting appears in the problem sections.

Solved Problems

FREQUENCY DISTRIBUTIONS, DISPLAYING DATA

1.1. Consider the following frequency distribution which gives the number f of students who got x correct answers on a 20-question exam:

x (correct answers)	9	10	12	13	14	15	16	17	18	19	20
f (number of students)	1	2	1	2	7	2	1	7	2	6	4

Display the data in a histogram and a frequency polygon.

The histogram appears in Fig. 1-23. The frequency polygon also appears in Fig. 1-23; it is obtained from the histogram by connecting the midpoints of the tops of the rectangles in the histogram.

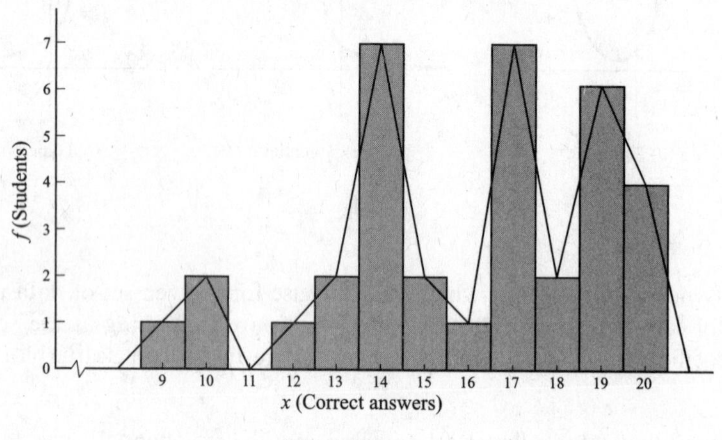

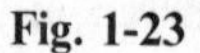

Fig. 1-23

1.2. Consider the following twenty data items:

$$\begin{array}{cccccccccc} 3 & 5 & 3 & 4 & 4 & 7 & 6 & 5 & 2 & 4 \\ 2 & 5 & 5 & 6 & 4 & 3 & 5 & 4 & 5 & 5 \end{array}$$

(*a*) Construct a frequency distribution (f) and a cumulative distribution (cf) of the data.
(*b*) Display the data in a histogram.

(*a*) Construct the table in Fig. 1-24(*a*). Here we show the *tally count*, which is used to find the frequency of each number. That is, as we run through the data list, we add a slash each time the number appears, and a line through four slashes indicates a fifth time the number appeared.

(*b*) The histogram is shown in Fig. 1-24(*b*).

1.3. The following scores were obtained in a statistics exam:

$$\begin{array}{cccccccccc} 74 & 80 & 65 & 85 & 95 & 72 & 76 & 72 & 93 & 84 \\ 75 & 75 & 60 & 74 & 75 & 63 & 78 & 87 & 90 & 70 \end{array}$$

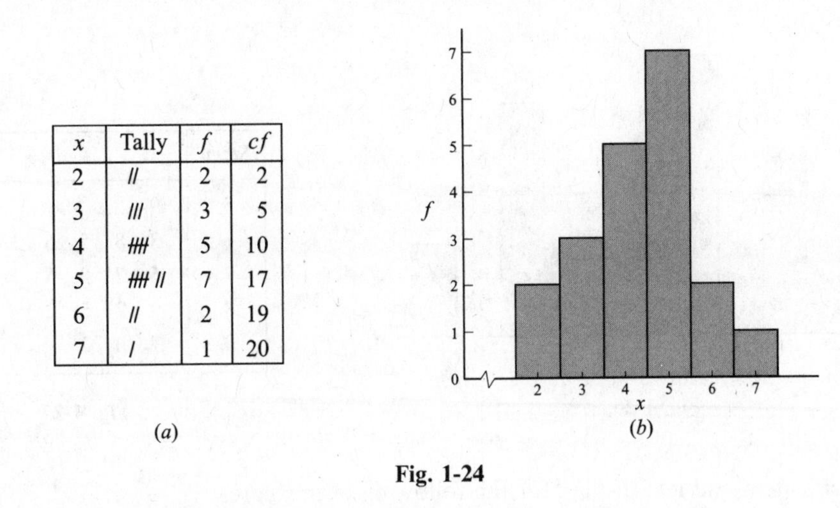

x	Tally	f	cf
2	//	2	2
3	///	3	5
4	~~////~~	5	10
5	~~////~~ //	7	17
6	//	2	19
7	/	1	20

Fig. 1-24

Find the frequency distribution when the data are classified into four classes 60–70, 70–80, 80–90, 90–100, and display the results in a histogram. (If a number falls on a class boundary, put it in the class to the right of the number.)

The frequency distribution, including the tally count, appears in Fig. 1-25(a) and the histogram appears in Fig. 1-25(b).

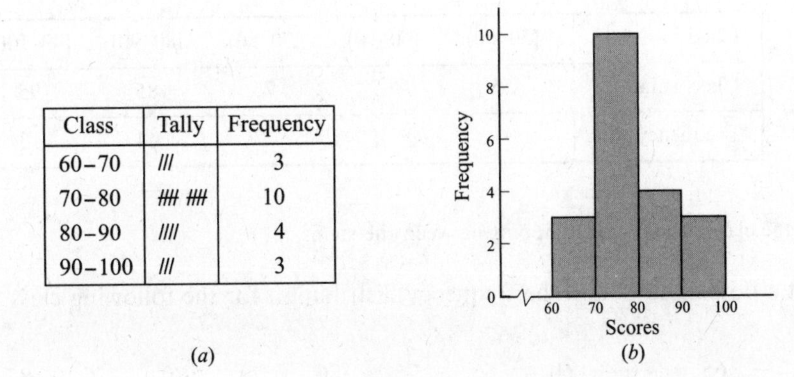

Class	Tally	Frequency
60–70	///	3
70–80	~~////~~ ~~////~~	10
80–90	////	4
90–100	///	3

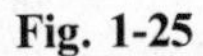

Fig. 1-25

1.4. The yearly rainfall, measured to the nearest tenth of a centimeter, for a 30-year period follows:

42.3	35.7	47.6	31.2	28.3	37.0	41.3	32.4	41.3	29.3
34.3	35.2	43.0	36.3	35.7	41.5	43.2	30.7	38.4	46.5
43.2	31.7	36.8	43.6	45.2	32.8	30.7	36.2	34.7	35.3

Classify the data into 10 classes, [28, 30), [30, 32), . . . , [44, 46), [46, 48), and display the results in a histogram.

The frequency distribution of the classification, where x_i denotes the class value and f_i denotes the frequency, follows:

Class	28–30	30–32	32–34	34–36	36–38	38–40	40–42	42–44	44–46	46–48
x_i	29	31	33	35	37	39	41	43	45	47
f_i	2	4	2	6	4	1	3	5	1	2

The histogram of the distribution appears in Fig. 1-26.

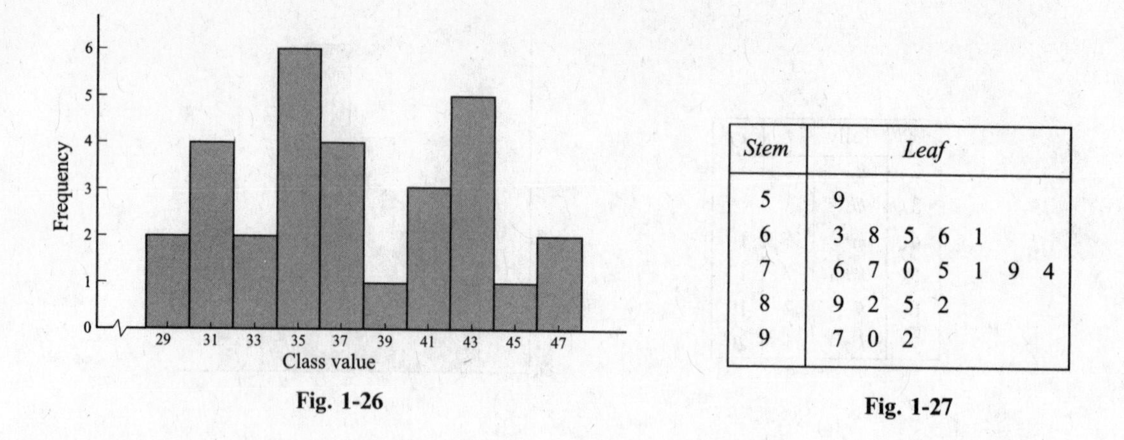

Stem	Leaf
5	9
6	3 8 5 6 1
7	6 7 0 5 1 9 4
8	9 2 5 2
9	7 0 2

Fig. 1-26 **Fig. 1-27**

1.5. Construct a stem-and-leaf display for the following exam scores:

63	68	59	66	76	82	70	71	74	85
97	65	89	90	77	61	75	79	92	82

The stem-and-leaf display appears in Fig. 1-27. Specifically, we use a place value, in this case the tens digit, as the "stem" and the unit digits as "leaves". The display gives the following frequency distribution:

Class interval	[50–60)	[60–70)	[70–80)	[80–90)	[99–100)
Class value x_i	55	65	75	85	95
Frequency f_i	1	5	7	4	3

Note that the class value is the midpoint between the stems.

1.6. Construct a dotplot to obtain the frequency distribution for the following class values of exam scores:

65	70	60	65	75	80	70	70	75	85
95	65	90	90	75	60	75	80	90	80

(The dotplot is sometimes used instead of a tally count.)

The dotplot appears in Fig. 1-28. Specifically, we mark off the class values on a horizontal axis, and then record the occurrence of a class value by a dot over the mark denoting the class value on the axis. The frequency distribution follows:

Class value	60	65	70	75	80	85	90	95
Frequency	2	3	3	4	3	1	3	1

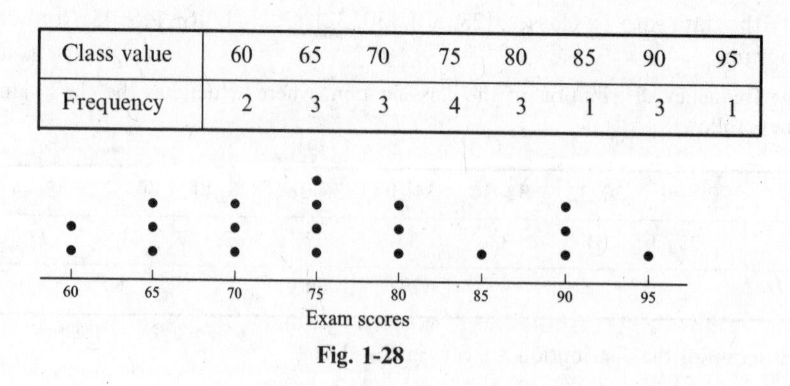

Fig. 1-28

MEAN, MEDIAN, VARIANCE, STANDARD DEVIATION

1.7. Find the sample mean $\bar{x}$, median $\tilde{x}$, variance s^2, and standard deviation s for the data:

$$4, \quad 6, \quad 6, \quad 7, \quad 9, \quad 10$$

Use formula (*1.3*) to obtain s^2.

There are $n = 6$ numbers. Hence,

$$\bar{x} = \frac{4+6+6+7+9+10}{6} = \frac{42}{6} = 7$$

The median is the average of the third and fourth numbers:

$$\tilde{x} = \frac{6+7}{2} = 6.5$$

By formula (*1.3*),

$$s^2 = \frac{(4-7)^2 + (6-7)^2 + (6-7)^2 + (7-7)^2 + (9-7)^2 + (10-7)^2}{5}$$

$$= \frac{9+1+1+0+4+9}{5} = \frac{24}{5} = 4.8$$

$$s = \sqrt{4.8} \approx 2.19$$

1.8. Find the sample mean $\bar{x}$, median $\tilde{x}$, variance s^2, and standard deviation s for the data:

$$8, \quad 7, \quad 12, \quad 5, \quad 6, \quad 7, \quad 4$$

Use formula (*1.5*) to obtain s^2.

There are $n = 7$ numbers. Hence,

$$\bar{x} = \frac{8+7+12+5+6+7+4}{7} = \frac{49}{7} = 7$$

To find the median, first arrange the numbers in increasing order:

$$4, \quad 5, \quad 6, \quad 7, \quad 7, \quad 8, \quad 12$$

The median is the fourth number: $\tilde{x} = 7$.

To apply formula (*1.5*) for s^2, we first construct the following table from the given data:

								Sum
x	8	7	12	5	6	7	4	49
x^2	64	49	144	25	36	49	16	383

Then, by formula (*1.5*),

$$s^2 = \frac{383 - (49)^2/7}{6} \approx 6.67$$

$$s = \sqrt{s^2} \approx 2.58$$

1.9. Find the sample mean $\bar{x}$, median $\tilde{x}$, variance s^2, and standard deviation s for the number x of correct scores in Problem 1.1.

First compute the table in Fig. 1-29. Then, by formula (*1.2*)

$$\bar{x} = \frac{\sum f_i x_i}{\sum f_i} = \frac{560}{35} = 16$$

The 35 scores x_i (including repetitions) are arranged in increasing order in Fig. 1-29. The median $\tilde{x}$ is the 18th score; hence $\tilde{x} = 17$. Using formula (*1.7*) for s^2 we get

$$s^2 = \frac{\sum f_i x_i^2 - (\sum f_i x_i)^2 / \sum f_i}{(\sum f_i) - 1} = \frac{9278 - (560)^2/35}{34} \approx 9.35$$

$$s = \sqrt{s^2} \approx 3.06$$

x_i	f_i	$f_i x_i$	x_i^2	$f_i x_i^2$
9	1	9	81	81
10	2	20	100	200
12	1	12	144	144
13	2	26	169	338
14	7	98	196	1372
15	2	30	225	450
16	1	16	256	256
17	7	119	289	2023
18	2	36	324	648
19	6	154	361	2166
20	4	80	400	1600
Sums	35	560		9278

Fig. 1-29

1.10. Find the sample mean $\bar{x}$, median $\tilde{x}$, variance s^2, and standard deviation s for the scores obtained in a statistics exam in Problem 1.3.

Letting x_i denote the class value of the ith class, compute the table in Fig. 1-30. Then, by formula (*1.2*),

$$\bar{x} = \frac{1570}{20} = 78.5$$

There are 20 scores; hence the median $\tilde{x}$ is the average of the 10th and 11th class values. Therefore

$$\tilde{x} = \frac{75 + 75}{2} = 75$$

Using formula (*1.7*) for s^2 we get

$$s^2 = \frac{124{,}900 - (1570)^2/20}{19} \approx 87.11, \quad s \approx 9.33$$

Class limits	Class value, x_i	f_i	$f_i x_i$	x_i^2	$f_i x_i^2$
60–70	65	3	195	4225	12,675
70–80	75	10	750	5625	56,250
80–90	85	4	340	7225	28,900
90–100	95	3	285	9025	27,075
	Sums	20	1570		124,900

Fig. 1-30

QUARTILES AND PERCENTILES

1.11. Consider the following data:

2, 7, 4, 4, 6, 1, 8, 15, 12, 7, 3, 16, 1, 2, 11, 5, 15, 4

(*a*) Find the first quartile Q_1, second quartile Q_2, and third quartile Q_3 for the data.

(*b*) Find the 5-number summary of the data.

(*a*) First arrange the data in numerical order:

1, 1, 2, 2, 3, 4, 4, 4, 5, 6, 7, 7, 8, 11, 12, 15, 15, 16

Q_2 is the median $\tilde{x}$, and since there are 18 values $\tilde{x}$ is the average of the 9th and 10th values. Thus

$$Q_2 = \frac{5+6}{2} = 5.5$$

Q_1 is the median of the values to the left of $\tilde{x}$. There are nine of these, so Q_1 is the fifth one. Thus $Q_1 = 3$.

Q_3 is the median of the values to the right of $\tilde{x}$. There are also nine of these, so Q_3 is the fifth one. Thus $Q_3 = 11$.

(*b*) The 5-number summary consists of the lowest value L, the quartiles Q_1, Q_2, Q_3, and the highest value H. Thus:

$$L = 1, \quad Q_1 = 3, \quad Q_2 = 5.5, \quad Q_3 = 11, \quad H = 16$$

1.12. Consider the following data:

5	6	7	7	9	10	12	15	15	20
21	22	25	27	28	32	34	34	35	40
41	48	51	56	57	65	75	76	78	80
81	84	88	88	89	90	91	92	93	97

Find the percentiles (*a*) P_{21}, (*b*) P_{40}, (*c*) P_{75}.

There are 40 data points, and they are arranged in numerical order. To determine the kth percentile, we first break up $kn/100$ into its integer and decimal parts.

(*a*) $n = 40$, $k = 21$. Thus

$$\frac{kn}{100} = \frac{21 \cdot 40}{100} = 8.4 = 8 + 0.4$$

The integer part is 8 and the decimal part is 0.4. Hence,

$$P_{21} = \text{9th value} = 15$$

(*b*) $n = 40$, $k = 40$. Thus

$$\frac{kn}{100} = \frac{40 \cdot 40}{100} = 16 = 16 + 0$$

Here, the integer part is 16 and the decimal part is zero. Hence,

$$P_{40} = \frac{\text{16th value} + \text{17th value}}{2} = \frac{32 + 34}{2} = 33$$

(*c*) $n = 40$, $k = 75$. Thus

$$\frac{kn}{100} = \frac{75 \cdot 40}{100} = 30 = 30 + 0$$

The integer part is 30 and the decimal part is zero, so

$$P_{75} = \frac{30\text{th value} + 31\text{st value}}{2} = \frac{80 + 81}{2} = 80.5$$

1.13. For the 40 data points in Problem 1.12, verify that

(*a*) $Q_1 = P_{25}$, (*b*) $Q_1 = P_{50}$, (*c*) $Q_3 = P_{75}$.

(*a*) Q_1 is the median of the first 20 data points, which is the average of the 10th and 11th values:

$$Q_1 = \frac{20 + 21}{2} = 20.5$$

For P_{25} we first compute $(25 \cdot 40)/100 = 10 = 10 + 0$. Hence P_{25} is also the average of the 10th and 11th values.

(*b*) Q_2 is the median of all 40 data points, which is the average of the 20th and 21st values:

$$Q_2 = \frac{40 + 41}{2} = 40.5$$

For P_{50} we first compute $(50 \cdot 40)/100 = 20 = 20 + 0$. Hence P_{50} is also the average of the 20th and 21st values.

(*c*) Q_3 is the median of the last 20 values, or the average of the 30th and 31st values:

$$Q_3 = \frac{80 + 81}{2} = 80.5$$

which is the value of P_{75} determined in Problem 1.12.

MISCELLANEOUS PROBLEMS INVOLVING ONE VARIABLE

1.14. Find the mode, range, and midrange of the data in:

(*a*) Problem 1.1, (*b*) Problem 1.4, (*c*) Problem 1.11

The mode is the value (or class value) which occurs most often (and more than once), the range is the difference between the largest and smallest values, and the midrange is the average of the smallest and largest values. Accordingly:

(*a*) Mode = 17, range = 20 − 9 = 11, midrange = (20 + 9)/2 = 14.5

(*b*) Mode = 35, range = 47 − 29 = 18, midrange = (47 + 29)/2 = 38

(*c*) Mode = 4, range = 16 − 1 = 15, midrange = (16 + 1)/2 = 8.5

1.15. An English class for foreign students consists of 20 French students, 25 Italian students, and 15 Spanish students. On an exam, the French students average 78, the Italian students 75, and the Spanish students 76. Find the mean grade for the class.

Here we use the formula for the grand mean $\bar{\bar{x}}$ (page 15) with

$$n_1 = 20,\ n_2 = 25,\ n_3 = 15,\ x_1 = 78,\ x_2 = 75,\ x_3 = 76$$

This yields

$$\bar{\bar{x}} = \frac{20(78) + 25(75) + 15(76)}{20 + 25 + 15} = \frac{4575}{60} = 76.25$$

That is, 76.25 is the mean grade for the class.

1.16. Student A received a score of 91 in a test whose scores had mean 82 and standard deviation 6. Student B received a score of 87 in another test whose scores had mean 80 and standard deviation 4. Which student got a "higher score"?

Transform the grades into standard units using

$$z = \frac{x - \bar{x}}{s}$$

This yields: $$z_A = \frac{91 - 82}{6} = \frac{9}{6} = 1.5 \quad \text{and} \quad z_B = \frac{87 - 80}{4} = \frac{7}{4} = 1.75$$

Thus, relatively speaking, B did better than A.

1.17. Suppose measurements of an item with a metric micrometer A yield a mean of 4.20 mm and a standard deviation of 0.015 mm, and suppose measurements of another item with an English micrometer B yield a mean of 1.10 inches and a standard deviation of 0.005 inches. Which micrometer is relatively "more" precise?

Calculate the two coefficients of variation. This yields:

$$V_A = \frac{0.015}{4.20}(100\%) = 0.36\% \quad \text{and} \quad V_B = \frac{0.005}{1.10}(100\%) = 0.45\%$$

Thus micrometer A is more precise.

BIVARIATE DATA

1.18. Estimate the correlation coefficient r for each data set shown in the scatterplots in Fig. 1.31.

The correlation coefficient r lies in the interval $[-1, 1]$. Moreover, r is close to 1 if the data are approximately linear with positive slope, r is close to -1 if the data are approximately linear with negative slope, and r is close to 0 if there is no relationship between the points. Accordingly:

(*a*) r is close to 1, say $r \approx 0.9$, since there appears to be a strong linear relationship between the points with positive slope.

(*b*) $r \approx 0$ since there appears to be no relationship between the points.

(*c*) r is close to -1, say $r \approx -0.9$, since there appears to be a strong linear relationship between the points but with negative slope.

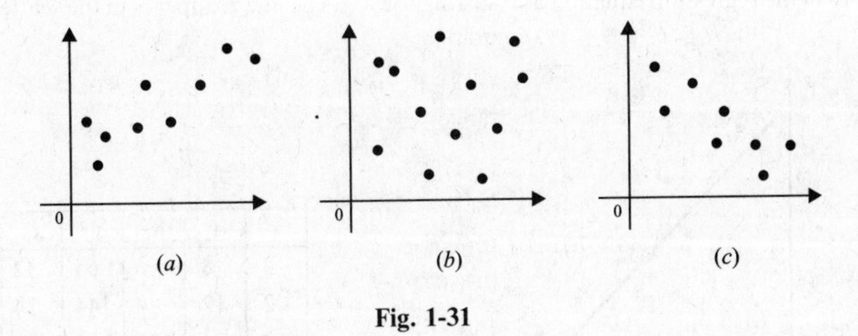

Fig. 1-31

1.19. Consider the following list of data values:

x	4	2	10	5	8
y	8	12	4	10	2

(*a*) Plot the data in a scatterplot.

(*b*) Compute the correlation coefficient r.

(*c*) Find L, the least-squares line $y = a + bx$.

(*d*) Graph L on the scatterplot in Part (*a*).

(*a*) The scatterplot (with L) is shown in Fig. 1-32(*a*).

(*b*) First complete the table in Fig. 1-32(*b*). Then, by formula (*1.11*), with $n = 5$,

$$r = \frac{162 - [(29)(36)]/5}{\sqrt{209 - (29)^2/5}\sqrt{328 - (36)^2/5}} = \frac{-46.8}{\sqrt{40.8}\sqrt{68.8}} = -8833$$

(*c*) First compute the standard deviations s_x and s_y of x and y, respectively. Using formulas (*1.4*) and (*1.5*), we get

$$s_x = \sqrt{\frac{209 - (29)^2/5}{4}} = 3.1937, \quad s_y = \sqrt{\frac{328 - (36)^2/5}{4}} = 4.1473$$

Substituting r, s_x, s_y into formula (*1.14*) for the slope b of the least-squares line L gives

$$b = \frac{rs_y}{s_x} = \frac{(-0.8833)(4.1473)}{3.1937} = -1.1470$$

To determine the y-intercept a of L, we first compute

$$\bar{x} = 29/5 = 5.8 \quad \text{and} \quad \bar{y} = 36/5 = 7.2$$

Then, by formula (*1.14*),

$$a = \bar{y} - b\bar{x} = 7.2 - (-1.1470)(5.8) = 13.8526$$

Hence L is the following equation:

$$y = 13.8526 - 1.1470x$$

Alternatively, we can find a and b using the normal equations in formula (*1.13*) with $n = 5$:

$$\begin{aligned} na + \textstyle\sum xb &= \textstyle\sum y \\ \textstyle\sum xa + \sum x^2 b &= \textstyle\sum xy \end{aligned} \quad \text{or} \quad \begin{aligned} 5a + \ \ 29b &= \ 36 \\ 29a + 209b &= 162 \end{aligned}$$

(These equations would be used if we did not also want $r, s_x, s_y, \bar{x}$, and $\bar{y}$.)

(*d*) To graph L, we find two points on L and draw the line through them. One of the two points is

$$(\bar{x}, \bar{y}) = (5.8, 7.2)$$

(which is on any least-squares line). Another point is (10, 2.3826), which is obtained by substituting $x = 10$ in the regression equation and solving for y. The line L appears in the scatterplot in Fig. 1.32(*a*).

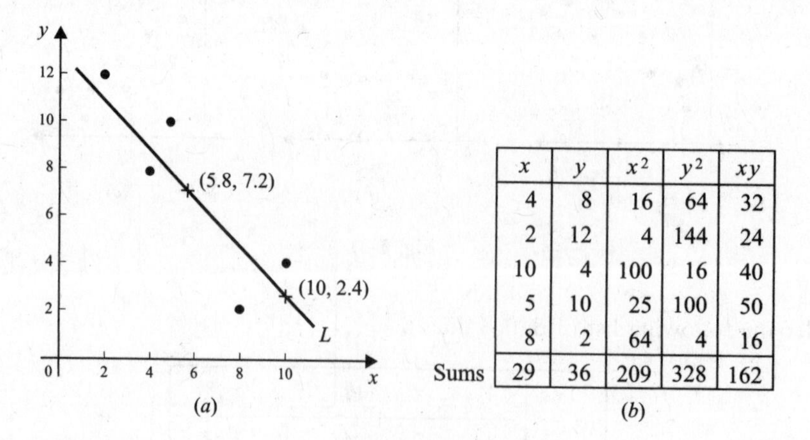

	x	y	x^2	y^2	xy
	4	8	16	64	32
	2	12	4	144	24
	10	4	100	16	40
	5	10	25	100	50
	8	2	64	4	16
Sums	29	36	209	328	162

(*b*)

Fig. 1-32

1.20. Repeat Problem 1.19 for the following data:

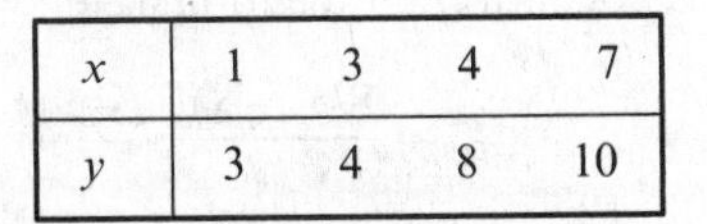

x	1	3	4	7
y	3	4	8	10

(*a*) The scatterplot (with L) is shown in Fig. 1-33(*a*).

(*b*) First complete the table in Fig. 1-33(*b*). Then, by formula (*1.11*), where $n = 4$,

$$r = \frac{117 - [(15)(25)]/4}{\sqrt{75 - (15)^2/4}\sqrt{189 - (25)^2/4}} = \frac{23.25}{\sqrt{18.75}\sqrt{32.75}} = 0.9382$$

(*c*) First compute the standard deviations s_x and s_y of x and y, respectively. Using formulas (*1.4*) and (*1.5*), we get

$$s_x = \sqrt{\frac{75 - (15)^2/4}{3}} = 2.5 \quad \text{and} \quad s_y = \sqrt{\frac{189 - (25)^2/4}{3}} = 3.304$$

Substituting r, s_x, s_y into formula (*1.14*) for the slope b of the least-squares line L gives

$$b = \frac{rs_y}{s_x} = \frac{(0.9675)(4.03)}{2.5} = 1.24$$

To determine the y-intercept a of L, we first compute

$$\bar{x} = \frac{15}{4} = 3.75 \quad \text{and} \quad \bar{y} = \frac{25}{4} = 6.25$$

Then, by formula (*1.14*),

$$a = \bar{y} - b\bar{x} = 6.25 - (1.24)(3.75) = 1.60$$

Hence L is the following equation:

$$y = 1.60 + 1.24x$$

Alternatively, we can find a and b by solving the normal equations in formula (*1.13*) with $n = 4$:

$$na + \sum xb = \sum y \qquad\qquad 4a + 15b = 25$$
$$\text{or}$$
$$\sum xa + \sum x^2 b = \sum xy \qquad\qquad 15a + 75b = 117$$

(These equations would be used if we did not also want $r, s_x, s_y, \bar{x}$, and $\bar{y}$.)

(*d*) To graph L, we find two points on L and draw the line through them. One point is $(\bar{x}, \bar{y}) = (3.75, 6.25)$. Another point is $(0, 1.60)$, the y-intercept. The line L appears in the scatterplot in Fig. 1-33(*a*).

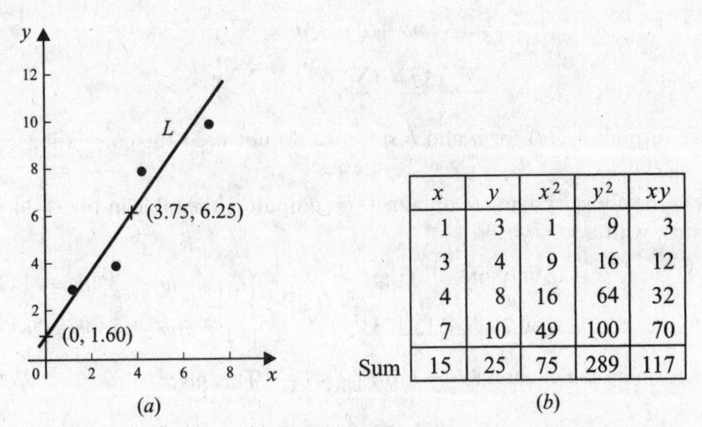

	x	y	x^2	y^2	xy
	1	3	1	9	3
	3	4	9	16	12
	4	8	16	64	32
	7	10	49	100	70
Sum	15	25	75	289	117

(*b*)

Fig. 1-33

1.21. Find the sample covariance s_{xy} of x and y for the data in (*a*) Problem 1.19, (*b*) Problem 1.20.

The sample covariance s_{xy} is obtained from the formula

$$s_{xy} = \frac{\Sigma(x_i - \bar{x})(y_i - \bar{y})}{n-1}$$

(*a*) We have:

$$\begin{aligned} s_{xy} &= [(4-5.8)(8-7.2) + (2-5.8)(12-7.2) + (10-5.8)(4-7.2) \\ &\quad + (5-5.8)(10-7.2) + (8-5.8)(2-7.2)]/4 \\ &= [-1.44 - 18.24 - 13.44 - 2.24 - 11.44]/4 \\ &= -46.8/4 = -11.7 \end{aligned}$$

We note that the variances s_x and s_y are always nonnegative, but the covariance s_{xy} can be negative, which indicates that y tends to decrease as x increases.

(*b*) We have:

$$\begin{aligned} s_{xy} &= [(1-3.75)(3-6.25) + (3-3.75)(4-6.75)(4-3.75)(6-8.25) + (7-3.75)(10-6.25)]/3 \\ &= [8.9375 + 2.0625 - 0.5625 + 12.1875]/3 \\ &= 22.625/3 = 7.5417 \end{aligned}$$

The covariance here is positive, which indicates that y tends to increase as x increases.

1.22. Let W denote the number of American women graduating with a doctoral degree in mathematics in a given year. Suppose that, for certain years, W has the following values:

Year	1980	1985	1990	1995
W	28	36	40	45

Assuming that the increase, year by year, is approximately linear and that it will increase linearly in the near future, estimate W for the years 2000, 2003, and 2005.

The estimation uses the least-squares line L, that is, the line $y = a + bx$ of best fit for the data (where x denotes the year and y denotes the value of W). The unknowns a and b will be determined by the following *normal equations* in formula (*1.13*):

$$na + (\textstyle\sum x)b = \sum y$$
$$(\textstyle\sum x)a + (\sum x^2)b = \sum xy$$

(We do not use formula (*1.14*) for a and b since we do not need the correlation coefficient r nor $s_x, s_y, \bar{x}$, and $\bar{y}$.)

The sums in the above system are obtained by computing the table in Fig. 1-34(*a*). Substitution in the normal equations, with $n = 4$, yields:

$$\begin{aligned} 4a + 350b &= 149 \\ 350a + 30{,}750b &= 13{,}175 \end{aligned} \quad \text{or} \quad \begin{aligned} E_1&: \quad 4a + 350b = 149 \\ E_2&: \quad 70a + 6150b = 2635 \end{aligned}$$

Eliminate a by forming the equation $E = -70E_1 + 4E_2$. This gives

$$100b = 110 \quad \text{or} \quad b = 1.1$$

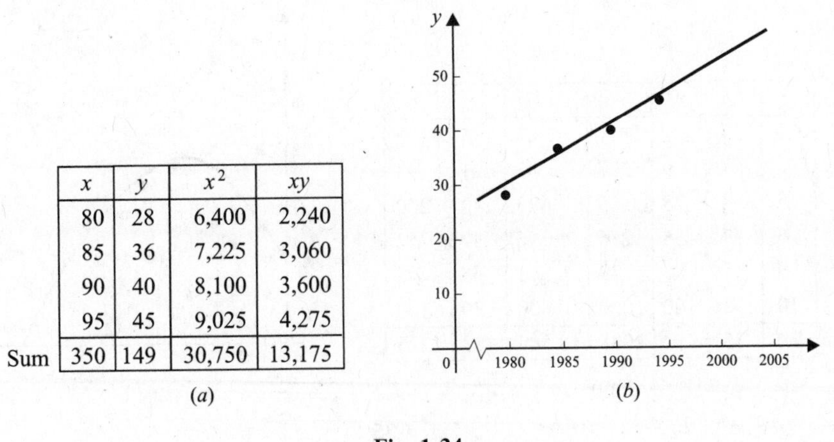

	x	y	x^2	xy
	80	28	6,400	2,240
	85	36	7,225	3,060
	90	40	8,100	3,600
	95	45	9,025	4,275
Sum	350	149	30,750	13,175

(a) (b)

Fig. 1-34

Substitute $b = 1.1$ in the first equation E_1 to obtain $a = -59$. Thus

$$y = -59 + 1.1x \tag{1}$$

is the line L. The original points and the line L are plotted in Fig. 1-34(b).

Substitute 100, 103, and 105 in (1) to obtain 51, 54.3, and 56.5, respectively. Thus one would expect that, approximately, $W = 51$, $W = 54$, and $W = 57$ women will receive doctoral degrees in the years 2000, 2003, and 2005, respectively.

1.23. Find the least-squares parabola C for the following data:

x	1	3	5	6	9	10
y	5	7	8	7	5	3

Plot C and the data points in the plane $\mathbf{R}^2$.

The parabola C has the form $y = a + bx + cx^2$ where the unknowns a, b, c are obtained from the following normal equations (which are analogous to the normal equations for the least-squares line L in formula (1.13)):

$$na + (\sum x)b + (\sum x^2)c = \sum y$$
$$(\sum x)a + (\sum x^2)b + (\sum x^3)c = \sum xy$$
$$(\sum x^2)a + (\sum x^3)b + (\sum x^4)c = \sum x^2 y$$

The sums in the system are obtained by computing the table in Fig. 1-35(a). Substitution in the normal equations, with $n = 6$, yields:

$$6a + \quad 34b + \quad 252c = \quad 35$$
$$34a + \ 252b + \ 2098c = \ 183$$
$$252a + 2098b + 18\,564c = 1225$$

Solving the system yields

$$a = \frac{12845}{3687} = 3.48, \qquad b = \frac{4179}{2458} = 1.70, \qquad c = -\frac{1279}{7374} = -0.173$$

Thus
$$y = 3.48 + 1.70x - 0.173x^2$$

is the required parabola C. The given data points and C are plotted in Fig. 1-35(b).

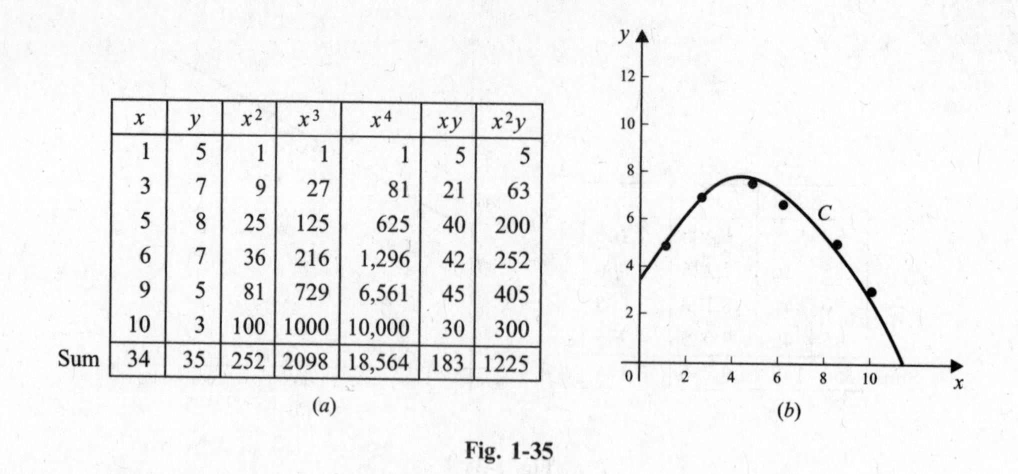

	x	y	x^2	x^3	x^4	xy	x^2y
	1	5	1	1	1	5	5
	3	7	9	27	81	21	63
	5	8	25	125	625	40	200
	6	7	36	216	1,296	42	252
	9	5	81	729	6,561	45	405
	10	3	100	1000	10,000	30	300
Sum	34	35	252	2098	18,564	183	1225

Fig. 1-35

1.24. Consider the following data which indicates exponential growth:

x	1	2	3	4	5	6
y	6	18	55	160	485	1460

Find the least-squares exponential curve C for the data, and plot the data points and C on the plane $\mathbf{R}^2$.

The curve C has the form $y = ab^x$ where a and b are unknowns. The logarithm (to base 10) of $y = ab^x$ yields

$$\log y = \log a + x \log b = a' + b'x$$

where $a' = \log a$ and $b' = \log b$. Thus we seek the least-squares line L for the following data:

x	1	2	3	4	5	6
$\log y$	0.7782	1.2553	1.7404	2.2041	2.6857	3.1644

Using the normal equations (*1.13*) for L, we get

$$a' = 0.3028, \qquad b' = 0.4767$$

The antiderivatives of a' and b' yield, approximately,

$$a = 2.0, \qquad b = 3.0$$

Thus $y = 2(3^x)$ is the required exponential curve C. The data points and C are plotted in Fig. 1-36.

1.25. Derive the normal equations (*1.13*) for the least-squares line L for n data points $P_i(x_i, y_i)$. [Our solution uses calculus.]

We want to minimize the least-squares error

$$D = \sum d_i^2 = \sum [y_i - (a + bx_i)]^2 = \sum [a + bx_i - y_i]^2$$

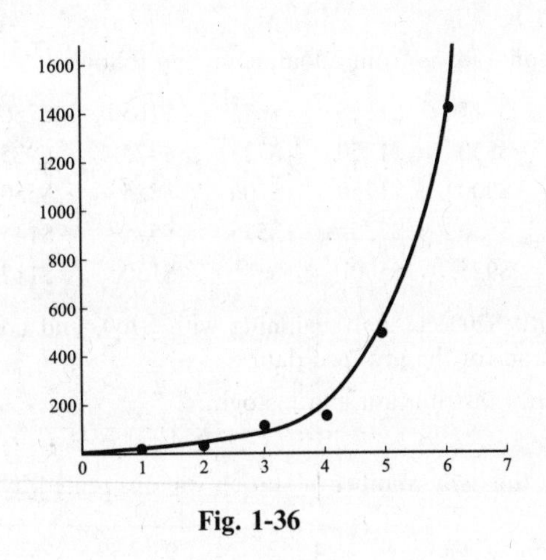

Fig. 1-36

which may be viewed as a function of a and b. The minimum may be obtained by setting the partial derivatives D_a and D_b of D with respect to a and b, respectively, equal to zero. The partial derivatives follow:

$$D_a = \sum 2(a + bx_i - y_i) \qquad \text{and} \qquad D_b = \sum 2(a + bx_i - y_i)x_i$$

Setting $D_a = 0$ and $D_b = 0$, we obtain the required equations

$$na + (\sum x_i)b = \sum y_i$$

$$(\sum x_i)a + (\sum x_i^2)b = \sum x_i y_i$$

Supplementary Problems

FREQUENCY DISTRIBUTIONS, DISPLAYING DATA

1.26. The following distribution gives the number of hours of overtime during one month for the employees of a company:

Overtime, h	0	1	2	3	4	5	6	7	8	9	10
Employees	10	2	4	2	6	4	2	4	6	2	8

Display the data in a histogram.

1.27. The frequency distribution of the weekly wages in dollars of a group of unskilled workers follows:

Weekly wages, $	140–160	160–180	180–200	200–220	220–240	240–260	260–280
Number of workers	18	24	32	20	8	6	2

Display the data in a (*a*) histogram, (*b*) frequency polygon.

1.28. The amounts of 45 personal loans from a loan company follow:

\$700	\$450	\$725	\$1125	\$675	\$1650	\$750	\$400	\$1050
\$500	\$750	\$850	\$1250	\$725	\$475	\$925	\$1050	\$925
\$850	\$625	\$900	\$1750	\$700	\$825	\$550	\$925	\$850
\$475	\$750	\$550	\$725	\$575	\$575	\$1450	\$700	\$450
\$700	\$1650	\$925	\$500	\$675	\$1300	\$1125	\$775	\$850

(*a*) Group the data into \$200 classes, beginning with \$400, and construct a frequency and cumulative frequency distribution for the grouped data.

(*b*) Display the frequency distribution in a histogram.

1.29. During a 30-day period, the daily number of station wagons rented by an automobile rental agency was as follows:

7	10	6	7	9	4	7	9	9	8	5	5	7	8	4
6	9	7	12	7	9	10	4	7	5	9	8	9	5	7

(*a*) Construct a dotplot (defined in Problem 1.6) of the data.

(*b*) Find its frequency and cumulative frequency distribution.

(*c*) Display the frequency distribution in a histogram.

1.30. A foreign automobile dealer sells English, French, German, Japanese, and Korean automobiles. The number of such automobiles sold in a month follow:

Country	English	French	German	Japanese	Korean
Number	5	3	12	20	10

Display the data in a (*a*) (horizontal) bar graph, (*b*) circular graph.

1.31. The following data are weights of the men (M) and women (W) in an exercise class.

122 (W)	117 (W)	117 (W)	167 (M)	114 (W)
195 (M)	145 (M)	158 (M)	158 (M)	190 (M)
110 (W)	134 (W)	165 (M)	104 (W)	132 (W)
107 (W)	105 (W)	181 (M)	142 (W)	123 (W)
155 (M)	155 (M)	172 (M)	149 (M)	120 (W)
140 (W)	163 (M)	125 (W)	130 (W)	150 (M)
187 (M)	147 (M)	118 (W)	159 (M)	160 (M)
115 (W)	175 (M)	125 (W)	177 (M)	121 (W)

(*a*) Construct a stem-and-leaf display (defined in Problem 1.5) of the data with the tens and hundreds digits as the stem and the units digit as the leaf.

(*b*) Construct a stem-and-leaf display of the data as in part (*a*), but put the leaves for the men's weights to the right of the stem and the leaves for the women's weights to the left of the stem.

MEAN, MEDIAN, MODE, MIDRANGE, VARIANCE, AND STANDARD DEVIATION

1.32. The prices of a pound of coffee in seven stores are:

\$5.58, \$6.18, \$5.84, \$5.75, \$5.67, \$5.95, \$5.62.

Find the (*a*) mean price, (*b*) median price.

1.33. For a given week, the average daily temperature was 35°, 33°, 30°, 36°, 40°, 37°, 38°. Find the (*a*) mean temperature, (*b*) median temperature.

1.34. During a given month, ten salespeople in an automobile dealership sold 13, 17, 10, 18, 17, 9, 17, 13, 15, 14 cars, respectively. Find the (*a*) mean, (*b*) median, (*c*) mode, (*d*) midrange.

1.35. Find the mean, median, mode, and midrange for the data in (*a*) Problem 1.26, (*b*) Problem 1.29.

1.36. Use the class value to find the mean, median, mode, and midrange for the data in (*a*) Problem 1.27, (*b*) Problem 1.28.

1.37. The students in a mathematics class are divided into four groups: (*a*) much greater than the median, (*b*) little above the median, (*c*) little below the median, (*d*) much below the median. On which group should the teacher concentrate in order to increase the median of the class? Mean of the class?

1.38. Find the variance s^2 and standard deviation s for the data in (*a*) Problem 1.26, (*b*) Problem 1.29.

1.39. Use the class value to find the variance s^2 and standard deviation s for the data in (*a*) Problem 1.27, (*b*) Problem 1.28.

1.40. Find the variance s^2 and standard deviation s for the data in (*a*) Problem 1.32, (*b*) Problem 1.33.

QUARTILES AND PERCENTILES

1.41. Find the quartiles Q_1, Q_2, Q_3 for the following data: 15, 17, 17, 20, 21, 21, 25, 27, 30, 31, 35.

1.42. Find the 5-number summary L, Q_1, Q_2, Q_3, H for the data in Problem 1.41.

1.43. Find the 5-number summary L, Q_1, Q_2, Q_3, H for the data in Problem 1.29.

1.44. Find P_{40}, P_{50}, and P_{85} for the following test scores:

55	60	68	73	76	84	88
57	62	70	75	77	84	90
58	64	71	75	79	85	91
58	66	71	76	80	87	93
58	66	72	76	82	88	95

1.45. With reference to the data in Problem 1.31, find P_{60}, P_{75}, and P_{93} for (*a*) the men's weights, (*b*) the women's weights, (*c*) the men's and women's weights combined.

MISCELLANEOUS PROBLEMS INVOLVING ONE VARIABLE

1.46. The students at a small school are divided into four groups: A, B, C, D. The number n of students in each group and the mean score $\bar{x}$ of each group on an exam follow:

A: $n = 80$, $\bar{x} = 78$; B: $n = 60$, $\bar{x} = 74$; C: $n = 85$, $\bar{x} = 77$; D: $n = 75$, $\bar{x} = 80$

Find the mean grade of the school.

1.47. Four students, A, B, C, D, received the following scores on exams with the following respective means $\bar{x}$ and standard deviations s:

A received 88, where $\bar{x} = 85, s = 4$; C received 90, where $\bar{x} = 86, s = 5$;
B received 85, where $\bar{x} = 82, s = 3$; D received 85, where $\bar{x} = 82, s = 2$

Rank the students by finding their respective standard scores z_A, z_B, z_C, z_D.

1.48. Three micrometers, A, B, C, yield the following respective means $\bar{x}$ and standard deviations s:

A: $\bar{x} = 23$, $s = 0.2$; B: $\bar{x} = 62$, $s = 0.5$; C: $\bar{x} = 48$, $s = 0.4$

Rank the micrometers by finding their respective coefficients of variation.

BIVARIATE DATA

1.49. The following table lists one person's oxygen utilization in units of liters per minute for times of t minutes into an exercise routine and t minutes following the routine:

t minutes	0	4	12	16	26
liters/minute during exercise	0.2	0.4	0.9	1.2	3.0
liters/minute following exercise	3.0	1.0	0.5	0.4	0.2

Let x = the oxygen rate during the exercise and y = the rate after the exercise. Find (*a*) the covariance S_{xy}, (*b*) the correlation coefficient r.

1.50. Consider the data in Problem 1.49. (*a*) Plot x against y in a scatterplot. (*b*) Find the least-squares line L for the data and graph L on the scatterplot in (*a*). (*c*) Find the least-squares hyperbolic curve C (which has the form $y = 1/(a + bx)$ or $1/y = a + bx$) for the data, and plot C on the scatterplot in (*a*). (Hint: Find the least-squares line for the data points $(x_i, 1/y_i)$.) (*d*) Which curve, L or C, best fits the data?

1.51. The following table lists average male weight in pounds and height in inches for certain ages which range from 1 to 21.

Age	1	3	6	10	13	16	21
Weight	20	30	45	60	95	140	155
Height	28	36	44	50	60	66	70

Find the correlation coefficient r for: (*a*) age and weight, (*b*) age and height, (*c*) weight and height.

1.52. Let x = weight, y = height for the data in Problem 1.51. (*a*) Plot x against y in a scatterplot. (*b*) Find the line L of best fit. (*c*) Graph L on the scatterplot in (*a*).

1.53. Repeat Problem 1.52, but let x = height and y = weight.

1.54. Find the least-squares exponential curve $y = ab^x$ for the following data:

x	1	2	3	4	5	6
y	6	12	24	50	95	190

1.55. Derive the following normal equations for the least-squares parabola $y = a + bx + cx^2$ for a set of n data points $P_i(x_i, y_i)$:

$$na + (\sum x)b + (\sum x^2)c = \sum y$$
$$(\sum x)a + (\sum x^2)b + (\sum x^3)c = \sum xy$$
$$(\sum x^2)a + (\sum x^3)b + (\sum x^4)c = \sum x^2 y$$

[This problem requires calculus as in Problem 1.25.]

Answers to Supplementary Problems

1.26. The histogram is shown in Fig. 1-37.

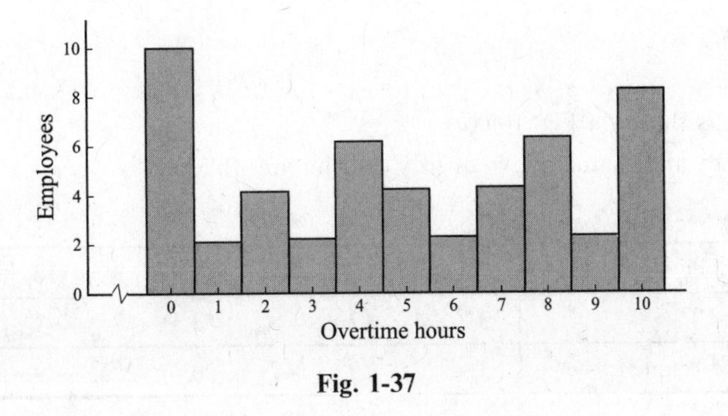

Fig. 1-37

1.27. The histogram and frequency polygon are shown in Fig. 1-38.

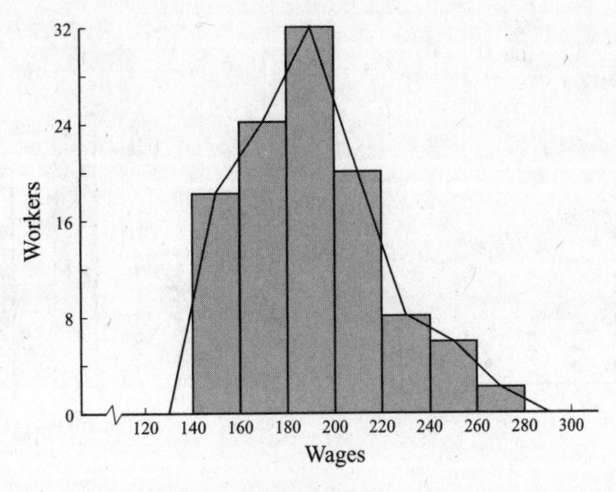

Fig. 1-38

1.28. (*a*) The frequency distribution (where the wage is divided by \$100 for notational convenience) follows:

Amount ÷ \$100	4–6	6–8	8–10	10–12	12–14	14–16	16–18
Number of loans	11	14	10	4	2	1	3

(*b*) The histogram is shown in Fig. 1-39.

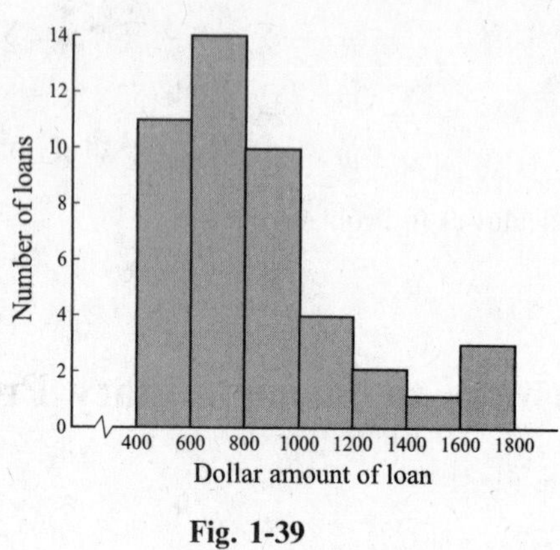

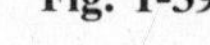
Fig. 1-39

1.29. (*a*) The dotplot is shown in Fig. 1-40(*a*).

(*b*) The frequency and cumulative frequency distributions follow:

Daily number of wagons	4	5	6	7	8	9	10	11	12
Frequency	3	4	2	8	3	7	2	0	1
Cumulative frequency	3	7	9	17	20	27	29	29	30

(*c*) The histogram is shown in Fig. 1-40(*b*).

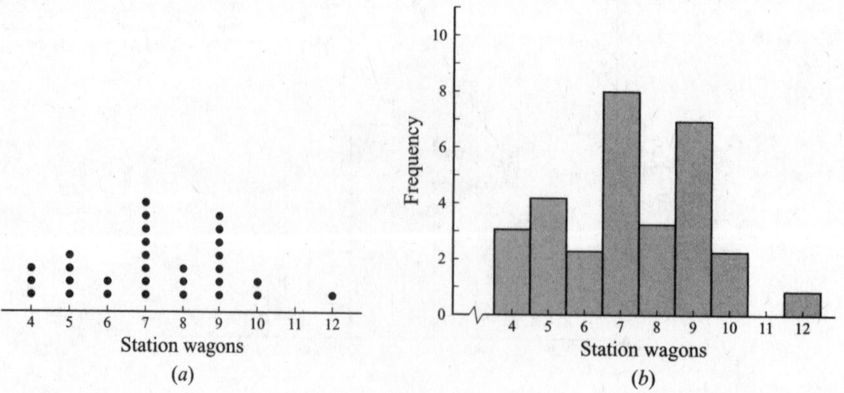

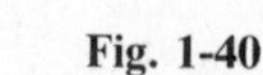
Fig. 1-40

1.30. See Fig. 1-41.

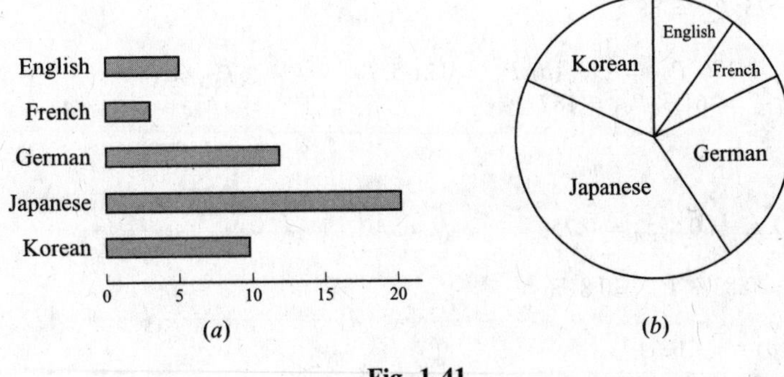

Fig. 1-41

1.31. See Fig. 1-42.

10	7 5 4
11	0 5 7 7 8 5 4
12	2 5 3 0 1
13	4 0 2
14	0 5 7 2 9
15	5 5 8 8 9 0
16	3 5 7 0
17	5 2 7
18	7 1
19	5 0

(a)

Women		Men
4 5 7	10	
4 5 8 7 7 5 0	11	
1 0 3 5 2	12	
2 0 4	13	
2 0	14	5 7 9
	15	5 5 8 8 9 0
	16	3 5 7 0
	17	5 2 7
	18	7 1
	19	5 0

(b)

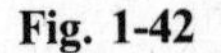

Fig. 1-42

1.32. (a) \$5.80, (b) \$5.75

1.33. (a) 35.67°, (b) 36.5°

1.34. (a) 14.3, (b) 14.5, (c) 17, (d) 13

1.35. (a) $\bar{x} = 4.92$, $\tilde{x} = 5$, mode = 0, midrange = 5; (b) $\bar{x} = 7.3$, $\tilde{x} = 7$, mode = 7, midrange = 8

1.36. (a) $\bar{x} = \$190.36$, $\tilde{x} = \$190$, mode = \$190, midrange = \$210;

(b) $\bar{x} = \$842.22$, $\tilde{x} = \$700$, mode = \$700, midrange = \$1100

1.37. Group (c) to increase the median. Likely (b) and (c) to increase the mean

1.38. (a) $s^2 = 12.97$ hours squared, $s = 3.60$ hours; (b) $s^2 = 4.00$, $s = 2.00$

1.39. (a) $s^2 = 858.58$ dollars squared, $s = \$29.30$; (b) $s^2 = 112{,}040.40$ dollars squared, $s = \$334.72$

1.40. (a) $s^2 = 0.0218$, $s = 0.1476$, (b) $s^2 = 2.3654$, $s = 1.538$

1.41. $Q_1 = 17, Q_2 = 21, Q_3 = 30$

1.42. $L = 15, Q_1 = 17, Q_2 = 21, Q_3 = 30, H = 35$

1.43. $L=4, Q_1 = 6, Q_2 = 7, Q_3 = 9, H = 12$

1.44. $P_{40} = 71.5, P_{50} = 75, P_{85} = 88$

1.45. (*a*) $P_{60} = 166, P_{75} = 176, P_{93} = 190$; (*b*) $P_{60} = 121.5, P_{75} = 127.5, P_{93} = 140$; (*c*) $P_{60} = 152.5, P_{75} = 161.5, P_{93} = 187$

1.46. $\bar{\bar{x}} = 77.4167$

1.47. $z_D = 1.5, z_B = 1.0, z_C = 0.8, z_A = 0.75$

1.48. $V_A = 0.87\%, V_C = 0.83\%, V_B = 0.81\%$

1.49. (*a*) $s_{xy} = -0.82$, (*b*) $r = -0.64$

1.50. (*a*) See Fig. 1-43, which also shows L and C. (*b*) $y = 1.78 - 0.66x$, (*c*) $y = 1/1.6x$, (*d*) C seems a better fit

1.51. (*a*) $r = 0.98$, (*b*) $r = 0.98$, (*c*) $r = 0.97$

1.52. (*a*) and (*c*) are shown in Fig. 1.44. (*b*) $y = 28.55 + 0.28x$

1.53. (*a*) and (*c*) are shown in Fig. 1-45. (*b*) $y = -88.98 + 3.30x$

1.54. $y = 3(2^x)$

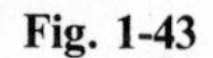

Fig. 1-43

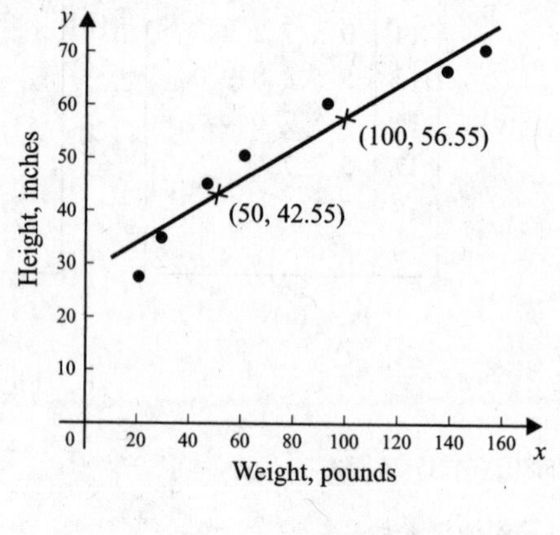

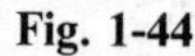

Fig. 1-44

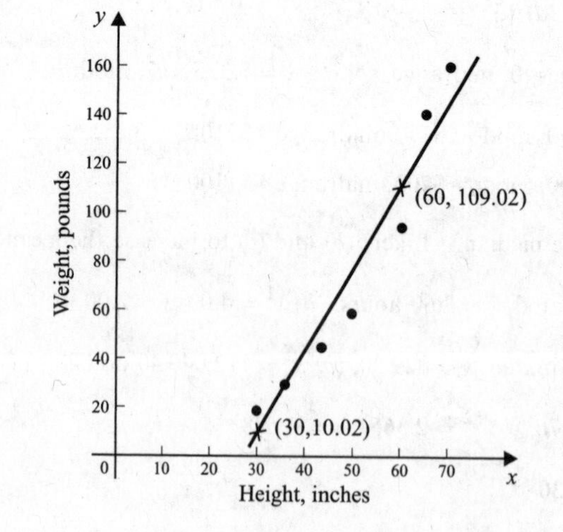

Fig. 1-45

Chapter 2

Sets and Counting

2.1 INTRODUCTION

The concept of a set lies at the foundations of mathematics and, in particular, probability and statistics. This concept formalizes the idea of grouping objects together and viewing them as a single entity. This chapter introduces this notion of a set and three basic operations on sets: union, intersection, and complement. We then discuss methods of counting the elements in a set or the logical possibilities of some event without necessarily enumerating each element or each case.

2.2 SETS AND ELEMENTS, SUBSETS

A set may be viewed as any well-defined collection of objects, called the *elements* or *members* of the set. We usually use capital letters; $A, B, X, Y, \ldots$ to denote sets, and lower-case letters, $a, b, x, y, \ldots$ to denote elements of sets. Synonyms for set are *class*, *collection*, and *family*.

The statement that an element a belongs to a set S is written

$$a \in S$$

(Here $\in$ is a symbol meaning "is an element of".) We also write $a, b \in S$ when both a and b belong to S. If every element of a set A also belongs to a set B, that is, if $a \in A$ implies $a \in B$, then A is called a *subset* of B, or A is said to be *contained* in B, written

$$A \subseteq B \quad \text{or} \quad B \supseteq A$$

Two sets are equal if they both have the same elements or, equivalently, if each is contained in the other. That is:

$$\boxed{A = B \quad \text{if and only if} \quad A \subseteq B \quad \text{and} \quad B \subseteq A}$$

The negations of $a \in A$, $A \subseteq B$, and $A = B$ are written $a \notin A$, $A \not\subseteq B$, and $A \neq B$, respectively.

Remark 1: It is common practice in mathematics to put a vertical line "|" or slanted line "/" through a symbol to indicate the opposite or negative meaning of the symbol.

Remark 2: The statement $A \subseteq B$ does not exclude the possibility that $A = B$. In fact, for any set A, we have $A \subseteq A$ since, trivially, every element in A belongs to A. However, if $A \subseteq B$ and $A \neq B$, then we say that A is a *proper subset* of A (sometimes written $A \subset B$).

Specifying Sets

There are essentially two ways to specify a particular set. One way, if possible, is to list its elements. For example,

$$A = \{1, 3, 5, 7, 9\}$$

means A is the set consisting of the numbers 1, 3, 5, 7, and 9. Note that the elements of the set are separated by commas and enclosed in braces { }. This is called the *tabular form* or *roster method* of a set.

The second way, called the *set-builder form* or *property method*, is to state those properties which characterize the elements in the set, that is, properties held by the members of the set but not by nonmembers. Consider, for example, the expression

$$B = \{x : x \text{ is an even integer}, x > 0\}$$

which is read:

"B is the set of x such that x is an even integer and $x > 0$"

It denotes the set B whose elements are the positive even integers. A letter, usually x, is used to denote a typical member of the set; the colon is read as "such that" and the comma as "and".

EXAMPLE 2.1

(*a*) The above set A can also be written as

$$A = \{x : x \text{ is an odd positive integer}, x < 10\}$$

We cannot list all the elements of the above set B, but we frequently specify the set by writing

$$B = \{2, 4, 6, \ldots\}$$

where we assume everyone knows what we mean. Observe that $9 \in A$ but $9 \notin B$. Also $6 \in B$, but $6 \notin A$.

(*b*) Consider the sets

$$A = \{1, 3, 5, 7, 9\}, \qquad B = \{1, 2, 3, 4, 5\}, \qquad C = \{3, 5\}$$

Then $C \subseteq A$ and $C \subseteq B$, since 3 and 5, the elements of C, are also members of A and B. On the other hand, $A \not\subseteq B$, since $7 \in A$, but $7 \notin B$, and $B \not\subseteq A$, since $2 \in B$ but $2 \notin A$.

(*c*) Suppose a die is tossed. The possible "number" or "points" which appear on the uppermost face of the die belongs to the set {1, 2, 3, 4, 5, 6}. Now suppose a die is tossed and an even number appears. Then the outcome is a member of the set {2, 4, 6} which is a (proper) subset of the set {1, 2, 3, 4, 5, 6} of all possible outcomes.

The following theorem applies.

Theorem 2.1: Let A, B, C be any sets. Then:

(i) $A \subseteq A$
(ii) If $A \subseteq B$ and $B \subseteq A$, then $A = B$
(iii) If $A \subseteq B$ and $B \subseteq C$, the $A \subseteq C$

Some sets occur very often in mathematics, so we have special symbols for them. The following special symbols will be used:

$\mathbf{P}$ = set of *counting numbers* or positive integers: $1, 2, 3, \ldots$

$\mathbf{N}$ = set of *natural numbers* or nonnegative integers: $0, 1, 2, \ldots$

$\mathbf{Z}$ = set of integers: $\ldots, -2, -1, 0, 1, 2, \ldots$

$\mathbf{R}$ = set of real numbers

Thus we have $\mathbf{P} \subseteq \mathbf{N} \subseteq \mathbf{Z} \subseteq \mathbf{R}$.

Universal Set, Empty Set

All sets under investigation in any application of set theory are assumed to be contained in some large fixed set called the *universal set* or *universe*. We denote this set by U unless otherwise specified.

Given a universal set U and a property P, there may be no elements in U which have the property P. The set with no elements is called the *empty set* or *null set*, and is denoted by $\varnothing$. There is only one empty set: If S and T are both empty, then $S = T$ since they have exactly the same elements, namely, none. The empty set $\varnothing$ is also regarded as a subset of every other set. Accordingly, we have

$$\varnothing \subseteq A \subseteq U$$

for any set A.

EXAMPLE 2.2

(*a*) In plane geometry, the universal set consists of all the points in the plane. In human population studies the universal set consists of all the people in the world.

(*b*) Consider the set

$$S = \{x : x \text{ is a positive integer, } x^2 = 3\}$$

Then S has no elements since no positive integer has the required property. Thus, $S = \varnothing$, the empty set.

Disjoint Sets

Two sets A and B are said to be *disjoint* if they have no elements in common. Consider, for example, the sets

$$A = \{1, 2\}, \qquad B = \{2, 4, 6\}, \qquad C = \{4, 5, 6, 7\}$$

Note that A and B are not disjoint, since each contains the element 2, and B and C are not disjoint since each contains the element 4, among others. On the other hand, A and C are disjoint since they have no element in common. We note that if A and B are disjoint, then neither is a subset of the other (unless one is the empty set).

2.3 VENN DIAGRAMS

A Venn diagram is a pictorial representation of sets where sets are represented by enclosed areas in the plane. The universal set U is represented by the points in a rectangle, and the other sets are represented by disks lying within the rectangle. If $A \subseteq B$, then the disk representing A will be entirely within the disk representing B, as in Fig. 2-1(*a*). If A and B are disjoint, i.e. have no elements in common, then the disk representing A will be separated from the disk representing B, as in Fig. 2-1(*b*).

On the other hand, if A and B are two arbitrary sets, it is possible that some elements are in A but not B, some elements are in B but not A, some are in both A and B, and some are in neither A nor B; hence, in general, we represent A and B as in Fig. 2-1(*c*).

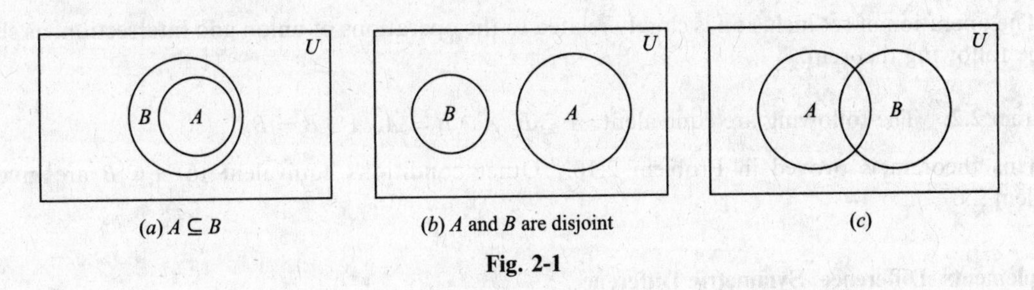

(*a*) $A \subseteq B$ (*b*) A and B are disjoint (*c*)

Fig. 2-1

2.4 SET OPERATIONS

This section defines a number of set operations, including the basic operations of union, intersection, and complement.

Union and Intersection

The *union* of two sets A and B, denoted by $A \cup B$, is the set of all elements which belong to A or to B; that is,

$$A \cup B = \{x : x \in A \quad \text{or} \quad x \in B\}$$

Here "or" is used in the sense of and/or. Figure 2-2(a) is a Venn diagram in which $A \cup B$ is shaded.

The *intersection* of two sets A and B, denoted by $A \cap B$, is the set of all elements which belong to both A and B; that is,

$$A \cap B = \{x : x \in A \quad \text{and} \quad x \in B\}$$

Figure 2-2(b) is a Venn diagram in which $A \cap B$ is shaded.

Recall that sets A and B are said to be disjoint if they have no elements in common or, using the above notation, if $A \cap B = \varnothing$, the empty set. If $S = A \cup B$ and $A \cap B = \varnothing$, then S is called the *disjoint union* of A and B.

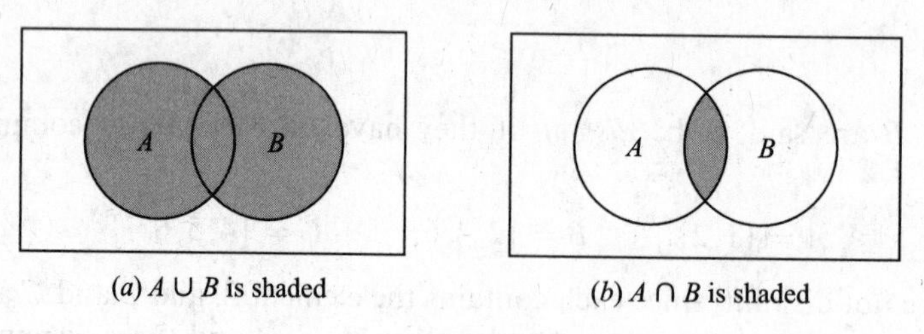

(a) $A \cup B$ is shaded (b) $A \cap B$ is shaded

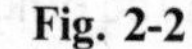

Fig. 2-2

EXAMPLE 2.3

(a) Let $A = \{1, 2, 3, 4\}$, $B = \{4, 5, 6\}$, $C = \{1, 3, 5, 7\}$. Then

$$A \cup B = \{1, 2, 3, 4, 5, 6\}, \qquad A \cup C = \{1, 2, 3, 4, 5, 7\}, \qquad B \cup C = \{1, 3, 4, 5, 6, 7\}$$
$$A \cap B = \{4\}, \qquad A \cap C = \{1, 3\}, \qquad B \cap C = \{5\}$$

(b) Let M and F denote, respectively, the set of male students and the set of female students in a college C. Then

$$M \cup F = C$$

since each student in C belongs to either M or F. Also,

$$M \cap F = \varnothing$$

since no student belongs to both M and F. Thus C is the disjoint union of M and F.

The operation of set inclusion is closely related to the operations of union and intersection, as shown by the following theorem.

Theorem 2.2: The following are equivalent: $A \subseteq B$, $A \cap B = A$, $A \cup B = B$.

This theorem is proved in Problem 2.10. Other conditions equivalent to $A \subseteq B$ are given in Problem 2.67.

Complements, Difference, Symmetric Difference

Recall that all sets under consideration at a particular time are subsets of a fixed universal set U. The *absolute complement* or, simply, *complement* of a set A, denoted by A^c is the set of elements in U which do not belong to A; that is

$$A^c = \{x : x \in U, \ x \notin A\}$$

Some texts denote the complement of A by A' or $\bar{A}$. Figure 2-3(*a*) is a Venn diagram in which A^c is shaded.

The *relative complement* of a set B with respect to a set A or, simply, the difference of A and B, denoted by $A \backslash B$, is the set of elements which belong to A but not to B; that is,

$$A \backslash B = \{x : x \in A, \ x \notin B\}$$

The set $A \backslash B$ is read "A minus B". Many texts denote $A \backslash B$ by $A - B$ or $A \sim B$. Figure 2-3(*b*) is a Venn diagram in which $A \backslash B$ is shaded.

The *symmetric difference* of sets A and B, denoted by $A \oplus B$, consists of those elements which belong to A or B but not both. That is,

$$A \oplus B = (A \cup B) \backslash (A \cap B) \quad \text{or, equivalently,} \quad A \oplus B = (A \backslash B) \cup (B \backslash A)$$

Figure 2-3(*c*) is a Venn diagram in which $A \oplus B$ is shaded.

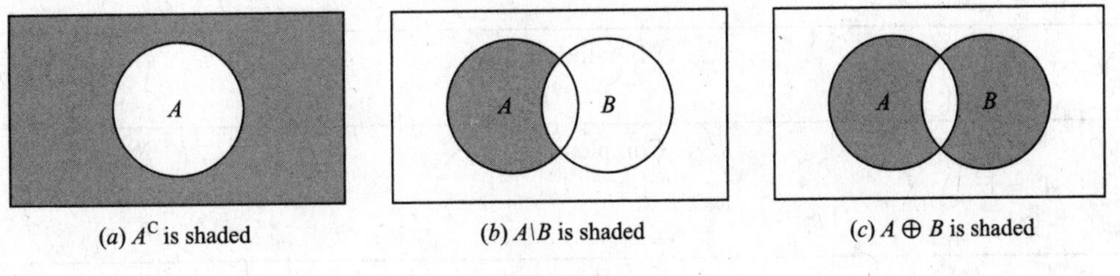

(*a*) A^C is shaded (*b*) $A \backslash B$ is shaded (*c*) $A \oplus B$ is shaded

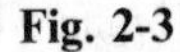

Fig. 2-3

EXAMPLE 2.4 Let $U = \mathbf{P} = \{1, 2, 3, \ldots\}$ be the universal set, and let

$$A = \{1, 2, 3, 4\}, \qquad B = \{3, 4, 5, 6, 7\}, \qquad C = \{6, 7, 8, 9\}, \qquad E = \{2, 4, 6, \ldots\}$$

(Here E is the set of even positive integers.) Then

$$A^c = \{5, 6, 7, \ldots\}, \qquad B^c = \{1, 2, 8, 9, 10, \ldots\}, \qquad E^c = \{1, 3, 5, \ldots\}$$

That is, E^c is the set of odd integers. Also,

$$A \backslash B = \{1, 2\}, \qquad B \backslash C = \{3, 4, 5\}, \qquad B \backslash A = \{5, 6, 7\}, \qquad C \backslash B = \{8, 9\}, \qquad C \backslash E = \{7, 9\}$$

Moreover, $A \oplus B = \{1, 2, 5, 6, 7\}$ and $B \oplus C = \{3, 4, 5, 8, 9\}$

Note that $A \oplus B = (A \backslash B) \cup (B \backslash A)$ and $B \oplus C = (B \backslash C) \cup (C \backslash B)$.

Algebra of Sets

Sets under the operations of union, intersection, and complement satisfy various laws (identities) which are listed in Table 2-1. In fact, we formally state this result:

Theorem 2.3: Sets satisfy the laws in Table 2-1.

Each of the laws in Table 2-1 follows from an equivalent logical law. Consider, for example, the proof of De Morgan's law:

$$(A \cup B)^c = \{x : x \notin (A \text{ or } B)\} = \{x : x \notin A \ \text{ and } \ x \notin B\} = A^c \cup B^c$$

Here we use the equivalent (De Morgan's) logical law:

$$\neg(p \vee q) \equiv \neg p \wedge \neg q$$

where $\neg$ means "not", $\vee$ means "or", and $\wedge$ means "and".

Sometimes Venn diagrams are used to illustrate the laws in Table 2-1 (*cf.* Problem 2.11).

Table 2-1 Laws of the algebra of sets

Idempotent Laws	
1*a*. $A \cup A = A$	1*b*. $A \cap A = A$
Associative Laws	
2*a*. $(A \cup B) \cup C = A \cup (B \cup C)$	2*b*. $(A \cap B) \cap C = A \cap (B \cap C)$
Commutative Laws	
3*a*. $A \cup B = B \cup A$	3*b*. $A \cap B = B \cap A$
Distributive Laws	
4*a*. $A \cup (B \cap C) = (A \cup B) \cap (A \cup C)$	4*b*. $A \cap (B \cup C) = (A \cap B) \cup (A \cap C)$
Identity Laws	
5*a*. $A \cup \varnothing = A$	5*b*. $A \cap U = A$
6*a*. $A \cup U = U$	6*b*. $A \cap \varnothing = \varnothing$
Involution Law	
7. $(A^c)^c = A$	
Complement Laws	
8*a*. $A \cup A^c = U$	8*b*. $A \cap A^c = \varnothing$
9*a*. $U^c = \varnothing$	9*b*. $\varnothing^c = U$
De Morgan's Laws	
10*a*. $(A \cup B)^c = A^c \cap B^c$	10*b*. $(A \cap B)^c = A^c \cup B^c$

Duality

The identities in Table 2-1 are arranged in pairs, as, for example, 2*a* and 2*b*. We now consider the principle behind this arrangement. Let E be an equation of set algebra. The *dual* E^* of E is the equation obtained by replacing each occurrence of $\cup$, $\cap$, U, $\varnothing$ in E by $\cap$, $\cup$, $\varnothing$, U, respectively. For example, the dual of

$$(U \cap A) \cup (B \cap A) = A \quad \text{is} \quad (\varnothing \cup A) \cap (B \cup A) = A$$

Observe that the pairs of laws in Table 2-1 are duals of each other. It is a fact of set algebra, called the *principle of duality*, that, if any equation E is an identity, then its dual E^* is also an identity.

2.5 FINITE AND COUNTABLE SETS

Sets can be finite or infinite. A set A is *finite* if it is empty or if it consists of exactly n elements, where n is a positive integer. Otherwise a set is said to be *infinite*.

EXAMPLE 2.5

(*a*) Let A denote the set of letters in the English alphabet. Then A is finite; it has 26 elements. Let D denote the set of the days of the week:

$$D = \{\text{Monday, Tuesday}, \ldots, \text{Sunday}\}$$

Then D is also finite; it has 7 elements.

(*b*) Let $R = \{x\text{: } x \text{ is a river on the earth}\}$. Although it may be difficult to count the number of rivers on the earth, R is still a finite set.

(*c*) Let Y denote the set of positive even integers, that is, $Y = \{2, 4, 6, \ldots\}$. Then Y is an infinite set.

(*d*) Let $\mathbf{I}$ be the *unit interval*, that is

$$\mathbf{I} = \{x : 0 \leq x \leq 1\}$$

Then $\mathbf{I}$ is also an infinite set.

Countable Sets

A set is *countable* if it is finite or if its elements can be listed in the form of a sequence, in which case it is said to be *countably infinite*; otherwise it is said to be *uncountable*. The above set Y of even integers is countably infinite, whereas it can be proven that the unit interval $\mathbf{I}$ is uncountable.

2.6 COUNTING ELEMENTS IN FINITE SETS, INCLUSION–EXCLUSION PRINCIPLE

The notation $n(S)$ or $|S|$ is used to denote the number of elements in a set S. Thus $n(A) = 26$, where A consists of the letters in the English alphabet and $n(D) = 7$ where D consists of the days in a week. Also, $n(\varnothing) = 0$, since the empty set has no elements.

The following lemma applies.

Lemma 2.4: Suppose A and B are finite disjoint sets. Then $A \cup B$ is finite and

$$n(A \cup B) = n(A) + n(B)$$

Proof: In counting the elements of $A \cup B$, first count those that are in A. There are $n(A)$ of these. The only other elements in $A \cup B$ are those that are in B but not in A. But since A and B are disjoint, no element of B is in A, so there are $n(B)$ elements in B which are not in A. Therefore, $n(A \cup B) = n(A) + n(B)$, as claimed.

Given any sets A and B, we note that A is the disjoint union of the sets $A \backslash B$ and $A \cap B$ (Problem 2.66). Thus Lemma 2.4 gives us the following useful result.

Theorem 2.5: Suppose A and B are finite sets. Then

$$n(A \backslash B) = n(A) - n(A \cap B)$$

That is, the number of elements in $A \backslash B$, that is, elements in A lying outside of B, is equal to the number of elements in A minus the number of elements in both A and B.

Inclusion–Exclusion Principles

There is also a formula for $n(A \cup B)$ even when they are not disjoint, called the inclusion–exclusion principle. Namely:

Theorem (Inclusion–Exclusion Principle) 2.6: Suppose A and B are finite sets. Then $A \cap B$ and $A \cup B$ are finite and

$$n(A \cup B) = n(A) + n(B) - n(A \cap B)$$

That is, we find the number of elements in A or B (or both) by first adding $n(A)$ and $n(B)$ (inclusion) and then subtracting $n(A \cap B)$ (exclusion), since its elements were counted twice.

We can apply this result to get a similar result for three sets.

Corollary 2.7: Suppose A, B, C are finite sets. Then $A \cup B \cup C$ is finite and

$$n(A \cup B \cup C) = n(A) + n(B) + n(C) - n(A \cap B) - n(A \cap C) - n(B \cap C) + n(A \cap B \cap C)$$

Mathematical induction (Section 2.9) may be used to further generalize this result to any finite number of finite sets.

EXAMPLE 2.6 Suppose list A contains the 30 students in a mathematics class and list B contains the 35 students in an English class, and suppose there are 20 names on both lists. Find the number of students:

(*a*) on list A or B,
(*b*) only on list A,
(*c*) only on list B,
(*d*) on exactly one of the two lists.

(*a*) We seek $n(A \cup B)$. By Theorem 2.6,

$$n(A \cup B) = n(A) + n(B) - n(A \cap B) = 30 + 35 - 20 = 45$$

In other words, we combine the two lists and then cross out the 20 names which appear twice.

(*b*) List A contains 30 names and 20 of them are on list B; hence $30 - 20 =$ names are only on list A. That is, by Theorem 2.5,

$$n(A \backslash B) = n(A) - n(A \cap B) = 30 - 20 = 10$$

(*c*) Similarly, there are $35 - 20 = 15$ names only on list B. That is,

$$n(B \backslash A) = n(B) - n(A \cap B) = 35 - 20 = 15$$

(*d*) By (*b*) and (*c*), there are $10 + 15 = 25$ names on exactly one of the two lists. In other words $n(A \oplus B) = 25$.

2.7 PRODUCT SETS

Let A and B be two sets. The *product set* of A and B, denoted by $A \times B$ (read: A cross B), consists of all ordered pairs (a, b) where $a \in A$ and $b \in B$; that is,

$$A \times B = \{(a, b) \colon a \in A,\ b \in B\}$$

The product of a set with itself, say $A \times A$, is denoted by A^2.

We note that two ordered pairs (a, b) and (c, d) are equal if and only if their *first* elements a and c are equal and their *second* elements b and d are equal. That is:

$$(a, b) = (c, d) \quad \text{if and only if} \quad a = c \quad \text{and} \quad b = d$$

EXAMPLE 2.7

(*a*) The reader is familiar with the cartesian plane $\mathbf{R}^2 = \mathbf{R} \times \mathbf{R}$ as discussed in Section 1.8. Here each point P in the plane represents an ordered pair (a, b) of real numbers, and vice versa.

(*b*) Let $A = \{1, 2, 3\}$ and $B = \{a, b\}$. Then

$$A \times B = \{(1, a),\ (1, b),\ (2, a),\ (2, b),\ (3, a),\ (3, b)\}$$

The following theorem applies.

Theorem 2.8: Suppose A and B are finite. Then $A \times B$ is finite and

$$n(A \times B) = n(A) \cdot n(B)$$

The proof follows from the fact that, for each $a \in A$, there will be $n(B)$ ordered pairs in $A \times B$, beginning with a. Hence altogether there will be $n(A) \times n(B)$ ordered pairs in $A \times B$. That is, $n(A \times B) = n(A) \cdot n(B)$, as claimed.

Observe that in Example 2.7(*b*) we have $n(A) = 3$, $n(B) = 2$, and, as expected from Theorem 2.8, $n(A \times B) = 3(2) = 6$.

The concept of a product set is extended to any finite number of sets in a natural way. The product set of sets $A_1, A_2, \ldots, A_m$, written

$$A_1 \times A_2 \times \cdots \times A_m \quad \text{or} \quad \prod_{i=1}^{m} A_i$$

is the set of all m-tuples $(a_1, a_2, \ldots, a_m)$ where $a_1 \in A_1$, $a_2 \in A_2, \ldots, a_m \in A_m$. Furthermore, the above Theorem 2.8 may easily be extended, by induction, to the product of m sets; that is,

$$n(A_1 \times A_2 \times \cdots \times A_m) = n(A_1)n(A_2)\ldots n(A_m)$$

2.8 CLASSES OF SETS, POWER SETS, PARTITIONS

Given a set S, we may wish to talk about some of its subsets. Thus we would be considering a "set of sets". Whenever such a situation arises, to avoid confusion, we will speak of a *class* of sets or a *collection* of sets. If we wish to consider some of the sets in a given class of sets, then we will use the term subclass or subcollection.

EXAMPLE 2.8 Suppose $S = \{1, 2, 3, 4\}$. Let $\mathscr{A}$ be the class of subsets of S which contain exactly three elements of S. Then

$$\mathscr{A} = [\{1,2,3\}, \quad \{1,2,4\}, \quad \{1,3,4\}, \quad \{2,3,4\}]$$

The elements of $\mathscr{A}$ are the sets {1, 2, 3}, {1, 2, 4}, {1, 3, 4}, and {2, 3, 4}.

Let $\mathscr{B}$ be the class of subsets of S which contain 2 and two other elements of S. Then

$$\mathscr{B} = [\{1,2,3\}, \quad \{1,2,4\}, \quad \{2,3,4\}]$$

The elements of $\mathscr{B}$ are {1, 2, 3}, {1, 2, 4}, and {2, 3, 4}. Thus $\mathscr{B}$ is a subclass of $\mathscr{A}$. (To avoid confusion, we will usually enclose the sets of a class in brackets instead of braces.)

Power Sets

For a given set S, we may consider the class of all subsets of S. This class is called the *power set* of S, and it will be denoted by $\mathscr{P}(S)$. If S is finite, then so is $\mathscr{P}(S)$. In fact, the number of elements in $\mathscr{P}(S)$ is 2 raised to the power of S; that is,

$$n(\mathscr{P}(S)) = 2^{n(S)}$$

(For this reason, the power set of S is sometimes denoted by 2^S.]

EXAMPLE 2.9 Suppose $S = \{1, 2, 3\}$. Then

$$\mathscr{P}(S) = [\varnothing, \{1\}, \{2\}, \{3\}, \{1,2\}, \{1,3\}, \{2,3\}, S]$$

Note that the empty set $\varnothing$ belongs to $\mathscr{P}(S)$, since $\varnothing$ is a subset of S. Similarly, S belongs to $\mathscr{P}(S)$. As expected from the above remark, $\mathscr{P}(S)$ has $2^3 = 8$ elements.

Partitions

Let S be a nonempty set. A partition of S is a subdivision of S into nonoverlapping, nonempty subsets. Precisely, a *partition* of S is a collection $\{A_i\}$ of nonempty subsets of S such that:

(i) Each a in S belongs to one of the A_i.

(ii) The sets of $\{A_i\}$ are mutually disjoint; that is, if

$$A_i \neq A_j \quad \text{the} \quad A_i \cap A_j = \varnothing$$

The subsets in a partition are called *cells*. Figure 2-4 is a Venn diagram of a partition of the rectangular set S of points into five cells, A_1, A_2, A_3, A_4, A_5.

EXAMPLE 2.10 Consider the following collections of subsets of $S = \{1, 2, 3, \ldots, 8, 9\}$:

(i) [{1, 3, 5}, {2, 6}, {4, 8, 9}]
(ii) [{1, 3, 5}, {2, 4, 6, 8}, {5, 7, 9}]
(iii) [{1, 3, 5}, {2, 4, 6, 8}, {7, 9}]

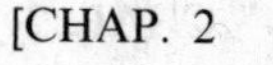

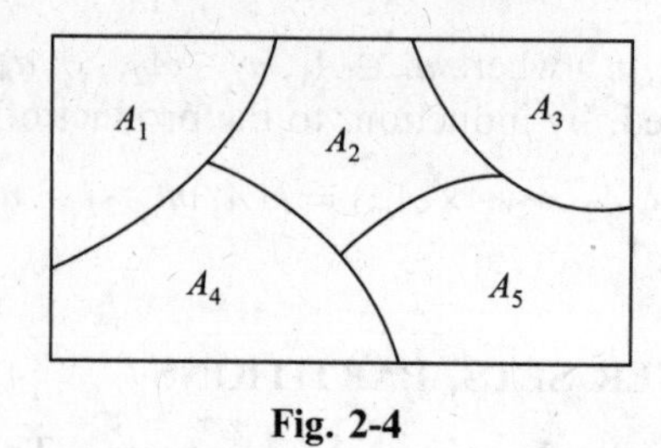

Fig. 2-4

Then (i) is not a partition of S since 7 in S does not belong to any of the subsets. Furthermore, (ii) is not a partition of S since {1, 3, 5} and {5, 7, 9} are not disjoint. On the other hand, (iii) is a partition of S.

Indexed Classes of Sets

When we speak of an *indexed class of sets* $\{A_i: i \in I\}$ or simply $\{A_i\}$, we mean that there is a set A_i assigned to each element $i \in I$. The set I is called the *indexing set* and the sets A_i are said to be indexed by I. When the indexing set is the set **P** of positive integers, the indexed class $\{A_1, A_2, \ldots\}$ is called a *sequence* of sets. By the *union* of these A_i, denoted by $\bigcup_{i \in I} A_i$ (or simply $\bigcup_i A_i$), we mean the set of elements each belonging to at least one of the A_i; and by the *intersection* of the A_i, denoted by $\bigcap_{i \in I} A_i$ (or simply $\bigcap_i A_i$), we mean the set of elements each belonging to every A_i. We also write

$$\bigcup_{i=1}^{\infty} A_i = A_i \cup A_2 \cup \cdots \qquad \text{and} \qquad \bigcap_{i=1}^{\infty} A_i = A_1 \cap A_2 \cap \cdots$$

for the union and intersection, respectively, of a sequence of sets.

Definition: A nonempty class $\mathscr{A}$ of subsets of U is called an *algebra* (*σ-algebra*) of sets if it has the following two properties:

(i) The complement of any set in $\mathscr{A}$ belongs to $\mathscr{A}$.

(ii) The union of any finite (countable) number of sets in $\mathscr{A}$ belongs to $\mathscr{A}$.

That is, $\mathscr{A}$ is closed under complements and finite (countable) unions.

It is simple to show (Problem 2.32) that any algebra (σ-algebra) of sets contains U and $\varnothing$ and is closed under finite (countable) intersections.

2.9 MATHEMATICAL INDUCTION

An essential property of the set $\mathbf{P} = \{1, 2, 3, \ldots\}$ of positive integers which is used in many proofs follows:

Principle of Mathematical Induction I: Let $A(n)$ be an assertion about the set **P** of positive integers, i.e. $A(n)$ is true or false for each integer $n \geq 1$. Suppose $A(n)$ has the following two properties:

(i) $A(1)$ is true.

(ii) $A(n+1)$ is true whenever $A(n)$ is true.

Then $A(n)$ is true for every positive integer.

We shall not prove this principle. In fact, this principle is usually given as one of the axioms when **P** is developed axiomatically.

EXAMPLE 2.11 Let $A(n)$ be the assertion that the sum of the first n odd numbers is n^2; that is:

$$A(n): \quad 1 + 3 + 5 + \cdots + (2n-1) = n^2$$

(The nth odd number is $2n - 1$ and the next odd number is $2n + 1$.) Observe that $A(n)$ is true for $n = 1$ since

$$A(1): \quad 1 = 1^2$$

Assuming $A(n)$ is true, we add $2n + 1$ to both sides of $A(n)$, obtaining

$$1 + 3 + 5 + \cdots + (2n - 1) + (2n + 1) = n^2 + (2n + 1) = (n + 1)^2$$

However, this is $A(n + 1)$. That is, $A(n + 1)$ is true assuming $A(n)$ is true. By the principle of mathematical induction, $A(n)$ is true for all $n \geq 1$.

There is another form of the principle of mathematical induction which is sometimes more convenient to use. Although it appears different, it is really equivalent to the above principle of induction.

Principle of Mathematical Induction II: Let $A(n)$ be an assertion about the set **P** of positive integers with the following two properties:

(i) $A(1)$ is true.

(ii) $A(n)$ is true whenever $A(k)$ is true for $1 \leq k < n$.

Then $A(n)$ is true for every positive integer.

Remark: Sometimes one wants to prove that an assertion A is true for a set of integers of the form

$$\{a, \ a + 1, \ a + 2, \ldots\}$$

where a is any integer, possibly 0. This can be done by simply replacing 1 by a in either of the above principles of mathematical induction.

2.10 COUNTING PRINCIPLES

Combinatorial analysis, which includes the study of permutations and combinations, is concerned with determining the number of logical possibilities of some event without necessarily identifying every case. There are two basic counting principles used throughout. One involves addition and the other multiplication.

Sum Rule Principle

The first counting principle follows:

> *Sum Rule Principle*: Suppose some event E can occur in m ways and a second event F can occur in n ways, and suppose both events cannot occur simultaneously. Then E or F can occur in $m + n$ ways.

This principle can be stated in terms of sets and it is simply a restatement of Lemma 2.4.

> *Sum Rule Principle*: Suppose A and B are disjoint sets. Then
>
> $$n(A \cup B) = n(A) + n(B)$$

Clearly, the principle can be extended to three or more events. Specifically, suppose an event E_1 can occur in n_1 ways, an event E_2 can occur in n_2 ways, an event E_3 can occur in n_3 ways, and so on, and suppose no two of the events can occur at the same time. Then one of the events can occur in $n_1 + n_2 + n_3 + \cdots$ ways.

Product Rule Principle

The second counting principle follows:

> *Product Rule Principle*: Suppose there is an event E which can occur in m ways and, independent of this event, there is a second event F which can occur in n ways. Then combinations of E and F can occur in mn ways.

This principle can also be stated in terms of sets and it is simply a restatement of Theorem 2.8.

> *Product Rule Principle*: Suppose A and B are finite sets. Then
> $$n(A \times B) = n(A) \cdot n(B)$$

Clearly, this principle can also be extended to three or more sets. Specifically, suppose an event E_1 can occur in n_1 ways, then an event E_2 can occur in n_2 ways, then an even E_3 can occur in n_3 ways, and so on. Then all of the events can occur in the order indicated in $n_1 \cdot n_2 \cdot n_3 \cdots$ ways.

EXAMPLE 2.12

(*a*) Suppose a college has 3 different history courses, 4 different literature courses, and 2 different science courses (with no prerequisites).
(1) There are $n = 3 + 4 + 2 = 9$ ways to choose 1 of the courses.
(2) There are $n = 3(4)(2) = 24$ ways to choose one of each of the courses.

(*b*) Suppose Airline A has three daily flights between Boston and Chicago, and Airline B has two daily flights between Boston and Chicago.
(1) There are $n = 3 + 2 = 5$ ways to fly from Boston to Chicago.
(2) There are $n = 3(2) = 6$ ways to fly Airline A from Boston to Chicago, and then Airline B from Chicago back to Boston.
(3) There are $n = 5(5) = 25$ ways to fly from Boston to Chicago, and then back again.

2.11 FACTORIAL NOTATION, BINOMIAL COEFFICIENTS

This section introduces some mathematical notation which is frequently used in combinatorics.

Factorial Notation

The product of the positive integers from 1 to n inclusive is denoted by $n!$ (read "n factorial"). That is,

$$n! = 1 \cdot 2 \cdot 3 \cdot \ldots \cdot (n-2)(n-1)n$$

In other words, $n!$ is defined by

$$1! = 1 \qquad \text{and} \qquad n! = n \cdot ((n-1)!)$$

It is also convenient to define $0! = 1$.

EXAMPLE 2.13

(*a*) $2! = 1 \cdot 2 = 2, \quad 3! = 1 \cdot 2 \cdot 3 = 6, \quad 4! = 1 \cdot 2 \cdot 3 \cdot 4 = 24,$

$5! = 5 \cdot 4! = 5 \cdot 24 = 120, \quad 6! = 6 \cdot 5! = 6 \cdot 120 = 720$

(*b*) $\dfrac{8!}{6!} = \dfrac{8 \cdot 7 \cdot 6!}{6!} = 8 \cdot 7 = 56, \quad 12 \cdot 11 \cdot 10 = \dfrac{12 \cdot 11 \cdot 10 \cdot 9!}{9!} = \dfrac{12!}{9!}, \quad \dfrac{12 \cdot 11 \cdot 10}{1 \cdot 2 \cdot 3} = 12 \cdot 11 \cdot 10 \cdot \dfrac{1}{3!} = \dfrac{12!}{3!\,9!}$

(c) $n(n-1)\cdots(n-r+1) = \dfrac{n(n-1)\cdots(n-r+1)(n-r)(n-r-1)\cdots 3\cdot 2\cdot 1}{(n-r)(n-r-1)\cdots 3\cdot 2\cdot 1} = \dfrac{n!}{(n-r)!}$,

$$\frac{n(n-1)\cdots(n-r+1)}{1\cdot 2\cdot 3\cdots(r-1)r} = n(n-1)\cdots(n-r+1)\cdot\frac{1}{r!} = \frac{n!}{(n-r)!}\cdot\frac{1}{r!} = \frac{n!}{r!\,(n-r)!}$$

Stirling's Approximation to *n*!

A direct evaluation of $n!$ when n is very large is impossible, even with modern-day computers. Accordingly, one frequently uses the approximation formula

$$n! \sim \sqrt{2\pi n}\, n^n e^{-n}$$

(Here $e = 2.71828\ldots$.) The symbol $\sim$ means that, as n gets larger and larger (that is, as $n \to \infty$), the ratio of both sides approaches 1.

Binomial Coefficients

The symbol $\binom{n}{r}$ (read "nCr" or "n choose r"), where r and n are positive integers with $r \le n$, is defined as follows:

$$\binom{n}{r} = \frac{n(n-1)(n-2)\cdots(n-r+1)}{1\cdot 2\cdot 3\cdots(r-1)r} \quad \text{or (by Example 2.13)} \quad \binom{n}{r} = \frac{n!}{r!(n-r)!}$$

But $n-(n-r) = r$; hence we have the following important relation:

$$\binom{n}{n-r} = \binom{n}{r} \quad \text{or, in other words, if } a+b=n \text{ then} \quad \binom{n}{a} = \binom{n}{b}$$

EXAMPLE 2.14

(a) $\binom{8}{2} = \dfrac{8\cdot 7}{1\cdot 2} = 28$ $\quad \binom{9}{4} = \dfrac{9\cdot 8\cdot 7\cdot 6}{1\cdot 2\cdot 3\cdot 4} = 126$ $\quad \binom{12}{5} = \dfrac{12\cdot 11\cdot 10\cdot 9\cdot 8}{1\cdot 2\cdot 3\cdot 4\cdot 5} = 792$

$\binom{10}{3} = \dfrac{10\cdot 9\cdot 8}{1\cdot 2\cdot 3} = 120$ $\quad \binom{13}{1} = \dfrac{13}{1} = 13$

Note that $\binom{n}{r}$ has exactly r factors in both the numerator and the denominator.

(b) Compute $\binom{10}{7}$. This can be done two ways:

$$\binom{10}{7} = \frac{10\cdot 9\cdot 8\cdot 7\cdot 6\cdot 5\cdot 4}{1\cdot 2\cdot 3\cdot 4\cdot 5\cdot 6\cdot 7} = 120 \quad \text{or} \quad \binom{10}{7} = \binom{10}{3} = \frac{10\cdot 9\cdot 8}{1\cdot 2\cdot 3} = 120$$

Observe that the second method (which uses $7+3=10$) saves space and time.

Binomial Coefficients and Pascal's Triangle

The number $\binom{n}{r}$ are called the *binomial coefficients*, since they appear as the coefficients in the expansion of $(a+b)^n$. Specifically, one can prove (Problem 2.59):

Theorem 2.9: $(a+b)^n = \displaystyle\sum_{k=0}^{n}\binom{n}{k} a^{n-k} b^k$

The coefficients of the successive powers of $a+b$ can be arranged in a triangular array of numbers, called Pascal's triangle, as pictured in Fig. 2-5. The numbers in Pascal's triangle have the following interesting properties:

(1) The first and last number in each row is 1.
(2) Every other number in the array can be obtained by adding the two numbers appearing directly above it. For example, $10 = 4+6$, $15 = 5+10$, $20 = 10+10$.

$$
\begin{aligned}
(a+b)^0 &= 1\\
(a+b)^1 &= a + b\\
(a+b)^2 &= a^2 + 2ab + b^2\\
(a+b)^3 &= a^3 + 3a^2b + 3ab^2 + b^3\\
(a+b)^4 &= a^4 + 4a^3b + 6a^2b^2 + 4ab^3 + b^4\\
(a+b)^5 &= a^5 + 5a^4b + 10a^3b^2 + 10a^2b^3 + 5ab^4 + b^5\\
(a+b)^6 &= a^6 + 6a^5b + 15a^4b^2 + 20a^3b^3 + 15a^2b^4 + 6ab^5 + b^6\\
&\cdots\cdots
\end{aligned}
$$

$$
\begin{array}{c}
1\\
1 \quad 1\\
1 \quad 2 \quad 1\\
1 \quad 3 \quad 3 \quad 1\\
1 \quad 4 \quad 6 \quad 4 \quad 1\\
1 \quad 5 \quad (10) \quad 10 \quad 5 \quad 1\\
1 \quad 6 \quad (15) \quad (20) \quad 15 \quad 6 \quad 1\\
\cdots\cdots
\end{array}
$$

Fig. 2-5

Since the numbers appearing in Pascal's triangle are the binomial coefficients, property (2) comes from the following theorem (proved in Problem 2.40):

Theorem 2.10: $\dbinom{n+1}{r} = \dbinom{n}{r-1} + \dbinom{n}{r}$

2.12 PERMUTATIONS

Any arrangement of a set of n objects in a given order is called a *permutation* of the objects (taken all at a time). Any arrangement of any $r \leq n$ of those objects in a given order is called an *r-permutation* or a *permutation of the n objects taken r at a time*. Consider, for example, the set of letters a, b, c, and d. Then:

(1) *bdca, dcba,* and *acdb* are permutations of the four letters (taken all at a time);
(2) *bad, adb, cbd,* and *bca* are permutations of the four letters taken three at a time;
(3) *ad, cb, da,* and *bd* are permutations of the four letters taken two at a time.

The number of permutations of n objects taken r at a time is denoted by

$$P(n,r), \quad {}_nP_r, \quad P_{n,r}, \quad P^n_r, \quad \text{or} \quad (n)_r$$

We shall use $P(n,r)$. Before we derive the general formula for $P(n,r)$ we consider a particular case.

EXAMPLE 2.15 Find the number of permutations of six objects, say A, B, C, D, E, F, taken three at a time. In other words, find the number of "three-letter words" using only the given six letters without repetitions.

Let the general three-letter word be represented by the following three boxes:

☐ ☐ ☐

Now the first letter can be chosen in six different ways; following this, the second letter can be chosen in five different ways; and, following this, the last letter can be chosen in four different ways. Write each number in its appropriate box as follows:

Thus by the fundamental principle of counting there are $6 \cdot 5 \cdot 4 = 120$ possible three-letter words without repetitions from the six letters, or there are 120 permutations of six objects taken three at a time:

$$P(6,3) = 120$$

Derivation of the Formula for $P(n, r)$

The derivation of the formula for the number of permutations of n objects taken r at a time, or the number of r-permutations of n objects, $P(n,r)$, follows the procedure in the preceding example. The first element in an r-permutation of n objects can be chosen in n different ways; following this, the second element in the permutation can be chosen in $n-1$ ways; and, following this, the third element in the permutation can be chosen in $n-2$ ways. Continuing in this manner, we have that the rth (last) element in the r-permutation can be chosen in $n-(r-1)=n-r+1$ ways. Thus, by the fundamental principle of counting, we have

$$P(n,r)=n(n-1)(n-2)\cdots(n-r+1)$$

By Example 2.13(*c*), we see that

$$n(n-1)(n-2)\cdots(n-r+1)=\frac{n(n-1)(n-2)\cdots(n-r+1)\cdot(n-r)!}{(n-r)!}=\frac{n!}{(n-r)!}$$

Thus we have proven:

Theorem 2.11: $P(n,r)=\dfrac{n!}{(n-r)!}$

In the special case in which $r=n$, we have

$$P(n,n)=n(n-1)(n-2)\cdots 3\cdot 2\cdot 1=n!$$

Accordingly,

Corollary 2.12: There are $n!$ permutations of n objects (taken all at a time).

For example, there are $3!=1\cdot 2\cdot 3=6$ permutations of the three letters a, b, and c. These are

$$abc,\quad acb,\quad bac,\quad bca,\quad cab,\quad cba$$

Permutations with repetitions

Frequently we want to know the number of permutations of a *multiset*; that is, a set of objects some of which are alike. We will let

$$P(n;n_1,n_2,\ldots,n_r)$$

denote the number of permutations of n objects of which n_1 are alike, n_2 are alike, ..., n_r are alike. The general formula follows:

Theorem 2.13: $P(n;n_1,n_2,\ldots,n_r)=\dfrac{n!}{n_1!\,n_2!\ldots n_r!}$

We indicate the proof of the above theorem by a particular example. Suppose we want to form all possible five-letter "words" using the letters from the word "BABBY". Now there are $5!=120$ permutations of the objects B_1, A, B_2, B_3, Y, where the three Bs are distinguished. Observe that the following six permutations

$$B_1B_2B_3AY,\quad B_2B_1B_3AY,\quad B_3B_1B_2AY,\quad B_1B_3B_2AY,\quad B_2B_3B_1AY,\quad B_3B_2B_1AY$$

produce the same word when the subscripts are removed. The 6 comes from the fact that there are $3!=3\cdot 2\cdot 1=6$ different ways of placing the three Bs in the first three positions in the permutation. This is true for each set of three positions in which the Bs can appear. Accordingly there are

$$P(5;3)=\frac{5!}{3!}=\frac{120}{6}=20$$

different five-letter words that can be formed using the letters from the word "BABBY".

EXAMPLE 2.16 Find the number m of seven-letter words that can be formed using the letters of the word "BENZENE".

We seek the number of permutations of seven objects of which three are alike (the three Es) and two are alike (the two Ns). By Theorem 2.13,

$$m = P(7;3,2) = \frac{7!}{3!\,2!} = \frac{7\cdot 6\cdot 5\cdot 4\cdot 3\cdot 2\cdot 1}{3\cdot 2\cdot 1\cdot 2\cdot 1} = 420$$

Ordered Samples

Many problems in combinatorial analysis and, in particular, probability and statistics are concerned with choosing an element from a set S containing n elements (or a card from a deck or a person from a population). When we choose one element after another from the set S, say r times, we call the choice an ordered sample of size r. We consider two cases:

(1) *Sampling with replacement*

Here the element is replaced in the set S before the next element is chosen. Since there are n different ways to choose each element (repetitions are allowed), the product rule principle tells us that there are

$$\overbrace{n\cdot n\cdot n\cdots n}^{r\text{ times}} = n^r$$

different ordered samples with replacement of size r.

(2) *Sampling without replacement*

Here the element is not replaced in the set S before the next element is chosen. Thus there are no repetitions in the ordered sample. Accordingly, an ordered sample of size r without replacement is simply an r-permutation of the elements in the set S with n elements. Thus there are

$$P(n,r) = n(n-1)(n-2)\cdots(n-r+1) = \frac{n!}{(n-r)!}$$

different ordered samples without replacement of size r from a population (set) with n elements. In other words, by the product rule, the first element can be chosen in n ways, the second in $n-1$ ways, and so on.

EXAMPLE 2.17 Three cards are chosen in succession from a deck with 52 cards. Find the number of ways this can be done (*a*) with replacement, (*b*) without replacement.

(*a*) Since each card is replaced before the next card is chosen, each card can be chosen in 52 ways. Thus there are

$$52(52)(52) = 52^3 = 140{,}608$$

different ordered samples of size $r = 3$ with replacement.

(*b*) Since there is no replacement, the first card can be chosen in 52 ways, the second card in 51 ways, and the last card in 50 ways. Thus there are

$$P(52,3) = 52(51)(50) = 132{,}600$$

different ordered samples of size $r = 3$ without replacement.

2.13 COMBINATIONS

Suppose we have a collection of n objects. A *combination* of these n objects taken r at a time is any selection of r of the objects where order doesn't count. In other words, an *r-combination* of a set of n objects is any subset of r elements. For example, the combinations of the letters a, b, c, d taken three at a time are

$$\{a,b,c\},\ \{a,b,d\},\ \{a,c,d\},\ \{b,c,d\} \quad \text{or simply} \quad abc,\ abd,\ acd,\ bcd$$

Observe that the following combinations are equal:

$$abc, \quad acb, \quad bac, \quad bca, \quad cab, \quad cba$$

That is, each denotes the same set $\{a, b, c\}$.

The number of combinations of n objects taken r at a time is denoted by

$$C(n, r)$$

The symbols ${}_nC_r$, $C_{n,r}$, and C_r^n also appear in various texts. Before we give the general formula for $C(n, r)$, we consider a special case.

EXAMPLE 2.18 Find the number of combinations of four objects, a, b, c, d, taken three at a time.

Each combination consisting of three objects determines $3! = 6$ permutations of the objects in the combination as pictured in Fig. 2-6. Thus the number of combinations multiplied by $3!$ equals the number of permutations. That is:

$$C(4, 3) \cdot 3! = P(4, 3) \quad \text{or} \quad C(4, 3) = \frac{P(4, 3)}{3!}$$

But $P(4, 3) = 4 \cdot 3 \cdot 2 = 24$ and $3! = 6$. Thus $C(4, 3) = 4$, which is noted in Fig. 2-6.

Combinations	Permutations
abc	*abc, acb, bac, bca, cab, cba*
abd	*abd, adb, bad, bda, dab, dba*
acd	*acd, adc, cad, cda, dac, dca*
bcd	*bcd, bdc, cbd, cdb, dbc, dcb*

Fig. 2-6

Formula for $C(n, r)$

Since any combination of n objects taken r at a time determines $r!$ permutations of the objects in the combination, we can conclude that

$$P(n, r) = r!\, C(n, r)$$

Thus we obtain

Theorem 2.14: $C(n, r) = \dfrac{P(n, r)}{r!} = \dfrac{n!}{r!\,(n-r)!}$

Recall that the binomial coefficient $\binom{n}{r}$ was defined to be $\dfrac{n!}{r!\,(n-r)!}$. Thus:

$$\boxed{C(n, r) = \binom{n}{r}}$$

We shall use $C(n, r)$ and $\binom{n}{r}$ interchangeably.

EXAMPLE 2.19

(*a*) Find the number m of committees of three that can be formed from eight people. Each committee is, essentially, a combination of the eight people taken three at a time. Thus

$$m = C(8, 3) = \binom{8}{3} = \frac{8 \cdot 7 \cdot 6}{1 \cdot 2 \cdot 3} = 56$$

(*b*) A farmer buys three cows, two pigs, and four hens from a man who has six cows, five pigs, and eight hens. How many choices does the farmer have?

The farmer can choose the cows in $\binom{6}{3}$ ways, the pigs in $\binom{5}{2}$ ways, and the hens in $\binom{8}{4}$ ways. Hence altogether he can choose the animals in

$$\binom{6}{3}\binom{5}{2}\binom{8}{4} = \frac{6 \cdot 5 \cdot 4}{1 \cdot 2 \cdot 3} \cdot \frac{5 \cdot 4}{1 \cdot 2} \cdot \frac{8 \cdot 7 \cdot 6 \cdot 5}{1 \cdot 2 \cdot 3 \cdot 4} = 20 \cdot 10 \cdot 70 = 14{,}000 \text{ ways}$$

EXAMPLE 2.20 Find the number m of ways that 9 toys can be divided between 4 children if the youngest is to receive 3 toys and each of the others 2 toys.

There are $C(9,3) = 84$ ways to first choose 3 toys for the youngest. Then there are $C(6,2) = 15$ ways to choose 2 of the remaining 6 toys for the oldest. Next, there are $C(4,2) = 6$ ways to choose 2 of the remaining 4 toys for the second oldest. The third oldest receives the remaining 2 toys. Thus, by the product rule,

$$m = 84(15)(6) = 7560$$

Alternately, by Problem 2.123,

$$m = \frac{9!}{3!\,2!\,2!\,2!} = 7560$$

2.14 TREE DIAGRAMS

A tree diagram is a device used to enumerate all the possible outcomes of a sequence of experiments or events where each event can occur in a finite number of ways. The construction of tree diagrams is illustrated in the following example.

EXAMPLE 2.21

(*a*) Find the product set $A \times B \times C$ where $A = \{1,2\}$, $B = \{a,b,c\}$, and $C = \{3,4\}$.

The tree diagram for the set $A \times B \times C$ appears in Fig. 2-7. Observe that the tree is constructed from left to right, and that the number of branches at each point corresponds to the number of possible outcomes of the next event. Each endpoint of the tree is labeled by the corresponding element of $A \times B \times C$. As expected from Theorem 2.8, $A \times B \times C$ contains $n = 2(3)(2) = 12$ elements.

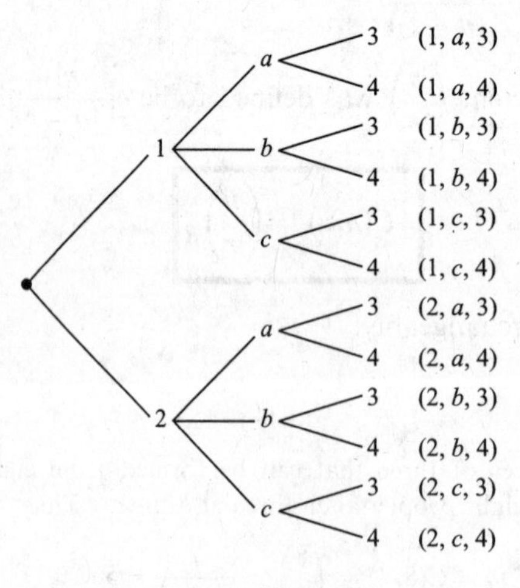

Fig. 2-7

(*b*) Marc and Erik are to play a tennis tournament. The first person to win two games in a row or who wins a total of three games wins the tournament. Find the number of ways the tournament can occur.

The tree diagram showing the possible outcomes of the tournament appears in Fig. 2-8. Specifically, there are 10 endpoints, which correspond to the 10 ways that the tournament can occur:

MM, MEMM, MEMEM, MEMEE, MEE, EMM, EMEMM, EMEME, EMEE, EE

The path from the beginning of the tree to the endpoint describes who won which game in the individual tournament.

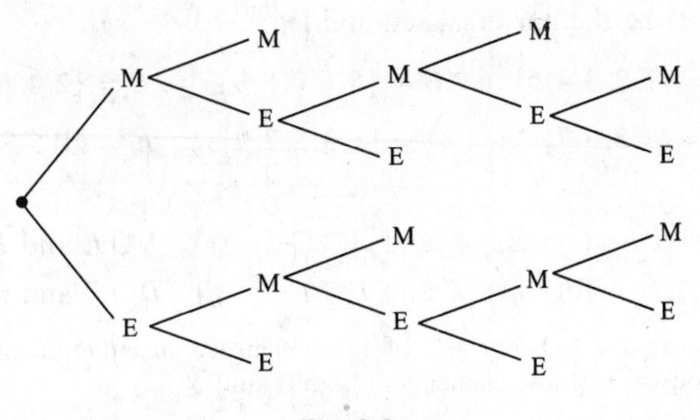

Fig. 2-8

Solved Problems

SETS, SUBSETS

2.1. List the elements of the following sets, where $\mathbf{P} = \{1, 2, 3, \ldots\}$:

(*a*) $A = \{x : x \in \mathbf{P},\ 3 < x < 7\}$,
(*b*) $B = \{x : x \in \mathbf{P},\ x \text{ is even},\ x < 9\}$,
(*c*) $C = \{x : x \in \mathbf{P}, x + 4 = 3\}$,
(*d*) $D = \{x : x \in \mathbf{P},\ x \text{ is a multiple of } 5\}$

(*a*) A consists of the positive integers between 3 and 7; hence $A = \{4, 5, 6\}$.

(*b*) B consists of the even positive integers less than 9; hence $B = \{2, 4, 6, 8\}$.

(*c*) There are no positive integers which satisfy the condition $x + 4 = 3$; hence C contains no elements. In other words $C = \varnothing$, the empty set.

(*d*) D is infinite, so we cannot list all its elements. However, sometimes we can write $D = \{5, 10, 15, 20, \ldots\}$ assuming everyone understands that we mean the multiples of 5.

2.2. Show that $A = \{2, 3, 4, 5\}$ is not a subset of $B = \{x : x \in \mathbf{P},\ x \text{ is even}\}$.

It is necessary to show that at least one element in A does not belong to B. Now $3 \in A$ and, since B consists of even numbers, $3 \notin B$; hence A is not a subset of B.

2.3. Show that $A = \{2, 3, 4, 5\}$ is a proper subset of $C = \{1, 2, 3, \ldots, 8, 9\}$.

Each element of A belongs to C so $A \subseteq C$. On the other hand, $1 \in C$ but $1 \notin A$. Hence $A \neq C$. Therefore A is a proper subset of C.

2.4. Prove Theorem 2.1(iii): If $A \subseteq B$ and $B \subseteq C$, then $A \subseteq C$.

We must show that each element in A also belongs to C. Let $x \in A$. Now $A \subseteq B$ implies $x \in B$. But $B \subseteq C$; hence $x \in C$. We have shown that $x \in A$ implies $x \in C$, that is, that $A \subseteq C$.

SET OPERATIONS

2.5. Let $U = \{1, 2, \ldots, 9\}$ be the universal set, and let

$$A = \{1,2,3,4,5\} \qquad C = \{5,6,7,8,9\} \qquad E = \{2,4,6,8\}$$
$$B = \{4,5,6,7\} \qquad D = \{1,3,5,7,9\} \qquad F = \{1,5,9\}$$

Find:
(*a*) $A \cup B$ and $A \cap B$ (*c*) $A \cup C$ and $A \cap C$ (*e*) $E \cup E$ and $E \cap E$
(*b*) $B \cup D$ and $B \cap D$ (*d*) $D \cup E$ and $D \cap E$ (*f*) $D \cup F$ and $D \cap F$

Recall that the union $X \cup Y$ consists of those elements in either X or Y (or both), and that the intersection $X \cap Y$ consists of those elements in both X and Y.

(*a*) $A \cup B = \{1,2,3,4,5,6,7\}$ $A \cap B = \{4,5\}$
(*b*) $B \cup D = \{1,3,4,5,6,7,9\}$ $B \cap D = \{5,7\}$
(*c*) $A \cup C = \{1,2,3,4,5,6,7,8,9\} = U$ $A \cap C = \{5\}$
(*d*) $D \cup E = \{1,2,3,4,5,6,7,8,9\} = U$ $D \cap E = \varnothing$
(*e*) $E \cup E = \{2,4,6,8\} = E$ $E \cap E = \{2,4,6,8\} = E$
(*f*) $D \cup F = \{1,3,5,7,9\} = D$ $D \cap F = \{1,5,9\} = F$

Observe that $F \subseteq D$; so by Theorem 2.2 we must have $D \cup F = D$ and $D \cap F = F$.

2.6. Consider the sets in the preceding Problem 2.5. Find:

(*a*) A^c, B^c, D^c, E^c (*b*) $A \backslash B, B \backslash A, D \backslash E, F \backslash D$ (*c*) $A \oplus B, C \oplus D, E \oplus F$

(*a*) The complement X^c consists of those elements in the universal set U which do not belong to X. Hence:

$$A^c = \{6,7,8,9\}, \quad B^c = \{1,2,3,8,9\}, \quad D^c = \{2,4,6,8\} = E, \quad E^c = \{1,3,5,7,9\} = D$$

(*b*) The difference $X \backslash Y$ consists of the elements in X which do not belong to Y. Hence:

$$A \backslash B = \{1,2,3\}, \quad B \backslash A = \{6,7\}, \quad D \backslash E = \{1,3,5,7,9\} = D, \quad F \backslash D = \varnothing$$

(*c*) The symmetric difference $X \oplus Y$ consists of the elements in X or Y but not in both X and Y. Hence:

$$A \oplus B = \{1,2,3,6,7\}, \quad C \oplus D = \{1,3,8,9\}, \quad E \oplus F = \{2,4,6,8,1,5,9\} = E \cup F$$

2.7. Show that we can have $A \cap B = A \cap C$ without $B = C$.

Let $A = \{1,2\}$, $B = \{2,3\}$, and $C = \{2,4\}$. Then $A \cap B = \{2\}$ and $A \cap C = \{2\}$. Accordingly,

$$A \cap B = A \cap C \text{ but } B \neq C$$

2.8. Prove: $B \backslash A = B \cap A^c$. Thus the set operation of difference can be written in terms of the operations of intersection and complementation.

$$B \backslash A = \{x : x \in B,\ x \notin A\} = \{x : x \in B,\ x \in A^c\} = B \cap A^c$$

2.9. Prove: $(A \cap B) \subseteq A \subseteq (A \cup B)$ and $(A \cap B) \subseteq B \subseteq (A \cup B)$.

Since every element in $A \cap B$ is in both A and B, it is certainly true that if $x \in (A \cap B)$ then $x \in A$; hence $(A \cap B) \subseteq A$. Furthermore, if $x \in A$, then $x \in (A \cup B)$ (by the definition of $A \cup B$), so $A \subseteq (A \cup B)$. Putting these together gives $(A \cap B) \subseteq A \subseteq (A \cup B)$. Similarly, $(A \cap B) \subseteq B \subseteq (A \cup B)$.

2.10. Prove Theorem 2.2: The following are equivalent: $A \subseteq B$, $A \cap B = A$, and $A \cup B = B$.

Suppose $A \subseteq B$ and let $x \in A$. Then $x \in B$, hence $x \in A \cap B$ and $A \subseteq A \cap B$. By Problem 2.9, $(A \cap B) \subseteq A$. Therefore $A \cap B = A$. On the other hand, suppose $A \cap B = A$ and let $x \in A$. Then $x \in (A \cap B)$; hence $x \in A$ and $x \in B$. Therefore, $A \subseteq B$. Both results show that $A \subseteq B$ is equivalent to $A \cap B = A$.

Suppose again that $A \subseteq B$. Let $x \in (A \cup B)$. Then $x \in A$ or $x \in B$. If $x \in A$, then $x \in B$ because $A \subseteq B$. In either case, $x \in B$. Therefore $A \cup B \subseteq B$. By Problem 2.9, $B \subseteq A \cup B$. Therefore $A \cup B = B$. Now suppose $A \cup B = B$ and let $x \in A$. Then $x \in A \cup B$ by definition of union of sets. Hence $x \in B = A \cup B$. Therefore $A \subseteq B$. Both results show that $A \subseteq B$ is equivalent to $A \cup B = B$.

Thus $A \subseteq B$, $A \cup B = A$ and $A \cup B = B$ are equivalent.

VENN DIAGRAMS, ALGEBRA OF SETS, DUALITY

2.11. Illustrate De Morgan's Law $(A \cup B)^c = A^c \cap B^c$ (proved in Section 2.5) using Venn diagrams.

Shade the area outside $A \cup B$ in a Venn diagram of sets A and B. This is shown in Fig. 2-9(*a*); hence the shaded area represents $(A \cup B)^c$. Now shade the area outside A in a Venn diagram of A and B with strokes in one direction (///), and then shade the area outside B with strokes in another direction (\\\). This is shown in Fig. 2-9(*b*); hence the cross-hatched area (area where both lines are present) represents the intersection of A^c and B^c, i.e. $A^c \cap B^c$. Both $(A \cup B)^c$ and $A^c \cap B^c$ are represented by the same area; thus the Venn diagrams indicate $(A \cup B)^c = A^c \cap B^c$. (We emphasize that a Venn diagram is not a formal proof, but it can indicate relationships between sets.)

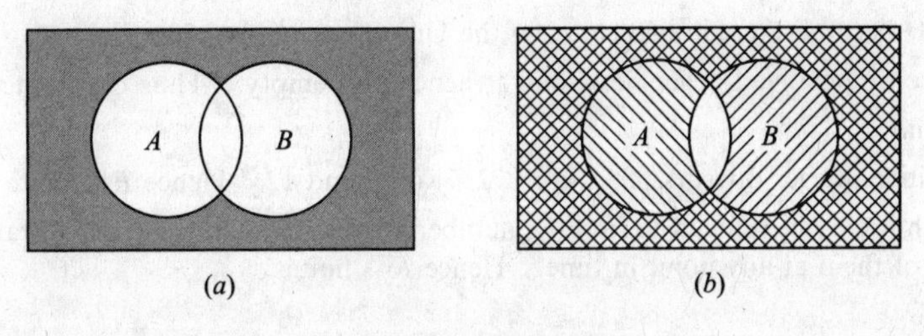

(*a*) (*b*)

Fig. 2-9

2.12. Prove the Distributive Law: $A \cap (B \cup C) = (A \cap B) \cup (A \cap C)$ (Theorem 2.3(4*b*)).

By the definitions of union and intersection,

$$A \cap (B \cup C) = \{x : x \in A,\ x \in B \cup C\}$$
$$= \{x : x \in A,\ x \in B \quad \text{or} \quad x \in A,\ x \in C\} = (A \cap B) \cup (A \cap C)$$

Here we use the analogous logical law $p \wedge (q \vee r) \equiv (p \wedge q) \vee (p \wedge r)$ where $\wedge$ denotes "and" and $\vee$ denotes "or".

2.13. Prove $(A \cup B) \backslash (A \cap B) = (A \backslash B) \cup (B \backslash A)$. (Thus either one may be used to define the symmetric difference $A \oplus B$.)

Using $X \backslash Y = X \cap Y^c$ and the laws in Table 2-1, including De Morgan's laws, we obtain:

$$\begin{aligned}(A \cup B) \backslash (A \cap B) &= (A \cup B) \cap (A \cap B)^c = (A \cup B) \cap (A^c \cup B^c) \\ &= (A \cup A^c) \cup (A \cap B^c) \cup (B \cap A^c) \cup (B \cap B^c) \\ &= \varnothing \cup (A \cap B^c) \cup (B \cap A^c) \cup \varnothing \\ &= (A \cap B^c) \cup (B \cap A^c) = (A \backslash B) \cup (B \backslash A)\end{aligned}$$

2.14. Write the dual of each set equation:

(*a*) $(U \cap A) \cup (B \cap A) = A$

(*b*) $(A \cup B \cup C)^c = (A \cup C)^c \cap (A \cup B)^c$

(*c*) $(A \cap U) \cap (\varnothing \cup A^c) = \varnothing$

(*d*) $(A \cap U)^c \cap A = \varnothing$

Interchange $\cup$ and $\cap$ and also U and $\varnothing$ in each set equation:

(*a*) $(\varnothing \cup A) \cap (B \cup A) = A$

(*b*) $(A \cap B \cap C)^c = (A \cap C)^c \cup (A \cap B)^c$

(*c*) $(A \cup \varnothing) \cup (U \cap A^c) = U$

(*d*) $(A \cup \varnothing)^c \cup A = U$

FINITE SETS AND THE COUNTING PRINCIPLE

2.15. Determine which of the following sets are finite:

(*a*) $A = \{\text{seasons in the year}\}$

(*b*) $B = \{\text{states in the Union}\}$

(*c*) $C = \{\text{positive integers less than 1}\}$

(*d*) $D = \{\text{odd integers}\}$

(*e*) $E = \{\text{positive integral divisors of 12}\}$

(*f*) $F = \{\text{cats living in the United States}\}$

(*a*) A is finite since there are four seasons in the year, i.e. $n(A) = 4$.

(*b*) B is finite because there are 50 states in the Union, i.e. $n(B) = 50$.

(*c*) There are no positive integers less than 1; hence C is empty. Thus C is finite and $n(C) = 0$.

(*d*) D is infinite.

(*e*) The positive integer divisors of 12 are 1, 2, 3, 4, 6, and 12. Hence E is finite and $n(E) = 6$.

(*f*) Although it may be difficult to find the number of cats living in the United States, there is still a finite number of them at any point in time. Hence F is finite.

2.16. Suppose 50 science students are polled to see whether or not they have studied French (F) or German (G) yielding the following data: 25 studied French, 20 studied German, 5 studied both. Find the number of the students who studied: (*a*) only French, (*b*) French or German, (*c*) neither language.

(*a*) Here 25 studied French, and 5 of them also studied German; hence $25 - 5 = 20$ students only studied French. That is, by Theorem 2.5,

$$n(F \backslash G) = n(F) - n(F \cap G) = 25 - 5 = 20$$

(*b*) By the inclusion–exclusion principle, Theorem 2.6,

$$n(F \cup G) = n(F) + n(G) - n(F \cap G) = 25 + 20 - 5 = 40$$

(*c*) Since 40 studied French or German, $50 - 40 = 10$ studied neither language.

2.17. In a survey of 60 people, it was found that:

25 read *Newsweek* magazine
26 read *Time*
26 read *Fortune*
9 read both *Newsweek* and *Fortune*
11 read both *Newsweek* and *Time*
8 read both *Time* and *Fortune*

3 read all three magazines

(*a*) Find the number of people who read at least one of the three magazines.

(*b*) Fill in the correct number of people in each of the eight regions of the Venn diagram in Fig. 2-10(*a*) where N, T, and F denote the set of people who read *Newsweek*, *Time*, and *Fortune*, respectively.

(*c*) Find the number of people who read exactly one magazine.

(*a*) We want $n(N \cup T \cup F)$, by Corollary 2.7:

$$n(N \cup T \cup F) = n(N) + n(T) + n(F) - n(N \cap T) - n(N \cap F) - n(T \cap F) + n(N \cap T \cap F)$$
$$= 25 + 26 + 26 - 11 - 9 - 8 + 3 = 52$$

(*b*) The required Venn diagram in Fig. 2-10(*b*) is obtained as follows:

3 read all three magazines

$11 - 3 = 8$ read *Newsweek* and *Time* but not all three magazines

$9 - 3 = 6$ read *Newsweek* and *Fortune* but not all three magazines

$8 - 3 = 5$ read *Time* and *Fortune* but not all three magazines

$25 - 8 - 6 - 3 = 8$ read only *Newsweek*

$26 - 8 - 5 - 3 = 10$ read only *Time*

$26 - 6 - 5 - 3 = 12$ read only *Fortune*

$60 - 52 = 8$ read no magazine at all

(*c*) $8 + 10 + 12 = 30$ read only one magazine.

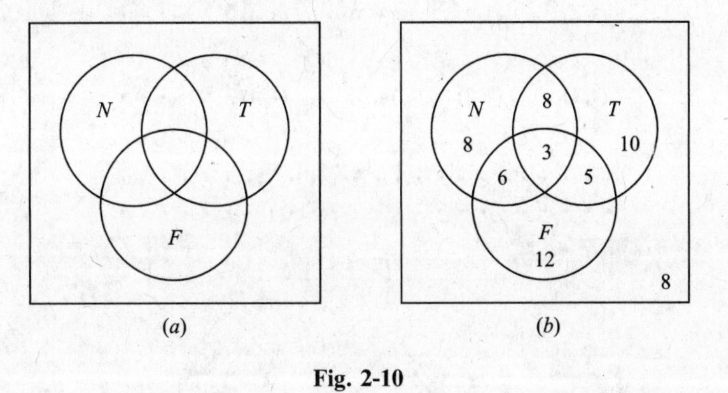

Fig. 2-10

2.18. Prove Theorem 2.6: If A and B are finite sets, then $A \cup B$ and $A \cap B$ are finite and $n(A \cup B) = n(A) + n(B) - n(A \cap B)$.

If A and B are finite, then clearly $A \cap B$ and $A \cup B$ are finite.

Suppose we count the element of A and then count the elements of B. Then every element in $A \cap B$ would be counted twice, once in A and once in B. Hence

$$n(A \cup B) = n(A) + n(B) - n(A \cap B)$$

Alternatively (Problem 2.66), A is the disjoint union of $A\backslash B$ and $A \cap B$, B is the disjoint union of $B\backslash A$ and $A \cap B$, and $A \cup B$ is the disjoint union of $A\backslash B$, $A \cap B$, and $B\backslash A$. Therefore, by Lemma 2.4,

$$\begin{aligned} n(A \cup B) &= n(A\backslash B) + n(A \cap B) + n(B\backslash A) \\ &= n(A\backslash B) + n(A \cap B) + n(B\backslash A) + n(A \cap B) - n(A \cap B) \\ &= n(A) + n(B) - n(A \cap B). \end{aligned}$$

2.19. Show that each set is countable: (*a*) set **Z** of integers, (*b*) **P** × **P**.

A set S is countable if (*a*) S is finite or (*b*) the element of S can be listed in the form of a sequence or, in other words, there is a one-to-one correspondence between the positive integers (counting numbers) $\mathbf{P} = \{1, 2, 3, \ldots\}$ and S. Neither set is finite.

(*a*) The following shows a one-to-one correspondence between **P** and **Z**:

Counting numbers **P**:	1	2	3	4	5	6	7	8	...
	↓	↓	↓	↓	↓	↓	↓	↓	
Integers **Z**:	0	1	−1	2	−2	3	−3	4	...

That is, $n \in \mathbf{P}$ corresponds to either $n/2$, when n is even, or $(1 - n)/2$, when n is odd. Thus **Z** is countable.

(*b*) Figure 2-11 shows that **P** × **P** can be written as an infinite sequence as follows:

$$(1, 1), \quad (2, 1), \quad (1, 2), \quad (1, 3), \quad (2, 2), \quad \ldots$$

Specifically, the sequence is determined by "following the arrows" in Fig. 2-11.

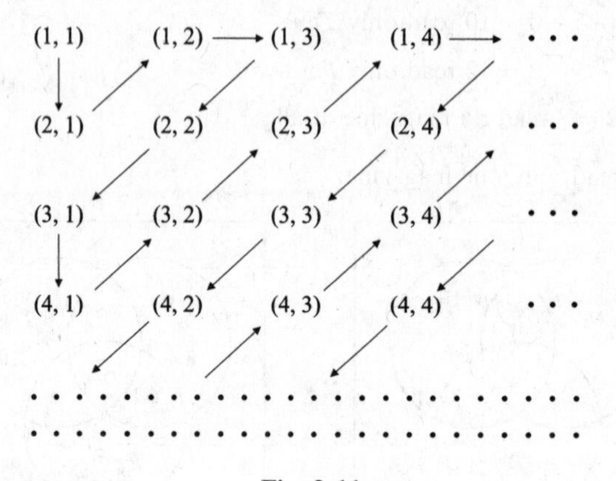

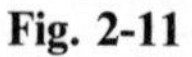
Fig. 2-11

PRODUCT SETS

2.20. Find x and y given that $(3x, x - 2y) = (6, -8)$.

Two ordered pairs are equal if and only if the corresponding components are equal. Hence we obtain the equations $3x = 6$ and $x - 2y = -8$ from which $x = 2$, $y = 5$.

2.21. Let $A = \{1, 2, 3\}$ and $B = \{a, b\}$. Find (*a*) $A \times B$, (*b*) $B \times A$.

(*a*) $A \times B$ consists of all ordered pairs with the first component from A and the second component from B. Thus

$$A \times B = \{(1, a),\ (1, b),\ (2, a),\ (2, b),\ (3, a),\ (3, b)\}$$

(b) Here the first component is from B and the second component is from A:

$$B \times A = \{(a, 1),\ (a, 2),\ (a, 3),\ (b, 1),\ (b, 2),\ (b, 3)\}$$

2.22. Let $A = \{a, b, c, d\}$ and $B = \{x, y, z\}$. Determine the number of elements in (a) $A \times B$, (b) $B \times A$, (c) A^3, (d) B^4.

Here $n(A) = 4$ and $n(B) = 3$. To obtain the number of elements in each product set, multiply the number of elements in each set in the product:

(a) $n(A \times B) = 4(3) = 12$

(b) $n(B \times A) = 3(4) = 12$

(c) $n(A^3) = 4(4)(4) = 64$

(d) $n(B^4) = 3^4 = 81$

2.23. Each toss of a coin will yield either a head or a tail. Let $C = \{H, T\}$ denote the set of outcomes. Find C^3, $n(C^3)$, and explain what C^3 represents.

Since $n(C) = 2$, we have $n(C^3) = 2^3 = 8$. Omitting certain commas and parentheses for notational convenience,

$$C^3 = \{HHH,\ HHT,\ HTH,\ HTT,\ THH,\ THT,\ TTH,\ TTT\}$$

C^3 represents all possible sequences of outcomes of three tosses of the coin.

2.24. Prove: $A \times (B \cap C) = (A \times B) \cap (A \times C)$.

$$\begin{aligned} A \times (B \cap C) &= \{(x, y) : x \in A,\ y \in B \cap C\} \\ &= \{(x, y) : x \in A,\ y \in B, y \in C\} \\ &= \{(x, y) : (x, y) \in A \times B,\ (x, y) \in A \times C\} \\ &= (A \times B) \cap (A \times C) \end{aligned}$$

CLASSES OF SETS, PARTITIONS

2.25. Find the elements of the set $A = [\{1, 2, 3\},\ \{4, 5\},\ \{6, 7, 8\}]$.

A is a class of sets; its elements are the sets $\{1, 2, 3\}$, $\{4, 5\}$, and $\{6, 7, 8\}$.

2.26. Determine the power set $\mathscr{P}(A)$ of $A = \{a, b, c, d\}$.

The elements of $\mathscr{P}(A)$ are the subsets of A. Hence

$$\mathscr{P}(A) = [A,\ \{a, b, c\},\ \{a, b, d\},\ \{a, c, d\},\ \{b, c, d\},\ \{a, b\},\ \{a, c\},\ \{a, d\},\ \{b, c\},\ \{b, d\},\ \{c, d\},\ \{a\},\ \{b\},\ \{c\},\ \{d\},\ \varnothing]$$

As expected, $\mathscr{P}(A)$ has $2^4 = 16$ elements.

2.27. Let $S = \{a, b, c, d, e, f, g\}$. Determine which of the following are partitions of S:

(a) $P_1 = [\{a, c, e\},\ \{b\},\ \{d, g\}]$

(b) $P_2 = [\{a, e, g\},\ \{c, d\},\ \{b, e, f\}]$

(c) $P_3 = [\{a, b, e, g\},\ \{c\},\ \{d, f\}]$

(d) $P_4 = [\{a, b, c, d, e, f, g\}]$

(*a*) P_1 is not a partition of S since $f \in S$ does not belong to any of the cells.

(*b*) P_2 is not a partition of S since $e \in S$ belongs to two of the cells.

(*c*) P_3 is a partition of S since each element in S belongs to exactly one cell.

(*d*) P_4 is a partition of S into one cell, S itself.

2.28. Find all partitions of $S = \{a, b, c, d\}$.

Note first that each partition of S contains either 1, 2, 3, or 4 distinct cells. The partitions are as follows:

(1) $[\{a, b, c, d\}]$

(2) $[\{a\}, \{b, c, d\}]$, $[\{b\}, \{a, c, d\}]$, $[\{c\}, \{a, b, d\}]$, $[\{d\}, \{a, b, c\}]$,
$[\{a, b\}, \{c, d\}]$, $[\{a, c\}, \{b, d\}]$, $[\{a, d\}, \{b, c\}]$

(3) $[\{a\}, \{b\}, \{c, d\}]$, $[\{a\}, \{c\}, \{b, d\}]$, $[\{a\}, \{d\}, \{b, c\}]$,
$[\{b\}, \{c\}, \{a, d\}]$, $[\{b\}, \{d\}, \{a, c\}]$, $[\{c\}, \{d\}, \{a, b\}]$,

(4) $[\{a\}, \{b\}, \{c\}, \{d\}]$.

There are 15 different partitions of S.

2.29. Let $\mathbf{P} = \{1, 2, 3, \ldots\}$ and, for each $n \in \mathbf{P}$, let

$$A_n = \{x : x \text{ is a multiple of } n\} = \{n,\ 2n,\ 3n, \ldots\}$$

Find (*a*) $A_3 \cap A_5$, (*b*) $A_4 \cap A_6$, (*c*) $\bigcup_{i \in Q} A_i$, where $Q = \{2, 3, 5, 7, 11, \ldots\}$ is the set of prime numbers.

(*a*) Those numbers which are multiples of both 3 and 5 are the multiples of 15; hence $A_3 \cap A_5 = A_{15}$.

(*b*) The multiples of 12 and no other numbers belong to both A_4 and A_6; hence $A_4 \cap A_6 = A_{12}$.

(*c*) Every positive integer except 1 is a multiple of at least one prime number; hence

$$\bigcup_{i \in Q} A_i = \{2, 3, 4, \ldots\} = \mathbf{P} \backslash \{1\}$$

2.30. Prove: Let $\{A_i : i \in I\}$ be an indexed class of sets and let $i_0 \in I$. Then

$$\bigcap_{i \in I} A_i \subseteq A_{i_0} \subseteq \bigcup_{i \in I} A_i$$

Let $x \in \bigcap_{i \in I} A_i$; then $x \in A_i$ for every $i \in I$. In particular, $x \in A_{i_0}$. Hence $\bigcap_{i \in I} A_i \subseteq A_{i_0}$. Now let $y \in A_{i_0}$. Since $i_0 \in I$, $y \in \bigcap_{i \in I} A_i$. Hence $A_{i_0} \subseteq \bigcup_{i \in I} A_i$.

2.31. Prove (De Morgan's law): For any indexed class $\{A_i : i \in I\}$, we have $(\bigcup_i A_i)^c = \bigcap_i A_i^c$.

Using the definitions of union and intersection of indexed classes of sets:

$$\begin{aligned}(\textstyle\bigcup_i A_i)^c &= \{x : x \notin \textstyle\bigcup_i A_i\} = \{x : x \notin A_i \text{ for every } i\}\\ &= \{x : x \in A_i^c \text{ for every } i\} = \textstyle\bigcap_i A_i^c\end{aligned}$$

2.32. Let $\mathscr{A}$ be an algebra (σ-algebra) of subsets of U. Show that: (*a*) U and $\varnothing$ belong to $\mathscr{A}$; and (*b*) $\mathscr{A}$ is closed under finite (countable) intersections.

Recall that $\mathscr{A}$ is closed under complements and finite (countable) unions.

(a) Since $\mathscr{A}$ is nonempty, there is a set $A \in \mathscr{A}$. Hence the complement $A^c \in \mathscr{A}$, and the union $U = A \cup A^c \in \mathscr{A}$. Also the complement $\varnothing = U^c \in \mathscr{A}$.

(b) Let $\{A_i\}$ be a finite (countable) class of sets belonging to $\mathscr{A}$. By De Morgan's law (Problem 2.31) $(\bigcup_i A_i^c)^c = \bigcap_i A_i^{cc} = \bigcap_i A_i$. Hence $\bigcap_i A_i$ belongs to $\mathscr{A}$, as required.

MATHEMATICAL INDUCTION

2.33. Prove the assertion $A(n)$ that the sum of the first n positive integers is $\frac{1}{2}n(n+1)$; that is,

$$A(n)\colon\ 1 + 2 + 3 + \cdots + n = \tfrac{1}{2}n(n+1)$$

The assertion holds for $n = 1$ since

$$A(1)\colon\ 1 = \tfrac{1}{2} \cdot 1(1+1) = 1$$

Assuming $A(n)$ is true, we add $n+1$ to both sides of $A(n)$, obtaining

$$\begin{aligned} 1 + 2 + 3 + \cdots + n + (n+1) &= \tfrac{1}{2}n(n+1 + (n+1) \\ &= \tfrac{1}{2}[n(n+1) + 2(n+1)] \\ &= \tfrac{1}{2}[(n+1)(n+2)] \end{aligned}$$

which is $A(n+1)$. That is, $A(n+1)$ is true whenever $A(n)$ is true. By the principle of induction, $A(n)$ is true for all n.

2.34. Prove the following assertion (for $n \geq 0$):

$$A(n)\colon\ 1 + 2 + 2^2 + 2^3 + \cdots + 2^n = 2^{n+1} - 1$$

$A(0)$ is true since $1 = 2^1 - 1$. Assuming $A(n)$ is true, we add 2^{n+1} to both sides of $A(n)$, obtaining

$$\begin{aligned} 1 + 2^1 + 2^2 + \cdots + 2^n + 2^{n+1} &= 2^{n+1} - 1 + 2^{n+1} \\ &= 2(2^{n+1}) - 1 \\ &= 2^{n+2} - 1 \end{aligned}$$

which is $A(n+1)$. Thus $A(n+1)$ is true whenever $A(n)$ is true. By the principle of induction, $A(n)$ is true for all $n \geq 0$.

FACTORIAL NOTATION, BINOMIAL COEFFICIENTS

2.35. Compute: (a) 4!, 5!, 6!, 7!, 8!, 9!, (b) 50!

(a) Use $(n+1)! = (n+1)n!$ after calculating 4! and 5!:

$$4! = 1 \cdot 2 \cdot 3 \cdot 4 = 24, \qquad 7! = 7(6!) = 7(720) = 5040$$

$$5! = 1 \cdot 2 \cdot 3 \cdot 4 \cdot 5 = 5(24) = 120, \qquad 8! = 8(7!) = 8(5040) = 40{,}320$$

$$6! = 6(5!) = 6(120) = 720, \qquad 9! = 9(8!) = 9(40{,}320) = 362{,}880$$

(b) Since n is very large, we use Stirling's approximation that $n! \sim \sqrt{2\pi n}\, n^n e^{-n}$ (where $e = 2.718$). Thus

$$50! \sim \sqrt{100\pi}\, 50^{50} e^{-50} = N$$

Evaluating N using a calculator, we get $N = 3.04 \times 10^{64}$ (which has 65 digits).

Alternatively, using (base 10) logarithms, we get

$$\begin{aligned}\log N &= \log(\sqrt{100\pi}50^{50}e^{-50})\\ &= \tfrac{1}{2}\log 100 + \tfrac{1}{2}\log\pi + 50\log 50 - 50\log e\\ &= \tfrac{1}{2}(2) + \tfrac{1}{2}(0.4972) + 50(1.6990) - 50(0.4343)\\ &= 64.4836\end{aligned}$$

The antilog yields $N = 3.04 \times 10^{64}$.

2.36. Compute: (a) $\dfrac{13!}{11!}$ (b) $\dfrac{7!}{10!}$

(a) $\dfrac{13!}{11!} = \dfrac{13\cdot 12\cdot 11\cdot 10\cdot 9\cdot 8\cdot 7\cdot 6\cdot 5\cdot 4\cdot 3\cdot 2\cdot 1}{11\cdot 10\cdot 9\cdot 8\cdot 7\cdot 6\cdot 5\cdot 4\cdot 3\cdot 2\cdot 1} = 13\cdot 12 = 156$

Alternatively, this could be solved as follows:

$$\frac{13!}{11!} = \frac{13\cdot 12\cdot 11!}{11!} = 13\cdot 12 = 156$$

(b) $\dfrac{7!}{10!} = \dfrac{7!}{10\cdot 9\cdot 8\cdot 7!} = \dfrac{1}{10\cdot 9\cdot 8} = \dfrac{1}{720}$

2.37. Compute: (a) $\dbinom{16}{3}$, (b) $\dbinom{12}{4}$

Recall that there are as many factors in the numerator as in the denominator.

(a) $\dbinom{16}{3} = \dfrac{16\cdot 15\cdot 14}{1\cdot 2\cdot 3} = 560$ (b) $\dbinom{12}{4} = \dfrac{12\cdot 11\cdot 10\cdot 9}{1\cdot 2\cdot 3\cdot 4} = 495$

2.38. Compute: (a) $\dbinom{8}{5}$, (b) $\dbinom{9}{7}$

(a) $\dbinom{8}{5} = \dfrac{8\cdot 7\cdot 6\cdot 5\cdot 4}{1\cdot 2\cdot 3\cdot 4\cdot 5} = 56$ or, since $8-5=3$, $\dbinom{8}{5} = \dbinom{8}{3} = \dfrac{8\cdot 7\cdot 6}{1\cdot 2\cdot 3} = 56$

(b) Since $9-7=2$, $\dbinom{9}{7} = \dbinom{9}{2} = \dfrac{9\cdot 8}{1\cdot 2} = 36$

2.39. Prove: $\dbinom{17}{6} = \dbinom{16}{5} + \dbinom{16}{6}$

Now $\dbinom{16}{5} + \dbinom{16}{6} = \dfrac{16!}{5!\,11!} + \dfrac{16!}{6!\,10!}$. Multiply the first fraction by $\dfrac{6}{6}$ and the second by $\dfrac{11}{11}$ to obtain the same denominator in both fractions; and then add:

$$\begin{aligned}\binom{16}{5} + \binom{16}{6} &= \frac{6\cdot 16!}{6\cdot 5!\cdot 11!} + \frac{11\cdot 16!}{6!\cdot 11\cdot 10!} = \frac{6\cdot 16!}{6!\cdot 11!} + \frac{11\cdot 16!}{6!\cdot 11!}\\ &= \frac{6\cdot 16! + 11\cdot 16!}{6!\cdot 11!} = \frac{(6+11)\cdot 16!}{6!\cdot 11!} = \frac{17\cdot 16!}{6!\cdot 11!} = \frac{17!}{6!\cdot 11!} = \binom{17}{6}\end{aligned}$$

2.40. Prove Theorem 2.10: $\dbinom{n+1}{r} = \dbinom{n}{r-1} + \dbinom{n}{r}$

(The technique in this proof is similar to that of the preceding problem.)

Now $\binom{n}{r-1}+\binom{n}{r}=\frac{n!}{(r-1)!\cdot(n-r+1)!}+\frac{n!}{r!\cdot(n-r)!}$. To obtain the same denominator in both fractions, multiply the first fraction by $\frac{r}{r}$ and the second fraction by $\frac{n-r+1}{n-r+1}$. Hence

$$\begin{aligned}\binom{n}{r-1}+\binom{n}{r} &= \frac{r\cdot n!}{r\cdot(r-1)!\cdot(n-r+1)!}+\frac{(n-r+1)\cdot n!}{r!\cdot(n-r+1)\cdot(n-r)!}\\ &= \frac{r\cdot n!}{r!\,(n-r+1)!}+\frac{(n-r+1)\cdot n!}{r!\,(n-r+1)!}\\ &= \frac{r\cdot n!+(n-r+1)\cdot n!}{r!\,(n-r+1)!}=\frac{[r+(n-r+1)]\cdot n!}{r!\,(n-r+1)!}\\ &= \frac{(n+1)n!}{r!\,(n-r+1)!}=\frac{(n+1)!}{r!\,(n-r+1)!}=\binom{n+1}{r}\end{aligned}$$

COUNTING PRINCIPLES

2.41. Suppose a bookcase shelf has 6 mathematics texts, 3 physics texts, 4 chemistry texts, and 5 computer science texts. Find the number n of ways a student can choose: (*a*) one of the texts, (*b*) one of each type of text.

(*a*) Here the sum rule applies; hence $n = 6 + 3 + 4 + 5 = 18$.

(*b*) Here the product rule applies; hence $n = 6\cdot 3\cdot 4\cdot 5 = 360$.

2.42. A restaurant has a menu with 3 appetizers, 4 entrées, and 2 desserts. Find the number n of ways a customer can order an appetizer, entrée, and dessert.

Here the product rule applies, since the customer orders one of each. Thus $n = 3\cdot 4\cdot 2 = 24$.

2.43. A history class contains 7 male students and 5 female students. Find the number n of ways that the class can elect: (*a*) a class representative, (*b*) two class representatives, one male and one female, (*c*) a president and a vice-president.

(*a*) Here the sum rule is used; hence $n = 7 + 5 = 12$.

(*b*) Here the product rule is used; hence $n = 7\cdot 5 = 35$.

(*c*) There are 12 ways to elect the president and then 11 ways to elect the vice-president. Thus $n = 12\cdot 11 = 132$.

2.44. There are four bus lines from city A to city B and three bus lines from city B to city C. Find the number n of ways a person can travel by bus: (*a*) from A to C by way of B, (*b*) round-trip from A to C by way of B, (*c*) round-trip from A to C by way of B, without using a bus line more than once.

(*a*) There are 4 ways to go from A to B, and 3 ways from B to C; hence, by the product rule, $n = 4\cdot 3 = 12$.

(*b*) There are 12 ways to go from A to C by way of B, and 12 ways to return. Thus, by the product rule, $n = 12\cdot 12 = 144$.

(*c*) The person will travel from A to B to C to B to A. Enter these letters with connecting arrows as follows:

$$\mathrm{A}\to\mathrm{B}\to\mathrm{C}\to\mathrm{B}\to\mathrm{A}$$

There are 4 ways to go from A to B and 3 ways to go from B to C. Since a bus line is not to be used more than once, there are only 2 ways to go from C back to B and only 3 ways to go from B back to A. Enter these numbers above the corresponding arrows as follows:

$$A \xrightarrow{4} B \xrightarrow{3} C \xrightarrow{2} B \xrightarrow{3} A$$

Thus, by the product rule, $n = 4 \cdot 3 \cdot 2 \cdot 3 = 72$.

PERMUTATIONS, ORDERED SAMPLES

2.45. State the essential difference between permutations and combinations, with examples.

Order counts with permutations, such as words, sitting in a row, or electing a president, vice-president, and treasurer. Order does not count with combinations, such as committees or teams (without counting positions). The product rule is usually used with permutations since the choice for each of the ordered positions may be viewed as a sequence of events.

2.46. A family has 3 boys and 2 girls. (*a*) Find the number of ways they can sit in a row. (*b*) How many ways are there if the boys and girls are each to sit together?

(*a*) The five children can sit in a row in $5 \cdot 4 \cdot 3 \cdot 2 \cdot 1 = 5! = 120$ ways.

(*b*) There are two ways to distribute them according to sex: BBBGG or GGGBB. In each case, the boys can sit in $3 \cdot 2 \cdot 1 = 3! = 6$ ways, and the girls can sit in $2 \cdot 1 = 2! = 2$ ways. Thus, altogether, there are $2 \cdot 3! \cdot 2! = 2 \cdot 6 \cdot 2 = 24$ ways.

2.47. Suppose repetitions are not allowed. (*a*) Find the number n of three-digit numbers that can be formed from the digits 2, 3, 5, 6, 7, and 9. (*b*) How many of them are even? (*c*) How many of them exceed 400?

There are 6 digits, and the three-digit number may be pictured by __, __, __. In each case, write down the number of ways that one can fill each of the positions.

(*a*) There are 6 ways to fill the first position, 5 ways to fill the second position, and 4 ways to fill the third position. This may be pictured by: $\underline{6}, \underline{5}, \underline{4}$. Thus $n = 6 \cdot 5 \cdot 4 = 120$.
Alternatively, n is the number of permutations of 6 things taken 3 at a time, so

$$n = P(6, 3) = 6 \cdot 5 \cdot 4 = 120$$

(*b*) Since the numbers must be even, the last digit must be either 2 or 4. Thus the third position is filled first and it can be done in 2 ways. Then there are now 5 ways to fill the middle position and 4 ways to fill the first position. This may be pictured by: $\underline{4}, \underline{5}, \underline{2}$. Thus $4 \cdot 5 \cdot 2 = 120$ of the numbers are even.

(*c*) Since the numbers must exceed 400, they must begin with 5, 6, 7, or 9. Thus we first fill the first position, which can be done in 4 ways. Then there are 5 ways to fill the second position and 4 ways to fill the third position. This may be pictured by: $\underline{4}, \underline{5}, \underline{4}$. Thus $4 \cdot 5 \cdot 4 = 80$ of the numbers exceed 400.

2.48. Find the number n of distinct permutations that can be formed from all the letters of each word: (*a*) THEM, (*b*) UNUSUAL, (*c*) SOCIOLOGICAL.

This problem concerns permutations with repetitions.

(*a*) $n = 4! = 24$, since there are 4 letters and no repetitions.

(*b*) $n = \frac{7!}{3!} = 840$, since there are 7 letters of which 3 are U.

(*c*) $n = \frac{12!}{3!\,2!\,2!\,2!}$, since there are 12 letters of which 3 are O, 2 are C, 2 are I, and 2 are L.

2.49. A class contains 8 students. Find the number of ordered samples of size 3: (*a*) with replacement, (*b*) without replacement.

(*a*) Each student in the ordered sample can be chosen in 8 ways; hence there are $8 \cdot 8 \cdot 8 = 8^3 = 512$ samples of size 3 with replacement.

(*b*) The first student in the sample can be chosen in 8 ways, the second in 7 ways, and the last in 6 ways. Thus there are $8 \cdot 7 \cdot 6 = 336$ samples of size 3 without replacement.

2.50. Find n if: (*a*) $P(n, 2) = 72$, (*b*) $2P(n, 2) + 50 = P(2n, 2)$.

(*a*) $P(n, 2) = n(n-1) = n^2 - n$; hence $n^2 - n = 72$ or $n^2 - n - 72 = 0$ or $(n-9)(n+8) = 0$. Since n must be positive, the only answer is $n = 9$.

(*b*) $P(n, 2) = n(n-1) = n^2 - n$ and $P(2n, 2) = 2n(2n-1) = 4n^2 - 2n$. Hence

$$2(n^2 - n) + 50 = 4n^2 - 2n \quad \text{or} \quad 2n^2 - 2n + 50 = 4n^2 - 2n \quad \text{or} \quad 50 = 2n^2 \quad \text{or} \quad n^2 = 25$$

Since n must be positive, the only answer is $n = 5$.

COMBINATIONS, PARTITIONS

2.51. A class contains 10 students with 6 men and 4 women. Find the number n of ways:

(*a*) a 4-member committee can be selected from the students,

(*b*) a 4-member committee with 2 men and 2 women can be selected,

(*c*) the class can elect a president, vice-president, treasurer, and secretary.

(*a*) This concerns combinations, not permutations, since order does not count. There are "10 choose 4" such committees. That is,

$$n = C(10, 4) = \binom{10}{4} = \frac{10 \cdot 9 \cdot 8 \cdot 7}{4 \cdot 3 \cdot 2 \cdot 1} = 210$$

(*b*) The 2 men can be chosen from the 6 men in $\binom{6}{2}$ ways, and the 2 women can be chosen from the 4 women in $\binom{4}{2}$ ways. Thus, by the product rule,

$$n = \binom{6}{2}\binom{4}{2} = \frac{6 \cdot 5}{2 \cdot 1} \cdot \frac{4 \cdot 3}{2 \cdot 1} = 15(6) = 90 \text{ ways}$$

(*c*) This concerns permutations, not combinations, since order does count. Thus

$$n = P(6, 4) = 6 \cdot 5 \cdot 4 \cdot 3 = 360$$

2.52. A box contains 7 blue socks and 5 red socks. Find the number n of ways two socks can be drawn from the box if: (*a*) they can be any color, (*b*) they must be the same color.

(*a*) There are "12 choose 2" ways to select 2 of the 12 socks. That is,

$$n = C(12, 2) = \binom{12}{2} = \frac{12 \cdot 11}{2 \cdot 1} = 66$$

(*b*) There are $C(7, 2) = 21$ ways to choose 2 of the 7 blue socks, and $C(5, 2) = 10$ ways to choose 2 of the 5 red socks. By the sum rule, $n = 21 + 10 = 31$.

2.53. Let $A, B, \ldots, J$ be 10 given points in the plane $\mathbf{R}^2$ such that no three of the points lie on the same line. Find the number n of:

(*a*) lines in $\mathbf{R}^2$ where each line contains two of the points,

(*b*) lines in $\mathbf{R}^2$ containing A and one of the other points,

(*c*) triangles whose vertices come from the given points,

(*d*) triangles whose vertices are A and two of the other points.

Since order does not count, this problem involves combinations.

(*a*) Each pair of points determines a line; hence

$$n = \text{"10 choose 2"} = C(10,2) = \binom{10}{2} = 66$$

(*b*) We need only choose one of the 9 remaining points; hence $n = 9$.

(*c*) Each triple of points determines a triangle; hence

$$n = \text{"10 choose 3"} = C(10,3) = \binom{10}{3} = 120$$

(*d*) We need only choose two of the 9 remaining points; hence $n = C(9,2) = 36$.

2.54. There are 12 students in a class. Find the number n of ways that 12 students can take three different tests if four students are to take each test.

There are $C(12,4) = 495$ ways to choose four students to take the first test; following this, there are $C(8,4) = 70$ ways to choose four students to take the second test. The remaining students take the third test. Thus $n = 70(495) = 34{,}650$.

2.55. Find the number n of ways 12 students can be partitioned into three teams A_1, A_2, A_3, so that each team contains four students. (Compare with preceding Problem 2.54.)

Let A denote one of the students. There are $C(11,3) = 165$ ways to choose three other students to be on the same team as A. Now let B be a student who is not on the same team as A. Then there are $C(7,3) = 35$ ways to choose three from the remaining students to be on the same team as B. The remaining four students form the third team. Thus $n = 35(165) = 5925$.

Alternatively, each partition $[A_1, A_2, A_3]$ can be arranged in $3! = 6$ ways as an ordered partition. By the preceding Problem 2.54, there are 34,650 such ordered partitions. Thus $n = 34\,650/6 = 5925$.

TREE DIAGRAMS

2.56. Construct the tree diagram that gives the permutations of $\{a, b, c\}$.

The tree diagram appears in Fig. 2-12. The six paths from the root of the tree yield the six permutations:

$$abc,\ acb,\ bac,\ bca,\ cab,\ cba$$

2.57. Jack has time to play roulette at most five times. At each play he wins or loses \$1. He begins with \$1 and will stop playing before the five times if he loses all his money or if he wins \$3, that is,

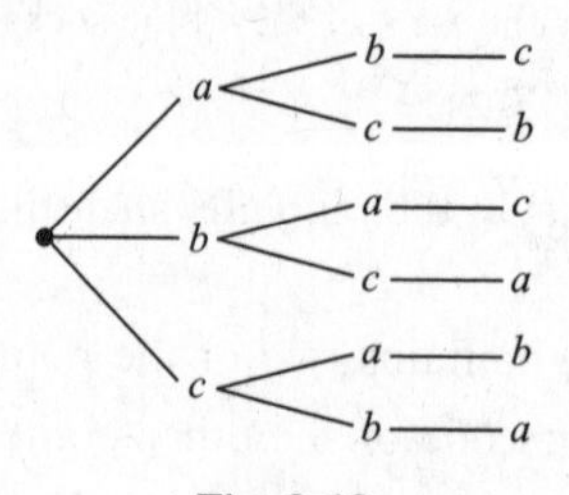

Fig. 2-12

if he has \$4. Find the number of ways the betting can occur, and find the number of times he will stop before betting five times.

Construct the appropriate tree diagram, as shown in Fig. 2-13. Each number in the diagram denotes the number of dollars he has at that moment of time. The betting can occur in 11 ways, and Jack will stop betting before the five times are up in only three of the cases.

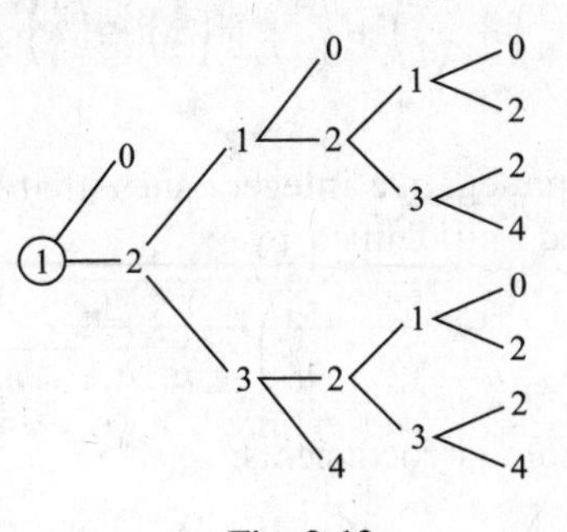

Fig. 2-13

MISCELLANEOUS PROBLEMS

2.58. Prove the binomial theorem 2.9: $(a+b)^n = \sum_{r=0}^{n} \binom{n}{r} a^{n-r} b^r$.

The theorem is true for $n = 1$, since

$$\sum_{r=0}^{1} \binom{1}{r} a^{1-r} b^r = \binom{1}{0} a^1 b^0 + \binom{1}{1} a^0 b^1 = a + b = (a+b)^1$$

We assume the theorem holds for $(a+b)^n$ and prove it is true for $(a+b)^{n+1}$

$$\begin{aligned} (a+b)^{n+1} &= (a+b)(a+b)^n \\ &= (a+b)\left[a^n + \binom{n}{1} a^{n-1} b + \cdots + \binom{n}{r-1} a^{n-r+1} b^{r-1} + \binom{n}{r} a^{n-r} b^r + \cdots + \binom{n}{1} a b^{n-1} + b^n\right] \end{aligned}$$

Now the term in the product which contains b^r is obtained from

$$\begin{aligned} b\left[\binom{n}{r-1} a^{n-r+1} b^{r-1}\right] + a\left[\binom{n}{r} a^{n-r} b^r\right] &= \binom{n}{r-1} a^{n-r+1} b^r + \binom{n}{r} a^{n-r+1} b^r \\ &= \left[\binom{n}{r-1} + \binom{n}{r}\right] a^{n-r+1} b^r \end{aligned}$$

But, by Theorem 2.10 $\binom{n}{r-1} + \binom{n}{r} = \binom{n+1}{r}$. Thus the term containing b^r is $\binom{n+1}{r} a^{n-r+1} b^r$. Note that $(a+b)(a+b)^n$ is a polynomial of degree $n+1$ in b. Consequently,

$$(a+b)^{n+1} = (a+b)(a+b)^n = \sum_{r=0}^{n+1} \binom{n+1}{r} a^{n-r+1} b^r$$

which was to be proved.

2.59. Prove: $\binom{4}{0} + \binom{4}{1} + \binom{4}{2} + \binom{4}{3} + \binom{4}{4} = 16$

Note $16 = 2^4 = (1+1)^4$. Expanding $(1+1)^4$, using the binomial theorem, yields:

$$16 = (1+1)^4 = \binom{4}{0}1^4 + \binom{4}{1}1^3 1^1 + \binom{4}{2}1^2 1^2 + \binom{4}{3}1^1 1^3 + \binom{4}{4}1^4$$
$$= \binom{4}{0} + \binom{4}{1} + \binom{4}{2} + \binom{4}{3} + \binom{4}{4}$$

2.60. Let n and $n_1, n_2, \ldots, n_r$ be nonnegative integers such that $n_1 + n_2 + \cdots + n_r = n$. The *multinomial coefficients* are denoted and defined by:

$$\binom{n}{n_1 \quad n_2 \quad \ldots \quad n_r} = \frac{n!}{n_1!\, n_2! \cdots n_r!}$$

Compute the following multinomial coefficients:

(*a*) $\binom{6}{3, 2, 1}$, (*b*) $\binom{8}{4, 2, 2, 0}$, (*c*) $\binom{10}{5, 3, 2, 2}$

(*a*) $\binom{6}{3, 2, 1} = \dfrac{6!}{3!\,2!\,1!} = \dfrac{6 \cdot 5 \cdot 4 \cdot 3 \cdot 2 \cdot 1}{3 \cdot 2 \cdot 1 \cdot 2 \cdot 1 \cdot 1} = 60$

(*b*) $\binom{8}{4, 2, 2, 0} = \dfrac{8!}{4!\,2!\,2!\,0!} = \dfrac{8 \cdot 7 \cdot 6 \cdot 5 \cdot 4 \cdot 3 \cdot 2 \cdot 1}{4 \cdot 3 \cdot 2 \cdot 1 \cdot 2 \cdot 1 \cdot 2 \cdot 1 \cdot 1} = 420$

(*c*) The expression $\binom{10}{5, 3, 2, 2}$ has no meaning, since $5 + 3 + 2 + 2 \neq 10$.

Supplementary Problems

SETS AND SUBSETS

2.61. List the elements of the following sets if the universal set is $U = \{a, b, c, \ldots, y, z\}$. Furthermore, identify which of the sets, if any, are equal.

$A = \{x : x \text{ is a vowel}\}$ $\qquad$ $C = \{x : x \text{ precedes } f \text{ in the alphabet}\}$

$B = \{x : x \text{ is a letter in the word "}\textit{little}\text{"}\}$ $\qquad$ $D = \{x : x \text{ is a letter in the word "}\textit{title}\text{"}\}$

2.62. Let $A = \{1, 2, \ldots, 8, 9\}$, $B = \{2, 4, 6, 8\}$, $C = \{1, 3, 5, 7, 9\}$, $D = \{3, 4, 5\}$, and $E = \{3, 5\}$. Which of the above sets can equal a set X under each of the following conditions?

(*a*) X and B are disjoint

(*b*) $X \subseteq D$ but $X \not\subseteq B$

(*c*) $X \subseteq A$ but $X \not\subseteq C$

(*d*) $X \subseteq C$ but $X \not\subseteq A$

SET OPERATIONS

Problems 2.63–2.66 refer to the universal set $U = \{1, 2, 3, \ldots, 8, 9\}$ and the sets:

$$A = \{1, 2, 5, 6\}, \qquad B = \{2, 5, 7\}, \qquad C = \{1, 3, 5, 7, 9\}$$

2.63. Find: (*a*) $A \cap B$ and $A \cap C$, (*b*) $A \cup B$ and $B \cup C$, (*c*) A^c and C^c.

2.64. Find: (*a*) $A \backslash B$ and $A \backslash C$, (*b*) $A \oplus B$ and $A \oplus C$.

2.65. Find: (*a*) $(A \cup C)\backslash B$, (*b*) $(A \cup B)^c$, (*c*) $(B \oplus C)\backslash A$.

2.66. Let A and B be any sets. Prove:

(*a*) A is the disjoint union of $A\backslash B$ and $A \cap B$.

(*b*) $A \cup B$ is the disjoint union of $A\backslash B$, $A \cap B$, and $B\backslash A$.

2.67. Prove the following:

(*a*) $A \subseteq B$ if and only if $A \cap B^c = \varnothing$.

(*b*) $A \subseteq B$ if and only if $A^c \cup B = U$.

(*c*) $A \subseteq B$ if and only if $B^c \subseteq A^c$.

(*d*) $A \subseteq B$ if and only if $A\backslash B = \varnothing$.

(Compare results with Theorem 2.2.)

2.68. Prove the absorption laws: (*a*) $A \cup (A \cap B) = A$, (*b*) $A \cap (A \cup B) = A$.

2.69. The formula $A\backslash B = A \cup B^c$ defines the difference operation in terms of the operations of intersection and complement. Find a formula that defines the union $A \cup B$ in terms of the operations of intersection and complement.

VENN DIAGRAMS, ALGEBRA OF SETS, DUALITY

2.70. The Venn diagram in Fig. 2-14 shows sets A, B, C. Shade the following sets:

(*a*) $A\backslash(B \cup C)$, (*b*) $A^c \cap (B \cap C)$, (*c*) $(A \cup C) \cap (B \cup C)$.

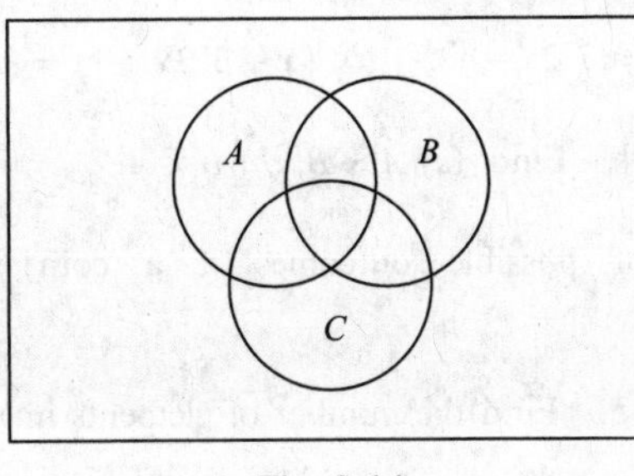

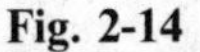

Fig. 2-14

2.71. Write the dual of each equation:

(*a*) $A \cup (A \cap B) = A$, (*b*) $(A \cap B) \cup (A^c \cap B) \cup (A \cap B^c) \cup (A^c \cap B^c) = U$

2.72. Use the laws in Table 2-1 to prove $(A \cap B) \cup (A \cap B^c) = A$.

FINITE SETS AND THE COUNTING PRINCIPLE

2.73. Determine which of the following sets are finite:

(*a*) lines parallel to the x-axis,

(*b*) letters in the English alphabet,

(*c*) animals living on the earth,

(*d*) circles through the origin (0, 0).

2.74. Given $n(U) = 20$, $n(A) = 12$, $n(B) = 9$, $n(A \cap B) = 4$, find:

(*a*) $n(A \cup B)$, (*b*) $n(A^c)$, (*c*) $n(B^c)$, (*d*) $n(A\backslash B)$, (*e*) $n(\varnothing)$.

2.75. Among 120 freshmen at a college, 40 take mathematics, 50 take English, and 15 take both mathematics and English. Find the number of freshmen who:

(*a*) do not take mathematics,
(*b*) take mathematics or English,
(*c*) take mathematics, but not English,
(*d*) take English, but not mathematics,
(*e*) take exactly one of the two subjects,
(*f*) take neither mathematics nor English.

2.76. A survey on a sample of 25 new cars being sold at a local auto dealer was conducted to see which of three popular options, air-conditioning (A), radio (R), and power windows (W), were already installed. The survey found:

15 had air-conditioning
12 had radio
11 had power windows
5 had air-conditioning and power windows
9 had air-conditioning and radio
4 had radio and power windows
3 had all three options

Find the number of cars that had: (*a*) only power windows, (*b*) only air-conditioning, (*c*) only radio, (*d*) radio and power windows but not air-conditioning, (*e*) air-conditioning and radio, but not power windows, (*f*) only one of the options, (*g*) none of the options.

2.77. Use Theorem 2.6 to prove Corollary 2.7: Suppose A, B, C are finite sets. Then $A \cup B \cup C$ is finite and

$$n(A \cup B \cup C) = n(A) + n(B) + n(C) - n(A \cap B) - n(A \cap C) - n(B \cap C) + n(A \cap B \cap C)$$

PRODUCT SETS

2.78. Find x and y if: (*a*) $(x+2, 3) = (7, 2x+y)$, (*b*) $(y-3, 2x+1) = (x+2, y+4)$.

2.79. Let $A = \{a, b\}$ and $B = \{1, 2, 3\}$. Find: (*a*) $A \times B$, (*b*) $B \times A$.

2.80. Let $C = \{H, T\}$, the set of possible outcomes if a coin is tossed. Find: (*a*) $C^2 = C \times C$, (*b*) $C^4 = C \times C \times C \times C$.

2.81. Suppose $n(A) = 3$, and $n(B) = 5$. Find the number of elements in: (*a*) $A \times B$, $B \times A$, (*b*) A^2, B^2, A^3, B^3; (*c*) $A \times A \times B \times A$.

CLASSES OF SETS, PARTITIONS

2.82. Find the power set $\mathscr{P}(A)$ of $A = \{a, b, c, d, e\}$.

2.83. Let $S = \{1, 2, 3, 4, 5, 6\}$. Determine whether each of the following is a partition of S:

(*a*) $[\{1, 3, 5\}, \{2, 4\}, \{3, 6\}]$
(*b*) $[\{1, 5\}, \{2\}, \{3, 6\}]$
(*c*) $[\{1, 5\}, \{2\}, \{4\}, \{3, 6\}]$
(*d*) $[\{1\}, \{3, 6\}, \{2, 4, 5\}, \{3, 6\}]$
(*e*) $[\{1, 2, 3, 4, 5, 6\}]$
(*f*) $[\{1\}, \{2\}, \{3\}, \{4\}, \{5\}, \{6\}]$

2.84. Find all partitions of $S = \{1, 2, 3\}$.

2.85. For each positive integer $n \in \mathbf{P}$, let $A_n = \{n, 2n, 3n, \ldots\}$, the multiples of n. Find: (*a*) $A_2 \cap A_7$, (*b*) $A_6 \cap A_8$, (*c*) $A_5 \cup A_{20}$, (*d*) $A_5 \cap A_{20}$, (*e*) $A_s \cup A_{st}$, where $s, t \in \mathbf{P}$, (*f*) $A_s \cap A_{st}$, where $s, t \in \mathbf{P}$.

2.86. Prove: If $J \subseteq \mathbf{P}$ is infinite, then $\cap(A_i \colon i \in J) = \varnothing$. (Here the A_i are the sets in Problem 2.85.)

2.87. Let $[A_1, A_2, \ldots, A_m]$ and $[B_1, B_2, \ldots, B_n]$ be partitions of S. Show that the collection of sets

$$[A_i \cap B_j; \quad i = 1, \ldots, m, \quad j = 1, \ldots, n] \backslash \varnothing$$

(where the empty set $\varnothing$ is deleted), is also a partition of S, called the *cross partition*.

2.88. Prove: For any indexed class of sets $\{A_i: i \in I\}$ and any set B: (*a*) $B \cup (\bigcap_i A_i) = \bigcap_i (B \cup A_i)$, (*b*) $B \cap (\bigcup_i A_i) = \bigcup_i (B \cap A_i)$.

2.89. Prove (De Morgan's law): $(\bigcap_i A_i)^c = \bigcup_i A_i^c$.

2.90. Show that each of the following is an algebra of subsets of U: (*a*) $\mathscr{A} = \{\varnothing, U\}$, (*b*) $\mathscr{B} = \{\varnothing, A, A^c, U\}$, (*c*) $\mathscr{P}(U)$, the power set of U.

2.91. Let $\mathscr{A}$ and $\mathscr{B}$ be algebras (σ-algebras) of subsets of U. Prove that the intersection $\mathscr{A} \cap \mathscr{B}$ is also an algebra (σ-algebra) of subsets of U.

MATHEMATICAL INDUCTION

2.92. Prove: $2 + 4 + 6 + \cdots + 2n = n(n + 1)$.

2.93. Prove: $1 + 4 + 7 + \cdots + (3n - 2) = 2n(3n - 1)$.

2.94. Prove: $1^2 + 2^2 + 3^2 + \cdots + n^2 = \dfrac{n(n+1)(2n+1)}{6}$.

2.95. Prove that, for $n \geq 4$: (*a*) $n! \geq 2^n$; (*b*) $2^n \geq n^2$; (*c*) $n^2 \geq 2n + 5$.

FACTORIAL NOTATION AND BINOMIAL COEFFICIENTS

2.96. Find: (*a*) 10!, 11!, 12!, (*b*) 60! (Hint: Use Stirling's approximation to $n!$.)

2.97. Simplify: (*a*) $\dfrac{(n+1)!}{n!}$, (*b*) $\dfrac{n!}{(n-2)!}$, (*c*) $\dfrac{(n-1)!}{(n+2)!}$, (*d*) $\dfrac{(n-r+1)!}{(n-r-1)!}$

2.98. Evaluate: (*a*) $\binom{5}{2}$, (*b*) $\binom{7}{3}$, (*c*) $\binom{14}{2}$, (*d*) $\binom{6}{4}$, (*e*) $\binom{20}{17}$, (*f*) $\binom{18}{15}$

2.99. Show that:

(*a*) $\binom{n}{0} + \binom{n}{1} + \binom{n}{2} + \binom{n}{3} + \cdots + \binom{n}{n} = 2^n$

(*b*) $\binom{n}{0} - \binom{n}{1} + \binom{n}{2} - \binom{n}{3} + \cdots + \binom{n}{n} = 0$

2.100. Evaluate the following multinomial coefficients (defined in Problem 2.58):

(*a*) $\binom{6}{2,3,1}$, (*b*) $\binom{8}{4,3,1,0}$, (*c*) $\binom{8}{3,3,2}$, (*d*) $\binom{9}{4,3,2,1}$

COUNTING PRINCIPLES, SUM AND PRODUCT RULES

2.101. A store sells clothes for men. It has 3 different kinds of jackets, 6 different kinds of shirts, and 4 different kinds of pants. Find the number of ways a person can buy: (*a*) one of the items for a present, (*b*) one of each of the items for a present.

2.102. A restaurant has, on its dessert menu, 4 kinds of cakes, 3 kinds of cookies, and 5 kinds of ice cream. Find the number of ways a person can select: (*a*) one of the desserts, (*b*) one of each kind of dessert.

2.103. A class contains 8 male students, and 6 female students. Find the number of ways that the class can elect: (*a*) a class representative, (*b*) two class representatives, one male and one female, (*c*) a president and a vice-president.

2.104. There are 6 roads between A and B and 4 roads between B and C. Find the number of ways a person can drive: (*a*) from A to C by way of B, (*b*) round-trip from A to C by way of B, (*c*) round-trip from A to C by way of B without using the same road more than once.

2.105. Suppose a code consists of two letters followed by a digit. Find the number of: (*a*) codes, (*b*) codes with distinct letters, (*c*) codes with the same letters.

PERMUTATIONS, ORDERED SAMPLES

2.106. Find the number n of ways a judge can award first, second, and third places in a contest with 18 contestants.

2.107. Find the number n of ways 6 people can ride a toboggan where: (*a*) anyone can drive, (*b*) one of three must drive.

2.108. Find the number n of permutations that can be formed from all the letters of each word: (*a*) QUEUE, (*b*) COMMITTEE, (*c*) PROPOSITION, (*d*) BASEBALL.

2.109. A box contains 10 light bulbs. Find the number n of ordered samples of size 3: (*a*) with replacement, (*b*) without replacement.

2.110. A class contains 6 students. Find the number n of ordered samples of size 4: (*a*) with replacement, (*b*) without replacement.

COMBINATIONS, PARTITIONS

2.111. A class contains 9 boys and 3 girls. Find the number of ways a teacher can select a committee of 4.

2.112. Repeat Problem 2.111, but where: (*a*) there are to be 2 boys and 2 girls, (*b*) there is to be exactly one girl, (*c*) there is to be at least one girl.

2.113. A box contains 6 blue socks and 4 white socks. Find the number of ways two socks can be drawn from the box where: (*a*) there are no restrictions, (*b*) they are different colors, (*c*) they are to be the same color.

2.114. A woman has 11 close friends. Find the number of ways she can invite 5 of them to dinner.

2.115. Repeat Problem 2.114, but where: (*a*) two of the friends are married and will not attend separately, (*b*) two of the friends are not on speaking terms and will not attend together.

2.116. A student is to answer 8 out of 10 questions on an exam. Find the number of choices.

2.117. Repeat Problem 2.116, but where: (*a*) the first three questions must be answered, (*b*) at least 4 of the first 5 questions must be answered.

2.118. There are 9 students in a class. Find the number of ways the students can take three tests if 3 students are to take each test.

2.119. There are 9 students in a class. Find the number of ways the students can be partitioned into three teams containing 3 students each. (Compare with Problem 2.118.)

2.120. Find the number of ways 9 toys may be divided among four children if the youngest is to receive 3 toys and each of the others 2 toys.

TREE DIAGRAMS

2.121. Teams A and B play in the World Series of baseball, where the team that first wins four games wins the series. Find the number n of ways the series can occur given that A wins the first game and that the team that wins the second game also wins the fourth game, and list the n ways the series can occur.

2.122. Suppose $A, B, \ldots, F$ in Fig. 2-15 denote islands, and the lines connecting them bridges. A man begins at A and walks from island to island. He stops for lunch when he cannot continue to walk without crossing the same bridge twice. (*a*) Construct the appropriate tree diagram, and find the number of ways he can take his walk before eating lunch. (*b*) At which islands can he eat his lunch?

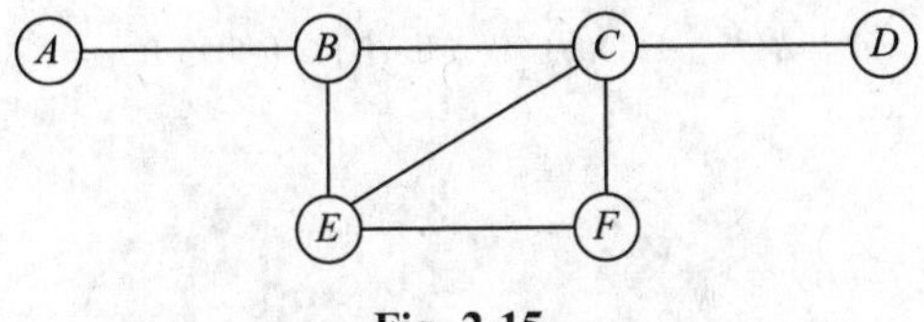

Fig. 2-15

MISCELLANEOUS PROBLEMS

2.123. Suppose n objects are partitioned into r ordered cells with $n_1, n_2, \ldots, n_r$ elements. Show that the number of such ordered partitions is

$$\frac{n!}{n_1!\,n_2!\,n_3!\ldots n_r!}$$

2.124. There are n married couples at a party. (*a*) Find the number of (unordered) pairs at the party. (*b*) Find the number of handshakes if each person shakes hands with every other person other than his or her spouse.

Answers to Supplementary Problems

2.61. $A = \{\text{a, e, i, o, u}\}$, $B = D = \{\text{l, i, t, e}\}$, $C = \{\text{a, b, c, d, e}\}$

2.62. (*a*) C, E; (*b*) D, E; (*c*) B; (*d*) none

2.63. (*a*) $A \cap B = \{2, 5\}$, $A \cap C = \{5\}$; (*b*) $A \cup B = \{1, 2, 5, 6, 7\}$, $B \cup C = \{1, 2, 3, 5, 7, 9\}$; (*c*) $A^c = \{3, 4, 7, 8, 9\}$, $C^c = \{2, 4, 6, 8\}$

2.64. (*a*) $A \backslash B = \{1, 6\}$, $A \backslash C = \{2, 6\}$; (*b*) $A \oplus B = \{1, 6, 7\}$, $A \oplus C = \{2, 6, 7, 9\}$

2.65. (*a*) $\{1, 3, 6, 9\}$, (*b*) $\{3, 4, 8, 9\}$, (*c*) $\{3, 9\}$

2.69. $A \cup B = (A^c \cap B^c)^c$

2.70. See Fig. 2-16.

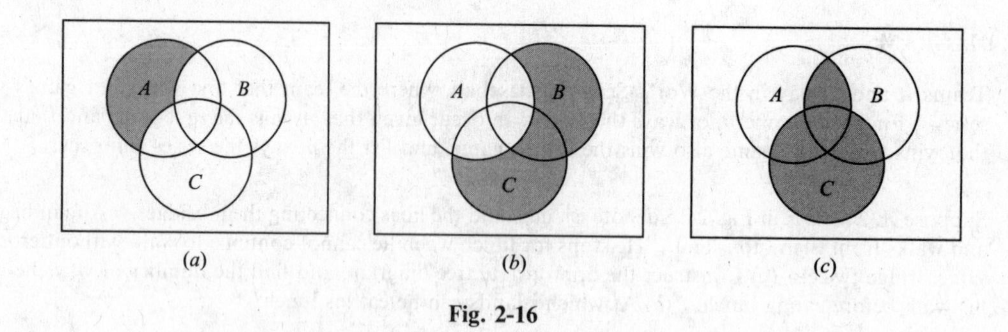

Fig. 2-16

2.71. (a) $A \cap (A \cup B) = A$, (b) $(A \cup B) \cap (A^c \cup B) \cap (A \cup B^c) \cap (A^c \cup B^c) = \varnothing$

2.73. (b), (d), (e)

2.74. (a) 17, (b) 8, (c) 11, (d) 8, (e) 0

2.75. (a) 80, (b) 75, (c) 25, (d) 35, (e) 60, (f) 45

2.76. (a) 5, (b) 4, (c) 2, (d) 1, (e) 6, (f) 11, (g) 2

2.78. (a) $x = 5$, $y = 7$; (b) $x = 8$, $y = 13$

2.79. $A \times B = \{a1, a2, a3, b1, b2, b3\}$, $B \times A = \{1a, 1b, 2a, 2b, 3a, 3b\}$

2.80. $C^2 = \{HH, HT, TH, TT\}$,
$C^4 = \{HHHH, HHHT, HHTH, HHTT, HTHH, HTHT, HTTH, HTTT, THHH, THHT, THTH, THTT, TTHH, TTHT, TTTH, TTTT\}$

2.81. (a) 15, 15, 9, 25; (b) 45, 27

2.82. $P(A) = [\varnothing, a, b, c, d, e, ab, ac, ad, ae, bc, bd, be, cd, ce, de, abc, abd, abe, acd, ace, ade, bcd, bce, bde, cde, abcd, abce, abde, acde, bcde, A]$. Note $n(P(a)) = 2^5 = 32$.

2.83. (a) and (b): no. Others: yes.

2.84. [S], [{1, 2}, {3}], [{1, 3}, {2}], [{2, 3}, {1}], [{1}, {2}, {3}]

2.85. (a) A_{14}, (b) A_{24}, (c) A_5, (d) A_{20}, (e) A_s, (f) A_{st}

2.96. (a) 3,628,800; 39,916,800; 479,001,600, (b) $\log(60!) = 81.92$ so $60! = 6.59 \times 10^{81}$

2.97. (a) $n+1$, (b) $n(n-1) = n^2 - n$, (c) $1/[n(n+1)(n+2)]$, (d) $(n-r)(n-r+1)$

2.98. (a) 10, (b) 35, (c) 91, (d) 15, (e) 1140, (f) 816

2.99. Hint: (a) expand $(1+1)^n$, (b) expand $(1-1)^n$

2.100. (a) 60, (b) 280, (c) 560, (d) not defined

2.101. (*a*) 13, (*b*) 72

2.102. (*a*) 12, (*b*) 60

2.103. (*a*) 14, (*b*) 48, (*c*) 182

2.104. (*a*) 24, (*b*) 576, (*c*) 360

2.105. (*a*) 6760, (*b*) 6500, (*c*) 260

2.106. $n = 18 \cdot 17 \cdot 16 = 4896$

2.107. (*a*) $6! = 720$, (*b*) $3 \cdot 5! = 360$

2.108. (*a*) 30, (*b*) $\dfrac{9!}{2!\,2!\,2!} = 45{,}360$, (*c*) $\dfrac{11!}{2!\,3!\,2!} = 1{,}663{,}200$, (*d*) $\dfrac{8!}{2!\,2!\,2!} = 5040$

2.109. (*a*) $10^3 = 1000$, (*b*) $10 \cdot 9 \cdot 8 = 720$

2.110. (*a*) $6^4 = 1296$, (*b*) $6 \cdot 5 \cdot 4 \cdot 3 = 360$

2.111. $C(12, 4) = 495$

2.112. (*a*) $C(9,2) \cdot C(3,2) = 108$, (*b*) $C(9,3) \cdot C(3,1) = 252$,
(*c*) $9 + 108 + 252 = 369$ or $C(12,4) - C(9,4) = 495 - 126 = 369$

2.113. (*a*) $C(10,2) = 45$, (*b*) $6 \cdot 4 = 24$, (*c*) $C(6,2) + C(4,2) = 21$ or $45 - 24 = 21$

2.114. $C(11, 5) = 462$

2.115. (*a*) 210, (*b*) 252

2.116. $C(10, 8) = C(10, 2) = 45$

2.117. (*a*) $C(7,5) = C(7,2) = 21$, (*b*) $25 + 10 = 35$

2.118. 1680

2.119. 280

2.120. 7560

2.121. Construct the appropriate tree diagram as in Fig. 2-17. Note that the tree begins at A, the winner of the first game, and that there is only one choice in the fourth game, the winner of the second game. The diagram shows that $n = 15$ and that the series can occur in the following 15 ways:

$$AAAA,\ AABAA,\ AABABA,\ AABABBA,\ AABABBB,\ ABABAA,\ ABABABA,\ ABABABB,$$
$$ABABBAA,\ ABABBAB,\ ABABBB,\ ABBBAAA,\ ABBBAAB,\ ABBBAB,\ ABBBB$$

2.122. (*a*) See Fig. 2-18. There are 11 ways to take his walk. (*b*) B, D, or E

2.124. (*a*) $C(2n, 2) = 2n(2n-1)/2$, (*b*) $C(2n, 2) - n = 2n(2n-1)/2 - n$

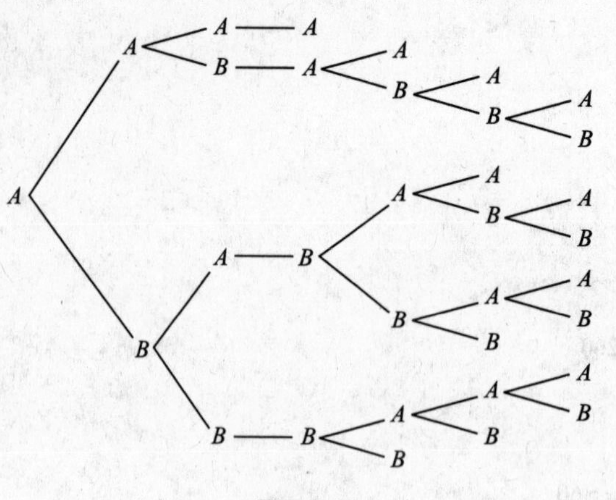

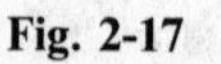

Fig. 2-17

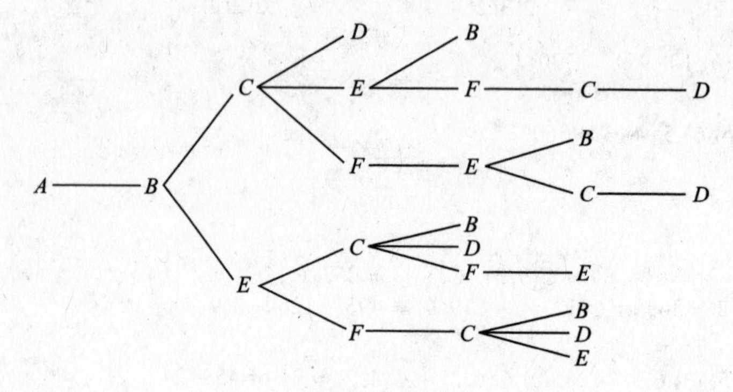

Fig. 2-18

Chapter 3

Basic Probability

3.1 INTRODUCTION

Probability theory is the mathematical modeling of the phenomenon of chance or randomness. If a coin is tossed in a random manner, it can land heads or tails, but we do not know which of these will occur on a single toss. However, suppose we let s be the number of times heads appears when the coin is tossed n times. As n increases, the ratio $f = s/n$, called the *relative frequency* of the outcome, becomes more stable. If the coin is perfectly balanced, then we expect that the coin will land heads approximately 50 percent of the time or, in other words, the relative frequency will approach 1/2. Alternatively, assuming the coin is perfectly balanced, we can arrive at the value 1/2 deductively. That is, any side of the coin is as likely to occur as the other; hence the chances of getting a head is one in two, which means the probability of getting a head is 1/2. Although the specific outcome on any one toss is unknown, the behavior over the long run is determined. This stable long-run behavior of random phenomena forms the basis of probability theory.

Consider another experiment, the tossing of a six-sided die (Fig. 3-1) and observing the number of dots, or pips, that appear on the top side. Suppose the experiment is repeated n times and let s be the number of times 4 dots appear on top. Again, as n increases, the *relative frequency* $f = s/n$ of the outcome 4 becomes more stable. Assuming the die is perfectly balanced, we would expect that the stable or long-run value of this ratio is 1/6, and we say the probability of getting a 4 is 1/6.

Alternatively, we can arrive at the value 1/6 deductively. That is, with a perfectly balanced die, any one side of the die is as likely as any other to occur on top. Thus the chance of getting a 4 is one in six or, in other words, the probability of getting a 4 is 1/6. Again, although the specific outcome on any one toss is unknown, the behavior over the long run is determined.

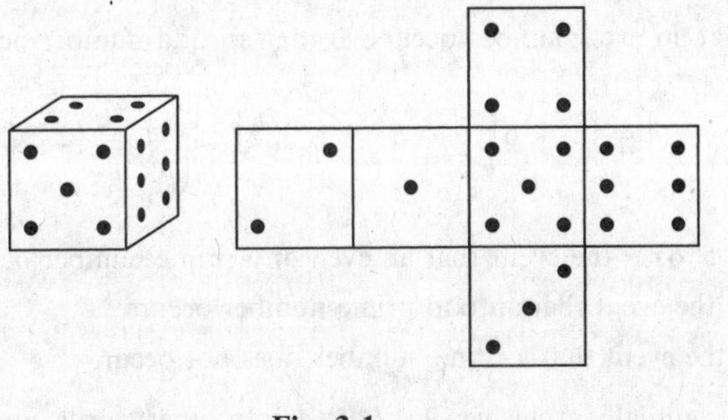

Fig. 3-1

The historical development of probability theory is similar to the above discussion. That is, letting E denote an event, an outcome of an experiment, there were two ways to obtain the probability p of E:

(*a*) *Classical (a priori) definition*: Suppose an event E can occur in s ways out of a total of n equally likely possible ways. Then $p = s/n$.

(*b*) *Frequency (a posteriori) definition*: Suppose after n repetitions, where n is very large, an event E occurs s times. Then $p = s/n$.

Both of the above definitions have serious flaws. The classical definition is essentially circular, since the idea of "equally likely" is the same as that of "with equal probability" which has not been defined. The frequency definition is not well-defined since "very large" has not been defined.

The modern treatment of probability theory is axiomatic, using set theory. Specifically, a mathematical model of an experiment is obtained by arbitrarily assigning probabilities to all the events, except that the assignments must satisfy certain axioms listed below. Naturally, the reliability of our mathematical model for a given experiment depends upon the closeness of the assigned probabilities to the actual limiting relative frequencies. This then gives rise to problems of testing and reliability, which form the subject matter of statistics.

3.2 SAMPLE SPACE AND EVENTS

The set S of all possible outcomes of a given experiment is called the *sample space*. A particular outcome, i.e. an element in S, is called a *sample point*. An *event* A is a set of outcomes or, in other words, a subset of the sample space S. In particular, the set $\{a\}$ consisting of a single sample point $a \in S$ is called an *elementary event*. Furthermore, the empty set $\varnothing$ and S itself are subsets of S and so are events; $\varnothing$ is sometimes called the *impossible event* or the *null event*.

Since an event is a set, we can combine events to form new events using the various set operations:

(1) $A \cup B$ is the event that occurs iff A occurs *or* B occurs (or both).
(2) $A \cap B$ is the event that occurs iff A occurs *and* B occurs.
(3) A^c, the complement of A, also written $\bar{A}$, is the event that occurs iff A does *not* occur.

Two events A and B are called *mutually exclusive* if they are disjoint, that is, if $A \cap B = \varnothing$. In other words, A and B are mutually exclusive iff they cannot occur simultaneously. Three or more events are mutually exclusive if every two of them are mutually exclusive.

EXAMPLE 3.1

(*a*) Experiment: Toss a die and observe the number (of dots) that appears on top.
The sample space S consists of the six possible numbers; that is,

$$S = \{1, 2, 3, 4, 5, 6\}$$

Let A be the event that an even number occurs, B that an odd number occurs, and C that a prime number occurs; that is, let

$$A = \{2, 4, 6\}, \qquad B = \{1, 3, 5\}, \qquad C = \{2, 3, 5\}.$$

Then

$A \cup C = \{2, 3, 4, 5, 6\}$ is the event that an even or a prime number occurs.

$B \cap C = \{3, 5\}$ is the event that an odd prime number occurs.

$C^c = \{1, 4, 6\}$ is the event that a prime number does not occur.

Note that A and B are mutually exclusive: $A \cap B = \varnothing$. In other words, an even number and an odd number cannot occur simultaneously.

(*b*) Experiment: Toss a coin three times and observe the sequence of heads (H) and tails (T) that appears.
The sample space S consists of the following eight elements:

$$S = \{HHH, HHT, HTH, HTT, THH, THT, TTH, TTT\}$$

Let A be the event that two or more heads appear consecutively, and B that all the tosses are the same; that is, let

$$A = \{HHH, HHT, THH\} \quad \text{and} \quad B = \{HHH, TTT\}$$

Then $A \cap B = \{HHH\}$ is the elementary event in which only heads appear. The event that five heads appear is the empty set $\varnothing$.

(*c*) Experiment: Toss a coin until a head appears, and then count the number of times the coin is tossed. The sample space of this experiment is $S = \{1, 2, 3, \ldots, \infty\}$. Here ∞ refers to the case when a head never appears, and the coin is tossed an infinite number of times. Since every positive integer is an element of S, the sample space is infinite. In fact, this is an example of a sample space which is *countably infinite*.

(*d*) Experiment: Let a pencil drop, head first, into a rectangular box and note the point at the bottom of the box that the pencil first touches. Here S consists of all the points on the bottom of the box. Let the rectangular area in Fig. 3-2 represent these points. Let A and B be the events that the pencil drops into the corresponding areas illustrated in Fig. 3-2.

Remark: The sample space S in Example 3.1(*d*) is an example of a continuous sample space. (A sample space S is *continuous* if it is an interval or a product of intervals.) In such a case, only special subsets (called *measurable* sets) will be events. On the other hand, if the sample space S is *discrete*, that is, if S is finite or countably infinite, then every subset of S is an event.

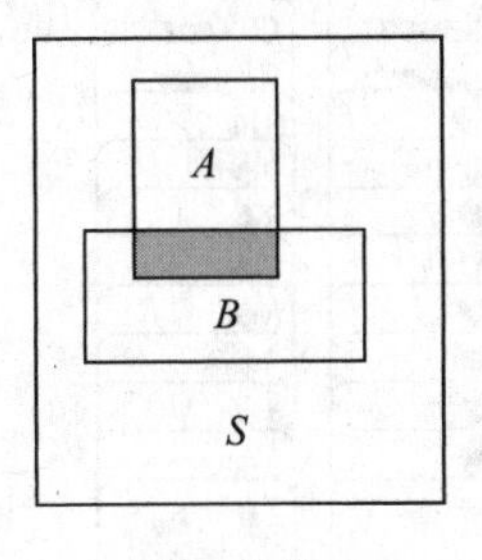

Fig. 3-2

EXAMPLE 3.2: Toss of a pair of dice A pair of dice is tossed and the two numbers appearing on the top are recorded. There are six possible numbers, $1, 2, \ldots, 6$, on each die. Thus S consists of the pairs of numbers from 1 to 6, and hence $n(S) = 6 \cdot 6 = 36$. Figure 3-3 shows these 36 pairs of numbers arranged in an array where the rows are labeled by the first die and the columns by the second die. Let A be the event that the sum of the two numbers is 6, and let B be the event that the largest of the two numbers is 4. That is, let

$$A = \{(1,5),\ (2,4),\ (3,3),\ (4,2),\ (5,1)\}, \qquad B = \{(1,4),\ (2,4),\ (3,4),\ (4,4),\ (4,3),\ (4,2),\ (4,1)\}$$

These events are pictured in Fig. 3-3. Then the event "A and B" consists of those pairs of integers whose sum is 6 and whose largest number is 4 or, in other words, the intersection of A and B. Thus

$$A \cap B = \{(2,4),\ (4,2)\}$$

Similarly, "A or B", the sum is 6 or the largest is 4, shaded in Fig. 3-3, is the union $A \cup B$.

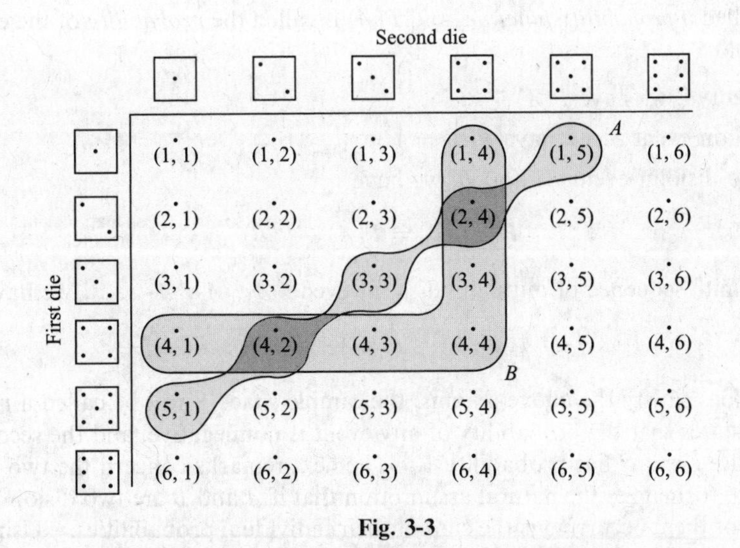

Fig. 3-3

EXAMPLE 3.3: Deck of cards A card is drawn from an ordinary deck of 52 cards which is pictured in Fig. 3-4(a). The sample space S consists of the four *suits*, clubs (C), diamonds (D), hearts (H), and spades (S), where each suit contains 13 cards which are numbered 2 to 10, and jack (J), queen (Q), king (K), and ace (A). The hearts (H) and diamonds (D) are red cards, and the spades (S) and clubs (C) are black cards. Figure 3-4(b) pictures 52 points which represent the deck S of cards in the obvious way. Let E be the event of a *picture card*, that is, a jack (J), queen (Q), or king (K), and let F be the event of a heart. Then $E \cap F = \{\text{JH}, \text{QH}, \text{KH}\}$, as shaded in Fig. 3-4($b$).

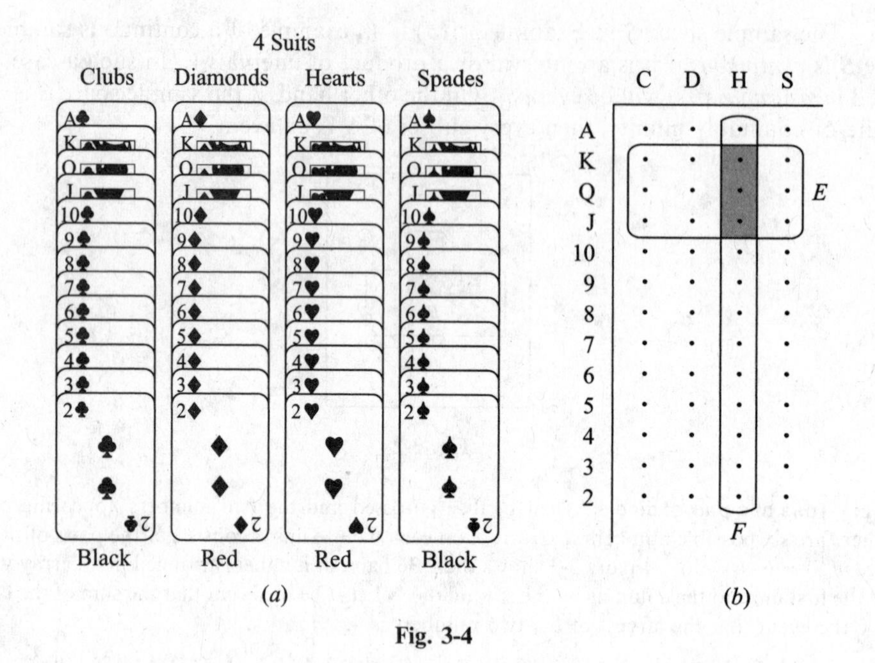

Fig. 3-4

3.3 AXIOMS OF PROBABILITY

Let S be a sample space, let $\mathscr{C}$ be the class of all events, and let P be a real-valued function defined on $\mathscr{C}$. Then P is called a *probability function*, and $P(A)$ is called the *probability* of the event A when the following axioms hold:

[P_1] For any event A, we have $P(A) \geq 0$.

[P_2] For the certain event S, we have $P(S) = 1$.

[P_3] For any two disjoint events A and B, we have

$$P(A \cup B) = P(A) + P(B)$$

[P_3'] For any infinite sequence of mutually disjoint events $A_1, A_2, A_3, \ldots$, we have

$$P(A_1 \cup A_2 \cup A_3 \cup \ldots) = P(A_1) + P(A_2) + P(A_3) + \ldots$$

Moreover, when P does satisfy the above axioms, the sample space S will be called a *probability space*.

The first axiom states that the probability of any event is nonnegative, and the second axiom states that the certain or sure event S has probability 1. The next remarks concern the two axioms [P_3] and [P_3']. The axiom [P_3] formalizes the natural assumption that if A and B are two disjoint events then the probability of either of them occurring is the sum of their individual probabilities. Using mathematical induction, we can then extend this *additive property* for two sets to any finite number of disjoint events; that is, for any mutually disjoint sets $A_1, A_2, \ldots, A_n$, we have

$$P(A_1 \cup A_2 \cup \cdots \cup A_n) = P(A_1) + P(A_2) + \cdots + P(A_n) \qquad (*)$$

We emphasize that $[P_3']$ does not follow from $[P_3]$, even though (*) is true for every positive integer n. However, if the sample space S is finite, then only $[P_3]$ is needed, that is, $[P_3']$ is superfluous.

Theorems on Probability Spaces

The following theorems follow directly from our axioms, and will be proved in Problems 3.20–3.25.

Theorem 3.1: The impossible event or, in other words, the empty set $\varnothing$, has probability zero; that is, $P(\varnothing) = 0$.

The next theorem, called the complement rule, formalizes our intuition that if we hit a target, say, $p = 1/3$ of the times, then we miss the target $1 - p = 2/3$ of the times. (Recall A^c denotes the complement of the set A.)

Theorem 3.2 (complement rule): For any event A, we have

$$P(A^c) = 1 - P(A)$$

The next theorem tells us that the probability of any event must lie between 0 and 1. That is:

Theorem 3.3: For any event A, we have $0 \le P(A) \le 1$.

The following theorem applies to the case that one event is a subset of another event.

Theorem 3.4: If $A \subseteq B$, then $P(A) \le P(B)$.

The following theorem concerns two arbitrary events.

Theorem 3.5: For any two events A and B, we have

$$P(A \backslash B) = P(A) - P(A \cap B)$$

The next theorem, called the *general addition rule*, or simply *addition rule*, is similar to the inclusion–exclusion principle for sets.

Theorem (addition rule) 3.6: For any two events A and B,

$$P(A \cup B) = P(A) + P(B) - P(A \cap B)$$

Applying the above theorem twice (Problem 3.26), we obtain:

Corollary 3.7: For any events, A, B, C, we have

$$P(A \cup B \cup C) = P(A) + P(B) + P(C) - P(A \cap B) - P(A \cap C) - P(B \cap C) + P(A \cap B \cap C)$$

Clearly, like the analogous inclusion–exclusion principle for sets, the addition rule can be extended to any finite number of sets.

3.4 FINITE PROBABILITY SPACES

Consider a sample space S and the class $\mathscr{C}$ of all events. (If S is finite then we assume, unless otherwise stated, that all subsets of S are events.) As noted above, S becomes a probability space by assigning probabilities to the events in $\mathscr{C}$ so that they satisfy the probability axioms. This section shows how this is usually done for finite sample spaces. The next section discusses infinite sample spaces.

Finite Equiprobable Spaces

Suppose S is a finite sample space with n elements, and suppose the physical characteristics of the experiment suggest that the various outcomes of the experiment be assigned equal probabilities. Then S becomes a probability space, called a *finite equiprobable space*, if each point in S is assigned the probability $1/n$ and if each event A containing r points is assigned the probability r/n. In other words,

$$P(A) = \frac{\text{number of elements in } A}{\text{number of elements in } S} = \frac{n(A)}{n(S)}$$

or
$$P(A) = \frac{\text{number of ways that the event } A \text{ can occur}}{\text{number of ways that the sample space } S \text{ can occur}}$$

We emphasize that the above formula for $P(A)$ can only be used with respect to an equiprobable space, and cannot be used in general.

We state the above result formally.

Theorem 3.8: Let S be a finite sample space and, for any subset A of S, let $P(A) = n(A)/n(S)$. Then P satisfies axioms $[P_1]$, $[P_2]$, and $[P_3]$.

The expression "at random" will be used only with respect to an equiprobable space; formally, the statement "choose a point at random from a set S" shall mean that S is an equiprobable space where each point in S has the same probability.

EXAMPLE 3.4

(*a*) A card may be selected at random from an ordinary deck of 52 playing cards (see Fig. 3-4). Consider the events:

$$A = \{\text{card is a heart}\} \text{ and } B = \{\text{card is a face card}\}$$

(A face card is a jack (J), queen (Q), or king (K).) We compute $P(A)$, $P(B)$, and $P(A \cap B)$. Since we have an equiprobable space,

$$P(A) = \frac{\text{number of hearts}}{\text{number of cards}} = \frac{13}{52} = \frac{1}{4}, \qquad P(B) = \frac{\text{number of face cards}}{\text{number of cards}} = \frac{12}{52} = \frac{3}{13}$$

$$P(A \cap B) = \frac{\text{number of heart face cards}}{\text{number of cards}} = \frac{3}{52}$$

Suppose we want the probability that the card is a heart or a face card, that is, suppose we want $P(A \cup B)$. We can count the number of such cards and use Theorem 3.8, or use the above data and Theorem 3.6, to obtain

$$P(A \cup B) = P(A) + P(B) - P(A \cap B) = \frac{1}{4} + \frac{3}{13} - \frac{3}{52} = \frac{22}{52} = \frac{11}{26}$$

(*b*) Suppose a student is selected at random from 80 students where 30 are taking mathematics, 20 are taking chemistry, and 10 are taking mathematics and chemistry. Find the probability p that the student is taking mathematics (M) or chemistry (C).

Since the space is equiprobable, we have:

$$P(M) = \frac{30}{80} = \frac{3}{8}, \qquad P(C) = \frac{20}{80} = \frac{1}{4}, \qquad P(M \text{ and } C) = P(M \cap C) = \frac{10}{80} = \frac{1}{8}$$

Thus, by the addition rule (Theorem 3.6),

$$p = P(M \text{ or } C) = P(M \cup C) = P(M) + P(C) - P(M \cap C) = \frac{3}{8} + \frac{1}{4} - \frac{1}{8} = \frac{1}{2}$$

Finite Probability Spaces

Let S be a finite sample space, say $S = \{a_1, a_2, \ldots, a_n\}$. A *finite probability space*, or *finite probability model*, is obtained by assigning to each point a_i in S a real number p_i, called the *probability* of a_i, satisfying the following properties:

(1) Each p_i is nonnegative, that is, $p_i \geq 0$.
(2) The sum of the p_i is 1, that is, $p_1 + p_2 + \cdots + p_n = 1$.

The *probability* $P(A)$ of an event A is defined to be the sum of the probabilities of the points in A. For notational convenience, we write $P(a_i)$ instead of $P(\{a_i\})$.

Sometimes the points in a finite sample space S and their assigned probabilities are given in the form of a table as follows:

Outcome	a_1	a_2	$\cdots$	a_n
Probability	p_1	p_2	$\cdots$	p_n

Such a table is called a *probability distribution*.

The fact that $P(A)$, the sum of the probabilities of the points in A, does define a probability space is stated formally below (and proved in Problem 3.30).

Theorem 3.9: The above function $P(A)$ satisfies the axioms $[P_1]$, $[P_2]$, and $[P_3]$.

EXAMPLE 3.5

(*a*) Experiment: Let three coins be tossed and the number of heads observed. (Compare with Example 3.1(*b*).) Then the sample space is $S = \{0, 1, 2, 3\}$. The following assignments on the elements of S defines a probability space:

Outcome	0	1	2	3
Probability	1/8	3/8	3/8	1/8

That is, each probability is nonnegative, and the sum of the probabilities is 1. Let A be the event that at least one head appears, and let B be the event that all heads or all tails appear; that is, let

$$A = \{1, 2, 3\} \quad \text{and} \quad B = \{0, 3\}$$

Then, by definition

$$P(A) = P(1) + P(2) + P(3) = \frac{3}{8} + \frac{3}{8} + \frac{1}{8} = \frac{7}{8}$$

and

$$P(B) = P(0) + P(3) = \frac{1}{8} + \frac{1}{8} = \frac{1}{4}$$

(*b*) Three horses A, B, C are in a race; A is twice as likely to win as B, and B is twice as likely to win as C. Find their respective probabilities of winning, that is find $P(A)$, $P(B)$, $P(C)$.

Let $P(C) = p$. Since B is twice as likely to win as C, $P(B) = 2p$; and since A is twice as likely to win as B, $P(A) = 2P(B) = 2(2p) = 4p$. Now the sum of the probabilities must be 1; hence

$$p + 2p + 4p = 1 \quad \text{or} \quad 7p = 1 \quad \text{or} \quad p = \frac{1}{7}$$

Accordingly,

$$P(A) = 4p = \frac{4}{7}, \qquad P(B) = 2p = \frac{2}{7}, \qquad P(C) = p = \frac{1}{7}$$

Question: What is the probability that B or C wins, that is, $P(\{B, C\})$? By definition

$$P(\{B, C\}) = P(B) + P(C) = \frac{2}{7} + \frac{1}{7} = \frac{3}{7}$$

3.5 INFINITE SAMPLE SPACES

This section considers infinite sample spaces S. There are two cases, the case where S is countably infinite, and the case where S is uncountable. We note that a finite or a countably infinite probability space S is said to be *discrete*. Moreover, an uncountable space S which consists of a continuum of points, such as an interval or product of intervals, is said to be *continuous*.

Countably Infinite Sample Spaces

Suppose S is a countably infinite sample space; say

$$S = \{a_1, a_2, a_3, \ldots\}$$

Then, as in the finite case, we obtain a probability space by assigning each $a_i \in S$ a real number p_i, called its probability, such that:

(1) Each p_i is nonnegative, that is, $p_i \geq 0$.
(2) The sum of the p_i is equal to 1, that is

$$p_1 + p_2 + p_3 + \cdots = \sum_{i=1}^{\infty} p_i = 1$$

The probability $P(A)$ of an event A is then the sum of the probabilities of its points.

EXAMPLE 3.6 Consider the sample space $S = \{1, 2, 3, \ldots, \infty\}$ of the experiment of tossing a coin until a head appears; here n denotes the number of times the coin is tossed. A probability space is obtained by setting

$$p(1) = \frac{1}{2}, \quad p(2) = \frac{1}{4}, \quad p(3) = \frac{1}{8}, \ldots, p(n) = \frac{1}{2^n}, \ldots, p(\infty) = 0$$

Consider the events:

$$A = \{n \text{ is at most } 3\} \quad \text{and} \quad B = \{n \text{ is even}\}$$

Then

$$P(A) = P(1, 2, 3) = \frac{1}{2} + \frac{1}{4} + \frac{1}{8} = \frac{7}{8}$$

$$P(B) = P(2, 4, 6, 8, \ldots) = \frac{1}{4} + \frac{1}{4^2} + \frac{1}{4^3} + \cdots$$

Note $P(B)$ is a geometric series with $a = 1/4$ and $r = 1/4$; hence

$$P(B) = \frac{a}{1-r} = \frac{1/4}{3/4} = \frac{1}{3}$$

Uncountable Spaces

The only uncountable sample spaces S which we will consider here are those with some finite geometrical measurement $m(S)$ such as length, area, or volume, and in which a point is selected at random. The probability of an event A, i.e. that the selected point belongs to A, is then the ratio of $m(A)$ to $m(S)$; that is,

$$P(A) = \frac{\text{length of } A}{\text{length of } S} \quad \text{or} \quad P(A) = \frac{\text{area of } A}{\text{area of } S} \quad \text{or} \quad P(A) = \frac{\text{volume of } A}{\text{volume of } S}$$

Such a probability space is said to be *uniform*.

EXAMPLE 3.7 On the real line **R**, points a and b are selected at random such that $-2 \le b \le 0$ and $0 \le a \le 3$, as shown in Fig. 3-5(*a*). Find the probability p that the distance d between a and b is greater than 3.

The sample space S consists of the ordered pairs (a, b) and so forms the rectangular region shown in Fig. 3-5(*b*). On the other hand, the set A of points (a, b) for which $d = a - b > 3$ consists of those points which lie below the line $x - y = 3$, and hence form the shaded region in Fig. 3-5(*b*). Thus

$$p = P(A) = \frac{\text{area of } A}{\text{area of } S} = \frac{2}{6} = \frac{1}{3}$$

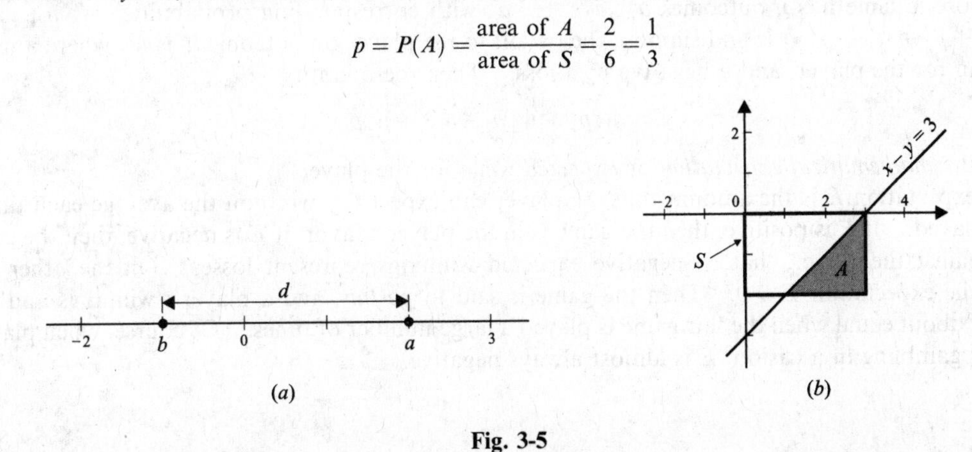

Fig. 3-5

3.6 CLASSICAL BIRTHDAY PROBLEM

The classical birthday problem concerns the probability p that n people have distinct birthdays. In solving this problem, we assume that a person's birthday can fall on any day with the same probability and that $n \le 365$.

Since there are n people and 365 different days, there are 365^n ways in which the n people can have their birthdays. On the other hand, if the n persons are to have distinct birthdays, then the first person can be born on any of the 365 days, the second person can be born on the remaining 364 days, the third person can be born on the remaining 363 days, etc. Thus there are $365 \cdot 364 \cdot 363 \cdot \cdots \cdot (365 - n + 1)$ ways the n persons can have distinct birthdays. The probability p that at least two people have the same birthday follows:

$$\begin{aligned} p &= 1 - \text{the probability that no two people have the same birthday} \\ &= 1 - \frac{365 \cdot 364 \cdot 363 \cdot \cdots \cdot (365 - n + 1)}{365^n} \end{aligned}$$

The values of p where n is a multiple of 10 up to 60 follow:

n	10	20	30	40	50	60
p	0.117	0.411	0.706	0.891	0.970	0.994

We also note that $p = 0.476$ for $n = 22$ and that $p = 0.507$ for $n = 23$. Accordingly:

> In a group of 23 people, it is more likely that at least two of them have the same birthday than that they all have distinct birthdays.

The above table also tells us that, in a group of 60 or more people, the probability that two or more of them have the same birthday exceeds 99 percent.

3.7 EXPECTATION

Games of chance involving probability are very popular throughout the world. Mathematical expectation, defined and discussed in this section, is the measure which decides the fairness of such a game of chance.

Suppose a game has n outcomes $a_1, a_2, \ldots, a_n$ with corresponding probabilities $p_1, p_2, \ldots, p_n$, where $p_1 + p_2 + \cdots + p_n = 1$, and suppose the payoff to the player on outcome a_i is w_i, where a positive w_i is a win for the player, and a negative w_i a loss. Then the quantity

$$E = w_1 p_1 + w_2 p_2 + \cdots + w_n p_n$$

is called the *mathematical expectation* or *expected value* for the player.

The expectation E is the amount that the player can expect to "win" on the average each time the game is played. If E is positive, then the game is in the player's favor; if E is negative, then the game is biased against the player, that is, negative expected winnings represent losses. On the other hand, suppose the expectation $E = 0$. Then the game is said to be *fair*, and a player's winnings and losses should be about equal when the fair game is played a large number of times. Of course, when playing a lottery or gambling in a casino, E is almost always negative.

EXAMPLE 3.8

(*a*) There are three envelopes containing \$100, \$200, and \$6000, respectively. A player selects an envelope and keeps what is in it. Find the expected winnings E of the player.

The player chooses an envelope at random; hence each envelope has probability of 1/3 of being chosen. Accordingly,

$$E = 100\left(\frac{1}{3}\right) + 200\left(\frac{1}{3}\right) + 6000\left(\frac{1}{3}\right) = 2{,}100$$

That is, the expected winnings of the player is \$2100.

(*b*) A player tosses two fair dice. If the sum is 7 or 11, the player wins \$7, otherwise the player loses \$2. Determine the expected value E of the game.

The sample space S consists of the 36 pairs of numbers pictured in Fig. 3-3. Eight of them will result in a sum of 7 or 11, namely,

$$(1,6),\ (2,5),\ (3,4),\ (4,3),\ (5,2),\ (6,1),\ (5,6),\ (6,5)$$

and $36 - 8 = 28$ will not. Thus

$$p(7 \text{ or } 11) = \frac{8}{36} \quad \text{and} \quad P\,(\text{neither 7 nor 11}) = \frac{28}{36}$$

Therefore,

$$E = 7\left(\frac{8}{36}\right) + (-2)\left(\frac{28}{36}\right) = 0$$

Thus the game is fair, and the player should break even over the long run.

Solved Problems

SAMPLE SPACES AND EVENTS

3.1. Let A and B be events. Find an expression and exhibit the Venn diagram for the events: (*a*) A but not B, (*b*) neither A nor B, (*c*) either A or B, but not both.

(*a*) Since A but not B occurs, shade the area of A outside of B, as in Fig. 3-6(*a*). Note that B^c, the complement of B, occurs, since B does not occur; hence A and B^c occur. In other words the event is $A \cap B^c$.

(*b*) "Neither A nor B" means "not A and not B" or $A^c \cap B^c$. By De Morgan's law, this is also the set $(A \cup B)^c$; hence shade the area outside of A and B, i.e. outside $A \cup B$, as in Fig. 3-6(*b*).

(*c*) Since A or B, but not both, occurs, shade the area of A and B, except where they intersect, as in Fig. 3-6(*c*). The event is equivalent to the occurrence of A but not B or B but not A. Thus the event is $(A \cap B^c) \cup (B \cap A^c)$.

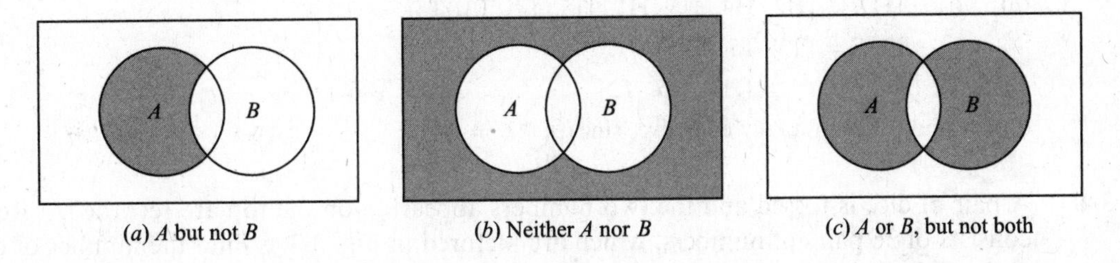

(*a*) A but not B (*b*) Neither A nor B (*c*) A or B, but not both

Fig. 3-6

3.2. Let A, B, C be events. Find an expression and exhibit the Venn diagram for the events: (*a*) A and B but not C occurs, (*b*) only A occurs.

(*a*) Since A and B but not C occurs, shade the intersection of A and B which lies outside of C, as in Fig. 3-7(*a*). The event consists of the elements in A, in B, and in C^c (not in C), that is, the event is the intersection $A \cap B \cap C^c$.

(*b*) Since only A is to occur, shade the area of A which lies outside of B and C, as in Fig. 3-7(*b*). The event consists of the elements in A, in B^c (not in B), and in C^c (not in C), that is, the event is the intersection $A \cap B^c \cap C^c$.

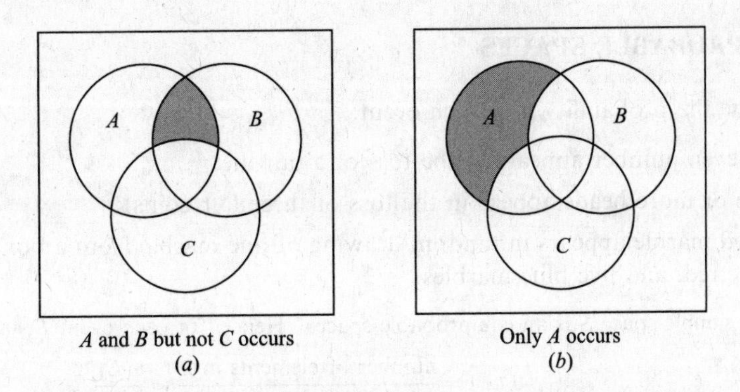

A and B but not C occurs (*a*) Only A occurs (*b*)

Fig. 3-7

3.3. Let a coin and a die be tossed; and let the sample space S consist of the 12 elements:

$$S = \{H1, H2, H3, H4, H5, H6, T1, T2, T3, T4, T5, T6\}$$

(*a*) Express explicitly the following events:

$$A = \{\text{heads and an even number appears}\},$$
$$B = \{\text{a prime number appears}\}, \quad C = \{\text{tails and an odd number appears}\}$$

(*b*) Express explicitly the events that: (i) A or B occurs, (ii) B and C occur, (iii) only B occurs.

(*c*) Which pair of the events A, B, and C are mutually exclusive?

(*a*) The elements of A are those elements of S consisting of an H and an even number:

$$A = \{\text{H2, H4, H6}\}$$

The elements of B are those points in S whose second component is a prime number:

$$B = \{\text{H2, H3, H5, T2, T3, T5}\}$$

The elements of C are those points in S consisting of a T and an odd number: $C = \{\text{T1, T3, T5}\}$.

(*b*) (i) $A \cup B = \{\text{H2, H4, H6, H3, H5, T2, T3, T5}\}$

(ii) $B \cap C = \{\text{T3, T5}\}$

(iii) $B \cap A^c \cap C^c = \{\text{H3, H5, T2}\}$

(*c*) A and C are mutually exclusive, since $A \cap C = \varnothing$.

3.4. A pair of dice is tossed and the two numbers appearing on the top are recorded. Recall that S consists of 36 pairs of numbers, which are pictured in Fig. 3-3. Find the number of elements in each of the following events:

(*a*) $A = \{\text{two numbers are equal}\}$ (*c*) $C = \{\text{5 appears on first die}\}$

(*b*) $B = \{\text{sum is 10 or more}\}$ (*d*) $D = \{\text{5 appears on at least one die}\}$

Use Fig. 3-3 to help count the number of elements in each of the events:

(*a*) $A = \{(1,1), (2,2), \ldots, (6,6)\}$, so $n(A) = 6$.

(*b*) $B = \{(6,4), (5,5), (4,6), (6,5), (5,6), (6,6)\}$, so $n(B) = 6$.

(*c*) $C = \{(5,1), (5,2), \ldots, (5,6)\}$, so $n(C) = 6$.

(*d*) There are six pairs with 5 as the first element, and six pairs with 5 as the second element. However, (5, 5) appears in both places. Hence

$$n(D) = 6 + 6 - 1 = 11$$

Alternatively, count the pairs in Fig. 3-3 which are in D to get $n(D) = 11$.

FINITE EQUIPROBABLE SPACES

3.5. Determine the probability p of each event:

(*a*) An even number appears in the toss of a fair die.

(*b*) One or more heads appear in the toss of three fair coins.

(*c*) A red marble appears in random drawing of one marble from a box containing four white, three red, and five blue marbles.

Each sample space S is an equiprobable space. Hence, for each event E, use

$$P(E) = \frac{\text{number of elements in } E}{\text{number of elements in } S} = \frac{n(E)}{n(S)}$$

(*a*) The event can occur in three ways, 2, 4, or 6, out of 6 cases; hence $p = 3/6 = 1/2$.

(*b*) Assuming the coins are distinguished, there are 8 cases:

$$\text{HHH, HHT, HTH, HTT, THH, THT, TTH, TTT}$$

Only the last case is not favorable; hence $p = 7/8$.

(*c*) There are $4 + 3 + 5 = 12$ marbles of which 3 are red; hence $p = 3/12 = 1/4$.

3.6. A single card is drawn from an ordinary deck S of 52 cards (see Fig. 3-4). Find the probability p that the card is: (*a*) a king, (*b*) a face card (jack, queen, or king), (*c*) a red card (heart or diamond), (*d*) a red face card.

Here $n(S) = 52$.

(*a*) There are four kings; hence $p = 4/52 = 1/13$.

(*b*) There are $4(3) = 12$ face cards; hence $p = 12/52 = 3/13$.

(*c*) There are 13 hearts and 13 diamonds; hence $p = 26/52 = 1/2$.

(*d*) There are six face cards which are red; hence $p = 6/52 = 3/26$.

3.7. Consider the sample space S and events A, B, C in Problem 3.3, where a coin and a die are tossed. Suppose the coin and die are fair; hence S is an equiprobable space. Find: (*a*) $P(A)$, $P(B)$, $P(C)$, (*b*) $P(A \cup B)$, $P(B \cap C)$, $P(B \cap A^c \cap C^c)$.

Since S is an equiprobable space, use $P(E) = n(E)/n(S)$. Here $n(S) = 12$. So we need only count the number of elements in the given set.

(*a*) $P(A) = 3/12$, $P(B) = 6/12$, $P(C) = 3/12$.

(*b*) $P(A \cup B) = 8/12$, $P(B \cap C) = 2/12$, $P(B \cap A^c \cap C^c) = 3/12$.

3.8. A box contains two white sox and two blue sox. Two sox are drawn at random. Find the probability p they are a match (same color).

There are $C(4, 2) = \binom{4}{2} = 6$ ways to draw two of the sox. Only two pairs will yield a match. Thus $p = 2/6 = 1/3$.

3.9. Five horses are in a race. Audrey picks two of the horses at random, and bets on them. Find the probability p that Audrey picked the winner.

There are $C(5, 2) = \binom{5}{2} = 10$ ways to pick two of the horses. Four of the pairs will contain the winner. Thus $p = 4/10 = 2/5$.

3.10. A class contains 10 men and 20 women, of which half the men and half the women have brown eyes. Find the probability p that a person chosen at random is a man or has brown eyes.

Let $A = \{\text{men}\}$, $B = \{\text{brown eyes}\}$. We seek $P(A \cup B)$. We have

$$P(A) = \frac{10}{30} = \frac{1}{3}, \qquad P(B) = \frac{15}{30} = \frac{1}{2}, \qquad P(A \cap B) = \frac{5}{30} = \frac{1}{6}$$

Thus, by Theorem 3.6 (addition rule),

$$P(A \cup B) = P(A) + P(B) - P(A \cap B) = \frac{1}{3} + \frac{1}{2} - \frac{1}{6} = \frac{2}{3}$$

FINITE PROBABILITY SPACES

3.11. A sample space S consists of four elements; that is, $S = \{a_1, a_2, a_3, a_4\}$. Under which of the following functions does S become a probability space?

(a) $P(a_1) = \frac{1}{2}$ $P(a_2) = \frac{1}{3}$ $P(a_3) = \frac{1}{4}$ $P(a_4) = \frac{1}{5}$

(b) $P(a_1) = \frac{1}{2}$ $P(a_2) = \frac{1}{4}$ $P(a_3) = -\frac{1}{4}$ $P(a_4) = \frac{1}{2}$

(c) $P(a_1) = \frac{1}{2}$ $P(a_2) = \frac{1}{4}$ $P(a_3) = \frac{1}{8}$ $P(a_4) = \frac{1}{8}$

(d) $P(a_1) = \frac{1}{2}$ $P(a_2) = \frac{1}{4}$ $P(a_3) = \frac{1}{4}$ $P(a_4) = 0$

(a) Since the sum of the values on the sample points is greater than one, the function does not define S as a probability space.

(b) Since $P(a_3)$ is negative, the function does not define S as a probability space.

(c) Since each value is nonnegative and the sum of the values is one, the function does define S as a probability space.

(d) The values are nonnegative and add up to one; hence the function does define S as a probability space.

3.12. A coin is weighted so that heads is twice as likely to appear as tails. Find $P(T)$ and $P(H)$.

Let $P(T) = p$; then $P(H) = 2p$. Now set the sum of the probabilities equal to one, that is, set $p + 2p = 1$. Then $p = 1/3$. Thus $P(H) = 1/3$ and $P(T) = 2/3$.

3.13. Suppose A and B are events with $P(A) = 0.6$, $P(B) = 0.3$ and $P(A \cap B) = 0.2$. Find the probability that:

(a) A does not occur. (c) A or B occurs.

(b) B does not occur. (d) Neither A nor B occurs.

(a) By the complement rule, $P(\text{not } A) = P(A^c) = 1 - P(A) = 0.4$

(b) By the complement rule, $P(\text{not } B) = P(B^c) = 1 - P(B) = 0.7$

(c) By the addition rule,

$$P(A \text{ or } B) = P(A \cup B) = P(A) + P(B) - P(A \cap B)$$
$$= 0.6 + 0.3 - 0.2 = 0.7$$

(d) Recall (Fig. 3-6(b)) that neither A nor B is the complement of $A \cup B$. Therefore,

$$P\text{ (neither } A \text{ nor } B) = P((A \cup B)^c) = 1 - P(A \cup B) = 1 - 0.7 = 0.3$$

3.14. A die is weighted so that the outcomes produce the following probability distribution:

Outcome	1	2	3	4	5	6
Probability	0.1	0.3	0.2	0.1	0.1	0.2

Consider the events:

$$A = \{\text{even number}\}, \quad B = \{2, 3, 4, 5\}, \quad C = \{x : x < 3\}, \quad D = \{x : x > 7\}$$

Find the following probabilities: (a) $P(A)$, (b) $P(B)$, (c) $P(C)$, (d) $P(D)$.

For any event E, find $P(E)$ by summing the probabilities of the elements in E.

(*a*) $A = \{2, 4, 6\}$, so $P(A) = 0.3 + 0.1 + 0.2 = 0.6$

(*b*) $P(B) = 0.3 + 0.2 + 0.1 + 0.1 = 0.7$

(*c*) $C = \{1, 2\}$, so $P(C) = 0.1 + 0.3 = 0.4$

(*d*) $D = \varnothing$, the empty set. Hence $P(D) = 0$

3.15. For the data in Problem 3.14, find: (*a*) $P(A \cap B)$, (*b*) $P(A \cup C)$, (*c*) $P(B \cap C)$.

First find the elements in the event, and then add the probabilities of the elements.

(*a*) $A \cap B = \{2, 4\}$, so $P(A \cap B) = 0.3 + 0.1 = 0.4$

(*b*) $A \cup C = \{1, 2, 3, 4, 5\} = \{6\}^c$, so $P(A \cup C) = 1 - 0.2 = 0.8$

(*c*) $B \cap C = \{2\}$, so $P(B \cap C) = 0.3$

3.16. Find the probability p of an event E if the odds that E occurs are a to b.

The odds that E occurs are given by the ratio $p : (1 - p)$. Hence

$$\frac{p}{1-p} = \frac{a}{b} \quad \text{or} \quad bp = a - ap \quad \text{or} \quad ap + bp = a \quad \text{or} \quad p = \frac{a}{a+b}$$

3.17. The odds that an event E occurs are 3 to 2. Find the probability of E.

Let $p = P(E)$. Set the odds equal to $p : (1 - p)$ to obtain

$$\frac{p}{1-p} = \frac{3}{2} \quad \text{or} \quad 2p = 3 - 3p \quad \text{or} \quad 5p = 3 \quad \text{or} \quad p = \frac{3}{7}$$

Alternatively, use the formula in Problem 3.16 to directly obtain $p = a/(a+b) = 3/(3+2) = 3/5$.

UNCOUNTABLE UNIFORM SPACES

3.18. A point is chosen at random inside a rectangle measuring 3 inches by 5 inches. Find the probability p that the point is at least one inch from the edge.

Let S denote the set of points inside the rectangle, and let A denote the set of points at least one inch from the edge. S and A are pictured in Fig. 3-8(*a*). Note that A is a rectangular area measuring 1 inch by 3 inches. Thus:

$$p = \frac{\text{area of } A}{\text{area of } S} = \frac{1 \cdot 3}{3 \cdot 5} = \frac{3}{15} = \frac{1}{5}$$

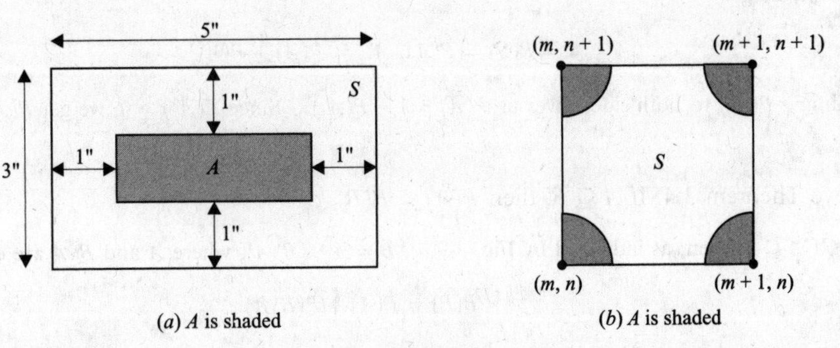

(*a*) A is shaded (*b*) A is shaded

Fig. 3-8

3.19. Consider the plane $\mathbf{R}^2$, and let X denote the subset of points with integer coordinates. A coin of radius 1/4 is tossed randomly on the plane. Find the probability that the coin covers a point of X.

Let S denote the set of points inside a square with corners

$$(m, n), \ (m, n+1), \ (m+1, n), \ (m+1, n+1)$$

where m and n are integers. Let A denote the set of points in S with distance less than 1/4 from any corner point, as pictured in Fig. 3-8(*b*). Note that the area of A is equal to the area inside a circle of radius 1/4. Suppose the center of the coin falls in S. Then the coin will cover a point in X if and only if its center falls in A. Accordingly,

$$p = \frac{\text{area of } A}{\text{area of } S} = \frac{\pi(1/4)^2}{1} = \frac{\pi}{16} \approx 0.2$$

(*Note*: We cannot take S to be all of $\mathbf{R}^2$, since the area of $\mathbf{R}^2$ is infinite.)

PROOFS OF THEOREMS

3.20. Prove Theorem 3.1: $P(\emptyset) = 0$

For any event A, we have $A \cup \emptyset = A$, where A and $\emptyset$ are disjoint. Using $[P_3]$, we get

$$P(A) = P(A \cup \emptyset) = P(A) + P(\emptyset)$$

Adding $-P(A)$ to both sides gives $P(\emptyset) = 0$.

3.21. Prove Theorem 3.2 (complement rule): $P(A^c) = 1 - P(A)$.

$S = A \cup A^c$ where A and A^c are disjoint. By $[P_2]$, $P(S) = 1$. Thus, using $[P_3]$, we get

$$1 = P(S) = P(A \cup A^c) = P(A) + P(A^c)$$

Adding $-P(A)$ to both sides gives us $P(A^c) = 1 - P(A)$.

3.22. Prove Theorem 3.3: $0 \le P(A) \le 1$.

By $[P_1]$, $P(A) \ge 0$. Hence we need only show that $P(A) \le 1$. Since $S = A \cup A^c$, where A and A^c are disjoint, we get

$$1 = P(S) = P(A \cup A^c) = P(A) + P(A^c)$$

Adding $-P(A^c)$ to both sides gives us $P(A) = 1 - P(A^c)$. Since $P(A^c) \ge 0$, we get $P(A) \le 1$, as required.

3.23. Prove Theorem 3.4: If $A \subseteq B$, then $P(A) \le P(B)$.

If $A \subseteq B$, then, as indicated by Fig. 3-9(*a*), $B = A \cup (B \backslash A)$, where A and $B \backslash A$ are disjoint. Hence

$$P(B) = P(A) + P(B \backslash A)$$

By $[P_1]$, we have $P(B \backslash A) \ge 0$; hence $P(A) \le P(B)$.

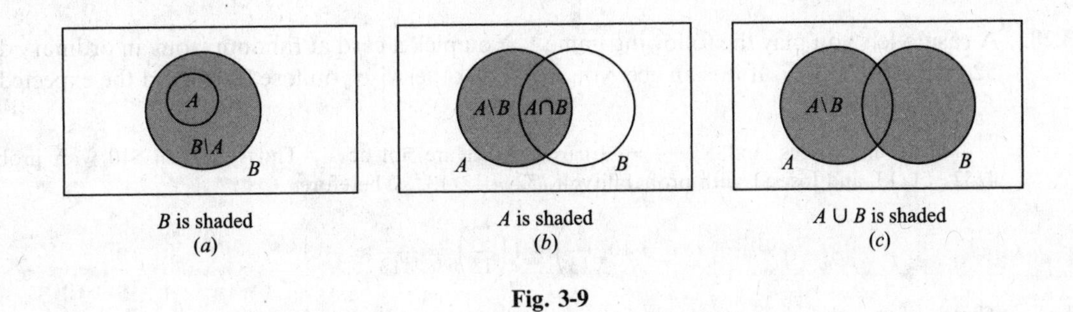

B is shaded
(*a*)

A is shaded
(*b*)

$A \cup B$ is shaded
(*c*)

Fig. 3-9

3.24. Prove Theorem 3.5: $P(A \backslash B) = P(A) - P(A \cap B)$.

As indicated by Fig. 3-9(*b*), $A = (A \backslash B) \cup (A \cap B)$, where $A \backslash B$ and $A \cap B$ are disjoint. Accordingly, by $[P_3]$,

$$P(A) = P(A \backslash B) + P(A \cap B)$$

from which our result follows.

3.25. Prove Theorem (addition rule) 3.6: For any events A and B,

$$P(A \cup B) = P(A) + P(B) - P(A \cap B)$$

As indicated by Fig. 3-9(*c*), $A \cup B = (A \backslash B) \cup B$, where $A \backslash B$ and B are disjoint sets. Thus, using Theorem 3.5,

$$\begin{aligned} P(A \cup B) &= P(A \backslash B) + P(B) = P(A) - P(A \cap B) + P(B) \\ &= P(A) + P(B) - P(A \cap B) \end{aligned}$$

3.26. Prove Corollary 3.7: For any events A, B, C,

$$P(A \cup B \cup C) = P(A) + P(B) + P(C) - P(A \cap B) - P(A \cap C) - P(B \cap C) + P(A \cap B \cap C)$$

Let $D = B \cup C$. Then $A \cap D = A \cap (B \cup C) = (A \cap B) \cup (A \cap C)$ and

$$P(A \cap D) = P(A \cap B) + P(A \cap C) - P(A \cap B \cap A \cap C) = P(A \cap B) + P(A \cap C) - P(A \cap B \cap C)$$

Thus

$$\begin{aligned} P(A \cup B \cup C) &= P(A \cup D) = P(A) + P(D) - P(A \cap D) \\ &= P(A) + P(B) + P(C) - P(B \cap C) - [P(A \cap B) + P(A \cap C) - P(A \cap B \cap C)] \\ &= P(A) + P(B) + P(C) - P(B \cap C) - P(A \cap B) - P(A \cap C) + P(A \cap B \cap C) \end{aligned}$$

EXPECTATION

3.27. A player tosses two fair coins. He wins \$2 if 2 heads occur, and \$1 if 1 head occurs. On the other hand, he loses \$3 if no heads occur. Find the expected value E of the game. Is the game fair? (The game is fair, favorable, or unfavorable to the player according as $E = 0$, $E > 0$, or $E < 0$.)

The sample space $S = \{\text{HH}, \text{HT}, \text{TH}, \text{TT}\}$ where each outcome has probability 1/4. The player wins \$2 in the first case, \$1 in the second and third cases, and loses \$3 in the last case. Thus

$$E = 2\left(\frac{1}{4}\right) + 1\left(\frac{1}{4}\right) + 1\left(\frac{1}{4}\right) - 3\left(\frac{1}{4}\right) = \frac{1}{4}$$

Thus the game is favorable to the player. (Specifically, he will win, on the average, 25 cents per play, e.g. if he plays 100 times, then he will likely win about \$25.)

3.28. A casino lets you play the following game. You pick a card at random from an ordinary deck of 52 cards (Fig. 3-4). If it is an ace, you win \$10; otherwise you lose \$1. Find the expected value E of your playing.

There are 4 aces, and $52 - 4 = 48$ cards which are not aces. Thus you win \$10 with probability $4/52 = 1/13$, and lose \$1 with probability $48/52 = 12/13$. Therefore,

$$E = 10\left(\frac{1}{13}\right) - 1\left(\frac{12}{13}\right) = -\frac{2}{13} \approx -0.15$$

That is, if you play the game many times, you will lose, on the average, 15 cents per game.

MISCELLANEOUS PROBLEMS

3.29. Show that axiom $[P_3]$ follows from axiom $[P_3']$.

First we show that $P(\emptyset) = 0$ using $[P_3']$ instead of $[P_3]$. We have $\emptyset = \emptyset + \emptyset + \emptyset + \ldots$ where the empty sets are disjoint. Say $P(\emptyset) = a$. Then, by $[P_3']$,

$$P(\emptyset) = P(\emptyset + \emptyset + \emptyset + \cdots) = P(\emptyset) + P(\emptyset) + P(\emptyset) + \cdots$$

However, zero is the only real number a satisfying

$$a = a + a + a + \cdots$$

Therefore, $P(\emptyset) = 0$.

Suppose A and B are disjoint. Then $A, B, \emptyset, \emptyset, \ldots$ are disjoint, and $A \cup B = A \cup B \cup \emptyset \cup \emptyset \cup \ldots$. Hence, by $[P_3']$,

$$P(A \cup B) = P(A \cup B \cup \emptyset \cup \emptyset \cup \cdots) = P(A) + P(B) + P(\emptyset) + P(\emptyset) + \cdots$$
$$= P(A) + P(B) + 0 + 0 + \cdots = P(A) + P(B)$$

which is $[P_3]$.

3.30. Prove Theorem 3.9. Suppose $S = \{a_1, a_2, \ldots, a_n\}$ and each a_i is assigned the probability p_i where: (i) $p_i \geq 0$, and (ii) $\sum p_i = 1$. For any event A, let

$$P(A) = \sum(p_j : a_j \in A)$$

Then P satisfies: (*a*) $[P_1]$, (*b*) $[P_2]$, (*c*) $[P_3]$.

(*a*) Each $p_j \geq 0$; hence $P(A) = \sum p_j \geq 0$.

(*b*) Every $a_j \in S$; hence $P(S) = p_1 + p_2 + \cdots + p_n = 1$.

(*c*) Suppose A and B are disjoint, and

$$P(A) = \sum(p_j : a_j \in A), \quad P(B) = \sum(p_k : a_k \in B)$$

Then the a_j's and a_k's are distinct. Therefore:

$$P(A \cup B) = \sum(p_t : p_t \in A \cup B)$$
$$= \sum(p_t : a_t \in A) + \sum(p_t : a_t \in B) = P(A) + P(B)$$

3.31. Let $S = \{a_1, a_2, \ldots, a_s\}$ and $T = \{b_1, b_2, \ldots, b_t\}$ be finite probability spaces. Let the number $p_{ij} = P(a_i)P(b_j)$ be assigned to the ordered pair (a_i, b_j) in the product set $S \times T = \{(s,t) : s \in S, t \in T\}$. Show that the p_{ij} define a probability space on $S \times T$; that is, show that: (i) the p_{ij} are nonnegative, and (ii) the sum of the p_{ij} equals one. This is called the *product probability space*. (We emphasize that this is not the only probability function that can be defined on the product set $S \times T$.)

Since $P(a_i)$, $P(b_j) \geq 0$, for each i and each j, $p_{ij} = P(a_i)P(b_j) \geq 0$. Furthermore,

$$\begin{aligned}
&p_{11} + p_{12} + \cdots + p_{1t} + p_{21} + p_{22} + \cdots + p_{2t} + \cdots + p_{s1} + p_{s2} + \cdots + p_{st} \\
&= P(a_1)P(b_1) + \cdots + P(a_1)P(b_t) + \cdots + P(a_s)P(b_1) + \cdots + P(a_s)P(b_t) \\
&= P(a_1)[P(b_1) + \cdots + P(b_t)] + \cdots + P(a_s)[P(b_1) + \cdots + P(b_t)] \\
&= P(a_1) \cdot 1 + \cdots + P(a_s) \cdot 1 = P(a_1) + \cdots + P(a_s) = 1
\end{aligned}$$

3.32. A die is tossed 100 times. The following table lists the six numbers and the frequency with which each number appeared:

Number	1	2	3	4	5	6
Frequency	14	17	20	18	15	16

(*a*) Find the relative frequency f of each of the following events:

$A = \{3 \text{ appears}\}$, $B = \{5 \text{ appears}\}$, $C = \{\text{even number appears}\}$

(*b*) Find a probability model for the data.

(*a*) The relative frequency $f = \dfrac{\text{number of successes}}{\text{total number of trials}}$. Thus:

$$f_A = \frac{20}{100} = 0.20, \qquad f_B = \frac{15}{100} = 0.15, \qquad f_C = \frac{17 + 18 + 16}{100} = 0.52$$

(*b*) The geometric symmetry of the die indicates that we first assume an equal probability space. Statistics is then used to decide whether or not the given data supports the assumption of a fair die.

Supplementary Problems

SAMPLE SPACES AND EVENTS

3.33. Let A and B be events. Find an expression and exhibit the Venn diagram for the event that: (*a*) A or not B occurs, (*b*) only A occurs.

3.34. Let A, B, and C be events. Find an expression and exhibit the Venn diagram for the event that:
(*a*) A and B but not C occur, (*c*) none of the events occurs,
(*b*) A or C, but not B, occur, (*d*) at least two of the events occur.

3.35. A penny, a dime, and a die are tossed.

(*a*) Describe a suitable sample space S, and find $n(S)$.

(*b*) Express explicitly the following events:

$A = \{\text{two heads and an even number}\}$, $B = \{2 \text{ appears}\}$

$C = \{\text{exactly one head and an odd number}\}$

(*c*) Express explicitly the events: (i) A and B, (ii) only B, (iii) B and C.

FINITE EQUIPROBABLE SPACES

3.36. Determine the probability of each event:

(*a*) An odd number appears in the toss of a fair die.

(*b*) One or more heads appear in the toss of four fair coins.

(*c*) One or both numbers exceed 4 in the toss of two fair die.

(*d*) A red or a face card appears when a card is randomly selected from a 52-card deck.

3.37. A student is chosen at random to represent a class with five freshmen, eight sophomores, three juniors, and two seniors. Find the probability that the student is: (*a*) a sophomore, (*b*) a junior, (*c*) a junior or a senior.

3.38. Of 10 girls in a class, 3 have blue eyes. Two of the girls are chosen at random. Find the probability that: (*a*) both have blue eyes, (*b*) neither has blue eyes, (*c*) at least one has blue eyes, (*d*) exactly one has blue eyes.

3.39. Three bolts and three nuts are in a box. Two parts are chosen at random. Find the probability that one is a bolt and one is a nut.

3.40. A box contains two white sox, two blue sox, and two red sox. Two sox are drawn at random. Find the probability they are a match (same color).

3.41. Of 120 students, 60 are studying French, 50 are studying Spanish, and 20 are studying both French and Spanish. A student is chosen at random. Find the probability that the student is studying: (*a*) French or Spanish, (*b*) neither French nor Spanish, (*c*) only French, (*d*) exactly one of the two languages.

FINITE PROBABILITY SPACES

3.42. Under which of the following functions does $S = \{a_1, a_2, a_3\}$ become a probability space?

(*a*) $P(a_1) = \frac{1}{4}, P(a_2) = \frac{1}{3}, P(a_3) = \frac{1}{2}$, (*c*) $P(a_1) = \frac{1}{6}, P(a_2) = \frac{1}{3}, P(a_3) = \frac{1}{2}$,

(*b*) $P(a_1) = \frac{2}{3}, P(a_2) = -\frac{1}{3}, P(a_3) = \frac{2}{3}$, (*d*) $P(a_1) = 0,\ P(a_2) = \frac{1}{3}, P(a_3) = \frac{2}{3}$.

3.43. A coin is weighted so that heads is three times as likely to appear as tails. Find $P(H)$ and $P(T)$.

3.44. Three students A, B, and C are in a swimming race. A and B have the same probability of winning and each is twice as likely to win as C. Find the probability that: (*a*) B wins, (*b*) C wins, (*c*) B or C wins.

3.45. Suppose A and B are events with $P(A) = 0.7$, $P(B) = 0.5$ and $P(A \cap B) = 0.4$. Find the probability that:

(*a*) A does not occur. (*c*) A but not B occurs.

(*b*) A or B occurs. (*d*) Neither A nor B occurs.

3.46. Consider the following probability distribution:

Outcome	1	2	3	4	5	6
Probability	0.1	0.4	0.1	0.1	0.2	0.1

Find the following probabilities, where:

$$A = \{\text{even number}\}, \qquad B = \{2, 3, 4, 5\}, \qquad C = \{1, 2\}$$

(*a*) $P(A)$, $P(B)$, $P(C)$, (*b*) $P(A \cap B)$, $P(A \cup C)$, $P(B \cap C)$.

3.47. Find the probability of an event E if the odds that it will occur are: (*a*) 2 to 1, (*b*) 5 to 11.

3.48. In a swimming race, the odds that A will win are 2 to 3 and the odds that B will win are 1 to 4. Find the probability p and the odds that: (*a*) A or B will win, (*b*) neither A nor B will win.

NONCOUNTABLE UNIFORM SPACES

3.49. A point is chosen at random inside a circle. Find the probability p that the point is closer to the center of the circle than to its radius.

3.50. A point A is selected at random inside an equilateral triangle whose side length is 3. Find the probability p that the distance of A from any corner is greater than 1.

3.51. A coin of diameter 1/2 is tossed randomly onto the plane $\mathbf{R}^2$. Find the probability p that the coin does not intersect any line of the form: (*a*) $x = k$, where k is an integer, (*b*) $x + y = k$, where k is an integer.

3.52. A point X is selected at random from a line segment AB with midpoint O. Find the probability p that the line segments AX, XB, and AO can form a triangle.

EXPECTATION

3.53. You have won a contest. Your prize is to select one of four envelopes and keep what is in it. One envelope contains a check for \$100, another for \$200, another for \$400, and another for \$2000. What is the mathematical expectation E of your winnings?

3.54. A game consists of tossing a fair die. A player wins if the number is even and loses if the number is odd. The winning or losing (dollar) payoff is equal to the number appearing. Find the player's mathematical expectation E.

3.55. A mathematics professor gives an "extra-credit" problem on a test. If it is done correctly, 15 points are added to the test score, and if it is done partially correctly, 5 points are added; otherwise 5 points are subtracted. Suppose a student's probability of getting the problem completely right is 1/4, and only partially correct is 1/2. Find the student's mathematical expectation E for extra credit.

3.56. A game consists of tossing a fair coin four times. A player wins \$3 if two or more heads appear; otherwise the player loses \$4. Find the expected value E of the game.

MISCELLANEOUS PROBLEMS

3.57. A die is tossed 50 times. The following table gives the six numbers and their frequency of occurrence:

Number	1	2	3	4	5	6
Frequency	7	9	8	7	9	10

Find the relative frequency of each event:
(*a*) 4 appears, (*b*) an odd number appears, (*c*) a prime number, 2, 3, or 5, appears.

3.58. Use mathematical induction to prove: For any events $A_1, A_2, \ldots, A_n$,

$$P(A_1 \cup \cdots \cup A_n) = \sum_i P(A_i) - \sum_{i<j} P(A_i \cap A_j) + \sum_{i<j<k} P(A_i \cap A_j \cap A_k) - \cdots \pm P(A_1 \cap \cdots \cap A_n)$$

Remark: This result generalizes Theorem 3.6 (addition rule) for two sets, and Corollary 3.7 for three sets.

Answers to Supplementary Problems

3.33. (*a*) $A \cup B^c$, (*b*) $A \cap B^c$

3.34. (*a*) $A \cap B \cap C^c$, (*b*) $(A \cup C) \cap B$, (*c*) $(A \cup B \cup B)^c = A^c \cap B^c \cap C^c$,
(*d*) $(A \cap B) \cup (A \cap C) \cup (B \cap C)$

3.35. (*a*) $n(S) = 24$, $S = \{H, T\} \times \{H, T\} \times \{1, 2, \ldots, 6\}$
(*b*) $A = \{HH2, HH4, HH6\}$, $B = \{HH2, HT2, TH2, TT2\}$, $C = \{HT1, HT3, HT5, TH1, TH3, TH5\}$
(*c*) (i) HH2, (ii) HT2, TH2, TT2, (iii) $\varnothing$

3.36. (*a*) 3/6, (*b*) 15/16, (*c*) 20/36, (*d*) 32/52

3.37. (*a*) 8/18, (*b*) 3/18, (*c*) 5/18

3.38. (*a*) 1/15, (*b*) 7/15, (*c*) 8/15, (*d*) 7/15

3.39. 3/5

3.40. 1/5

3.41. (*a*) 3/4, (*b*) 1/4, (*c*) 1/3, (*d*) 7/12

3.42. (*c*) and (*d*)

3.43. $P(H) = 3/4$, $P(T) = 1/4$

3.44. (*a*) 2/5, (*b*) 1/5, (*c*) 3/5

3.45. (*a*) 0.3, (*b*) 0.8, (*c*) 0.2, (*d*) 0.2

3.46. (*a*) 0.6, 0.8, 0.5, (*b*) 0.5, 0.7, 0.4

3.47. (*a*) 2/3, (*b*) 5/16

3.48. (*a*) $p = 2/5$; odds 3 to 2, (*b*) $p = 3/5$; odds 2 to 3

3.49. 1/4

3.50. $1 - 2\pi/(9\sqrt{3}) = 1 - 2\sqrt{3}\pi/27$

3.51. (*a*) 1/2; (*b*) $1 - \sqrt{2}/2$

3.52. 1/2

3.53. $675

3.54. $0.50

3.55. 5

3.56. $13/16 \approx \$0.81$

3.57. (*a*) 7/50, (*b*) 24/50, (*c*) 26/50

Chapter 4

Conditional Probability and Independence

4.1 INTRODUCTION

The notions of conditional probability and independence will be motivated by two well-known examples.

Gender Gap

Suppose candidate A receives 52 percent of the entire vote, but only 46 percent of the female vote. Let $P(A)$ denote the probability that a random person voted for A, but let $P(A|W)$ denote the probability that a random woman voted for A. Then

$$P(A) = 0.52 \qquad \text{but} \qquad P(A|W) = 0.46$$

$P(A|W)$ is called the condition probability of A given W; note that $P(A|W)$ only looks at the reduced sample space consisting of women. The fact that $P(A) \neq P(A|W)$ is called the "gender gap" in politics. On the other hand, suppose $P(A) = P(A|W)$; then we say that voting for A is independent of the gender of the voter.

Insurance Rates

Auto insurance rates usually depend on the probability that a random person will be involved in an accident. It is well known that male drivers under 25 years old get into more accidents than the general public. That is, letting $P(A)$ denote the probability of an accident and letting E denote male drivers under 25 years old, the data tells us that

$$P(A) < P(A|E)$$

Again we use the notation $P(A|E)$ to denote the probability of an accident A given that the driver E is male and under 25 years old.

This chapter formally defines conditional probability and independence. We also cover finite stochastic processes, Bayes' Theorem, and independent repeated trials.

4.2 CONDITIONAL PROBABILITY

Suppose E is an event in a sample space S with $P(E) > 0$. The probability that an event A occurs once E has occurred or, specifically, the *conditional probability of A given E*, written $P(A|E)$, is defined as follows:

$$P(A|E) = \frac{P(A \cap E)}{P(E)}$$

As pictured in the Venn diagram in Fig. 4-1, $P(A|E)$ measures, in a certain sense, the relative probability of A with respect to the reduced space E.

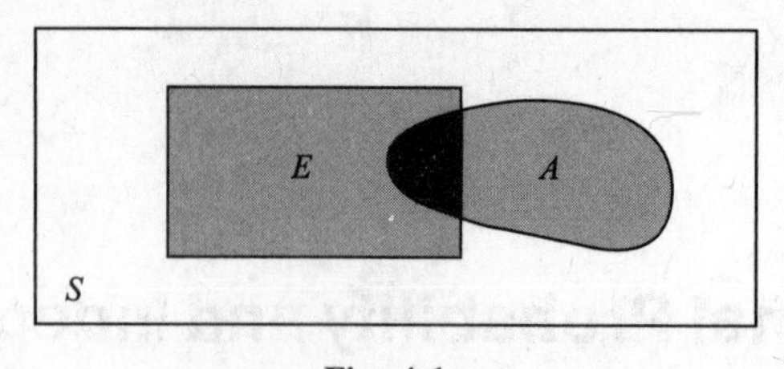

Fig. 4-1

Now suppose S is an equiprobable space, and we let $n(A)$ denote the number of elements in the event A. Then

$$P(A \cap E) = \frac{n(A \cap E)}{n(S)}, \qquad P(E) = \frac{n(E)}{n(S)}, \qquad \text{so} \qquad P(A|E) = \frac{P(A \cap E)}{P(E)} = \frac{n(A \cap E)}{n(E)}$$

We state this result formally.

Theorem 4.1: Suppose S is an equiprobable space and A and B are events. Then

$$P(A|E) = \frac{\text{number of elements in } A \cap E}{\text{number of elements in } E} = \frac{n(A \cap E)}{n(E)}$$

EXAMPLE 4.1

(*a*) A pair of fair dice is tossed. The sample space S consists of the 36 ordered pairs (a, b), where a and b can be any of the integers from 1 to 6 (see Fig. 3-3). Thus the probability of any point is 1/36. Find the probability that one of the die is 2 if the sum is 6. That is, find $P(A|E)$ where

$$E = \{\text{sum is } 6\} \qquad \text{and} \qquad A = \{2 \text{ appears on at least one die}\}$$

Also find $P(A)$.

Now E consists of five elements, specifically

$$E = \{(1,5), (2,4), (3,3), (4,2), (5,1)\}$$

Two of them, $(2,4)$ and $(4,2)$, belong to A; that is, $A \cap E = \{(2,4), (4,2)\}$. By Theorem 4.1, $P(A|E) = 2/5$.

On the other hand, A consists of 11 elements, specifically,

$$A = \{(2,1), (2,2), (2,3), (2,4), (2,5), (2,6), (1,2), (3,2), (4,2), (5,2), (6,2)\}$$

and S consists of 36 elements; hence $P(A) = 11/36$.

(*b*) A couple has two children; the sample space is $S = \{dd, bg, gb, gg\}$ with probability 1/4 for each point. Find the probability p that both children are boys if it is known that: (i) at least one of the children is a boy, (ii) the older child is a boy.

(i) Here the reduced sample space consists of three elements, $\{bb, bg, gb\}$; hence $p = 1/3$.

(ii) Here the reduced sample space consists of two elements, $\{bb, bg\}$; hence $p = 1/2$.

Multiplication Theorem for Conditional Probability

Suppose A and B are events in a sample space S with $P(A) > 0$. By definition of conditional probability,

$$P(B|A) = \frac{P(A \cap B)}{P(A)}$$

Multiplying both sides by $P(A)$ gives us the following useful result:

Theorem 4.2 (multiplication theorem for conditional probability):

$$P(A \cap B) = P(A)P(B|A)$$

The multiplication theorem gives us a formula for the probability that events A and B both occur. It can be extended to three or more events. For three events, we get:

Corollary 4.3: $P(A \cap B \cap C) = P(A)P(B|A)P(C|A \cap B)$

That is, the probability that A, B, and C occur is equal to the product of (i) the probability that A occurs, (ii) the probability that B occurs, assuming that A occurred, and (iii) the probability that C occurs, assuming that A and B have occurred.

EXAMPLE 4.2 A lot contains 12 items, of which 4 are defective. Three items are drawn at random from the lot, one after the other. Find the probability p that all three are nondefective.

The probability that the first item is nondefective is $\frac{8}{12}$, since eight of 12 items are nondefective. If the first item is nondefective, then the probability that the next item is nondefective is $\frac{7}{11}$, since only seven of the remaining 11 items are nondefective. If the first two items are nondefective, then the probability that the last item is nondefective is $\frac{6}{10}$, since only 6 of the remaining 10 items are now nondefective. Thus by the multiplication theorem,

$$p = \frac{8}{12} \cdot \frac{7}{11} \cdot \frac{6}{10} = \frac{14}{55} \approx 0.25$$

4.3 FINITE STOCHASTIC PROCESSES AND TREE DIAGRAMS

Consider a (*finite*) *stochastic process*, that is, a finite sequence of experiments where each experiment has a finite number of outcomes with given probabilities. A convenient way of describing such a process is by means of a *labeled tree diagram*, as illustrated below. The multiplication theorem (Theorem 4.2) can then be used to compute the probability of an event which is represented by a given path of the tree.

EXAMPLE 4.3

(*a*) Suppose the following three boxes are given:

Box X has 10 light bulbs, of which four are defective.

Box Y has 6 light bulbs, of which one is defective.

Box Z has 8 light bulbs, of which three are defective.

A box is chosen at random, and then a bulb is randomly selected from the chosen box. Find the probability p that the bulb is nondefective.

Here we perform a sequence of two experiments:

(i) Select one of the three boxes.

(ii) Select a bulb which is either defective (D) or nondefective (N).

The tree diagram in Fig. 4-2 describes this process and gives the probability of each branch of the tree. The multiplication theorem tells us that the probability of a given path of the tree is the product of the probabilities

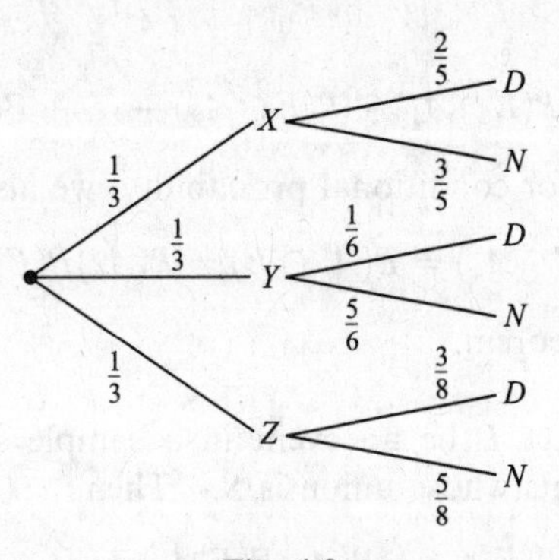

Fig. 4-2

of each branch of the path. For example, the probability of selecting box X and then a nondefective bulb N from box X is

$$\frac{1}{3}\cdot\frac{3}{5}=\frac{1}{5}$$

Since there are three disjoint paths which lead to a nondefective bulb N, the sum of the probabilities of these paths gives us the required probability. Namely,

$$p = P(N) = \frac{1}{3}\cdot\frac{3}{5}+\frac{1}{3}\cdot\frac{5}{6}+\frac{1}{3}\cdot\frac{5}{8}=\frac{247}{360}\approx 0.686$$

(*b*) Consider the stochastic process in part (*a*). If a nondefective bulb N is chosen, find the probability that the bulk came from box Z. In other words, find $P(Z|N)$, the conditional probability of box Z given a nondefective bulb N.

Now box Z and a nondefective bulb N can only occur on the bottom path, which has probability $\frac{1}{3}\cdot\frac{5}{8}=\frac{5}{24}$, that is, $P(Z\cap N)=\frac{5}{24}$. Furthermore, by (*a*) we have $P(N)=\frac{247}{360}$. Thus, by the definition of conditional probability,

$$P(Z|N)=\frac{P(Z\cap N)}{P(N)}=\frac{5/24}{247/360}=\frac{75}{247}=0.304$$

In other words, we divide the probability of the successful path by the probability of the reduced sample space consisting of all the paths leading to N.

4.4 TOTAL PROBABILITY AND BAYES' FORMULA

Suppose a set S is the union of mutually disjoint subsets $A_1, A_2, \ldots, A_n$, and suppose E is any subset of S. Then, as illustrated in Fig. 4-3 for the case $n = 3$,

$$E = E\cap S = E\cap(A_1\cup A_2\cup\cdots\cup A_n) = (E\cap A_1)\cup(E\cap A_2)\cup\cdots\cup(E\cap A_n)$$

Moreover, the n subsets on the right in the above equation are also mutually disjoint.

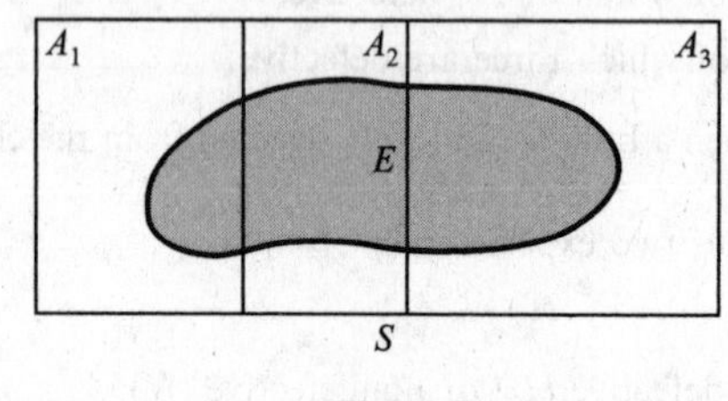

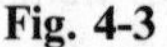

Fig. 4-3

Now suppose S is a sample space and the above subsets $A_1, A_2, \ldots, A_n$, E are events. Since the $E\cap A_k$ are disjoint, we get

$$P(E) = P(E\cap A_1)+P(E\cap A_2)+\cdots+P(E\cap A_n)$$

Using the multiplication theorem for conditional probability, we also get

$$P(E\cap A_k) = P(A_k\cap E) = P(A_k)P(E|A_k)$$

Thus we arrive at the following theorem.

Theorem 4.4 (total probability): Let E be an event in a sample space S, and let $A_1, A_2, \ldots, A_n$ be mutually disjoint events whose union is S. Then

$$P(E) = P(A_1)P(E|A_1)+P(A_2)P(E|A_2)+\cdots+P(A_n)P(E|A_n)$$

The equation in Theorem 4.4 is called the *law of total probability*. Note that the sets $A_1, A_2, \ldots, A_n$ are pairwise disjoint and their union is all of S. That is, the As form a *partition* of S.

EXAMPLE 4.4 A factory uses three machines X, Y, Z to produce certain items. Suppose:

(1) Machine X produces 50 percent of the items, of which 3 percent are defective.

(2) Machine Y produces 30 percent of the items, of which 4 percent are defective.

(3) Machine Z produces 20 percent of the items, of which 5 percent are defective.

Find the probability p that a randomly selected item is defective.

Let D denote the event that an item is defective. Then, by the law of total probability,

$$\begin{aligned} P(D) &= P(X)P(D|X) + P(Y)P(D|Y) + P(Z)P(D|Z) \\ &= (0.50)(0.03) + (0.30)(0.04) + (0.20)(0.05) = 0.037 = 3.7 \text{ percent} \end{aligned}$$

Bayes' Theorem

Suppose the events $A_1, A_2, \ldots, A_n$ do form a partition of the sample space S, and E is any event. Then, for $k = 1, 2, \ldots, n$, the multiplication theorem for conditional probability tells us that $P(A_k \cap E) = P(A_k)P(E|A_k)$. Therefore,

$$P(A_k|E) = \frac{P(A_k \cap E)}{P(E)} = \frac{P(A_k)P(E|A_k)}{P(E)}$$

Using the law of total probability (Theorem 4.4) for the denominator $P(E)$, we arrive at the next theorem.

Theorem 4.5 (Bayes' formula): Let E be an event in a sample space S, and let $A_1, A_2, \ldots, A_n$ be disjoint events whose union is S. Then, for $k = 1, 2, \ldots, n$,

$$P(A_k|E) = \frac{P(A_k)P(E|A_k)}{P(A_1)P(E|A_1) + P(A_2)P(E|A_2) + \cdots + P(A_n)P(E|A_n)}$$

The above equation is called *Bayes' rule* or *Bayes' formula*, after the English mathematician Thomas Bayes (1702–61). If we think of the events $A_1, A_2, \ldots, A_n$ as possible causes of the event E, then Bayes' formula enables us to determine the probability that a particular one of the As occurred, given that E occurred.

EXAMPLE 4.5 Consider the factory in Example 4.4. Suppose a defective item is found among the output. Find the probability that it came from each of the machines, that is, find $P(X|D)$, $P(Y|D)$, and $P(Z|D)$.

Recall that $P(D) = P(X)P(D|X) + P(Y)P(D|Y) + P(Z)P(D|Z) = 0.037$. Therefore, by Bayes' formula,

$$P(X|D) = \frac{P(X)P(D|X)}{P(D)} = \frac{(0.50)(0.03)}{0.037} = \frac{15}{37} = 40.5 \text{ percent}$$

$$P(Y|D) = \frac{P(Y)P(D|Y)}{P(D)} = \frac{(0.30)(0.04)}{0.037} = \frac{12}{37} = 32.5 \text{ percent}$$

$$P(Z|D) = \frac{P(Z)P(D|Z)}{P(D)} = \frac{(0.20)(0.05)}{0.037} = \frac{10}{37} = 27.0 \text{ percent}$$

Stochastic Interpretation of Total Probability and Bayes' Formula

Frequently, problems involving the total probability law and Bayes' formula can be interpreted as a two-step stochastic process. Figure 4-4 gives the stochastic tree corresponding to Fig. 4-3, where the first step in the tree involves the events A_1, A_2, A_3, which partition S, and the second step involves the arbitrary event E.

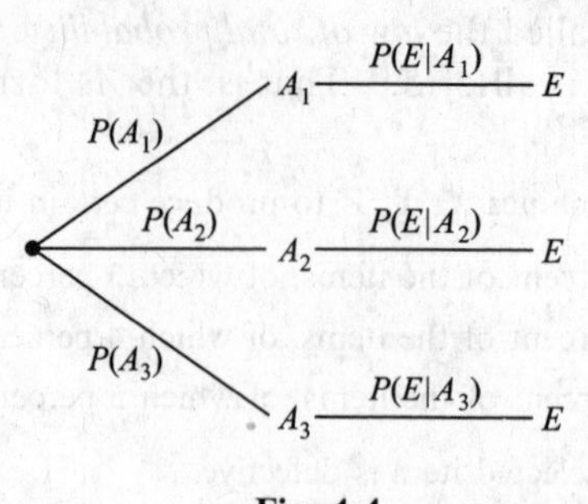

Fig. 4-4

Suppose we want $P(E)$. Using the tree diagram, we obtain

$$P(E) = P(A_1)P(E|A_1) + P(A_2)P(E|A_2) + P(A_3)P(E|A_3)$$

Furthermore, for $k = 1, 2, 3$,

$$P(A_k|E) = \frac{P(A_k \cap E)}{P(E)} = \frac{P(A_k)P(E|A_k)}{P(E)}$$

$$= \frac{P(A_k)P(E|A_k)}{P(A_1)P(E|A_1) + P(A_2)P(E|A_2) + P(A_3)P(E|A_3)}$$

Observe that the above two formulas are simply the total probability law and Bayes' formula, for the case $n = 3$. The stochastic approach also applies to any positive integer n. (See Problem 4.12.)

4.5 INDEPENDENT EVENTS

Events A and B in probability space S are said to be *independent* if the occurrence of one of them does not influence the occurrence of the other. More specifically, B is independent of A if $P(B)$ is the same as $P(B|A)$. Now, substituting $P(B)$ for $P(B|A)$ in the multiplication Theorem 4.2, that is, $P(A \cap B) = P(A)P(B|A)$, yields:

$$\boxed{P(A \cap B) = P(A)P(B)}$$

We formally use the above equation as our definition of independence.

Definition: Events A and B are *independent* if $P(A \cap B) = P(A)P(B)$; otherwise they are *dependent*.

We emphasize that independence is a symmetric relation. In particular:

$$\boxed{P(A \cap B) = P(A)P(B) \qquad \text{implies both} \qquad P(B|A) = P(B) \qquad \text{and} \qquad P(A|B) = P(A)}$$

Note also that disjoint (mutually exclusive) events are not independent unless one of them has zero probability. That is, suppose $A \cap B = \emptyset$ and A and B are independent. Then

$$P(A)P(B) = P(A \cap B) = 0 \qquad \text{and so} \qquad P(A) = 0 \qquad \text{or} \qquad P(B) = 0$$

EXAMPLE 4.6 A fair coin is tossed three times, yielding the equiprobable space

$$S = \{\text{HHH, HHT, HTH, HTT, THH, THT, TTH, TTT}\}$$

Consider the events:

$$A = \{\text{first toss is heads}\} = \{\text{HHH, HHT, HTH, HTT}\}$$
$$B = \{\text{second toss is heads}\} = \{\text{HHH, HHT, THH, THT}\}$$
$$C = \{\text{exactly two heads in a row}\} = \{\text{HHT, THH}\}$$

Clearly A and B are independent events; this fact is verified below. On the other hand, the relationship between A and C and between B and C is not obvious. We claim that A and C are independent, but that B and C are dependent. Note that:

$$P(A) = \frac{4}{8} = \frac{1}{2}, \qquad P(B) = \frac{4}{8} = \frac{1}{2}, \qquad P(C) = \frac{2}{8} = \frac{1}{4}$$

Also,

$$P(A \cap B) = P(\{\text{HHH, HHT}\}) = \frac{1}{4}, \qquad P(A \cap C) = P(\{\text{HHT}\}) = \frac{1}{8}, \qquad P(B \cap C) = P(\{\text{HHT, THH}\}) = \frac{1}{4}$$

Accordingly,

$$P(A)P(B) = \frac{1}{2} \cdot \frac{1}{2} = \frac{1}{4} = P(A \cap B), \qquad \text{so } A \text{ and } B \text{ are independent}$$

$$P(A)P(C) = \frac{1}{2} \cdot \frac{1}{4} = \frac{1}{8} = P(A \cap C), \qquad \text{so } A \text{ and } C \text{ are independent}$$

$$P(B)P(C) = \frac{1}{2} \cdot \frac{1}{4} = \frac{1}{8} \neq P(B \cap C), \qquad \text{so } B \text{ and } C \text{ are dependent}$$

Frequently, we will postulate that two events are independent, or the experiment itself will imply that two events are independent.

EXAMPLE 4.7 The probability that A hits a target is $\frac{1}{4}$, and the probability that B hits the target is $\frac{2}{5}$. Both shoot at the target. Find the probability that at least one of them hits the target, i.e. that A or B (or both) hit the target.

Here $P(A) = \frac{1}{2}$ and $P(B) = \frac{2}{5}$, and we seek $P(A \cup B)$. Furthermore, we assume that A and B are independent events; that is, that the probability that A or B hits the target is not influenced by what the other does. Therefore:

$$P(A \cap B) = P(A)P(B) = \frac{1}{4} \cdot \frac{2}{5} = \frac{1}{10}$$

Accordingly, by the addition rule, Theorem 3.6,

$$P(A \cup B) = P(A) + P(B) - P(A \cap B) = \frac{1}{4} + \frac{2}{5} - \frac{1}{10} = \frac{11}{20}$$

Independence of Three or More Events

Three events A, B, C are *independent* if the following two conditions hold:

(1) They are pairwise independent; that is,

$$P(A \cap B) = P(A)P(B), \qquad P(A \cap C) = P(A)P(C), \qquad P(B \cap C) = P(B)P(C)$$

(2) $P(A \cap B \cap C) = P(A)P(B)P(C)$

Problem 4.17 shows that pairwise independence does not imply independence, that is, (1) does not imply (2); and Problem 4.18 shows that (2) does not imply (1).

Independence of more than three events is defined analogously. Namely, the events $A_1, A_2, \ldots, A_n$ are *independent* if any proper subset of them is independent and

$$P(A_1 \cap A_2 \cap \cdots \cap A_n) = P(A_1)P(A_2) \cdots P(A_n)$$

Observe that induction is used in this definition.

4.6 INDEPENDENT REPEATED TRIALS

Previously we discussed probability spaces which were associated with an experiment repeated a finite number of times, such as the tossing of a coin three times. This concept of repetition is formalized as follows:

Definition: Let S be a finite probability space. By the space of n *independent repeated trials*, we mean the probability space S_n consisting of ordered n-tuples of elements of S, with the probability of an n-tuple defined to be the product of the probabilities of its components:

$$P((s_1, s_2, \ldots, s_n)) = P(s_1)P(s_2)\cdots P(s_n)$$

EXAMPLE 4.8 Suppose that whenever three horses a, b, c race together, their respective probabilities of winning are 1/2, 1/3, and 1/6. In other words:

$$S = \{a, b, c\}, \quad \text{with} \quad P(a) = \frac{1}{2}, \quad P(b) = \frac{1}{3}, \quad \text{and} \quad P(c) = \frac{1}{6}$$

If the horses race twice, then the sample space S_2 of the two repeated trials follows:

$$S_2 = \{aa,\ ab,\ ac,\ ba,\ bb,\ bc,\ ca,\ cb,\ cc\}$$

For notational convenience, we have written ac for the ordered pair (a, c). The probability of each point in S_2 follows:

$$P(aa) = P(a)P(a) = \frac{1}{2}\left(\frac{1}{2}\right) = \frac{1}{4} \qquad P(ba) = \frac{1}{6} \qquad P(ca) = \frac{1}{12}$$

$$P(ab) = P(a)P(b) = \frac{1}{2}\left(\frac{1}{3}\right) = \frac{1}{6} \qquad P(bb) = \frac{1}{9} \qquad P(cb) = \frac{1}{18}$$

$$P(ac) = P(a)P(c) = \frac{1}{2}\left(\frac{1}{6}\right) = \frac{1}{12} \qquad P(bc) = \frac{1}{18} \qquad P(cc) = \frac{1}{36}$$

Thus the probability that c wins the first race and a wins the second race is $P(ca) = \frac{1}{12}$.

Repeated Trials as a Stochastic Process

A repeated-trials process may also be viewed as a stochastic process whose tree diagram has the following properties:

(i) Each branch point has the same outcomes.

(ii) All branches leading to the same outcome have the same probability.

For example, the tree diagram for the repeated-trials process in Example 4.8 appears in Fig. 4-5. Observe that:

(i) Each branch point has outcomes a, b, c.

(ii) All branches leading to outcome a have probability $\frac{1}{2}$, to outcome b have probability $\frac{1}{3}$, and to outcome c have probability $\frac{1}{6}$.

These two properties are expected, as noted above.

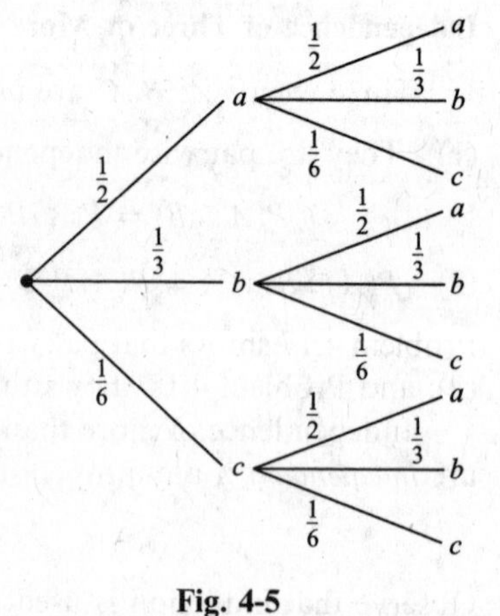

Fig. 4-5

Solved Problems

CONDITIONAL PROBABILITY

4.1. Three fair coins, a penny, a nickel, and a dime, are tossed. Find the probability p that they are all heads if: (*a*) the penny is heads, (*b*) at least one of the coins is heads, (*c*) the dime is tails.

The sample space has eight elements:

$$S = \{\text{HHH, HHT, HTH, HTT, THH, THT, TTH, TTT}\}$$

(*a*) If the penny is heads, the reduced sample space is $A = \{\text{HHH, HHT, HTH, HTT}\}$. Since the coins are all heads in 1 of 4 cases, $p = 1/4$.

(*b*) If one or more of the coins is heads, the reduced sample space is

$$B = \{\text{HHH, HHT, HTH, HTT, THH, THT, TTH}\}$$

Since the coins are all heads in 1 of 7 cases, $p = 1/7$.

(*c*) If the dime is tails, the reduced sample space is $C = \{\text{HTH, HTT, TTH, TTT}\}$. None contains all heads; hence $p = 0$.

4.2. A pair of fair dice is thrown. Find the probability p that the sum is 10 or greater if: (*a*) 5 appears on the first die, (*b*) 5 appears on at least one die.

Figure 3-3 shows the 36 ways the pair of dice can be thrown.

(*a*) If a 5 appears on the first die, then the reduced sample space has six elements:

$$A = \{(5,1), (5,2), (5,3), (5,4), (5,5), (5,6)\}$$

The sum is 10 or greater on two of the six outcomes: $(5,5)$, $(5,6)$. Hence $p = \frac{2}{6} = \frac{1}{3}$.

(*b*) If a 5 appears on at least one of the dice, then the reduced sample space has eleven elements:

$$B = \{(5,1), (5,2), (5,3), (5,4), (5,5), (5,6), (1,5), (2,5), (3,5), (4,5), (6,5)\}$$

The sum is 10 or greater on three of the eleven outcomes: $(5,5)$, $(5,6)$, $(6,5)$. Hence $p = \frac{3}{11}$.

4.3. In a certain college town, 25 percent of the students failed mathematics, 15 percent failed chemistry, and 10 percent failed both mathematics and chemistry. A student is selected at random.

(*a*) If the student failed chemistry, what is the probability that he or she failed mathematics?
(*b*) If the student failed mathematics, what is the probability that he or she failed chemistry?
(*c*) What is the probability that the student failed mathematics or chemistry?
(*d*) What is the probability that the student failed neither mathematics nor chemistry?

(*a*) The probability that a student failed mathematics, given that he or she failed chemistry, is

$$P(M|C) = \frac{P(M \cap C)}{P(C)} = \frac{0.10}{0.15} = \frac{2}{3}$$

(*b*) The probability that a student failed chemistry, given that he or she failed mathematics is

$$P(C|M) = \frac{P(C \cap M)}{P(M)} = \frac{0.10}{0.25} = \frac{2}{5}$$

(*c*) By the addition rule (Theorem 3.6),

$$P(M \cup C) = P(M) + P(C) - P(M \cap C) = 0.25 + 0.15 - 0.10 = 0.30$$

(*d*) Students who failed neither mathematics nor chemistry form the complement of the set $M \cup C$; that is, form the set $(M \cup C)^c$. Hence

$$P((M \cup C)^c) = 1 - P(M \cup C) = 1 - 0.30 = 0.70$$

4.4. Let A and B be events with $P(A) = 0.6$, $P(B) = 0.3$, and $P(A \cap B) = 0.2$. Find: (*a*) $P(A|B)$ and $P(B|A)$, (*b*) $P(A \cup B)$, (*c*) $P(A^c)$ and $P(B^c)$.

(*a*) By definition of conditional probability,

$$P(A|B) = \frac{P(A \cap B)}{P(B)} = \frac{0.2}{0.3} = \frac{2}{3}, \qquad P(B|A) = \frac{P(A \cap B)}{P(A)} = \frac{0.2}{0.6} = \frac{1}{3}$$

(*b*) By the addition rule, Theorem 3.6,

$$P(A \cup B) = P(A) + P(B) - P(A \cap B) = 0.6 + 0.3 - 0.2 = 0.7$$

(*c*) By the complement rule,

$$P(A^c) = 1 - P(A) = 1 - 0.6 = 0.4 \qquad \text{and} \qquad P(B^c) = 1 - P(B) = 1 - 0.3 = 0.7$$

4.5. Consider the data in Problem 4.4. Find: (*a*) $P(A^c|B^c)$, (*b*) $P(B^c|A^c)$.

First compute $P(A^c \cap B^c)$. By De Morgan's law, $(A \cup B)^c = A^c \cap B^c$. Hence, by the complement rule,

$$P(A^c \cap B^c) = P((A \cup B)^c) = 1 - P(A \cup B) = 1 - 0.7 = 0.3$$

(*a*) $P(A^c|B^c) = \dfrac{P(A^c \cap B^c)}{P(B^c)} = \dfrac{0.3}{0.7} = \dfrac{3}{7}$

(*b*) $P(B^c|A^c) = \dfrac{P(A^c \cap B^c)}{P(A^c)} = \dfrac{0.3}{0.4} = \dfrac{3}{4}$

4.6. A class has 12 boys and 4 girls. Suppose three students are selected at random from the class. Find the probability p that they are all boys.

The probability that the first student selected is a boy is 12/16 since there are 12 boys out of 16 students. If the first student is a boy, then the probability that the second is a boy is 11/15, since there are 11 boys left out of 15 students. Finally, if the first two students selected were boys, then the probability that the third student is a boy is 10/14, since there are 10 boys left out of 14 students. Thus, by the multiplication theorem, the probability that all three are boys is

$$p = \frac{12}{16} \cdot \frac{11}{15} \cdot \frac{10}{14} = \frac{11}{28}$$

Another Method

There are $C(16, 3) = 560$ ways to select 3 students out of the 16 students, and $C(12, 3) = 220$ ways to select 3 boys out of 12 boys; hence

$$p = \frac{220}{560} = \frac{11}{28}$$

Another Method

If the students are selected one after the other, then there are $16 \cdot 15 \cdot 14$ ways to select three students, and $12 \cdot 11 \cdot 10$ ways to select three boys; hence

$$p = \frac{12 \cdot 11 \cdot 10}{16 \cdot 15 \cdot 14} = \frac{11}{28}$$

4.7. Find $P(B|A)$ if: (*a*) A is a subset of B, (*b*) A and B are mutually exclusive (disjoint). (Assume $P(A) > 0$.)

(*a*) If A is a subset of B (as pictured in Fig. 4-6(*a*)), then whenever A occurs B must occur; hence $P(B|A) = 1$. Alternatively, if A is a subset of B, then $A \cap B = A$; hence

$$P(B|A) = \frac{P(A \cap B)}{P(A)} = \frac{P(A)}{P(A)} = 1$$

(*b*) If A and B are mutually exclusive, i.e. disjoint (as pictured in Fig. 4-6(*b*)), then whenever A occurs B cannot occur; hence $P(B|A) = 0$. Alternatively, if A and B are disjoint, then $A \cap B = \varnothing$; hence

$$P(B|A) = \frac{P(A \cap B)}{P(A)} = \frac{P(\varnothing)}{P(A)} = \frac{0}{P(A)} = 0$$

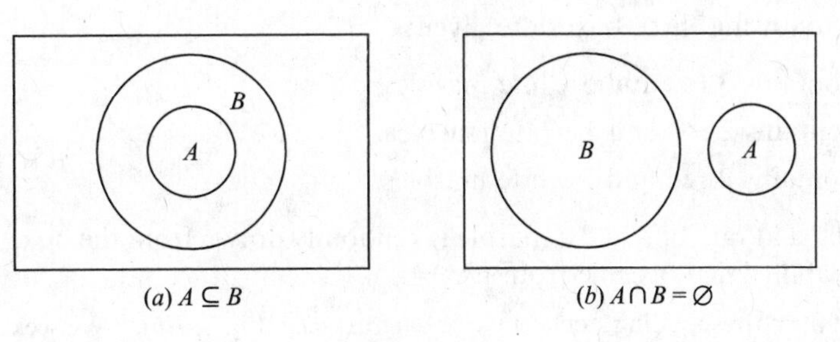

Fig. 4-6

FINITE STOCHASTIC PROCESSES

4.8. Let X, Y, Z be three coins in a box. Suppose X is a fair coin, Y is two-headed, and Z is weighted so that the probability of heads is 1/3. A coin is selected at random and is tossed.
(*a*) Find the probability that heads appears, that is, find $P(H)$.
(*b*) If heads appears, find the probability that it is the fair coin X, that is, find $P(X|h)$.
(*c*) If tails appears, find the probability it is the coin Z, that is, find $P(Z|T)$.

Construct the corresponding two-step stochastic tree diagram in Fig. 4-7(*a*).

(*a*) Heads appears along three of the paths; hence

$$P(H) = \frac{1}{3} \cdot \frac{1}{2} + \frac{1}{3} \cdot 1 + \frac{1}{3} \cdot \frac{1}{3} = \frac{11}{18}$$

(*b*) Note X and heads H appear only along the top path in Fig. 4-7(*a*); hence $P(X \cap H) = (1/3)(1/2) = 1/6$. Thus

$$P(X|H) = \frac{P(X \cap H)}{P(H)} = \frac{1/6}{11/18} = \frac{3}{11}$$

(*c*) $P(T) = 1 - P(H) = 1 - 11/18 = 7/18$. Alternatively, tails appears along two of the paths and so

$$P(T) = \frac{1}{3} \cdot \frac{1}{2} + \frac{1}{3} \cdot \frac{2}{3} = \frac{7}{18}$$

Note that Z and tails T appear only along the bottom path in Fig. 4-7(*a*); accordingly, we have $P(Z \cap T) = (1/3)(2/3) = 2/9$. Thus

$$P(Z|T) = \frac{P(Z \cap T)}{P(T)} = \frac{2/9}{7/18} = \frac{4}{7}$$

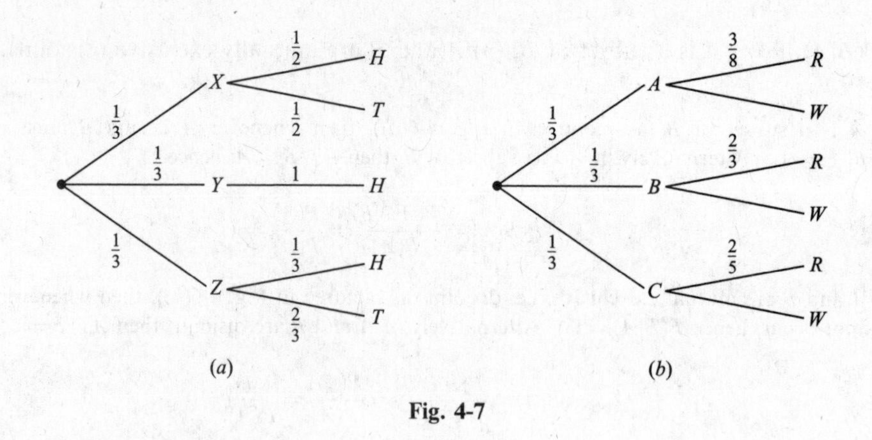

Fig. 4-7

4.9. Suppose the following three boxes are given:

Box A contains 3 red and 5 white marbles.

Box B contains 2 red and 1 white marbles.

Box C contains 2 red and 3 white marbles.

A box is selected at random, and a marble is randomly drawn from the box. If the marble is red, find the probability that it came from box A.

Construct the corresponding stochastic tree diagram as in Fig. 4-7(b). We seek $P(A|R)$, the probability that A was selected, given that the marble is red. Thus it is necessary to find $P(A \cap R)$ and $P(R)$. Note that A and R only occur on the top path; hence $P(A \cap R) = (1/3)(3/8) = 1/8$. There are three paths leading to a red marble R; hence

$$P(R) = \frac{1}{3} \cdot \frac{3}{8} + \frac{1}{3} \cdot \frac{2}{3} + \frac{1}{3} \cdot \frac{2}{5} = \frac{173}{360} \approx 0.48$$

Thus

$$P(A|R) = \frac{P(A \cap R)}{P(R)} = \frac{1/8}{173/360} = \frac{45}{173} \approx 0.26$$

4.10. Suppose the following two boxes are given:

Box A contains 3 red and 2 white marbles.

Box B contains 2 red and 5 white marbles.

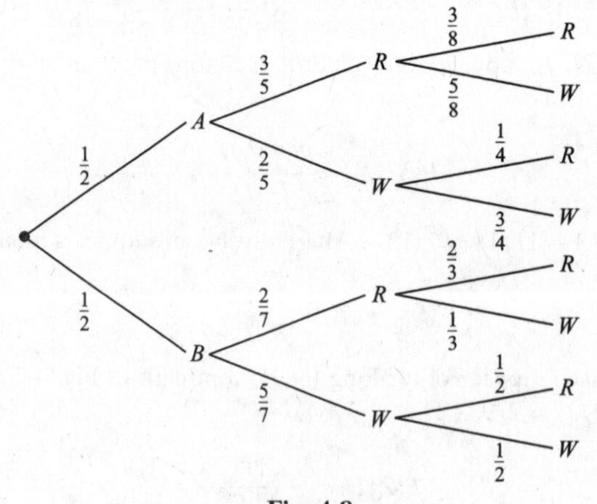

Fig. 4-8

A box is selected at random; a marble is drawn and put into the other box; then a marble is drawn from the second box. Find the probability p that both marbles drawn are of the same color.

Construct the corresponding stochastic tree diagram as in Fig. 4-8. Note that this is a three-step stochastic process: (1) choosing a box, (2) choosing a marble, (3) choosing a second marble. Note that if box A is selected and a red marble R is drawn and put into box B, then box B will have 3 red marbles and 5 white marbles.

There are four paths which lead to two marbles of the same color; hence

$$p = \frac{1}{2}\cdot\frac{3}{5}\cdot\frac{3}{8} + \frac{1}{2}\cdot\frac{2}{5}\cdot\frac{3}{4} + \frac{1}{2}\cdot\frac{2}{7}\cdot\frac{2}{3} + \frac{1}{2}\cdot\frac{5}{7}\cdot\frac{1}{2} = \frac{901}{1680} \approx 0.536$$

LAW OF TOTAL PROBABILITY, BAYES' RULE

4.11. In a certain city, 40 percent of the people consider themselves Conservatives (C), 35 percent consider themselves to be Liberals (L), and 25 percent consider themselves to be Independents (I). During a particular election, 45 percent of the Conservatives voted, 40 percent of the Liberals voted, and 60 percent of the Independents voted. If a randomly selected person voted, find the probability that the voter is (*a*) Conservative, (*b*) Liberal, (*c*) Independent.

Let V denote the event that a person voted. We need $P(V)$. By the law of total probability,

$$\begin{aligned} P(V) &= P(C)P(V|C) + P(L)P(V|L) + P(I)P(V|I) \\ &= (0.40)(0.45) + (0.35)(0.40) + (0.25)(0.60) = 0.47 \end{aligned}$$

By Bayes' rule:

(*a*) $P(C|V) = \dfrac{P(C)P(V|C)}{P(V)} = \dfrac{(0.40)(0.45)}{0.47} = \dfrac{18}{47} \approx 38.3\%$

(*b*) $P(L|V) = \dfrac{P(L)P(V|L)}{P(V)} = \dfrac{(0.35)(0.40)}{0.47} = \dfrac{14}{47} \approx 29.8\%$

(*c*) $P(I|V) = \dfrac{P(I)P(V|I)}{P(V)} = \dfrac{(0.25)(0.60)}{0.47} = \dfrac{15}{47} \approx 31.9\%$

4.12. Suppose a student dormitory in a college consists of the following:

(1) 30 percent are freshmen of whom 10 percent own a car
(2) 40 percent are sophomores of whom 20 percent own a car
(3) 20 percent are juniors of whom 40 percent own a car
(4) 10 percent are seniors of whom 60 percent own a car

A student is randomly selected from the dormitory.

(*a*) Find the probability that the student owns a car.
(*b*) If the student owns a car, find the probability that the student is a junior.

Let A, B, C, D denote, respectively, the sets of freshmen, sophomores, juniors, and seniors, and let E denote the set of students who own a car. Figure 4-9 is a stochastic tree describing the given data.

(*a*) We seek $P(E)$. By the law of total probability

$$\begin{aligned} P(E) &= (0.30)(0.10) + (0.40)(0.20) + (0.20)(0.40) + (0.10)(0.60) \\ &= 0.03 + 0.08 + 0.08 + 0.06 = 0.25 = 25\% \end{aligned}$$

(Alternatively, using Fig. 4-9, add the four paths to E to obtain $P(E) = 0.25 = 25\%$.)

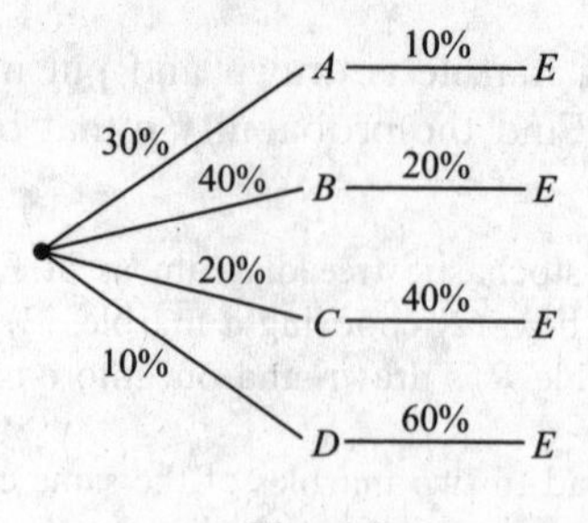

Fig. 4-9

(*b*) We seek $P(C|E)$. By Bayes' formula:

$$P(C|E) = \frac{P(C)P(E|C)}{P(E)} = \frac{(0.20)(0.40)}{0.25} = \frac{8}{25} = 32\%$$

Alternatively, using Fig. 4-9, divide the successful path containing C and E with probability

$$(20\%)\,(40\%) = 8\%$$

by the sum of the paths to E with probability 25 percent to obtain $P(C|E) = 32$ percent.

4.13. In a certain college, 4 percent of the men and 1 percent of the women are taller than 6 feet. Furthermore, 60 percent of the students are women. Suppose a randomly selected student is taller than 6 feet. Find the probability that the student is a woman.

Let $A = \{\text{students taller than 6 feet}\}$. We seek $P(W|A)$, the probability that a student is a woman given that the student is taller than 6 feet. By Bayes' formula,

$$P(W|A) = \frac{P(A|W)P(W)}{P(A|W)P(W) + P(A|M)P(M)} = \frac{1\% \cdot 60\%}{1\% \cdot 60\% + 4\% \cdot 40\%} = \frac{3}{11} \approx 0.27$$

INDEPENDENT EVENTS

4.14. The probability that A hits a target is $\frac{1}{3}$ and the probability that B hits a target is $\frac{1}{5}$. They both fire at the target. Find the probability that:

(*a*) A does not hit the target (*c*) One of them hits the target

(*b*) Both hit the target (*d*) Neither hits the target

We are given $P(A) = \frac{1}{3}$ and $P(B) = \frac{1}{5}$ (and we assume the events are independent).

(*a*) $P(\text{not } A) = P(A^c) = 1 - P(A) = 1 - \frac{1}{3} = \frac{2}{3}$.

(*b*) Since the events are independent,

$$P(A \text{ and } B) = P(A \cap B) = P(A) \cdot P(B) = \frac{1}{3} \cdot \frac{1}{5} = \frac{1}{15}$$

(*c*) By the addition rule, Theorem 3.6,

$$P(A \text{ or } B) = P(A \cup B) = P(A) + P(B) - P(A \cap B) = \frac{1}{3} + \frac{1}{5} - \frac{1}{15} = \frac{7}{15}$$

(*d*) By De Morgan's law, "Neither A nor B" is the complement of $A \cup B$. Hence

$$P(\text{neither } A \text{ nor } B) = P((A \cup B)^c) = 1 - P(A \cup B) = 1 - \frac{7}{15} = \frac{8}{15}$$

4.15. Consider the following events for a family with children:

$$A = (\text{children of both sexes}), \qquad B = \{\text{at most one boy}\}$$

(a) Show that A and B are independent events if a family has three children.

(b) Show that A and B are dependent events if a family has only two children.

(a) We have the equiprobable space $S = \{bbb, bbg, bgb, bgg, gbb, gbg, ggb, ggg\}$. Here

$$A = \{bbg, bgb, bgg, gbb, gbg, ggb\} \quad \text{so} \quad P(A) = \tfrac{6}{8} = \tfrac{3}{4}$$

$$B = \{bgg, gbg, ggb, ggg\} \quad \text{so} \quad P(B) = \tfrac{4}{8} = \tfrac{1}{2}$$

$$A \cap B = \{bgg, gbg, ggb\} \quad \text{so} \quad P(A \cap B) = \tfrac{3}{8}$$

Since $P(A)P(B) = \frac{3}{4} \cdot \frac{1}{2} = \frac{3}{8} = P(A \cap B)$, A and B are independent.

(b) We have the equiprobable space $S = \{bb, bg, gb, gg\}$. Here

$$A = \{bg, gb\} \quad \text{so} \quad P(A) = \tfrac{1}{2}$$

$$B = \{bg, gb, gg\} \quad \text{so} \quad P(B) = \tfrac{3}{4}$$

$$A \cap B = \{bg, gb\} \quad \text{so} \quad P(A \cap B) = \tfrac{1}{2}$$

Since $P(A)P(B) \neq P(A \cap B)$, A and B are dependent.

4.16. Box A contains 5 red marbles and 3 blue marbles, and Box B contains 3 red and 2 blue. A marble is drawn at random from each box.

(a) Find the probability p that both marbles are red.

(b) Find the probability p that one is red and one is blue.

(a) The probability of choosing a red marble from A is $\frac{5}{8}$ and from B is $\frac{3}{5}$. Since the events are independent,

$$p = \frac{5}{8} \cdot \frac{3}{5} = \frac{3}{8}$$

(b) The probability of choosing a red marble from A and a blue marble from B is $p_1 = \frac{5}{8} \cdot \frac{2}{5} = \frac{1}{4}$. The probability of choosing a blue marble from A and a red marble from B is $p_2 = \frac{3}{8} \cdot \frac{3}{5} = \frac{9}{40}$. Hence

$$p = p_1 + p_2 = \frac{1}{4} + \frac{9}{40} = \frac{19}{40}$$

4.17. Consider an equiprobable space $S = \{a, b, c, d\}$; hence each elementary event has the same probability $p = \frac{1}{4}$. Consider the events $A = \{a, d\}$, $B = \{b, d\}$, $C = \{c, d\}$.
(a) Show that A, B, C are pairwise independent. (b) Show that A, B, C are not independent.

(a) Here $P(A) = P(B) = P(C) = \frac{1}{2}$. Since $A \cap B = \{d\}$,

$$P(A \cap B) = P(\{d\}) = \frac{1}{4} = P(A)P(B)$$

Hence A and B are independent. Similarly, A and C are independent; and B and C are independent.

(b) Here $A \cap B \cap C = \{d\}$, so $P(A \cap B \cap C) = \frac{1}{4}$. Therefore,

$$P(A)P(B)P(C) = \frac{1}{8} \neq P(A \cap B \cap C)$$

Accordingly, A, B, C are not independent.

4.18. Consider an equiprobable space $S = \{1, 2, 3, 4, 5, 6, 7, 8\}$; hence each elementary event has probability 1/8. Consider the events:

$$A = \{1, 2, 3, 4\}, \qquad B = \{2, 3, 4, 5\}, \qquad C = \{4, 6, 7, 8\}$$

(*a*) Show that $P(A \cap B \cap C) = P(A)P(B)P(C)$.

(*b*) Show that:

(i) $P(A \cap B) \neq P(A)P(B)$

(ii) $P(A \cap C) \neq P(A)P(C)$

(iii) $P(B \cap C) \neq P(B)P(C)$

(*a*) Here $P(A) = P(B) = P(C) = \frac{4}{8} = \frac{1}{2}$. Since $A \cap B \cap C = \{4\}$,

$$P(A \cap B \cap C) = \tfrac{1}{8} = P(A)P(B)P(C)$$

(*b*) (i) $A \cap B = \{3, 4, 5\}$, so $P(A \cap B) = \frac{3}{8}$. But $P(A)P(B) = \frac{1}{4}$; hence $P(A \cap B) \neq P(A)P(B)$.

(ii) $A \cap C = \{4\}$, so $P(A \cap C) = \frac{1}{8}$. But $P(A)P(C) = \frac{1}{4}$; hence $P(A \cap C) \neq P(A)P(C)$.

(iii) $B \cap C = \{4\}$, so $P(B \cap C) = \frac{1}{8}$. But $P(B)P(C) = \frac{1}{4}$; hence $P(B \cap C) \neq P(B)P(C)$.

4.19. Prove: If A and B are independent events, then A^c and B^c are independent events.

Let $P(A) = x$ and $P(B) = y$. Then $P(A^c) = 1 - x$ and $P(B^c) = 1 - y$. Since A and B are independent, $P(A \cap B) = P(A)P(B) = xy$. Furthermore,

$$P(A \cup B) = P(A) + P(B) - P(A \cap B) = x + y - xy$$

By De Morgan's law, $(A \cup B)^c = A^c \cap B^c$; hence

$$P(A^c \cap B^c) = P((A \cup B)^c) = 1 - P(A \cup B) = 1 - x - y + xy$$

On the other hand,

$$P(A^c)P(B^c) = (1 - x)(1 - y) = 1 - x - y + xy$$

Thus $P(A^c \cap B^c) = P(A^c)P(B^c)$, and so A^c and B^c are independent.

In similar fashion, we can show that A and B^c, as well as A^c and B, are independent.

REPEATED TRIALS

4.20. A fair coin is tossed three times. Find the probability that there will appear: (*a*) three heads, (*b*) exactly two heads, (*c*) exactly one head, (*d*) no heads.

Let H denote a head and T a tail on any toss. The three tosses can be modeled as an equiprobable space in which there are eight possible outcomes:

$$S = \{\text{HHH, HHT, HTH, HTT, THH, THT, TTH, TTT}\}$$

However, since the result on any one toss does not depend on the result of any other toss, the three tosses may be modeled as three independent trials in which $P(\text{H}) = \frac{1}{2}$ and $P(\text{T}) = \frac{1}{2}$ on any one trial. Then:

(*a*) $P(\text{three heads}) = P(\text{HHH}) = \frac{1}{2} \cdot \frac{1}{2} \cdot \frac{1}{2} = \frac{1}{8}$.

(*b*) $P(\text{exactly two heads}) = P(\text{HHT or HTH or THH})$

$$= \tfrac{1}{2} \cdot \tfrac{1}{2} \cdot \tfrac{1}{2} + \tfrac{1}{2} \cdot \tfrac{1}{2} \cdot \tfrac{1}{2} + \tfrac{1}{2} \cdot \tfrac{1}{2} \cdot \tfrac{1}{2} = \tfrac{3}{8}$$

(*c*) As in (*b*), $P(\text{exactly one head}) = P(\text{exactly two tails}) = \frac{3}{8}$.

(*d*) As in (*a*), $P(\text{no heads}) = P(\text{TTT}) = \frac{1}{8}$.

4.21. Whenever horses a, b, c, d race together, their respective probabilities of winning are 0.2, 0.5, 0.1, 0.2. That is, $S = \{a, b, c, d\}$, where $P(a) = 0.2$, $P(b) = 0.5$, $P(c) = 0.1$, and $P(d) = 0.2$. They race three times.

(*a*) Describe and find the number of elements in the product probability space S_3.

(*b*) Find the probability that the same horse wins all three races.

(*c*) Find the probability that a, b, c each win one race.

(*a*) By definition, $S_3 = S \times S \times S = \{(x, y, z) : x, y, z \in S\}$ and

$$P((x, y, z)) = P(x)P(y)P(z)$$

Thus, in particular, S_3 contains $4^3 = 64$ elements.

(*b*) Writing xyz for (x, y, z), we seek the probability of the event

$$A = \{aaa, bbb, ccc, ddd\}$$

By definition

$$P(aaa) = (0.2)^3 = 0.008, \qquad P(ccc) = (0.1)^3 = 0.001$$

$$P(bbb) = (0.5)^3 = 0.125, \qquad P(ddd) = (0.2)^3 = 0.008$$

Thus $P(A) = 0.0008 + 0.125 + 0.001 + 0.008 = 0.142$.

(*c*) We seek the probability of the event

$$B = \{abc, acb, bac, bca, cab, cba\}$$

Every element in B has the same probability

$$p = (0.2)(0.5)(0.1) = 0.01. \qquad \text{Hence} \qquad P(B) = 6(0.01) = 0.06.$$

4.22. A certain soccer team wins (W) with probability 0.6, loses (L) with probability 0.3, and ties (T) with probability 0.1. The team plays three games over the weekend. (*a*) Determine the elements of the event A that the team wins at least twice and does not lose; and find $P(A)$. (*b*) Determine the elements of the event B that the team wins, loses, and ties in some order; and find $P(B)$.

(*a*) A consists of all ordered triples with at least two Ws and no Ls. Thus

$$A = \{\text{WWW, WWT, WTW, TWW}\}$$

Furthermore,

$$\begin{aligned} P(A) &= P(\text{WWW}) + P(\text{WWT}) + P(\text{WTW}) + P(\text{TWW}) \\ &= (0.6)(0.6)(0.6) + (0.6)(0.6)(0.1) + (0.6)(0.1)(0.6) + (0.1)(0.6)(0.6) \\ &= 0.216 + 0.036 + 0.036 + 0.036 = 0.324 \end{aligned}$$

(*b*) Here $B = \{\text{WLT, WTL, LWT, LTW, TWL, TLW}\}$. Every element in B has probability

$$p = (0.6)(0.3)(0.1) = 0.018; \qquad \text{hence} \qquad P(B) = 6(0.018) = 0.108$$

4.23. A certain type of missile hits its target with probability $p = 0.3$. Find the number of missiles that should be fired so that there is at least an 80 percent probability of hitting the target.

The probability of missing the target is $q = 1 - p = 0.7$. Hence the probability that n missiles miss the target is $(0.7)^n$. Thus we seek the smallest n for which

$$1 - (0.7)^n > 0.8 \qquad \text{or equivalently} \qquad (0.7)^n < 0.2$$

Compute:

$$(0.7)^1 = 0.7, \qquad (0.7)^2 = 0.49, \qquad (0.7)^3 = 0.343, \qquad (0.7)^4 = 0.2401, \qquad (0.7)^5 = 0.16807$$

Thus at least five missiles should be fired.

4.24. The probability that a man hits a target is 1/3. He fires at the target $n = 6$ times. (*a*) Describe and find the number of elements in the sample space S. (*b*) Let E be the event that he hits the target exactly $k = 2$ times. List the elements of E and find the number $n(E)$ of elements in E. (*c*) Find $P(E)$.

(*a*) S consists of all 6-element sequences consisting of S's (successes) and F's (failures); hence S contains $2^6 = 64$ elements.

(*b*) E consists of all sequences with two S's and four F's; hence E consists of the following elements:

SSFFFF, SFSFFF, SFFSFF, SFFFSF, SFFFFS, FSSFFF, FSFSFF, FSFFSF,

FSFFFS, FFSSFF, FFSFSF, FFSFFS, FFFSSF, FFFSFS, FFFFSS

Observe that the list contains 15 elements. (This is expected since we are distributing $k = 2$ letters S among the $n = 6$ positions in the sequence, and $C(6, 2) = 15$.) Thus $n(E) = 15$.

(*c*) Here $P(\text{S}) = 1/3$, so $P(\text{F}) = 1 - P(\text{S}) = 2/3$. Thus each of the above sequences occurs with the same probability

$$p = (1/3)^2(2/3)^4 = 16/729$$

Hence $P(E) = 15(16/729) = 80/243 \approx 33\%$.

Supplementary Problems

CONDITIONAL PROBABILITY

4.25. A fair die is tossed. Consider events $A = \{2, 4, 6\}$, $B = \{1, 2\}$, $C = \{1, 2, 3, 4\}$. Find:

(*a*) $P(A \text{ and } B)$ and $P(A \text{ or } C)$ (*c*) $P(A|C)$ and $P(C|A)$

(*b*) $P(A|B)$ and $P(B|A)$ (*d*) $P(B|C)$ and $P(C|B)$

4.26. A pair of fair dice is tossed. If the faces appearing are different, find the probability that:

(*a*) The sum is even, (*b*) The sum exceeds 9.

4.27. Let A and B be events with $P(A) = 0.6$, $P(B) = 0.3$, $P(A \cap B) = 0.2$. Find:

(*a*) $P(A \cup B)$, (*b*) $P(A|B)$, (*c*) $P(B|A)$.

4.28. Let A and B be events with $P(A) = \frac{1}{3}$, $P(B) = \frac{1}{4}$, and $P(A \cup B) = \frac{1}{2}$.

(*a*) Find $P(A|B)$ and $P(B|A)$. (*b*) Are A and B independent?

4.29. Two marbles are selected one after the other without replacement from a box containing 3 white marbles and 2 red marbles. Find:

(*a*) P(2 white), (*b*) P(2 white | first is white), (*c*) P(2 red), (*d*) P(2 red | second is red).

4.30. Two marbles are selected one after the other with replacement from a box containing 3 white marbles and 2 red marbles. Find:

(*a*) P(2 white), (*c*) P(2 white | first is white), (*b*) P(2 red), (*d*) P(2 red | second is red).

4.31. Two different digits are selected at random from the digits 1 through 5.

(*a*) If the sum is odd, what is the probability that 2 is one of the numbers selected?

(*b*) If 2 is one of the digits, what is the probability that the sum is odd?

4.32. Three cards are drawn in succession (without replacement) from a 52-card deck. Find:

(a) $P(3 \text{ aces} \mid \text{first card is an ace})$, ($b$) $P(3 \text{ aces} \mid \text{first two cards are aces})$

4.33. A die is weighted to yield the following probability distribution:

Number	1	2	3	4	5	6
Probability	0.2	0.1	0.1	0.3	0.1	0.2

Let $A = \{1, 2, 3\}$, $B = \{2, 3, 5\}$, $C = \{2, 4, 6\}$. Find:

(a) $P(A)$, $P(B)$, $P(C)$ (b) $P(A^c)$, $P(B^c)$, $P(C^c)$

(c) $P(A|B)$, $P(B|A)$ (d) $P(A|C)$, $P(C|A)$

(e) $P(B|C)$, $P(C|B)$

4.34. In a country club, 65 percent of the members play tennis, 40 percent play golf, and 20 percent play both tennis and golf. A member is chosen at random.

(a) Find the probability that he plays neither tennis nor golf.

(b) If he plays tennis, find the probability that he plays golf.

(c) If he plays golf, find the probability that he plays tennis.

4.35. In a certain college town, 25 percent of the boys and 10 percent of the girls are studying mathematics. The girls constitute 60 percent of the student body. If a student is chosen at random and is studying mathematics, determine the probability that the student is a girl.

FINITE STOCHASTIC PROCESSES

4.36. Two boxes are given as follows:

Box A contains 5 red marbles, 3 white marbles, and 8 blue marbles.

Box B contains 3 red marbles, and 5 white marbles.

A box is selected at random and a marble is randomly chosen. Find the probability that the marble is: (a) red, (b) white, (c) blue.

4.37. Refer to Problem 4.36. Find the probability that box A was selected if the marble is: (a) red, (b) white, (c) blue.

4.38. Two boxes are given as follows:

Box A contains 5 red marbles, 3 white marbles, and 8 blue marbles.

Box B contains 3 red marbles and 5 white marbles.

A fair die is tossed; if a 3 or 6 appears, a marble is randomly chosen from A, otherwise a marble is chosen from B. Find the probability that the marble is: (a) red, (b) white, (c) blue.

4.39. Refer to Problem 4.38. Find the probability that box A was selected if the marble is: (a) red, (b) white, (c) blue.

4.40. A box contains three coins, two of them fair and one two-headed. A coin is randomly selected and tossed twice. If heads appear both times, what is the probability that the coin is two-headed?

4.41. Two boxes are given as follows:

Box A contains 5 red marbles, and 3 white marbles.

Box B contains 1 red marble, and 2 white marbles.

A fair die is tossed; if a 3 or 6 appears, a marble is randomly chosen from B and put into A and a marble is drawn from A, otherwise a marble is chosen from A and put into B and a marble is drawn from B. Find the probability that both marbles are: (*a*) red, (*b*) white.

TOTAL PROBABILITY AND BAYES' FORMULA

4.42. A city is partitioned into districts A, B, C having 20 percent, 40 percent, and 40 percent of the registered voters, respectively. The registered voters listed as Democrats are 50 percent in A, 25 percent in B, and 75 percent in C.

(*a*) If a registered voter is chosen randomly in the city, find the probability that the voter is listed as a Democrat.

(*b*) A registered voter from the city is chosen at random and found to be listed as a Democrat. Find the probability that the voter came from district B.

4.43. Refer to Problem 4.42. Suppose a district is chosen at random, and then a registered voter is randomly chosen from the district.

(*a*) Find the probability that the voter is listed as a Democrat.

(*b*) If the voter is listed as a Democrat, what is the probability that the voter came from district A?

4.44. Women in City College constitute 60 percent of the freshmen, 40 percent of the sophomores, 40 percent of the juniors, and 45 percent of the seniors. The school population is 30 percent freshmen, 25 percent sophomores, 25 percent juniors, and 20 percent seniors.

(*a*) If a student from City College is chosen at random, find the probability that the student is a woman.

(*b*) If a student is a woman, what is the probability that she is a sophomore?

4.45. Refer to Problem 4.44. Suppose one of the classes is chosen, and then a student is randomly chosen from the class.

(*a*) Find the probability that the student is a woman.

(*b*) If the student is a woman, what is the probability that she is a sophomore?

4.46. A company produces light bulbs at three factories A, B, C.

Factory A produces 40 percent of the total number of bulbs, of which 2 percent are defective.
Factory B produces 35 percent of the toal number of bulbs, of which 4 percent are defective.
Factory C produces 25 percent of the total number of bulbs, of which 3 percent are defective.

If a defective bulb is found among the total output, find the probability that it came from:
(*a*) factory A, (*b*) factory B, (*c*) factory C.

4.47. Refer to Problem 4.46. Suppose a factory is chosen at random, and one of its bulbs is randomly selected. If the bulb is defective, find the probability that it came from:
(*a*) factory A, (*b*) factory B, (*c*) factory C.

4.48. A test for Alzheimer's disease is 95 percent effective in detecting the disease when it is present, but also gives a positive result 10 percent of the time when it is not present (false positive). Suppose 4 percent of the population over 65 years have Alzheimer's disease.

(*a*) What is the probability that a person over 65 years chosen at random will test positively for the disease?

(*b*) Suppose a person over 65 tests positively. What is the probability that the person has the disease?

(*c*) Suppose a person over 65 tests negatively. What is the probability that the person has the disease?

INDEPENDENT EVENTS

4.49. Let A and B be independent events with $P(A) = 0.3$ and $P(B) = 0.4$. Find: (*a*) $P(A \cap B)$ and $P(A \cup B)$, (*b*) $P(A|B)$ and $P(B|A)$.

4.50. Box A contains 5 red marbles and 3 blue marbles, and Box B contains 2 red and 3 blue. A marble is drawn at random from each box.

(*a*) Find the probability p that both marbles are red.

(*b*) Find the probability p that one is red and one is blue.

4.51. Let A and B be events with $P(A) = 0.3$, $P(A \cup B) = 0.5$, and $P(B) = p$. Find p if:
(*a*) A and B are disjoint, (*b*) A and B are independent, (*c*) A is a subset of B.

4.52. The probability that A hits a target is $\frac{1}{4}$ and the probability that B hits a target is $\frac{1}{3}$. They each fire once at the target. (*a*) Find the probability that they both hit the target. (*b*) Find the probability that the target is hit exactly once. (*c*) If the target is hit only once, what is the probability that A hit the target?

4.53. The probability that A hits a target is $\frac{1}{4}$ and the probability that B hits a target is $\frac{1}{3}$. They each fire twice. Find the probability that the target will be hit at least once?

4.54. The probabilities that three men hit a target are respectively 0.3, 0.5, and 0.4. Each fires once at the target. (As usual, assume that the three events that each hits the target are independent.)

(*a*) Find the probability that they all hit the target.

(*b*) Find the probability that exactly one of them hits the target.

(*c*) If only one hits the target, what is the probability that it was the first man?

4.55. Three fair coins are tossed. Consider the events:

$$A = \{\text{all heads or all tails}\}, \qquad B = \{\text{at least two heads}\}, \qquad C = \{\text{at most two heads}\}$$

Of the pairs (A, B), (A, C), and (B, C), which are independent?

4.56. Suppose A and B are independent events. Show that A and B^c are independent, and that A^c and B are independent.

4.57. Suppose A, B, C are independent events. Show that:
(*a*) A^c, B, C are independent; (*b*) A^c, B^c, C are independent; (*c*) A^c, B^c, C^c are independent.

4.58. Suppose A, B, C are independent events. Show that A and $B \cup C$ are independent.

REPEATED TRIALS

4.59. Whenever horses a, b, and c race together, their respective probabilities of winning are 0.3, 0.5, and 0.2. They race three times.

(*a*) Find the probability that the same horse wins all three races.

(*b*) Find the probability that a, b, c each win one race.

4.60. A team wins (W) with probability 0.5, loses (L) with probability 0.3, and ties (T) with probability 0.2. The team plays twice. (*a*) Determine the sample space S and the probability of each elementary event. (*b*) Find the probability that the team wins at least once.

4.61. A certain type of missile hits its target with probability $p = \frac{1}{3}$. (*a*) If 3 missiles are fired, find the probability that the target is hit at least once. (*b*) Find the number of missiles that should be fired so that there is at least a 90 percent probability of hitting the target.

4.62. In any game, the probability that the Hornets (H) will defect the Rockets (R) is 0.6. Find the probability that the Hornets will win a best-out-of-three series. (Assume no ties.)

4.63. The batting average of a baseball player is .300. He comes to bat 4 times. Find the probability that he will get: (*a*) exactly two hits, (*b*) at least one hit.

Answers to Supplementary Problems

4.25. (*a*) $\frac{1}{6}, \frac{4}{6}$; (*b*) $\frac{1}{2}, \frac{1}{3}$; (*c*) $\frac{1}{2}, \frac{2}{3}$; (*d*) $\frac{1}{2}$, 1

4.26. (*a*) $\frac{12}{30}$, (*b*) $\frac{4}{30}$

4.27. (*a*) 0.7, (*b*) $\frac{2}{3}$, (*c*) $\frac{1}{3}$

4.28. (*a*) $\frac{1}{3}, \frac{1}{4}$; (*b*) No

4.29. (*a*) $\frac{3}{10}$, (*b*) $\frac{1}{2}$, (*c*) $\frac{1}{10}$, (*d*) $\frac{1}{4}$

4.30. (*a*) $\frac{9}{25}$, (*b*) $\frac{3}{5}$, (*c*) $\frac{4}{25}$, (*d*) $\frac{2}{5}$

4.31. (*a*) $\frac{1}{3}$, (*b*) $\frac{3}{4}$

4.32. (*a*) $1/1275 = 0.08$ percent, (*b*) $1/50 = 2$ percent

4.33. (*a*) 0.4, 0.3, 0.6, (*b*) 0.6, 0.7, 0.4, (*c*) $\frac{2}{3}, \frac{1}{2}$, (*d*) $\frac{1}{6}, \frac{1}{4}$, (*e*) $\frac{1}{6}, \frac{1}{3}$

4.34. (*a*) 15 percent, (*b*) $20/65 \approx 30.1$ percent, (*c*) $\frac{1}{2} = 50$ percent

4.35. $6/16 = 37.5$ percent

4.36. (*a*) 11/32, (*b*) 13/32, (*c*) 8/32

4.37. (*a*) 5/11, (*b*) 3/13, (*c*) 1

4.38. (*a*) $17/48 \approx 35.4$ percent, (*b*) $23/48 \approx 47.9$ percent, (*c*) $8/48 \approx 16.7$ percent

4.39. (*a*) $5/17 \approx 29.4$ percent, (*b*) $3/23 \approx 13.0$ percent, (*c*) 1

4.40. 2/3

4.41. (*a*) $61/216 \approx 28.2$ percent, (*b*) $499/1296 \approx 38.5$ percent

4.42. (*a*) 50 percent, (*b*) 20 percent

4.43. (*a*) 50 percent, (*b*) $\frac{1}{3}$

4.44. (*a*) 47 percent, (*b*) $10/47 \approx 21.3$ percent

4.45. (*a*) 46.25 percent, (*b*) 21.6 percent

4.46. (*a*) $80/295 \approx 27.1$ percent, (*b*) $140/295 \approx 47.5$ percent, (*c*) $75/295 \approx 25.574$ percent

4.47. (*a*) $\frac{2}{9}$, (*b*) $\frac{4}{9}$, (*c*) $\frac{3}{9}$

4.48. (*a*) 13.4 percent, (*b*) 28.36 percent, (*c*) 0.23 percent

4.49. (*a*) 0.12, 0.58, (*b*) 0.3, 0.4

4.50. (*a*) $\frac{1}{4}$, (*b*) $\frac{21}{40}$

4.51. (*a*) 0.2, (*b*) $\frac{2}{7}$, (*c*) 0.5

4.52. (*a*) $\frac{1}{12}$, (*b*) $\frac{5}{12}$, (*c*) $\frac{2}{5}$

4.53. $1 - \frac{1}{4} = \frac{3}{4}$

4.54. (*a*) 6 percent, (*b*) 44 percent, (*c*) $9/44 \approx 20.45$ percent

4.55. Only A and B are independent.

4.59. (*a*) $P(aaa \text{ or } bbb \text{ or } ccc) = 0.26$, (*b*) $6(0.03) = 0.18$

4.60. (*a*) $S = \{$WW, WL, WT, LW, LL, LT, TW, TL, TT$\}$; 0.25, 0.15, 0.10, 0.15, 0.09, 0.06, 0.10, 0.06, 0.04

(*b*) $1 - 0.25 = 0.75$

4.61. (*a*) $1 - (2/3)^3 = 19/27$, (*b*) $(2/3)^n < 10$ percent so $n > 6$

4.62. $P(\text{HH or HRH or RHH}) = 64.8$ percent

4.63. (*a*) $6(0.44) = 26.5$ percent, (*b*) $1 - P(\text{MMMM}) \approx 76$ percent

Chapter 5

Random Variables

5.1 INTRODUCTION

The topic of random variables is fundamental to probability and statistics. This chapter formally defines a random variable and presents its basic properties. We end the chapter with the Law of Large Numbers, on which much of probability and statistics is based.

A random variable is a special kind of function, so we recall some notation and definitions about functions. Let S and T be sets. Suppose to each $s \in S$ there is assigned a unique element of T; the collection f of such assignments is called a *function* from S into T, and is written $f : S \to T$. We write $f(s)$ for the element of T that f assigns to $s \in S$, and call $f(s)$ the *image* of s under f or the *value* of f at s. The *image* $f(A)$ of any subset A of S, and the *preimage* $f^{-1}(B)$ of any subset B of T are defined by:

$$f(A) = \{f(s) : s \in A\} \qquad \text{and} \qquad f^{-1}(B) = \{s : f(s) \in B\}$$

In words, $f(A)$ consists of the images of points in A, and $f^{-1}(B)$ consists of those points whose images belong to B. In particular, the set $f(S)$ of all the image points is called the *range* (or *image*) of the function f.

5.2 RANDOM VARIABLES

Let S be the sample space of an experiment. Frequently, we wish to assign a specific number to each outcome of the experiment, e.g. the sum of the numbers on a toss of a pair of dice, the number of aces in a bridge hand, or the time (in hours) it takes for a light bulb to burn out. Such an assignment of numerical values is called a random variable. Namely:

Definition: A *random variable* X on a sample space S is a rule that assigns a numerical value to each outcome of S or, in other words, a function from S into the set R of real numbers.

Remark: If S is uncountable, then certain real-valued functions on S are not random variables. Specifically, X is a random variable if the preimage of every interval of R is an event of S. On the other hand, if S is a sample space in which every subset is an event, then every real-valued function on S is a random variable.

The notation R_X will be used to denote the set of numbers assigned by a random variable X, and we refer to R_X as the *range space*. This chapter will mainly investigate *discrete* random variables, where the range space R_X is finite or countable. *Continuous* random variables, where the range space is a continuum of numbers, such as an interval or a union of intervals, and which sometimes requires calculus, will be treated near the end of the chapter.

EXAMPLE 5.1

(*a*) A fair coin is tossed three times and the sequence of heads (H) and tails (T) is observed. The sample space S consists of the following eight elements:

$$S = \{\text{HHH, HHT, HTH, HTT, THH, THT, TTH, TTT}\}$$

Let X assign to each point in S the largest number of successive heads that occurs. Thus:

$$X(\text{TTT}) = 0, \qquad X(\text{HTH}) = X(\text{HTT}) = X(\text{THT}) = X(\text{TTH}) = 1$$
$$X(\text{HHT}) = X(\text{THH}) = 2, \qquad X(\text{HHH}) = 3$$

Then X is a random variable with range space

$$R_X = \{0, 1, 2, 3\}$$

(*b*) A pair of fair dice is tossed. The sample space S (pictured in Fig. 3-3) consists of the 36 ordered pairs (a, b), where a and b can be any integers between 1 and 6; that is,

$$S = \{(1,1),\ (1,2),\ \ldots,\ (6,6)\}$$

Let X assign to each point (a, b) in S the maximum of the numbers, that is, $X(a, b) = \max(a, b)$. For example,

$$X(1,1) = 1, \qquad X(2,3) = 3, \qquad X(4,4) = 4, \qquad X(6,5) = 6, \qquad X(6,6) = 6$$

Then X is a random variable and any number between 1 and 6 can occur. Therefore,

$$R_X = \{1, 2, 3, 4, 5, 6\}$$

Now let Y assign to each point (a, b) in S the sum of the numbers, that is, $Y(a, b) = a + b$. For example,

$$Y(1,1) = 2, \qquad Y(2,3) = 5, \qquad Y(4,4) = 8, \qquad Y(6,5) = 11, \qquad Y(6,6) = 12$$

Then Y is a random variable with range space

$$R_Y = \{2, 3, \ldots, 12\}$$

That is, no sum can be less than 2 and no sum can exceed 12.

(*c*) A point is chosen at random in a circle C with radius r. Let X denote the distance of the point from the center of the circle. Then X is a random variable and its range space is the closed interval with endpoints 0 and r, that is,

$$R_X = [0, r]$$

Here X is a continuous random variable.

Sums and Products of Random Variables

Let X and Y be random variables on the same sample space S. Then $X + Y$, $X + k$, kX, and XY are the functions on S defined by

$$(X + Y)(s) = X(s) + Y(s), \qquad (kX)(s) = kX(s)$$
$$(X + k)(s) = X(s) + k, \qquad (XY)(s) = X(s)Y(s)$$

More generally, for any polynomial or exponential function $h(x)$, we define $h(X)$ to be the function on S defined by

$$[h(X)](s) = h[X(s)]$$

It can be shown that these are also random variables. (This is trivial in the case that every subset of S is an event.)

The short notation $P(X = a)$ and $P(a \le X \le b)$ will be used, respectively, for the probability that "X maps into a" and "X maps into the interval $[a, b]$", that is:

$$P(X = a) \equiv P(\{s \in S : X(s) = a\}$$

and

$$P(a \le X \le b) \equiv P(\{s \in S : a \le X(s) \le b\}$$

Analogous meanings are given to

$$P(X \le a), \qquad P(X = a, Y = b), \qquad P(a \le X \le b, c \le Y \le d)$$

and so on.

5.3 PROBABILITY DISTRIBUTION OF A FINITE RANDOM VARIABLE

Suppose a random variable X assigns only a finite number of values to a sample space S, say

$$R_X = \{x_1, x_2, \ldots, x_n\}$$

(We assume $x_1 < x_2 < \cdots < x_n$.) Then X induces a function f which assigns probabilities to the points in R_X by

$$f(x_k) \equiv P(X = x_k)$$

The set of ordered pairs $[x_i, f(x_i)]$ is usually given by means of a table as follows:

x	x_1	x_2	$\cdots$	x_n
$f(x)$	$f(x_1)$	$f(x_2)$	$\cdots$	$f(x_n)$

This function f is called the *probability distribution* or, simply, the *distribution* of the random variable X; it has the following two properties:

$$\text{(i) } f(x_k) \geq 0, \qquad \text{(ii) } \sum_k x_k = 1$$

Thus R_X with the above assignment of probabilities is a probability space.

Suppose S is a finite equiprobable space. Then the following theorem (proved in Problem 5.23) applies.

Theorem 5.1: Let S be a finite equiprobable space, and let f be the distribution of a random variable X on S with range space $R_X = \{x_1, x_2, \ldots, x_n\}$. Then:

$$f(x_k) = \frac{\text{number of points in } S \text{ whose image is } x_k}{\text{number of points in } S}$$

Remark: It is convenient sometimes to extend a probability distribution f to all real numbers by defining

$$f(x) \equiv P(X = x)$$

For $x = x_k$, this reduces to the above, whereas for other values of x we get $f(x) = 0$. Furthermore, we can now write

$$\text{(i) } f(x) \geq 0, \qquad \text{(ii) } \sum_x f(x) = 1$$

where the sum in (ii) may be viewed as taking place over all values of x. A graph of $f(x)$ is called a *probability graph.*

Notation: Sometimes we will use the pair $[x_i, p_i]$ or $[x_i, P(x_i)]$ or $[x, P(X = x)]$ to denote a probability distribution instead of using the functional notation $[x, f(x)]$.

EXAMPLE 5.2 A coin is tossed three times yielding the sample space

$$S = \{\text{HHH}, \text{HHT}, \text{HTH}, \text{HTT}, \text{THH}, \text{THT}, \text{TTH}, \text{TTT}\}$$

Let X be the random variable which assigns to each point in S its largest number of successive heads as discussed in Example 5.1(a). Then the range space is $R_X = \{0, 1, 2, 3\}$. In particular, there exist:

(i) one point TTT, where $X = 0$,
(ii) four points, HTH, HTT, THT, TTH, where $X = 1$,
(iii) two points HHT and THH, where $X = 2$,
(iv) one point HHH, where $X = 3$.

(*a*) Suppose the coin is fair. Then S is an 8-element equiprobable space. Hence we can use Theorem 5.1 to obtain the following distribution f of X:

x	0	1	2	3
$f(x)$	$\frac{1}{8}$	$\frac{4}{8}$	$\frac{2}{8}$	$\frac{1}{8}$

There are two ways to present the probability graph of X. One is by the *bar chart* shown in Fig. 5-1(*a*), and the other is by the *histogram* shown in Fig. 5-1(*b*). Observe that the sum of the lengths of the bars in the bar chart is 1, whereas the sum of the areas of the rectangles in the histogram is 1. One may view the histogram as making the random variable continuous, where $X = 1$ means X lies between 0.5 and 1.5.

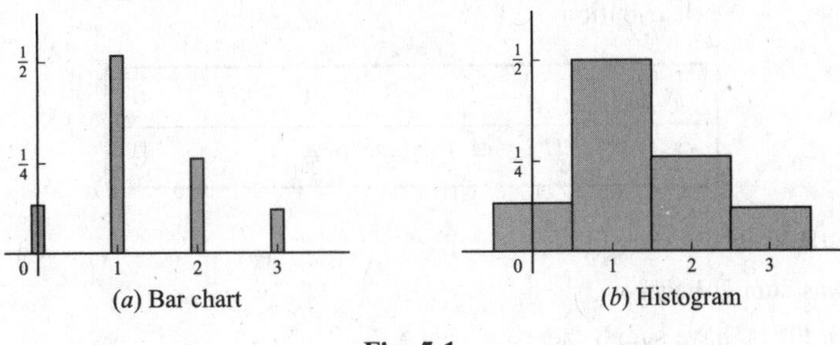

(*a*) Bar chart (*b*) Histogram

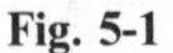
Fig. 5-1

(*b*) Suppose the coin is weighted so that $P(\mathrm{H}) = 2/3$ and $P(\mathrm{H}) = 1/3$. Then S is not an equiprobable space. Specifically, the probabilities of the points in S are as follows:

$$P(\mathrm{HHH}) = \frac{2}{3}\cdot\frac{2}{3}\cdot\frac{2}{3} = \frac{8}{27} \qquad P(\mathrm{THH}) = \frac{1}{3}\cdot\frac{2}{3}\cdot\frac{2}{3} = \frac{4}{27}$$

$$P(\mathrm{HHT}) = \frac{2}{3}\cdot\frac{2}{3}\cdot\frac{1}{3} = \frac{4}{27} \qquad P(\mathrm{THT}) = \frac{1}{3}\cdot\frac{2}{3}\cdot\frac{1}{3} = \frac{2}{27}$$

$$P(\mathrm{HTH}) = \frac{2}{3}\cdot\frac{1}{3}\cdot\frac{2}{3} = \frac{4}{27} \qquad P(\mathrm{TTH}) = \frac{1}{3}\cdot\frac{1}{3}\cdot\frac{2}{3} = \frac{2}{27}$$

$$P(\mathrm{HTT}) = \frac{2}{3}\cdot\frac{1}{3}\cdot\frac{1}{3} = \frac{2}{27} \qquad P(\mathrm{TTT}) = \frac{1}{3}\cdot\frac{1}{3}\cdot\frac{1}{3} = \frac{1}{27}$$

Since S is not an equiprobable space, we cannot use Theorem 5.1 to find the distribution f of X. Thus we find f directly:

$$f(0) = P(\mathrm{TTT}) = \frac{1}{27}$$

$$f(1) = P(\{\mathrm{HTH}, \mathrm{HTT}, \mathrm{THT}, \mathrm{TTH}\}) = \frac{4}{27} + \frac{2}{27} + \frac{2}{27} + \frac{2}{27} = \frac{10}{27}$$

$$f(2) = P(\{\mathrm{HHT}, \mathrm{THH}\}) = \frac{4}{27} + \frac{4}{27} = \frac{8}{27}$$

$$f(3) = P(\mathrm{HHH}) = \frac{8}{27}$$

Accordingly, the following is the distribution f of X:

x	0	1	2	3
$f(x)$	1/27	10/27	8/27	8/27

EXAMPLE 5.3 Let S be the sample space when a pair of fair dice is tossed, and let X and Y be the random variable on S in Example 5.1(*b*); namely, X denotes the maximum of the numbers appearing, i.e. $X(a,b) = \max(a,b)$, and Y denotes the sum of the numbers, i.e. $Y(a,b) = a+b$. Find the distribution f of X and the distribution g of Y. Also exhibit their probability graphs.

Here S is an equiprobable space with 36 points, so we can use Theorem 5.1 and simply count the number of points with the given numerical value.

First we compute the distribution f of X:

(1) One point (1, 1) has maximum value 1; hence $f(1) = \frac{1}{36}$.
(2) Three points, (1, 2), (2, 2), (2, 1), have maximum value 2; hence $F(2) = \frac{3}{36}$.
(3) Five points, (1, 3), (2, 3), (3, 3), (3, 2), (3, 1), have maximum value 3; hence $f(3) = \frac{5}{36}$.

Similarly,

$$f(4) = \tfrac{7}{36}, \qquad f(5) = \tfrac{9}{36}, \qquad f(6) = \tfrac{11}{36}$$

Accordingly, the following is the distribution f of X:

x	1	2	3	4	5	6
$f(x)$	$\frac{1}{36}$	$\frac{3}{36}$	$\frac{5}{36}$	$\frac{7}{36}$	$\frac{9}{36}$	$\frac{11}{36}$

Now we compute the distribution g of Y:

(1) One point (1, 1) has sum 2; hence $g(2) = \frac{1}{36}$.
(2) Two points, (1, 2), (2, 1), have sum 3; hence $g(3) = \frac{2}{36}$.
(3) Three points, (1, 3), (2, 2), (3, 1), have sum 4; hence $g(4) = \frac{3}{36}$.

Similarly,

$$g(5) = \tfrac{4}{36}, \qquad g(6) = \tfrac{5}{36}, \qquad g(7) = \tfrac{6}{36}, \ldots, g(12) = \tfrac{1}{36}$$

Accordingly, the following is the distribution g of Y:

y	2	3	4	5	6	7	8	9	10	11	12
$g(y)$	$\frac{1}{36}$	$\frac{2}{36}$	$\frac{3}{36}$	$\frac{4}{36}$	$\frac{5}{36}$	$\frac{6}{36}$	$\frac{5}{36}$	$\frac{4}{36}$	$\frac{3}{36}$	$\frac{2}{36}$	$\frac{1}{36}$

The probability bar charts of X and Y are pictured in Fig. 5-2.

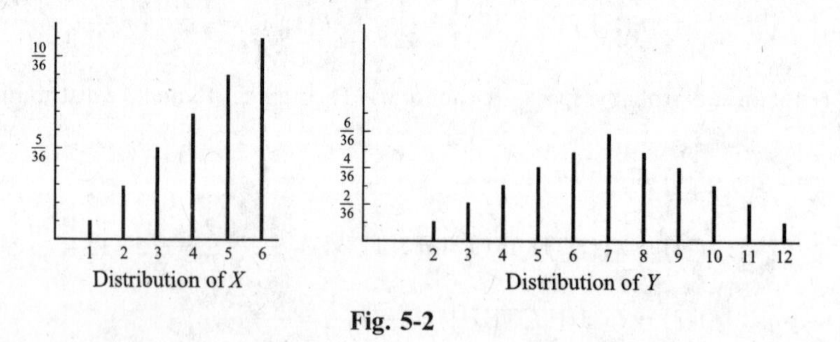

Fig. 5-2

5.4 EXPECTATION OF A FINITE RANDOM VARIABLE

Suppose X is a random variable whose distribution f is as follows:

x	x_1	x_2	x_3	...	x_n
$f(x)$	$f(x_1)$	$f(x_2)$	$f(x_3)$	...	$f(x_n)$

Then the *mathematical expectation* or *expected value* or, simply, the *expectation* of X, denoted by $E(X)$ or simply E, is defined by

$$E = E(X) = x_1 f(x_1) + x_2 f(x_2) + \cdots + x_n f(x_n) = \sum_i x_i f(x_i)$$

Equivalently, when the notation $[x_i, p_i]$ is used instead of $[x, f(x)]$,

$$E = E(X) = x_1 p_1 + x_2 p_2 + \cdots + x_n p_n = \sum_i x_i p_i$$

Roughly speaking, if the x_i are numerical outcomes of an experiment, then E is the expected value of the experiment.

EXAMPLE 5.4 A coin is tossed three times. Let X denote the largest number of successive heads.

(*a*) Suppose the coin is fair. The distribution of X appears in Example 5.2(*a*). Using this distribution we get

$$E = E(X) = 0\left(\frac{1}{8}\right) + 1\left(\frac{4}{8}\right) + 2\left(\frac{2}{8}\right) + 3\left(\frac{1}{8}\right) = \frac{11}{8} = 1.375$$

is the expected maximum number of successive heads.

(*b*) Suppose the coin is weighted so that $P(H) = \frac{2}{3}$ and $P(T) = \frac{1}{3}$. The distribution of X appears in Example 5.2(*b*). Using this distribution we get

$$E = E(X) = 0\left(\frac{1}{27}\right) + 1\left(\frac{10}{27}\right) + 2\left(\frac{8}{27}\right) + 3\left(\frac{8}{27}\right) = \frac{50}{27} \approx 1.852$$

is the expected maximum number of successive heads.

EXAMPLE 5.5 A pair of fair dice is tossed. Let X denote the maximum of the numbers appearing, i.e. $X(a, b) = \max(a, b)$, and let Y denote the sum of the numbers appearing, i.e. $Y(a, b) = a + b$. The distribution f of X is given in Example 5.3. Using the distribution of X, the expectation of X is computed as follows:

$$E(X) = \sum x_i f(x_i) = 1\left(\frac{1}{36}\right) + 2\left(\frac{3}{36}\right) + 3\left(\frac{5}{36}\right) + 4\left(\frac{7}{36}\right) + 5\left(\frac{9}{36}\right) + 6\left(\frac{11}{36}\right) = \frac{161}{36} \approx 4.5$$

The distribution g of Y is also given in Example 5.3. Using the distribution of Y, the expectation of Y is computed as follows:

$$E(Y) = \sum y_i g(y_i) = 2\left(\frac{1}{36}\right) + 3\left(\frac{2}{36}\right) + 4\left(\frac{3}{36}\right) + \cdots + 12\left(\frac{1}{36}\right) = \frac{252}{36} = 7$$

EXAMPLE 5.6 A fair coin is tossed 6 times. Let X denote the number of heads occurring. One can show that the distribution of X is as follows:

x	0	1	2	3	4	5	6
$f(x)$	$\frac{1}{64}$	$\frac{6}{64}$	$\frac{15}{64}$	$\frac{20}{64}$	$\frac{15}{64}$	$\frac{6}{64}$	$\frac{1}{64}$

Then the expected number of heads is

$$E = 0\left(\frac{1}{64}\right) + 1\left(\frac{6}{64}\right) + 2\left(\frac{15}{64}\right) + 3\left(\frac{20}{64}\right) + 4\left(\frac{15}{64}\right) + 5\left(\frac{6}{64}\right) + 6\left(\frac{1}{64}\right) = 3$$

This agrees with our intuition that, when a fair coin is repeatedly tossed, about half of them should be heads.

EXAMPLE 5.7 A sample of size 3 is selected at random from a box containing 12 items, of which 3 are defective. Let X denote the number of defective items in the sample. Find the expected number $E(X)$ of defective items.

The sample space S consists of $C(12, 3) = 220$ distinct equally likely samples of size 3. We note that there are:

$$\begin{aligned} C(9,3) &= \ \ 84 \text{ samples with 0 defective items,} \\ 3 \cdot C(9,2) &= 108 \text{ samples with 1 defective item,} \\ C(3,2) \cdot 9 &= \ \ 27 \text{ samples with 2 defective items,} \\ C(3,3) &= \ \ \ 1 \text{ sample with 3 defective items.} \end{aligned}$$

Since S is an equiprobable space, we can use Theorem 5.1 to obtain the following distribution f of X:

x	0	1	2	3
$f(x)$	$\frac{84}{220}$	$\frac{108}{220}$	$\frac{27}{220}$	$\frac{1}{220}$

Accordingly, the expected number of defective items in a sample is

$$E = 0\left(\frac{84}{220}\right) + 1\left(\frac{108}{220}\right) + 2\left(\frac{27}{220}\right) + 3\left(\frac{1}{220}\right) = \frac{165}{220} = 0.75$$

The following theorems and corollary (proved in Problems 5.26–5.28) relate the notion of expectation to operations on random variables defined in Section 5.2.

Theorem 5.2: Let X be a random variable and let k be a real number. Then:

$$\text{(i)}\quad E(kX) = kE(X), \qquad \text{(ii)}\quad E(X + k) = E(X) + k$$

Theorem 5.3: Let X and Y be random variables on the same sample space S. Then $E(X + Y) = E(X) + E(Y)$.

A simple induction argument yields:

Corollary 5.4: Let $X_1, X_2, \ldots, X_n$ be random variables on S. Then:

$$E(X_1 + X_2 + \cdots + X_n) = E(X_1) + E(X_2) + \cdots + E(X_n)$$

Expectation and Games of Chance

Consider a game of chance with n outcomes $a_1, a_2, \ldots, a_n$ with corresponding probabilities $p_1, p_2, \ldots, p_n$, and suppose the payoff to the player for outcome a_i is w_i (where a positive w_i is a win for the player, and a negative w_i a loss). Recall from Section 3.7 that the expectation of the player was the quantity

$$E = w_1 p_1 + w_2 p_2 + \cdots + w_n p_n$$

The assignment of the number w_i to a_i may be viewed as a random variable X, and the expected value $E(X)$ of X is the expectation E of the game. The game is *fair* if $E = 0$, *favorable* to the player if E is positive, and *unfavorable* to the player if E is negative.

EXAMPLE 5.8 A player tosses a fair die. If a prime number, 2, 3, or 5, occurs the player wins that number of dollars, but if a nonprime number occurs the player loses that number of dollars. The distribution of the game follows:

x	2	3	5	-1	-4	-6
$f(x)$	$\frac{1}{6}$	$\frac{1}{6}$	$\frac{1}{6}$	$\frac{1}{6}$	$\frac{1}{6}$	$\frac{1}{6}$

The negative numbers -1, -4, and -6 correspond to the fact that the player loses if a nonprime number occurs. The expected value of the game is

$$E = 2\left(\frac{1}{6}\right) + 3\left(\frac{1}{6}\right) + 5\left(\frac{1}{6}\right) - 1\left(\frac{1}{6}\right) - 4\left(\frac{1}{6}\right) - 6\left(\frac{1}{6}\right) = -\left(\frac{1}{6}\right)$$

Thus the game is unfavorable to the player, since the expected value E is negative.

Mean and Expected Value

Suppose X is a random variable on an equiprobable space $S = \{a_1, a_2, \ldots, a_n\}$, where X assigns the value x_i to a_i and all the x_i's are distinct. Then each x_i occurs with the same probability $p_i = 1/n$. Thus

$$E(x) = x_1\left(\frac{1}{n}\right) + x_2\left(\frac{1}{n}\right) + \cdots + x_n\left(\frac{1}{n}\right) = \frac{x_1 + x_2 + \cdots + x_n}{n}$$

which is the average or mean value of the numbers $x_1, x_2, \ldots, x_n$. In general, $E(X)$ is the *weighted average* of the possible values of X, where each value is weighted by its probability. For this reason $E(X)$ is also called the *mean* of the random variable X. Recall the mean was denoted by the Greek letter μ. Thus we use the following notation for the expectation of X:

$$\boxed{\mu = \mu_X = E(X)}$$

The mean is an important parameter for a probability distribution, and in Section 5.5 we introduce another important parameter, called the *standard deviation* of X.

5.5 VARIANCE AND STANDARD DEVIATION

The mean of a random variable X measures, in a certain sense, the "average" value of X. The concepts in this section, variance and standard deviation, measure the "spread" or "dispersion" of X.

Consider a random variable X with mean $\mu = E(X)$ and probability distribution

x	x_1	x_2	x_3	$\ldots$	x_n
$f(x)$	$f(x_1)$	$f(x_2)$	$f(x_3)$	$\ldots$	$f(x_n)$

The *variance* of X, denoted by $\mathrm{Var}(X)$, is defined by:

$$\mathrm{Var}(X) = (x_1 - \mu)^2 f(x_1) + (x_2 - \mu)^2 f(x_2) + \cdots + (x_n - \mu)^2 f(x_n)$$
$$= \sum (x_i - \mu)^2 f(x_i) = E((X - \mu)^2)$$

The *standard deviation* of X, denoted by σ_X or simply σ, is the nonnegative square root of $\mathrm{Var}(X)$; that is

$$\sigma_X = \sqrt{\mathrm{Var}(X)}$$

Accordingly, $\mathrm{Var}(X) = \sigma_X^2$. Both $\mathrm{Var}(X)$ and σ_X^2 or simply σ^2 are used to denote the variance of a random variable X.

The next theorem gives us an alternate formula for calculating the variance of X.

Theorem 5.5: $\text{Var}(X) = x_1^2 f(x_1) + x_2^2 f(x_2) + \cdots + x_n^2 f(x_n) - \mu^2$
$= \sum x_i^2 f(x_i) - \mu^2 = E(X^2) - \mu^2$

Proof: Using $\sum x_i f(x_i) = \mu$ and $\sum f(x_i) = 1$, we have

$$\begin{aligned}\sum (x_i - \mu)^2 f(x_i) &= \sum (x_i^2 - 2\mu x_i + \mu^2) f(x_i) \\ &= \sum x_i^2 f(x_i) - 2\mu \sum x_i f(x_i) + \mu^2 \sum f(x_i) \\ &= \sum x_i^2 f(x_i) - 2\mu^2 + \mu^2 = \sum x_i^2 f(x_i) - \mu^2\end{aligned}$$

which proves the theorem.

Remark: Both the variance $\text{Var}(X) = \sigma^2$ and the standard deviation σ measure the weighted spread of the values x_i about the mean μ; however, the standard deviation σ has the same units as μ.

EXAMPLE 5.9

(*a*) Let X denote the number of times heads occurs when a fair coin is tossed six times. The distribution of X appears in Example 5.6, where its mean $\mu = 3$ is computed. The variance of X is computed as follows:

$$\text{Var}(X) = (0-3)^2 \frac{1}{64} + (1-3)^2 \frac{6}{64} + (2-3)^2 \frac{15}{64} + \cdots + (6-3)\frac{1}{64} = 1.5$$

Alternatively: $\text{Var}(X) = 0^2 \frac{1}{64} + 1^2 \frac{6}{64} + 2^2 \frac{15}{64} + 3^2 \frac{20}{64} + 4^2 \frac{15}{64} + 5^2 \frac{6}{64} + 6^2 \frac{1}{64} - 3^2 = 1.5$

Thus the standard deviation is $\sigma = \sqrt{1.5} \approx 1.225$ (heads).

(*b*) Consider the random variable X in Example 5.7 where its mean $\mu = 0.75$ is computed. The variance of X is computed as follows:

$$\text{Var}(X) = 0^2 \frac{84}{220} + 1^2 \frac{108}{220} + 2^2 \frac{27}{220} + 3^2 \frac{1}{220} - (0.75)^2 = 0.46$$

Thus the standard deviation is

$$\sigma = \sqrt{\text{Var}(X)} = \sqrt{0.46} = 0.66$$

EXAMPLE 5.10 A pair of fair dice is tossed. Let X denote the maximum of the numbers appearing, i.e. $X(a, b) = \max(a, b)$, and let Y denote the sum of the numbers appearing, i.e. $Y(a, b) = a + b$. The distributions of X and Y appear in Example 5.3, and their expectations were computed in Example 5.5, yielding

$$\mu_X = E(X) = 4.5 \quad \text{and} \quad \mu_Y = 7$$

Find the variance and standard deviation of (*a*) X, (*b*) Y.

(*a*) First we compute $E(X^2)$ as follows:

$$\begin{aligned}E(X^2) = \sum x_i^2 f(x_i) &= 1^2\left(\frac{1}{36}\right) + 2^2\left(\frac{3}{36}\right) + 3^2\left(\frac{5}{36}\right) + 4^2\left(\frac{7}{36}\right) + 5^2\left(\frac{9}{36}\right) + 6^2\left(\frac{11}{36}\right) \\ &= \frac{791}{36} = 21.97\end{aligned}$$

Hence $\quad \text{Var}(X) = E(X^2) - \mu_X^2 = 21.97 - 19.98 = 1.99 \quad$ and $\quad \sigma_X = \sqrt{1.99} = 1.4$

(*b*) First we compute $E(Y^2)$ as follows:

$$E(Y^2) = \sum y_i^2 g(y_i) = 2^2\left(\frac{1}{36}\right) + 3^2\left(\frac{2}{36}\right) + \cdots + 12^2\left(\frac{1}{36}\right) = \frac{1974}{36} = 54.8$$

Hence $\quad \text{Var}(Y) = E(Y^2) - \mu_Y^2 = 54.8 - 49 = 5.8 \quad$ and $\quad \sigma_Y = \sqrt{5.8} = 2.4$

Remark: There are physical interpretations of the mean and variance. Suppose the x-axis is a thin wire and at each point x_i there is a unit with mass p_i. Then, if a fulcrum or pivot is placed at the point μ (Fig. 5-3(a)), the system will balance. Hence, μ is called the *center of mass* of the system of points x_i. On the other hand, if the system were rotating about an axis through the mean μ (Fig. 5-3(b)), then the variance σ^2 is a measure of the system's resistance to stopping. In technical terms, σ^2 is called the *moment of inertia* of the system.

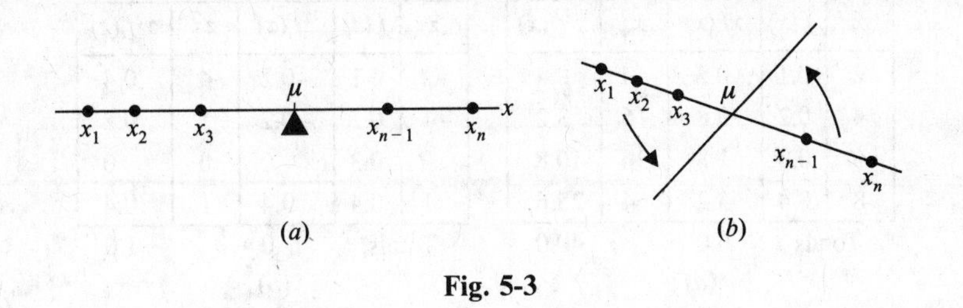

Fig. 5-3

A basic property of the variance and standard deviation is given in the following theorem (proved in Problem 5.29).

Theorem 5.6: Let X be a random variable, and let a and b be constants. Then:

$$\mathrm{Var}(aX + b) = a^2\,\mathrm{Var}(X) \quad \text{and} \quad \sigma_{aX+b} = |a|\sigma_X$$

In particular, we have the following special cases, where k is a real number:

(i) $\mathrm{Var}(X + k) = \mathrm{Var}(X)$ and hence $\sigma_{X+k} = \sigma_X$.

(ii) $\mathrm{Var}(kX) = k^2\,\mathrm{Var}(X)$ and hence $\sigma_{kX} = |k|\sigma_X$.

Standardized Random Variable

Let X be a random variable with mean μ and standard deviation $\sigma > 0$. Then the *standardized random variable* Z is defined by

$$Z = \frac{X - \mu}{\sigma}$$

Important properties of Z are contained in the next theorem (proved in Problem 5.31).

Theorem 5.7: The standardized random variable Z has mean $\mu_Z = 0$ and standard deviation $\sigma_Z = 1$.

EXAMPLE 5.11 Suppose a random variable X has the following distribution:

x	2	4	6	8
$f(x)$	0.1	0.2	0.3	0.4

(a) Compute the mean μ and standard deviation σ of X.

(b) Find the probability distribution of $Z = (X - \mu)/\sigma$, and show that $\mu_Z = 0$ and $\sigma_Z = 1$, as predicted by Theorem 5.7.

(a) First construct a data table as in Fig. 5-4(a). The total in the third column is the expected value of X; that is, $\mu = E(X) = \sum x_i f(x_i) = 6$. Similarly, the total in the fifth column is the expected value of X^2; that is $E(X^2) = \sum x_i^2 f(x_i) = 40$. Thus, by Theorem 5.5,

$$\sigma^2 = E(X^2) - \mu^2 = 40 - 6^2 = 4 \quad \text{and} \quad \sigma = 2$$

(*b*) Using $z = (x - 6)/2$ and $f(z) = f(x)$, construct a data table for the random variable $Z = (X - 6)/2$ as in Fig. 5-4(*b*). The first two columns of the table form the distribution of Z. The total in the third column is the expected value of Z; hence $\mu_Z = 0$. The total in the fifth column is the expected value of Z^2; hence $E(Z^2) = 1.0$. Thus, by Theorem 5.5,

$$\sigma_Z^2 = E(Z^2) - \mu_Z^2 = 1 - 0^2 = 1 \quad \text{and} \quad \sigma_Z = 1$$

x	$f(x)$	$xf(x)$	x^2	$x^2f(x)$
2	0.1	0.2	4	0.4
4	0.2	0.8	16	3.2
6	0.3	1.8	36	10.8
8	0.4	3.2	64	25.6
Totals		6.0		40.0

(*a*)

z	$f(z)$	$zf(z)$	z^2	$z^2f(z)$
−2	0.1	−0.2	4	0.4
−1	0.2	−0.2	1	0.2
0	0.3	0	0	0
1	0.4	0.4	1	0.4
Totals		0		1.0

(*b*)

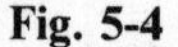

Fig. 5-4

5.6 JOINT DISTRIBUTION OF RANDOM VARIABLES

Let X and Y be random variables on the same sample space S with respective range spaces

$$R_X = \{x_1, x_2, \ldots, x_n\} \quad \text{and} \quad R_Y = \{y_1, y_2, \ldots, y_m\}$$

The *joint distribution* or *joint probability function* of X and Y is the function h on the product space $R_X \times R_Y$ defined by

$$h(x_i, y_j) \equiv P(X = x_i, Y = y_j) \equiv P(\{s \in S : X(s) = x_i, Y(s) = y_j\})$$

The function h is usually given in the form of a table as in Fig. 5-5. The function h has the properties:

(i) $h(x_i, y_j) \geq 0$, (ii) $\sum_i \sum_j h(x_i, y_j) = 1$

Thus h defines a probability space on the product $R_X \times R_Y$.

X \ Y	y_1	y_2	...	y_m	Sum
x_1	$h(x_1, y_1)$	$h(x_1, y_2)$	...	$h(x_1, y_m)$	$f(x_1)$
x_2	$h(x_2, y_1)$	$h(x_2, y_2)$	...	$h(x_2, y_m)$	$f(x_2)$
...	...	...	...	...	...
x_n	$h(x_n, y_1)$	$h(x_n, y_2)$	...	$h(x_n, y_m)$	$f(x_n)$
Sum	$g(y_1)$	$g(y_2)$	...	$g(y_m)$	

Fig. 5-5

The functions f and g on the right side and the bottom side, respectively, of the joint distribution table in Fig. 5-5 are defined by

$$f(x_i) = \sum_j h(x_i, y_j) \quad \text{and} \quad g(y_j) = \sum_i h(x_i, y_j)$$

That is, $f(x_i)$ is the sum of the entries in the ith row and $g(y_j)$ is the sum of the entries in the jth column. They are called the *marginal distributions*, and are, in fact, the (individual) distributions of X and Y respectively (Problem 5.13).

Now if X and Y are random variables with the above joint distribution (and respective means μ_X and μ_Y), then the *covariance* of X and Y, denoted by $\mathrm{Cov}(X, Y)$, is defined by

$$\mathrm{Cov}(X, Y) = \sum_{i,j}(x_i - \mu_X)(y_j - \mu_Y)h(x_i, y_j) = E[(X - \mu_X)(Y - \mu_Y)]$$

or equivalently (see Problem 5.30) by

$$\mathrm{Cov}(X, Y) = \sum_{i,j} x_i y_j h(x_i, y_j) - \mu_X\mu_Y = E(XY) - \mu_X\mu_Y$$

The *correlation* of X and Y, denoted by $\rho(X, Y)$, is defined by

$$\rho(X, Y) = \frac{\mathrm{Cov}(X, Y)}{\sigma_X\sigma_Y}$$

The correlation ρ is dimensionless and has the following properties:

(i) $\rho(X, Y) = \rho(Y, X)$

(ii) $-1 \le \rho \le 1$

(iii) $\rho(X, X) = 1,\ \rho(X, -X) = -1$

(iv) $\rho(aX + b, cY + d) = \rho(X, Y)$, if $a, c \ne 0$

We show below (Example 5.13) that pairs of random variables with identical (individual) distributions can have distinct covariances and correlations. Thus $\mathrm{Cov}(X, Y)$ and $\rho(X, Y)$ are measurements of the way that X and Y are interrelated.

EXAMPLE 5.12 Let S be the sample space when a pair of fair dice is tossed, and let X and Y be the random variables on S in Example 5.1(*b*); namely, X assigns the maximum of the numbers and Y assigns the sum of the numbers to each point in S. The joint distribution of X and Y appear in Fig. 5-6. The entry $h(3, 5) = \frac{2}{36}$ comes from the fact that (3, 2) and (2, 3) are the only points in S whose maximum number is 3 and whose sum is 5; that is,

$$h(3, 5) \equiv P(X = 3, Y = 5) = P(\{(3, 2), (2, 3)\}) = 2/36$$

The other entries are obtained in a similar manner.

X \ Y	2	3	4	5	6	7	8	9	10	11	12	Sum
1	$\frac{1}{36}$	0	0	0	0	0	0	0	0	0	0	$\frac{1}{36}$
2	0	$\frac{2}{36}$	$\frac{1}{36}$	0	0	0	0	0	0	0	0	$\frac{3}{36}$
3	0	0	$\frac{2}{36}$	$\frac{2}{36}$	$\frac{1}{36}$	0	0	0	0	0	0	$\frac{5}{36}$
4	0	0	0	$\frac{2}{36}$	$\frac{2}{36}$	$\frac{2}{36}$	$\frac{1}{36}$	0	0	0	0	$\frac{7}{36}$
5	0	0	0	0	$\frac{2}{36}$	$\frac{2}{36}$	$\frac{2}{36}$	$\frac{2}{36}$	$\frac{1}{36}$	0	0	$\frac{9}{36}$
6	0	0	0	0	0	$\frac{2}{36}$	$\frac{2}{36}$	$\frac{2}{36}$	$\frac{2}{36}$	$\frac{2}{36}$	$\frac{1}{36}$	$\frac{11}{36}$
Sum	$\frac{1}{36}$	$\frac{2}{36}$	$\frac{3}{36}$	$\frac{4}{36}$	$\frac{5}{36}$	$\frac{6}{36}$	$\frac{5}{36}$	$\frac{4}{36}$	$\frac{3}{36}$	$\frac{2}{36}$	$\frac{1}{36}$	

Fig. 5-6

Observe that the right side sum column does give the distribution f of X and the bottom sum row does give the distribution of Y in Example 5.3.

We compute the covariance and correlation of X and Y. First we compute $E(XY)$ as follows:

$$\begin{aligned} E(XY) &= \sum x_i y_j h(x_i, y_j) \\ &= 1(2)\left(\frac{1}{36}\right) + 2(3)\left(\frac{2}{36}\right) + 2(4)\left(\frac{1}{36}\right) + \cdots + 6(12)\left(\frac{1}{36}\right) = \frac{1232}{36} \approx 34.2 \end{aligned}$$

By Example 5.5, $\mu_X = 4.47$ and $\mu_Y = 7$, and by Example 5.10, $\sigma_X = 1.4$ and $\sigma_Y = 2.4$; hence

$$\text{Cov}(X, Y) = E(XY) - \mu_X\mu_Y = 34.2 - (4.47)(7) = 2.9$$

and
$$\rho(X, Y) = \frac{\text{Cov}(X, Y)}{\sigma_X\sigma_Y} = \frac{2.9}{(1.4)(2.4)} = 0.86$$

EXAMPLE 5.13 Let X and Y be random variables with joint distribution in Fig. 5-7(*a*), and let X' and Y' be random variables with joint distribution in Fig. 5-7(*b*). Observe that X and X' have the same (individual) distribution, and Y and Y' have the same distribution as follows:

x	1	3
$f(x)$	$\frac{1}{2}$	$\frac{1}{2}$

Distribution of X and X'

y	4	10
$g(y)$	$\frac{1}{2}$	$\frac{1}{2}$

Distribution of Y and Y'

Note $\mu_X = \mu_{X'} = 2$ and $\mu_Y = \mu_{Y'} = 7$.

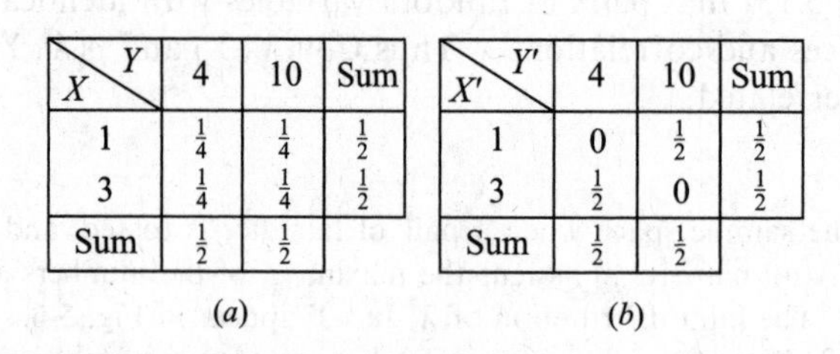

$X \backslash Y$	4	10	Sum
1	$\frac{1}{4}$	$\frac{1}{4}$	$\frac{1}{2}$
3	$\frac{1}{4}$	$\frac{1}{4}$	$\frac{1}{2}$
Sum	$\frac{1}{2}$	$\frac{1}{2}$	

(*a*)

$X' \backslash Y'$	4	10	Sum
1	0	$\frac{1}{2}$	$\frac{1}{2}$
3	$\frac{1}{2}$	0	$\frac{1}{2}$
Sum	$\frac{1}{2}$	$\frac{1}{2}$	

(*b*)

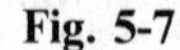

Fig. 5-7

We show that $\text{Cov}(X, Y) \neq \text{Cov}(X', Y')$ and hence $\rho(X, Y) \neq \rho(X', Y')$. We first compute $E(XY)$ and $E(X'Y')$ as follows:

$$E(XY) = 1\cdot 4\cdot \tfrac{1}{4} + 1\cdot 10\cdot \tfrac{1}{4} + 3\cdot 4\cdot \tfrac{1}{4} + 3\cdot 10\cdot \tfrac{1}{4} = 14$$
$$E(X'Y') = 1\cdot 4\cdot 0 + 1\cdot 10\cdot \tfrac{1}{2} + 3\cdot 4\cdot \tfrac{1}{2} + 3\cdot 10\cdot 0 = 11$$

Since $\mu_X = \mu_{X'} = 2$ and $\mu_Y = \mu_{Y'} = 7$,

$$\text{Cov}(X, Y) = E(XY) - \mu_X\mu_Y = 0 \quad \text{and} \quad \text{Cov}(X', Y') = E(X'Y') - \mu_{X'}\mu_{Y'} = -3$$

Remark: The notion of a joint distribution h is extended to any finite number of random variables $X, Y, \ldots, Z$ in the obvious way; that is, h is a function on the product set $R_X \times R_Y \times \cdots \times R_Z$ defined by

$$h(x_i, y_j, \ldots, z_k) \equiv P(X = x_i, Y = y_j, \ldots, Z = z_k)$$

5.7 INDEPENDENT RANDOM VARIABLES

A finite number of random variables $X, Y, \ldots, Z$ on a sample space S are said to be *independent* if

$$P(X = x_i,\ Y = y_j, \ldots, Z = z_k) \equiv P(X = x_i)P(Y = y_j)\cdots P(Z = z_k)$$

for any values $x_i, y_j, \ldots, z_k$. In particular, X and Y are independent if

$$P(X = x_i,\ Y = y_j) \equiv P(X = x_i)P(Y = y_j)$$

Now if X and Y have respective distributions f and g, and joint distribution h, then the above equation can be written as

$$h(x_i, y_j) = f(x_i)g(y_j)$$

In other words, X and Y are independent if each entry $h(x_i, y_j)$ is the product of its marginal entries.

EXAMPLE 5.14 Let X and Y be random variables with joint distribution in Fig. 5-8. Then X and Y are independent random variables since each entry in the joint distribution can be obtained by multiplying its marginal entries. For example,

$$P(1, 2) = P(X = 1)P(Y = 2) = (0.30)(0.20) = 0.06$$
$$P(1, 3) = P(X = 1)P(Y = 3) = (0.30)(0.50) = 0.15$$
$$P(1, 4) = P(X = 1)P(Y = 4) = (0.30)(0.30) = 0.09$$

and so on.

X \ Y	2	3	4	Sum
1	0.06	0.15	0.09	0.30
2	0.14	0.35	0.21	0.70
Sum	0.20	0.50	0.30	

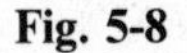

Fig. 5-8

EXAMPLE 5.15 A fair coin is tossed twice giving the equiprobable sample space $S = \{HH, HT, TH, TT\}$. Let X and Y be random variables on S as follows.

(*a*) Let $X = 1$ if the first toss is a head, $X = 0$ otherwise; let $Y = 1$ if both tosses are heads, $Y = 0$ otherwise. The joint distribution of X and Y appear in Fig. 5-9(*a*). X and Y are not independent. For example, $P(X = 0) = \frac{1}{2}$ and $P(Y = 0) = \frac{3}{4}$, but $P(0, 0) = \frac{1}{2} \neq P(X = 0)P(Y = 0)$.

(*b*) Again let $X = 1$ if the first toss is a head, $X = 0$ otherwise; but now let $Y = 1$ if the second toss is a head, $Y = 0$ otherwise. The joint distribution of X and Y appear in Fig. 5-9(*b*). Now X and Y are independent. Specifically,

$$P(X = x, Y = y) = P(X = x)P(Y = y)$$

for all four entries.

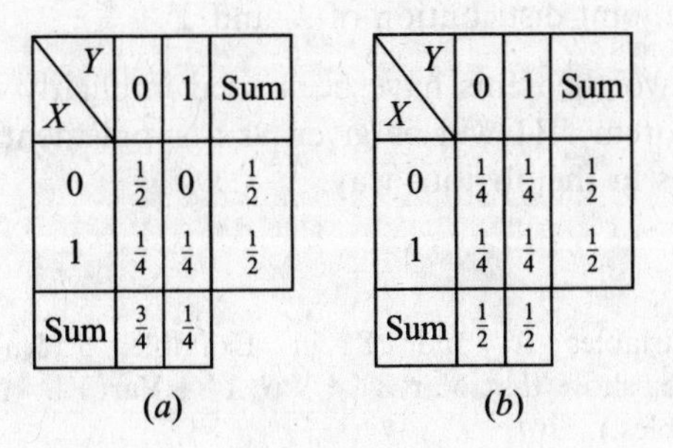

X \ Y	0	1	Sum
0	$\frac{1}{2}$	0	$\frac{1}{2}$
1	$\frac{1}{4}$	$\frac{1}{4}$	$\frac{1}{2}$
Sum	$\frac{3}{4}$	$\frac{1}{4}$	

(*a*)

X \ Y	0	1	Sum
0	$\frac{1}{4}$	$\frac{1}{4}$	$\frac{1}{2}$
1	$\frac{1}{4}$	$\frac{1}{4}$	$\frac{1}{2}$
Sum	$\frac{1}{2}$	$\frac{1}{2}$	

(*b*)

Fig. 5-9

The following theorems (proved in Problems 5.32–5.33) give important properties of independent random variables which do not hold in general.

Theorem 5.8: Let X and Y be independent random variables. Then:

(i) $E(XY) = E(X)E(Y)$

(ii) $\text{Var}(X + Y) = \text{Var}(X) + \text{Var}(Y)$

(iii) $\text{Cov}(X, Y) = 0$

Part (ii) in the above theorem generalizes as follows.

Theorem 5.9: Let $X_1, X_2, \ldots, X_n$ be independent random variables. Then:

$$\text{Var}(X_1 + \cdots + X_n) = \text{Var}(X_1) + \cdots + \text{Var}(X_n)$$

5.8 FUNCTIONS OF A RANDOM VARIABLE

Let X and Y be random variables on the same sample space S. Then Y is said to be a function of X if Y can be represented $Y = \Phi(X)$ for some real-valued function Φ of a real variable; that is, if $Y(s) = \Phi[X(s)]$ for every $s \in S$. For example, kX, X^2, $X + k$, and $(X + k)^2$ are all functions of X with $\Phi(x) = kx$, x^2, $x + k$, and $(x + k)^2$, respectively. We have the following fundamental result (proved in Problem 5.25).

Theorem 5.10: Let X and Y be random variables on the same sample space S with $Y = \Phi(X)$. Then

$$E(Y) = \sum_{i=1}^{n} \Phi(x_i) f(x_i)$$

where f is the distribution function of X.

Similarly, a random variable Z is said to be a function of X and Y if Z can be represented $Z = \Phi(X, Y)$, where Φ is a real-valued function of two real variables; that is, if

$$Z(s) = \Phi[X(s),\ Y(s)]$$

for every $s \in S$. For example, $X + Y$ is a function of X and Y with $\Phi(x, y) = x + y$.

Corresponding to the above theorem we have the following analogous result.

Theorem 5.11: Let X, Y, Z be random variables on the same sample space S with $Z = \Phi(X, Y)$. Then

$$E(Z) = \sum_{i,j} \Phi(x_i, y_j) h(x_i, y_j)$$

where h is the joint distribution of X and Y.

We note that the above two theorems have been used implicitly in the preceding discussion and theorems. The proof of Theorem 5.11 will be given as a supplementary problem; it generalizes to a function of n random variables in the obvious way.

EXAMPLE 5.16

(*a*) Consider the random variables X and Y in Example 5.15(*a*). Let $Z = X + Y$. Show that $E(Z) = E(X) + E(Y)$. Also, show that $\text{Var}(Z) \neq \text{Var}(X) + \text{Var}(Y)$. (Thus Theorem 5.8 need not hold for dependent random variables.)

Use the right marginal distribution in Fig. 5-9(*a*) for the distribution of X to obtain:

$$\mu_X = E(X) = 0\left(\frac{1}{2}\right) + 1\left(\frac{1}{2}\right) = \left(\frac{1}{2}\right) \quad \text{and} \quad E(X^2) = 0^2\left(\frac{1}{2}\right) + 1^2\left(\frac{1}{2}\right) = \frac{1}{2}$$

$$\text{Var}(X) = E(X^2) - \mu_X^2 = \frac{1}{2} - \frac{1}{4} = \frac{1}{4}$$

Use the left marginal distribution in Fig. 5-9(*a*) for the distribution of Y to obtain:

$$\mu_Y = E(Y) = 0\left(\frac{3}{4}\right) + 1\left(\frac{1}{4}\right) = \frac{1}{4} \quad \text{and} \quad E(Y^2) = 0^2\left(\frac{3}{4}\right) + 1^2\left(\frac{1}{4}\right) = \frac{1}{4}$$

$$\text{Var}(Y) = E(Y^2) - \mu_Y^2 = \frac{1}{4} - \frac{1}{16} = \frac{3}{16}$$

The random variable $Z = X + Y$ assumes the values 0, 1, 2 with respective probabilities $\frac{1}{2}, \frac{1}{4}, \frac{1}{4}$. Thus

$$\mu_Z = E(Z) = 0\left(\frac{1}{2}\right) + 1\left(\frac{1}{4}\right) + 2\left(\frac{1}{4}\right) = \frac{3}{4} \quad \text{and} \quad E(Z^2) = 0^2\left(\frac{1}{2}\right) + 1^2\left(\frac{1}{4}\right) + 2^2\left(\frac{1}{4}\right) = \frac{5}{4}$$

$$\text{Var}(Z) = E(Z^2) - \mu_Z^2 = \frac{5}{4} - \frac{9}{16} = \frac{11}{16}$$

Therefore,

$$E(X) + E(Y) = \frac{1}{2} + \frac{1}{4} = \frac{3}{4} = E(Z)$$

but

$$\text{Var}(X) + \text{Var}(Y) = \frac{1}{4} + \frac{3}{16} = \frac{7}{16} \neq \frac{11}{16} = \text{Var}(Z)$$

(*b*) Consider the random variables X and Y in Example 5.15(*b*). Let $Z = X + Y$. Show that $E(Z) = E(X) + E(Y)$. Also, show that $\text{Var}(Z) = \text{Var}(X) + \text{Var}(Y)$, which is expected since X and Y are independent.

The marginal distributions in Fig. 5-9(*b*) give the distributions of X and Y and they are identical. Thus

$$\mu_X = \mu_Y = E(X) = E(Y) = 0\left(\frac{1}{2}\right) + 1\left(\frac{1}{2}\right) = \frac{1}{2}$$

$$E(X^2) = E(Y^2) = 0^2\left(\frac{1}{2}\right) + 1^2\left(\frac{1}{2}\right) = \frac{1}{2}$$

$$\text{Var}(X) = \text{Var}(Y) = E(X^2) - \mu_X^2 = \frac{1}{2} - \frac{1}{4} = \frac{1}{4}$$

The random variable $Z = X + Y$ assumes the values 0, 1, 2 but now with respective probabilities $\frac{1}{4}, \frac{1}{2}, \frac{1}{4}$. Thus

$$\mu_Z = E(Z) = 0\left(\frac{1}{4}\right) + 1\left(\frac{1}{2}\right) + 2\left(\frac{1}{4}\right) = 1 \quad \text{and} \quad E(Z^2) = 0^2\left(\frac{1}{4}\right) + 1^2\left(\frac{1}{2}\right) + 2^2\left(\frac{1}{4}\right) = \frac{3}{2}$$

$$\text{Var}(Z) = E(Z^2) - \mu_Z^2 = \frac{3}{3} - 1 = \frac{1}{2}$$

Therefore,

$$E(X) + E(Y) = \frac{1}{2} + \frac{1}{2} = 1 = E(Z)$$

and

$$\text{Var}(X) + \text{Var}(Y) = \frac{1}{4} + \frac{1}{4} = \frac{1}{2} = \text{Var}(Z)$$

5.9 DISCRETE RANDOM VARIABLES IN GENERAL

Now suppose X is a random variable on a sample space S with a countable infinite range space, say $R_S = \{x_1, x_2, \ldots\}$. As in the finite case, X induces a function f on R_X, called the *distribution* of X, defined by

$$f(x_i) \equiv P(X = x_i)$$

The distribution is frequently presented in a table as follows:

x	x_1	x_2	x_3	$\ldots$
$f(x)$	$f(x_1)$	$f(x_2)$	$f(x_3)$	$\ldots$

The distribution f has the following two properties:

$$\text{(i)}\quad f(x_i) \geq 0 \qquad\qquad \text{(ii)}\quad \textstyle\sum_{i=1}^{\infty} f(x_i) = 1$$

Thus R_X with the above assignment of probabilities is a probability space.

The *expectation* $E(X)$ and *variance* $\text{Var}(X)$ of the above random variable X are defined by

$$E(X) = x_1 f(x_1) + x_2 f(x_2) + \cdots + \sum_{i=1}^{\infty} x_i f(x_i)$$

$$\text{Var}(X) = (x_1 - \mu)^2 f(x_1) + (x_2 - \mu)^2 f(x_2) + \cdots = \sum_{i=1}^{\infty} (x_i - \mu)^2 f(x_i)$$

when the relevant series converge absolutely. It can be shown that $\text{Var}(X)$ exists if and only if $\mu = E(X)$ and $E(X^2)$ both exist and that in this case the formula

$$\text{Var}(X) = E(X^2) - \mu^2$$

is valid just as in the finite case. When $\text{Var}(X)$ exists, the *standard deviation* σ_X is defined as in the finite case by

$$\sigma_X = \sqrt{\text{Var}(X)}$$

The notions of joint distribution, independent random variables, and functions of random variables carry over directly to the general case. It can be shown that if X and Y are defined on the same sample space S and if $\text{Var}(X)$ and $\text{Var}(Y)$ both exist, then the series

$$\text{Cov}(X, Y) = \sum_{i,j} (x_i - \mu_X)(y_j - \mu_Y) h(x_i, y_j)$$

converges absolutely and the relation

$$\text{Cov}(X, Y) = \sum_{i,j} x_i y_j h(x_i, y_j) - \mu_X \mu_Y = E(XY) - \mu_X \mu_Y$$

holds just as in the finite case.

Remark: To avoid technicalities we will establish many theorems in this chapter only for finite random variables.

5.10 CONTINUOUS RANDOM VARIABLES

Suppose that X is a random variable on a sample space S whose range space R_X is a continuum of numbers, such as an interval. Recall from the definition of a random variable that the set $\{a \leq X \leq b\}$ is an event in S and therefore the probability $P(a \leq X \leq b)$ is well defined. We assume there is a piecewise continuous function f: $\mathbf{R} \to \mathbf{R}$ such that $P(a \leq X \leq b)$ is equal to the area under the graph of f between $x = a$ and $x = b$, as shown in Fig. 5-10. In the language of calculus

$$P(a \leq X \leq b) = \int_a^b f(x)\, dx$$

In this case X is said to be a *continuous random variable*. The function f is called the *distribution* or the *continuous probability function* (or: *density function*) of X; it satisfies the conditions

$$\text{(i)}\quad f(x) \geq 0 \qquad \text{and} \qquad \text{(ii)}\quad \int_{-\infty}^{\infty} f(x)\, dx \equiv \int_{\mathbf{R}} f(x)\, dx = 1$$

That is, f is nonnegative and the total area under its graph is 1.

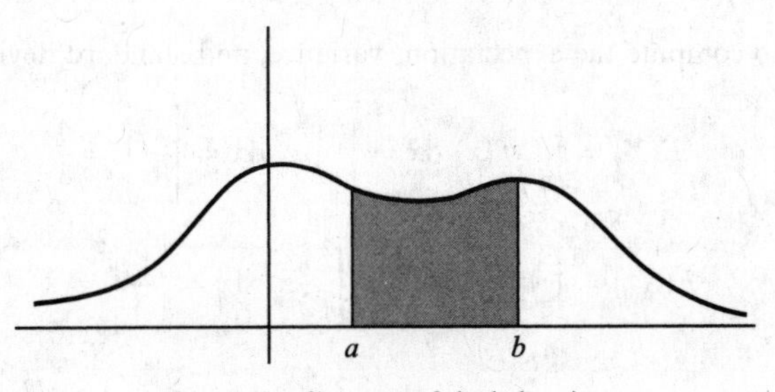

$P(a \leq X \leq b)$ = area of shaded region

Fig. 5-10

The *expectation* $E(X)$ for a continuous random variable X is defined by

$$E(X) = \int_{\mathbf{R}} xf(x)\, dx$$

when it exists. Functions of random variables are defined just as in the discrete case; and it can be shown that if $Y = \Phi(X)$. Then

$$E(Y) = \int_{\mathbf{R}} \Phi(x) f(x)\, dx$$

when the right side exists. The *variance* $\operatorname{Var}(X)$ is defined by

$$\operatorname{Var}(X) = E((X - \mu)^2) = \int_{\mathbf{R}} (x - \mu)^2 f(x)\, dx$$

when it exists. Just as in the discrete case, it can be shown that $\operatorname{Var}(X)$ exists if and only if $\mu = E(X)$ and $E(X^2)$ both exist, and then

$$\operatorname{Var}(X) = E(X^2) - \mu^2 = \int_{\mathbf{R}} x^2 f(x)\, dx - \mu^2$$

The *standard deviation* σ_X is defined by $\sigma_X = \sqrt{\operatorname{Var}(X)}$ when $\operatorname{Var}(X)$ exists.

We have already remarked that we will establish many results for finite random variables and take them for granted in the general discrete case and in the continuous case.

EXAMPLE 5.17 Let X be a random variable with the following distribution function f:

$$f(x) = \begin{cases} \frac{1}{2}x & \text{if } 0 \leq x \leq 2 \\ 0 & \text{elsewhere} \end{cases}$$

The graph of f appears in Fig. 5-11. Then

$$P(1 \leq X \leq 1.5) = \text{area of shaded region in diagram} = \frac{5}{16}$$

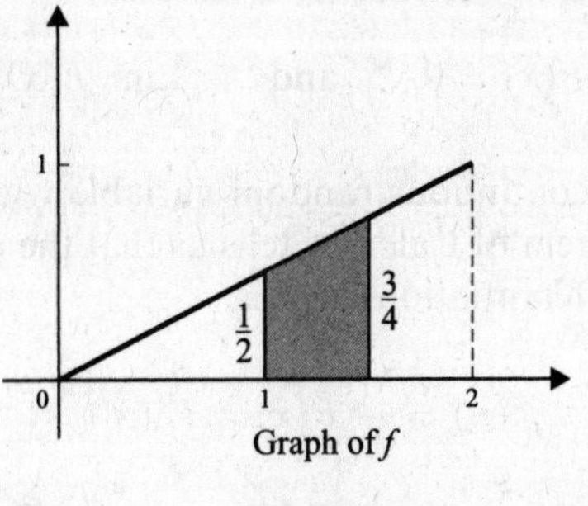

Graph of f

Fig. 5-11

Using calculus we are able to compute the expectation, variance, and standard deviation of X as follows:

$$E(X) = \int_{\mathbf{R}} xf(x)\,dx = \int_0^2 \frac{1}{2}x^2\,dx = \left[\frac{x^3}{6}\right]_0^2 = \frac{4}{3}$$

$$E(X^2) = \int_{\mathbf{R}} x^2 f(x)\,dx = \int_0^2 \frac{1}{2}x^3\,dx = \left[\frac{x^4}{8}\right]_0^2 = 2$$

$$\mathrm{Var}(X) = E(X^2) - \mu^2 = 2 - \frac{16}{9} = \frac{2}{9} \qquad \text{and} \qquad \sigma_X = \sqrt{\frac{2}{9}} = \frac{1}{3}\sqrt{2}$$

Independent Continuous Random Variables

A finite number of continuous random variables, say $X, Y, \ldots, Z$, are said to be *independent* if for any intervals $[a, a'], [b, b'], \ldots, [c, c']$,

$$P(a \le X \le a', b \le Y \le b', \ldots, c \le Z \le c') = P(a \le X \le a')P(b \le Y \le b') \cdots P(c \le Z \le c')$$

Observe that intervals play the same role in the continuous case as points did in the discrete case.

5.11 CUMULATIVE DISTRIBUTION FUNCTION

Let X be a random variable (discrete or continuous). The *cumulative distribution function* F of X is the function $F: \mathbf{R} \to \mathbf{R}$ defined by

$$F(a) = P(X \le a)$$

If X is a discrete random variable with distribution f, then F is the "step function" defined by

$$F(x) = \sum_{x_1 \le x} f(x_i)$$

If X is a continuous random variable with distribution f, then

$$F(x) = \int_{-\infty}^{x} f(t)\,dt$$

In either case, F is monotonic increasing, that is,

$$F(a) \le F(b) \qquad \text{whenever} \qquad a \le b$$

and the limit of F to the left is 0 and to the right is 1, that is,

$$\lim_{x \to -\infty} F(x) = 0 \qquad \text{and} \qquad \lim_{x \to -\infty} F(x) = 1$$

On the other hand, suppose X is a continuous random variable with cumulative distribution function $F(x)$. Then the Fundamental Theorem of Calculus tells us that the probability density function $f(x)$ of X can be obtained from $F(x)$ by differentiation, that is,

$$f(x) = \frac{d}{dx}F(x) = F'(X)$$

wherever the derivative exists.

EXAMPLE 5.18

(*a*) Let X be a discrete random variable with the following distribution function f:

x	-2	1	2	4
$f(x)$	$\frac{1}{4}$	$\frac{1}{8}$	$\frac{1}{2}$	$\frac{1}{8}$

The graph of the cumulative distribution function F of X appears in Fig. 5-12. Observe that F is a "step function" with a step at x_i with height $f(x_i)$.

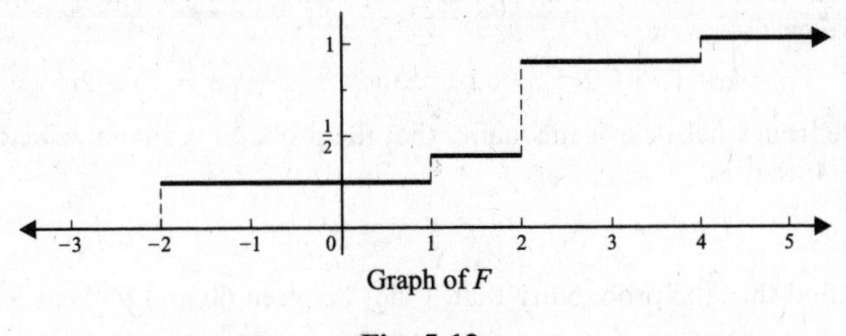

Graph of F

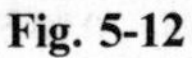

Fig. 5-12

(*b*) Let X be a continuous random variable with the following distribution function f:

$$f(x) = \begin{cases} \frac{1}{2}x & \text{if } 0 \le x \le 2 \\ 0 & \text{elsewhere} \end{cases}$$

The cumulative distribution function F of X follows:

$$F(x) = \begin{cases} 0 & \text{if } x < 0 \\ \frac{1}{4}x^2 & \text{if } 0 \le x \le 2 \\ 1 & \text{if } x > 2 \end{cases}$$

Here we use the fact that, for $0 \le x \le 2$,

$$F(x) = \int_0^x \frac{1}{2} t \, dt = \frac{1}{4} x^2$$

The graphs of f and F appear in Fig. 5-13(*a*) and (*b*), respectively.

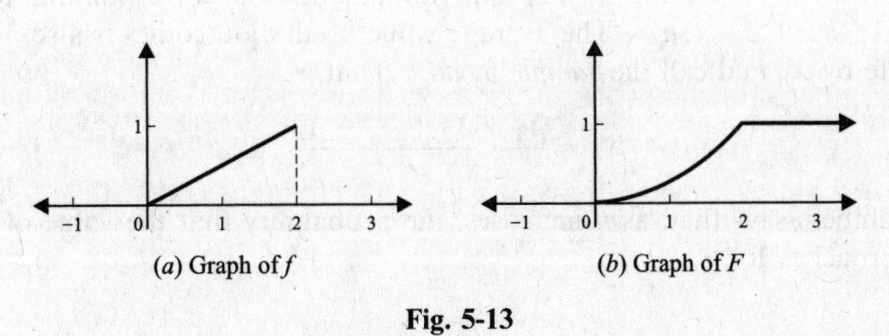

(*a*) Graph of f (*b*) Graph of F

Fig. 5-13

5.12 CHEBYSHEV'S INEQUALITY AND THE LAW OF LARGE NUMBERS

The standard deviation σ of a random variable X measures the weighted spread of the values of X about the mean μ. Therefore, for smaller σ, we would expect that X will be closer to μ. A more precise statement of this expectation is given by the following inequality, named after the Russian mathematician P. L. Chebyshev (1821–94).

Theorem 5.12 **(Chebyshev's inequality):** Let X be a random variable with mean μ and standard deviation σ. Then for any positive number k, the probability that a value of X lies in the interval $[\mu - k\sigma, \mu + k\sigma]$ is at least $1 - 1/k^2$. That is,

$$P(\mu - k\sigma \leq X \leq \mu + k\sigma) \geq 1 - \frac{1}{k^2}$$

A proof of this important theorem is given in Problem 5.34. We illustrate the use of the inequality in the next example.

EXAMPLE 5.19 Suppose X is a random variable with mean $\mu = 75$ and standard deviation $\sigma = 5$.

(*a*) What conclusion about X can be drawn from Chebyshev's inequality for $k = 2$ and $k = 3$?

Setting $k = 2$, we get

$$\mu - k\sigma = 75 - 2(5) = 65 \qquad \text{and} \qquad \mu + k\sigma = 75 + 2(5) = 85$$

Thus we can conclude from Chebyshev's inequality that the probability that a value of X lies between 65 and 85 is at least $1 - \frac{1}{2}^2 = 3/4$; that is,

$$P(65 \leq X \leq 85) \geq \frac{3}{4}$$

By letting $k = 3$, we find that the probability that X lies between 60 and 90 is at least $1 - \frac{1}{3}^2 = 8/9$.

(*b*) Estimate the probability that X lies between $75 - 20 = 55$ and $75 + 20 = 95$.

Set $k\sigma = 20$ and solve for k. Since $\sigma = 5$, we get $k \cdot 5 = 20$ and hence $k = 4$. Thus, by Chebyshev's inequality,

$$P(55 \leq X \leq 95) \geq 1 - \frac{1}{4^2} = \frac{15}{16} \approx 0.94$$

That is, the probability that X lies between 55 and 95 is at least 94 percent.

(*c*) Determine an interval $[a, b]$ about the mean for which the probability that X lies in the interval is at least 99 percent.

Set $1 - 1/k^2 = 0.99$ and solve for k. We get

$$1 - 0.99 = 1/k^2 \qquad \text{or} \qquad k^2 = 1/0.01 = 100 \qquad \text{or} \qquad k = 10$$

Thus the interval is $[75 - 10(5), 75 + 10(5)] = [25, 125]$.

Sample Mean and the Law of Large Numbers

The notion of n independent trials of a probability experiment was defined in Section 4.6. If X is a random variable with mean μ, then we can consider the numerical outcome of each particular trial to be a random variable with the same mean as X. The random variable corresponding to the ith outcome will be denoted by X_i $(i = 1, 2, \ldots, n)$. The average value of all n outcomes is also a random variable, which we will denote by $\bar{X}_n$ and call the *sample mean*. That is,

$$\bar{X}_n = \frac{X_1 + X_2 + \cdots + X_n}{n}$$

The law of large numbers says that, as n increases, the probability that the value of the sample mean $\bar{X}_n$ is close to μ approaches 1.

EXAMPLE 5.20 Suppose a die were tossed 5 times with outcomes

$$x_1 = 3, \qquad x_2 = 4, \qquad x_3 = 6, \qquad x_4 = 1, \qquad x_5 = 4$$

Then the corresponding value of the sample mean $\bar{X}_5$ is

$$\bar{x}_5 = \frac{3 + 4 + 6 + 1 + 4}{5} = 3.6$$

For a fair die, the mean $\mu = 3.5$. The law of large numbers tells us that, as n gets larger, there is a greater likelihood that $\bar{X}_n$ will get close to 3.5.

A more technical statement of the law of large numbers follows.

Theorem 5.13 (law of large numbers) For any positive number α, no matter how small, the probability that the sample mean $\bar{X}_n$ has a value in the interval $[\mu - \alpha, \mu + \alpha]$ approaches 1 as n approaches infinity. That is,

$$P(\mu - \alpha \leq \bar{X}_n \leq \mu + \alpha) \to 1 \quad \text{as} \quad n \to \infty$$

A proof of Theorem 5.13, based on Chebyshev's inequality, is sketched in Problem 5.35. A stronger version of the law of large numbers is given in more advanced treatments of probability theory.

The (strong) law of large numbers can also be used to show that if an event A occurs with probability p in a given model, then the average number of occurrences of A approaches p as the number of (independent) trials increases.

Solved Problems

RANDOM VARIABLES AND EXPECTED VALUE

5.1. Suppose a random variable X takes on the values -3, -1, 2, and 5 with respective probabilities

$$\frac{2k-3}{10}, \quad \frac{k+1}{10}, \quad \frac{k-1}{10}, \quad \frac{k-2}{10}$$

Determine the distribution of X.

Set the sum of the probabilities equal to 1, and solve for k, obtaining $k = 3$. Then put $k = 3$ into the above probabilities, yielding 0.3, 0.4, 0.2, 0.1. Thus the distribution of X follows:

x	-3	-1	2	5
$P(X = x)$	0.3	0.4	0.2	0.1

5.2. A fair coin is tossed four times. Let X denote the number of heads occurring. Find: (*a*) the distribution f of X, (*b*) $E(X)$, (*c*) the probability graph of X.

The sample space S is an equiprobable space consisting of $2^4 = 16$ sequences with H's and T's.

(*a*) Since X is the number of heads, and each sequence consists of four elements, X takes on the values of 0, 1, 2, 3, 4; that is, $R_X = \{0, 1, 2, 3, 4\}$.

(i) One point, TTTT, has no heads; hence $f(0) = \frac{1}{16}$.
(ii) Four points, HTTT, THTT, TTHT, TTTH, have one head; hence $f(1) = \frac{4}{16}$.
(iii) Six points, HHTT, HTHT, HTTH, THHT, THTH, TTHH, have two heads; hence $f(2) = \frac{6}{16}$.
(iv) Four points, HHHT, HHTH, HTHH, THHH, have one head; hence $f(1) = \frac{4}{16}$.
(v) One point, HHHH, has four heads; hence $f(4) = \frac{1}{16}$.

The distribution f of X follows:

x	0	1	2	3	4
$f(x)$	$\frac{1}{16}$	$\frac{4}{16}$	$\frac{6}{16}$	$\frac{4}{16}$	$\frac{1}{16}$

(*b*) The expected value $E(X)$ is obtained by multiplying each value of X by its probability and taking the sum. Hence

$$E(X) = 0\left(\frac{1}{16}\right) + 1\left(\frac{4}{16}\right) + 2\left(\frac{6}{16}\right) + 3\left(\frac{4}{16}\right) + 4\left(\frac{1}{16}\right) = 2$$

This agrees with our intuition that, when a fair coin is repeatedly tossed, about half of them should be heads.

(*c*) The probability bar chart of X appears in Fig. 5-14(*a*), and the probability histogram appears in Fig. 5-14(*b*). One may view the histogram as making the random variable continuous where $X = 1$ means X lies between 0.5 and 1.5.

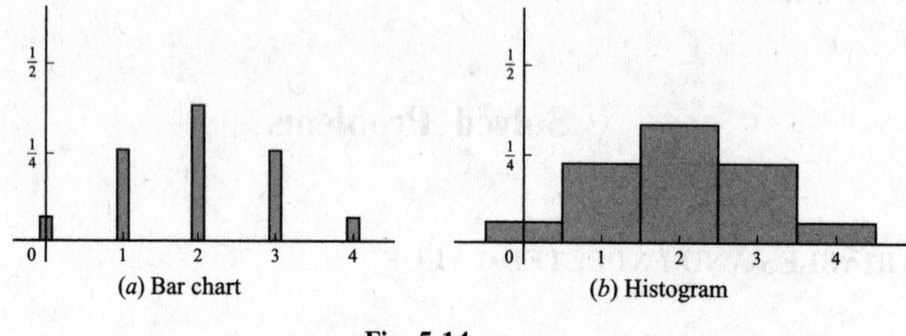

(*a*) Bar chart (*b*) Histogram

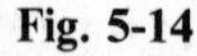
Fig. 5-14

5.3. A fair coin is tossed until a head or five tails occurs. Find the expected number E of tosses of the coin.

The sample space S consists of the six points

H, TH, TTH, TTTH, TTTTH, TTTTH, TTTTT

with respective probabilities (independent trials)

$$\frac{1}{2}, \quad \left(\frac{1}{2}\right)^2 = \frac{1}{4}, \quad \left(\frac{1}{2}\right)^3 = \frac{1}{8}, \quad \left(\frac{1}{2}\right)^4 = \frac{1}{16}, \quad \left(\frac{1}{2}\right)^5 = \frac{1}{32}, \quad \left(\frac{1}{2}\right)^5 = \frac{1}{32}$$

The random variable X of interest is the number of tosses in each outcome. Thus

$$X(\text{H}) = 1, \qquad X(\text{TTH}) = 3, \qquad X(\text{TTTTH}) = 5$$
$$X(\text{TH}) = 2, \qquad X(\text{TTTH}) = 4, \qquad X(\text{TTTTT}) = 5$$

These X values are assigned the following probabilities:

$$P(1) \equiv P(\text{H}) = \frac{1}{2}, \qquad P(2) \equiv P(\text{TH}) = \frac{1}{4}, \qquad P(3) \equiv P(\text{TTH}) = \frac{1}{8},$$
$$P(4) \equiv P(\text{TTTH}) = \frac{1}{16}, \qquad P(5) \equiv P(\{\text{TTTTH, TTTTT}\}) = \frac{1}{32} + \frac{1}{32} = \frac{1}{16}$$

Accordingly

$$E = E(X) = 1\left(\frac{1}{2}\right) + 2\left(\frac{1}{4}\right) + 3\left(\frac{1}{8}\right) + 4\left(\frac{1}{16}\right) + 5\left(\frac{1}{16}\right) \approx 1.9$$

5.4. A random sample with replacement of size $n = 2$ is chosen from the set {1, 2, 3}, yielding the 9-element equiprobable space

$$S = \{(1,1),\ (1,2),\ (1,3),\ (2,1),\ (2,2),\ (2,3),\ (3,1),\ (3,2),\ (3,3)\}$$

(*a*) Let X denote the sum of the two numbers. Find the distribution f of X, and find the expected value $E(X)$.

(*b*) Let Y denote the minimum of the two numbers. Find the distribution g of Y, and find the expected value $E(Y)$.

(*a*) The random variable X assumes the values 2, 3, 4, 5, 6; that is, $R_X = \{2, 3, 4, 5, 6\}$. We compute the distribution f of X:

(i) One point (1, 1) has sum 2; hence $f(2) = \frac{1}{9}$.
(ii) Two points, (1, 2), (2, 1), have sum 3; hence $f(3) = \frac{2}{9}$.
(iii) Three points, (1, 3), (2, 2), (1, 3), have sum 4; hence $f(4) = \frac{3}{9}$.
(iv) Two points, (2, 3), (3, 2), have sum 5; hence $f(5) = \frac{2}{9}$.
(v) One point (3, 3) has sum 6; hence $f(6) = \frac{1}{9}$.

Thus the following is the distribution f of X:

x	2	3	4	5	6
$f(x)$	$\frac{1}{9}$	$\frac{2}{9}$	$\frac{3}{9}$	$\frac{2}{9}$	$\frac{1}{9}$

The expected value $E(X)$ is obtained by multiplying each value of x by its probability and taking the sum. Hence

$$E(X) = 2\left(\frac{1}{9}\right) + 3\left(\frac{2}{9}\right) + 4\left(\frac{3}{9}\right) + 5\left(\frac{2}{9}\right) + 6\left(\frac{1}{9}\right) = 4$$

(*b*) The random variable Y only assumes the values 1, 2, 3; that is, $R_Y = \{1, 2, 3\}$. We compute the distribution g of X:

(i) Five points, (1, 1), (1, 2), (1, 3), (2, 1), (3, 1), have minimum 1; hence $g(1) = \frac{5}{9}$.
(ii) Three points, (2, 2), (2, 3), (3, 2), have minimum 2; hence $g(2) = \frac{3}{9}$.
(iii) One point (3, 3) has minimum 3; hence $g(3) = \frac{1}{9}$.

Thus the following is the distribution g of Y:

y	1	2	3
$g(y)$	$\frac{5}{9}$	$\frac{3}{9}$	$\frac{1}{9}$

The expected value $E(Y)$ is obtained by multiplying each value of y by its probability and taking the sum. Hence

$$E(Y) = 1\left(\frac{5}{9}\right) + 2\left(\frac{3}{9}\right) + 3\left(\frac{1}{9}\right) = \frac{12}{9} \approx 1.33$$

5.5. A player tosses two fair coins. The player wins \$2 if 2 heads occur, and \$1 if 1 head occurs. On the other hand, the player loses \$3 if no heads occur. Find the expected value E of the game. Is the game fair? (The game is fair, favorable, or unfavorable to the player according as $E = 0$, $E > 0$ or $E < 0$.)

The sample space is $S = \{\text{HH}, \text{HT}, \text{TH}, \text{TT}\}$ and each sample point has probability $\frac{1}{4}$. Letting X denote the player's gain, we have

$$X(\text{HH}) = \$2, \qquad X(\text{HT}) = X(\text{TH}) = \$1, \qquad X(\text{TT}) = -\$3$$

Thus the distribution of X and its expectation E are as follows:

x	2	1	-3
$P(X=x)$	$\frac{1}{4}$	$\frac{2}{4}$	$\frac{1}{4}$

$$E = E(X) = 2\left(\frac{1}{4}\right) + 1\left(\frac{2}{4}\right) - 3\left(\frac{1}{4}\right) = \frac{1}{4} = \$0.25$$

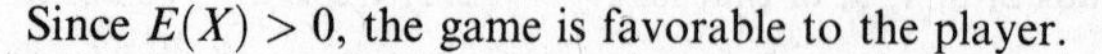

Since $E(X) > 0$, the game is favorable to the player.

5.6. A box contains eight light bulbs of which three are defective. A bulb is selected from the box and tested. If it is defective, another bulb is selected and tested, until a nondefective bulb is chosen. Find the expected number E of bulbs chosen.

Writing D for defective and N for nondefective, the sample space S has the four elements

$$\text{N}, \qquad \text{DN}, \qquad \text{DDN}, \qquad \text{DDDN}$$

with respective probabilities

$$\frac{5}{8}, \qquad \frac{3}{8}\cdot\frac{5}{7}=\frac{15}{56}, \qquad \frac{3}{8}\cdot\frac{2}{7}\cdot\frac{5}{6}=\frac{5}{56}, \qquad \frac{3}{8}\cdot\frac{2}{7}\cdot\frac{1}{6}\cdot\frac{5}{5}=\frac{1}{56}$$

The number X of bulbs chosen has the values

$$X(\text{N}) = 1, \qquad X(\text{DN}) = 2, \qquad X(\text{DDN}) = 3, \qquad X(\text{DDDN}) = 4$$

with the above respective probabilities. Hence:

$$E = E(X) = 1\left(\frac{5}{8}\right) + 2\left(\frac{15}{56}\right) + 3\left(\frac{5}{56}\right) + 4\left(\frac{1}{56}\right) = \frac{3}{2}$$

5.7. Concentric circles of radius 1 and 3 inches are drawn on a circular target of radius 5, as pictured in Fig. 5-15. A man receives 10, 5, or 3 points according to whether he hits the target inside the smaller circle, inside the middle annular region or inside the outer annular region, respectively. Suppose the man hits the target with probability $\frac{1}{2}$ and then is just as likely to hit one point of the target as the other. Find the expected number E of points he scores each time he fires.

The probability of scoring 10, 5, 3 or 0 points follows:

$$f(10) = \frac{1}{2}\cdot\frac{\text{area of 10 points}}{\text{area of target}} = \frac{1}{2}\cdot\frac{\pi(1)^2}{\pi(5)^2} = \frac{1}{50}$$

$$f(5) = \frac{1}{2}\cdot\frac{\text{area of 5 points}}{\text{area of target}} = \frac{1}{2}\cdot\frac{\pi(3)^2 - \pi(1)^2}{\pi(5)^2} = \frac{8}{50}$$

$$f(3) = \frac{1}{2}\cdot\frac{\text{area of 3 points}}{\text{area of target}} = \frac{1}{2}\cdot\frac{\pi(5)^2 - \pi(3)^2}{\pi(5)^2} = \frac{16}{50}$$

$$f(0) = \frac{1}{2}$$

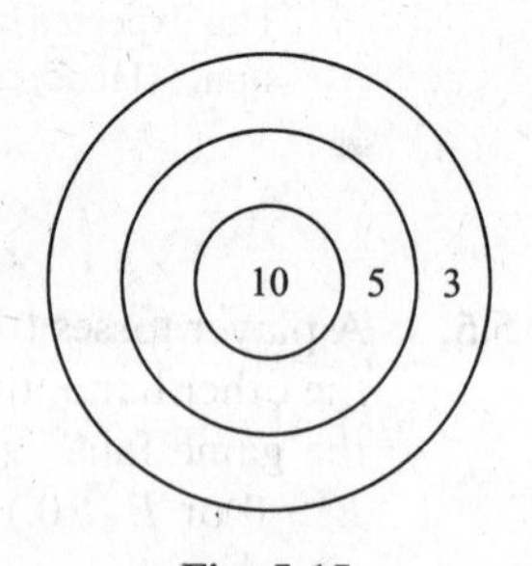

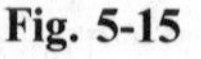
Fig. 5-15

Thus $E = 10\left(\frac{1}{50}\right) + 8\left(\frac{8}{50}\right) + 3\left(\frac{16}{50}\right) + 0\left(\frac{1}{2}\right) = \frac{98}{50} = 1.96.$

MEAN, VARIANCE, AND STANDARD DEVIATION

5.8. Find the mean $\mu = E(X)$, variance $\sigma^2 = \text{Var}(X)$, and standard deviation $\sigma = \sigma_X$ of each distribution:

(*a*)

x	3	8	12
$f(x)$	$\frac{1}{3}$	$\frac{1}{2}$	$\frac{1}{6}$

(*b*)

x	1	3	4	5
$f(x)$	0.4	0.1	0.2	0.3

Use the formulas,

$$\mu = E(X) = x_1 f(x_1) + x_2 f(x_2) + \cdots + x_m f(x_m) = \sum x_i f(x_i)$$
$$E(X^2) = x_1^2 f(x_1) + x_2^2 f(x_2) + \cdots + x_m^2 f(x_m) = \sum x_i^2 f(x_i)$$

Alternatively, form a data table with columns labeled by x, $f(x)$, $xf(x)$, x^2, $x^2 f(x)$. The sum of the third column is $\mu = E(X)$ and the sum of the fifth column is $E(X^2)$. Then use the formulas

$$\sigma^2 = \text{Var}(X) = E(X^2) - \mu^2 \qquad \text{and} \qquad \sigma = \sigma_X = \sqrt{\text{Var}(X)}$$

to obtain $\sigma^2 = \text{Var}(X)$ and σ.

(*a*) Form the data table in Fig. 5-16(*a*) to get $\mu = E(X) = 7$ and $E(X^2) = 59$. Alternatively, use the formulas directly to obtain

$$\mu = \sum x_i f(x_i) = 3\left(\frac{1}{3}\right) + 8\left(\frac{1}{2}\right) + 12\left(\frac{1}{6}\right) = 7$$
$$E(X^2) = \sum x_i^2 f(x_i) = 3^2\left(\frac{1}{3}\right) + 8^2\left(\frac{1}{2}\right) + 12^2\left(\frac{1}{6}\right) = 59$$

Then:

$$\sigma^2 = \text{Var}(X) = E(X^2) - \mu^2 = 59 - 7^2 = 10$$
$$\sigma = \sqrt{\text{Var}(X)} = \sqrt{10} = 3.2$$

(*b*) Form the data table in Fig. 5-16(*b*) to get $\mu = E(X) = 3$ and $E(X^2) = 12$. Alternatively, use the formulas directly to obtain

$$\mu = \sum x_i f(x_i) = 1(0.4) + 3(0.1) + 4(0.2) + 5(0.3) = 3$$
$$E(X^2) = \sum x_i^2 f(x_i) = 1(0.4) + 9(0.1) + 16(0.2) + 25(0.3) = 12$$

Then:

$$\sigma^2 = \text{Var}(X) = E(X^2) - \mu^2 = 12 - 9 = 3$$
$$\sigma = \sqrt{\text{Var}(X)} = \sqrt{3} = 1.7$$

x	$f(x)$	$xf(x)$	x^2	$x^2f(x)$
3	1/3	1	9	3
8	1/2	4	64	32
12	1/6	2	144	24
	Sums	7		59

(*a*)

x	$f(x)$	$xf(x)$	x^2	$x^2f(x)$
1	0.4	0.4	1	0.4
3	0.1	0.3	9	0.9
4	0.2	0.8	16	3.2
5	0.3	1.5	25	7.5
	Sums	3		12

(*b*)

Fig. 5-16

5.9. Find the mean $\mu = E(X)$, variance $\sigma^2 = \text{Var}(X)$, and standard deviation $\sigma = \sigma_X$ of each distribution:

(a)

x_i	-6	-4	3	12
p_i	$\frac{1}{4}$	$\frac{1}{8}$	$\frac{1}{2}$	$\frac{1}{8}$

(b)

x_i	2	3	5	8
p_i	0.3	0.1	0.4	0.2

Here the distribution is presented using x_i and p_i instead of x and $f(x)$. The following are the analogous formulas:

$$\mu = E(X) = x_1 p_1 + x_2 p_2 + \cdots + x_m p_m = \sum x_i p_i$$

$$E(X^2) = x_1^2 p_1 + x_2^2 p_2 + \cdots + x_m^2 p_m = \sum x_i^2 p_i$$

Then, as before,

$$\sigma^2 = \text{Var}(X) = E(X^2) - \mu^2 \quad \text{and} \quad \sigma = \sigma_X = \sqrt{\text{Var}(X)}$$

(a)
$$\mu = E(X) = \sum x_i p_i = -6\left(\frac{1}{4}\right) - 4\left(\frac{1}{8}\right) + 3\left(\frac{1}{2}\right) + 12\left(\frac{1}{8}\right) = 1$$

$$E(X^2) = \sum x_i^2 p_i = 36\left(\frac{1}{4}\right) + 16\left(\frac{1}{8}\right) + 9\left(\frac{1}{2}\right) + 144\left(\frac{1}{8}\right) = 33.5$$

Then:

$$\sigma^2 = \text{Var}(X) = E(X^2) - \mu^2 = 33.5 - 1^2 = 32.5$$

$$\sigma = \sqrt{\text{Var}(X)} = \sqrt{32.5} = 5.7$$

(b)
$$\mu = E(X) = \sum x_i p_i = 2(0.3) + 3(0.1) + 5(0.4) + 8(0.2) = 4.5$$

$$E(X^2) = \sum x_i^2 p_i = 2^2(0.3) + 3^2(0.1) + 5^2(0.4) + 8^2(0.2) = 24.9$$

Then:

$$\sigma^2 = \text{Var}(X) = E(X^2) - \mu^2 = 14.9 - (4.5)^2 = 5.35$$

$$\sigma = \sqrt{\text{Var}(X)} = \sqrt{5.35} = 2.31$$

5.10. A fair die is tossed. Let X denote twice the number appearing, and let Y be 1 or 3 according as an odd or even number appears. Find the distribution and expectation of: (a) X, (b) Y.

The sample space is $S = \{1, 2, 3, 4, 5, 6\}$ and each sample point has probability $\frac{1}{6}$.

(a) The images of the sample points are:

$$X(1) = 2, \quad X(2) = 4, \quad X(3) = 6, \quad X(4) = 8, \quad X(5) = 10, \quad X(6) = 12$$

As these are distinct, the distribution of X is

x_i	2	4	6	8	10	12
$P(x_i)$	$\frac{1}{6}$	$\frac{1}{6}$	$\frac{1}{6}$	$\frac{1}{6}$	$\frac{1}{6}$	$\frac{1}{6}$

Thus

$$E(X) = \sum x_i P(x_i) = \frac{2}{6} + \frac{4}{6} + \frac{6}{6} + \frac{8}{6} + \frac{10}{6} + \frac{12}{6} = 7$$

(b) The images of the sample points are:

$$Y(1) = 1, \quad Y(2) = 3, \quad Y(3) = 1, \quad Y(4) = 3, \quad Y(5) = 1, \quad Y(6) = 3$$

The two Y-values, 1 and 3, are each assumed at three sample points. Hence we have the distribution

y_i	1	3
$P(y_i)$	3/6	3/6

Thus

$$E(Y) = \sum y_i P(y_i) = \frac{3}{6} + \frac{9}{6} = 2$$

5.11. Let X and Y be the random variable in Problem 5.10. Recall that $Z = X + Y$ and $W = XY$ are random variables defined by

$$Z(s) = X(s) + Y(s) \quad \text{and} \quad W(s) = X(s)Y(s)$$

(*a*) Find the distribution and expectation of $Z = X + Y$. Verify that

$$E(X + Y) = E(X) + E(Y)$$

(*b*) Find the distribution and expectation of $W = XY$.

The sample space is still $S = \{1, 2, 3, 4, 5, 6\}$ and each sample point still has probability $\frac{1}{6}$.

(*a*) Use $Z(s) = (X + Y)(s) = X(s) + Y(s)$ and the values of X and Y from Problem 5.10 to obtain:

$$Z(1) = X(1) + Y(1) = 2 + 1 = 3 \qquad Z(4) = X(4) + Y(4) = 8 + 3 = 11$$
$$Z(2) = X(2) + Y(2) = 4 + 3 = 7 \qquad Z(5) = X(5) + Y(5) = 10 + 1 = 11$$
$$Z(3) = X(3) + Y(3) = 6 + 1 = 7 \qquad Z(6) = X(6) + Y(6) = 12 + 3 = 15$$

The image set is {3, 7, 11, 5}. The values 3 and 15 are each assumed at only one sample point and hence have probability $\frac{1}{6}$; the values 7 and 11 are each assumed at two sample points and hence have probability $\frac{2}{6}$. Thus the distribution of $Z = X + Y$ is:

z_i	3	7	11	15
$P(z_i)$	$\frac{1}{6}$	$\frac{2}{6}$	$\frac{2}{6}$	$\frac{1}{6}$

Thus

$$E(X + Y) = E(Z) = \sum z_i P(z_i) = \frac{3}{6} + \frac{14}{6} + \frac{22}{6} + \frac{15}{6} = 9$$

Moreover, $E(X + Y) = 9 = 7 + 2 = E(X) + E(Y)$.

(*b*) Use $W(s) = XY(s) = X(s)Y(s)$ to obtain:

$$W(1) = X(1)Y(1) = 2(1) = 2 \qquad W(4) = X(4)Y(4) = 8(3) = 24$$
$$W(2) = X(2)Y(2) = 4(3) = 12 \qquad W(5) = X(5)Y(5) = 10(1) = 10$$
$$W(3) = X(3)Y(3) = 6(1) = 6 \qquad W(6) = X(6)Y(6) = 12(3) = 36$$

Each value of $W = XY$ is assumed at just one sample point; hence the distribution of W is:

w_i	2	6	10	12	24	36
$P(w_i)$	$\frac{1}{6}$	$\frac{1}{6}$	$\frac{1}{6}$	$\frac{1}{6}$	$\frac{1}{6}$	$\frac{1}{6}$

Thus

$$E(XY) = E(W) = \sum w_i P(w_i) = \frac{2}{6} + \frac{6}{6} + \frac{10}{6} + \frac{12}{6} + \frac{24}{6} + \frac{36}{6} = 15$$

(Note: $E(XY) = 15 \neq (7)(2) = E(X)E(Y)$.)

5.12. Let X be a random variable with distribution:

x	1	2	3
$P(X = x)$	0.3	0.5	0.2

Find the mean, variance, and standard deviation of X. Then find the distribution, mean, variance, and standard deviation of the random variable $Y = \Phi(X)$, where: (*a*) $\Phi(x) = x^3$, (*b*) $\Phi(x) = 2^x$, (*c*) $\Phi(x) = x^2 + 3x + 4$.

The formulas for μ_X and $E(X^2)$ yield:

$$\mu_X = E(X) = \sum x_i P(x_i) = 1(0.3) + 2(0.5) + 3(0.2) = 1.9$$

$$E(X^2) = \sum x_i^2 P(x_i) = 1^2(0.3) + 2^2(0.5) + 3^2(0.2) = 4.1$$

Then:

$$\sigma^2 = \text{Var}(X) = E(X^2) - \mu^2 = 4.1 - (1.9)^2 = 0.49$$

$$\sigma = \sqrt{\text{Var}(X)} = \sqrt{0.49} = 0.7$$

Generally speaking, the distribution of $Y = \Phi(X)$ is as follows, where $P(y) = P(x)$:

y	$\Phi(1)$	$\Phi(2)$	$\Phi(3)$
$P(y)$	0.3	0.5	0.2

(*a*) Using $1^3 = 1$, $2^3 = 8$, $3^3 = 27$, the distribution of $Y = X^3$ is as follows:

y	1	8	27
$P(y)$	0.3	0.5	0.2

Therefore:

$$\mu_Y = E(Y) = \sum \Phi(x_i)P(x_i) = \sum y_i P(y_i) = 1(0.3) + 8(0.5) + 27(0.2) = 9.7$$

$$E(Y^2) = \sum y_i^2 P(y_i) = 1^2(0.3) + 8^2(0.5) + 27^2(0.2) = 178.1$$

Then

$$\sigma^2 = \text{Var}(Y) = E(Y^2) - \mu^2 = 178.1 - (9.7)^2 = 84.0$$

$$\sigma = \sqrt{\text{Var}(Y)} = \sqrt{84.0} = 9.17$$

(*b*) Using $2^1 = 2$, $2^2 = 4$, $2^3 = 8$, the distribution of $Y = 2^X$ is as follows:

y	2	4	8
$P(y)$	0.3	0.5	0.2

Therefore:

$$\mu_Y = E(Y) = \sum y_i P(y_i) = 2(0.3) + 4(0.5) + 8(0.2) = 4.2$$
$$E(Y^2) = \sum y_i^2 P(y_i) = 2^2(0.3) + 4^2(0.5) + 8^2(0.2) = 41.2$$

Then:

$$\sigma^2 = \text{Var}(Y) = E(Y^2) - \mu^2 = 41.2 - (4.2)^2 = 23.6$$
$$\sigma = \sqrt{\text{Var}(Y)} = \sqrt{23.6} = 4.86$$

(*c*) Substitute $x = 1, 2, 3$ in $\Phi(x) = x^2 + 3x + 4$ to obtain $\Phi(1) = 8$, $\Phi(2) = 14$, $\Phi(3) = 22$. Then the distribution of $Y = X^2 + 3X + 4$ is as follows:

y	8	14	22
$P(y)$	0.3	0.5	0.2

Therefore:

$$\mu_Y = E(Y) = \sum y_i P(y_i) = 8(0.3) + 14(0.5) + 22(0.2) = 13.9$$
$$E(Y^2) = \sum y_i^2 P(y_i) = 8^2(0.3) + 14^2(0.5) + 22^2(0.2) = 214$$

Then:

$$\sigma^2 = \text{Var}(Y) = E(Y^2) - \mu^2 = 214 - (13.9)^2 = 20.8$$
$$\sigma = \sqrt{\text{Var}(Y)} = \sqrt{20.8} = 4.56$$

JOINT DISTRIBUTIONS

5.13. Let X and Y be random variables with the joint distribution in Fig. 5-17.

(*a*) Find the distributions of X and Y.
(*b*) Find Cov(X, Y), i.e. the covariance of X and Y.
(*c*) Find $\rho(X, Y)$, i.e. the correlation of X and Y.
(*d*) Are X and Y independent random variables?

X \ Y	−3	2	4	Sum
1	0.1	0.2	0.2	0.5
3	0.3	0.1	0.1	0.5
Sum	0.4	0.3	0.3	

Fig. 5-17

(*a*) The marginal distribution on the right is the distribution of X, and the marginal distribution on the bottom is the distribution of Y. Namely,

x_i	1	3
$f(x_i)$	0.5	0.5

Distribution of X

y_j	−3	2	4
$g(y_j)$	0.4	0.3	0.3

Distribution of Y

(*b*) First compute μ_X and μ_Y as follows:

$$\mu_X = \sum x_i f(x_i) = (1)(0.5) + (3)(0.5) = 2$$
$$\mu_Y = \sum y_j g(y_j) = (-3)(0.4) + (2)(0.3) + (4)(0.3) = 0.6$$

Next compute $E(XY)$ as follows:

$$E(XY) = \sum x_i y_j h(x_i, y_j)$$
$$= (1)(-3)(0.1) + (1)(2)(0.2) + (1)(4)(0.2) + (3)(-3)(0.3) + (3)(2)(0.1) + (3)(4)(0.1) = 0$$

Then $\quad \mathrm{Cov}(X, Y) = E(XY) - \mu_X\mu_Y = 0 - (2)(0.6) = -1.2$

(*c*) First compute σ_X and σ_Y as follows:

$$E(X^2) = \sum x_i^2 f(x_i) = (1)(0.5) + (9)(0.5) = 5$$
$$\sigma_X^2 = \mathrm{Var}(X) = E(X^2) - \mu_X^2 = 5 - (2)^2 = 1$$
$$\sigma_x = \sqrt{1} = 1$$

and

$$E(Y^2) = \sum y_j^2 g(y_j) = (9)(0.4) + (4)(0.3) + (16)(0.3) = 9.6$$
$$\sigma_Y^2 = \mathrm{Var}(Y) = E(Y^2) - \mu_Y^2 = 9.6 - (0.6)^2 = 9.24$$
$$\sigma_Y = \sqrt{9.24} = 3.0$$

Then

$$\rho(X, Y) = \frac{\mathrm{Cov}(X, Y)}{\sigma_X \sigma_Y} = \frac{-1.2}{(1)(3.0)} = -0.4$$

(*d*) X and Y are not independent, since

$$P(X = 1, Y = -3) \neq P(X = 1)P(Y = -3)$$

i.e. the entry $h(1, -3) = 0.1$ is not equal to $f(1)g(-3) = (0.5)(0.4) = 0.2$, the product of its marginal entries.

5.14. Let X and Y be independent random variables with the following distributions:

x_i	1	2
$f(x_i)$	0.6	0.4

Distribution of X

y_j	5	10	15
$g(y_j)$	0.2	0.5	0.3

Distribution of Y

Find the joint distribution h of X and Y.

Since X and Y are independent, the joint distribution h can be obtained from the marginal distributions f and g. Specifically, first construct the joint distribution table with only the marginal distributions, as shown in Fig. 5-18(*a*). Then multiply the marginal entries to obtain the other entries; that is, set $h(x_i, y_j) = f(x_i)g(y_j)$. This yields the joint distribution of X and Y appearing in Fig. 5-18(*b*).

X \ Y	5	10	15	Sum
1				0.6
2				0.4
Sum	0.2	0.5	0.3	

(*a*)

X \ Y	5	10	15	Sum
1	0.12	0.30	0.18	0.6
2	0.08	0.20	0.12	0.4
Sum	0.2	0.5	0.3	

(*b*)

Fig. 5-18

5.15. A fair coin is tossed three times. Let X equal 0 or 1 according as a head or a tail occurs on the first toss, and let Y equal the total number of heads that occur.

(*a*) Find the distributions of X and Y.

(*b*) Find the joint distribution h of X and Y.

(*c*) Determine whether X and Y are independent.

(*d*) Find $\text{Cov}(X, Y)$.

(*a*) The sample space S consists of the following eight points, each with probability $\frac{1}{8}$:

$$S = \{\text{HHH, HHT, HTH, HTT, THH, THT, TTH, TTT}\}$$

We have

$$X(\text{HHH}) = 0, \quad X(\text{HHT}) = 0, \quad X(\text{HTH}) = 0, \quad X(\text{HTT}) = 0;$$
$$X(\text{THH}) = 1, \quad X(\text{THT}) = 1, \quad X(\text{TTH}) = 1, \quad X(\text{TTT}) = 1$$

and

$$Y(\text{HHH}) = 3; \quad Y(\text{HHT}) = 2, \quad Y(\text{HTH}) = 2, \quad Y(\text{THH}) = 2;$$
$$Y(\text{HTT}) = 1, \quad Y(\text{THT}) = 1, \quad Y(\text{TTH}) = 1; \quad Y(\text{TTT}) = 0$$

Thus the distributions of X and Y are as follows:

x_i	0	1
$f(x_i)$	$\frac{1}{2}$	$\frac{1}{2}$

Distribution of X

y_j	0	1	2	3
$g(y_j)$	$\frac{1}{8}$	$\frac{3}{8}$	$\frac{3}{8}$	$\frac{1}{8}$

Distribution of Y

(*b*) The joint distribution h of X and Y appears in Fig. 5-19. We obtain, for example, the entry $h(0, 2)$ using

$$h(0, 2) \equiv P(X = 0, Y = 2) = P(\{\text{HTH, HHT}\}) = \frac{2}{8}$$

The other entries are obtained similarly.

X \ Y	0	1	2	3	Sum
0	0	$\frac{1}{8}$	$\frac{2}{8}$	$\frac{1}{8}$	$\frac{1}{2}$
1	$\frac{1}{8}$	$\frac{2}{8}$	$\frac{1}{8}$	0	$\frac{1}{2}$
Sum	$\frac{1}{8}$	$\frac{3}{8}$	$\frac{3}{8}$	$\frac{1}{8}$	

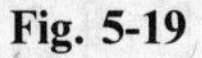
Fig. 5-19

(*c*) From the joint distribution, $P(0, 0) = 0$; but

$$P(X = 0) = \frac{1}{2} \quad \text{and} \quad P(Y = 0) = \frac{1}{8}$$

Since $0 \neq \left(\frac{1}{2}\right)\left(\frac{1}{8}\right)$, X and Y are not independent.

(*d*) We have:

$$\mu_X = \sum x_i f(x_i) = 0\left(\frac{1}{2}\right) + 1\left(\frac{1}{2}\right) = \frac{1}{2}$$

$$\mu_Y = \sum y_j g(y_j) = 0\left(\frac{1}{8}\right) + 1\left(\frac{3}{8}\right) + 2\left(\frac{3}{8}\right) + 3\left(\frac{1}{8}\right) = \frac{3}{2}$$

$$E(XY) = \sum x_i y_j h(x_i, y_j) = 1(1)\left(\frac{2}{8}\right) + 1(2)\left(\frac{1}{8}\right) + \text{terms with a factor } 0 = \frac{1}{2}$$

$$\text{Cov}(X, Y) = E(XY) - \mu_X \mu_Y = \frac{1}{2} - \frac{1}{2}\left(\frac{3}{2}\right) = -\frac{1}{4}$$

5.16. Let X be the random variable with the following distribution, and let $Y = X^2$:

x	-2	-1	1	2
$f(x)$	$\frac{1}{4}$	$\frac{1}{4}$	$\frac{1}{4}$	$\frac{1}{4}$

(*a*) Find the distribution of Y.

(*b*) Find the joint distribution of X and Y.

(*c*) Find $\text{Cov}(X, Y)$ and $\rho(X, Y)$.

(*d*) Determine whether X and Y are independent.

(*a*) Since $Y = X^2$, the random variable Y can only take on the values 4 and 1. Letting g denote the distribution of Y, we have:

$$g(4) = P(Y = 4) = P(X = 2 \text{ or } X = -2) = P(X = 2) + P(X = -2) = \frac{1}{4} + \frac{1}{4} = \frac{1}{2}$$

Similarly, $g(1) = \frac{1}{2}$. Thus the distribution g of Y is as follows:

y	1	4
$g(y)$	$\frac{1}{2}$	$\frac{1}{2}$

(*b*) The joint distribution h of X and Y appears in Fig. 5-20. Note that if $X = -2$, then $Y = 4$; hence $h(-2, 1) = 0$ and $h(-2, 4) = f(-2) = \frac{1}{4}$. The other entries are obtained in a similar way.

X \ Y	1	4	Sum
-2	0	$\frac{1}{4}$	$\frac{1}{4}$
-1	$\frac{1}{4}$	0	$\frac{1}{4}$
1	$\frac{1}{4}$	0	$\frac{1}{4}$
2	0	$\frac{1}{4}$	$\frac{1}{4}$
Sum	$\frac{1}{2}$	$\frac{1}{2}$	

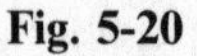

Fig. 5-20

(*c*) We have:

$$\mu_X = E(X) = \sum x_i f(x_i) = -2\left(\frac{1}{4}\right) - 1\left(\frac{1}{4}\right) + 1\left(\frac{1}{4}\right) + 2\left(\frac{1}{4}\right) = 0$$

$$\mu_Y = E(Y) = \sum y_j g(y_j) = 1\left(\frac{1}{2}\right) + 4\left(\frac{1}{2}\right) = \frac{5}{2}$$

$$E(XY) = \sum x_i y_j h(x_i, y_j) = -8\left(\frac{1}{4}\right) - 1\left(\frac{1}{4}\right) + 1\left(\frac{1}{4}\right) + 8\left(\frac{1}{4}\right) = 0$$

$$\text{Cov}(X, Y) = E(XY) - \mu_X \mu_Y = 0 - 0 \cdot \frac{5}{2} = 0 \quad \text{and so} \quad \rho(X, Y) = 0$$

(*d*) From the joint distribution, $P(-2, 1) = 0$; but $P(X = -2) = \frac{1}{4}$ and $P(Y = 1) = \frac{1}{2}$. Since $0 \neq \left(\frac{1}{4}\right)\left(\frac{1}{2}\right)$, X and Y are not independent.

Remark: Although Y is a function of X and X and Y are not independent, this example shows that it is still possible for the covariance and correlation to be 0, as always in the case when X and Y are independent.

CHEBYSHEV'S INEQUALITY

5.17. Suppose a random variable X has mean $\mu = 25$ and standard deviation $\sigma = 2$. Use Chebyshev's inequality to estimate: (*a*) $P(X \leq 35)$, (*b*) $P(X \geq 20)$.

(*a*) Recall Chebyshev's inequality states:

$$P(\mu - k\sigma \leq X \leq \mu + k\sigma) \geq 1 - \frac{1}{k^2}$$

Substitute $\mu = 25$, $\sigma = 2$ in $\mu + k\sigma$ and solve the equation $25 + 2k = 35$ for k, getting $k = 5$. Then

$$1 - \frac{1}{k^2} = 1 - \frac{1}{25} = \frac{24}{25} = 0.96$$

Since $\mu - k\sigma = 25 - 10 = 15$, Chebyshev's inequality gives

$$P(15 \leq X \leq 35) \geq 0.96$$

The event corresponding to $X \leq 35$ contains as a subset the event corresponding to $15 \leq X \leq 35$. Therefore,

$$P(X \leq 35) \geq P(15 \leq X \leq 35) \geq 0.96$$

Thus the probability that X is less than or equal to 35 is at least 96 percent.

(*b*) Substitute $\mu = 25$, $\sigma = 2$ in $\mu - k\sigma$, and solve the equation $25 - 2k = 20$ for k, getting $k = 2.5$. Then

$$1 - \frac{1}{k^2} = 1 - \frac{1}{6.25} = 0.84$$

Since $\mu + 2\sigma = 25 + 5 = 30$, Chebyshev's inequality gives

$$P(20 \leq X \leq 30) \geq 0.84$$

The event corresponding to $X \geq 20$ contains as a subset the event corresponding to $20 \leq X \leq 30$. Therefore,

$$P(X \geq 20) \geq P(20 \leq X \leq 30) \geq 0.84$$

which says that the probability that X is greater than or equal to 20 is at least 84 percent.

Remark: This problem illustrates that Chebyshev's inequality can be used to estimate $P(X \leq b)$ when $b \geq \mu$, and to estimate $P(X \geq a)$ when $a \leq \mu$.

5.18. Let X be a random variable with mean $\mu = 40$ and standard deviation $\sigma = 5$. Use Chebyshev's inequality to find a value b for which $P(40 - b \leq X \leq 40 + b) \geq 0.95$.

First solve $1 - \frac{1}{k^2} = 0.95$ for k as follows:

$$0.05 = \frac{1}{k^2} \quad \text{or} \quad k^2 = \frac{1}{0.05} = 20 \quad \text{or} \quad k = \sqrt{20} = 2\sqrt{5}$$

Then, by Chebyshev's inequality, $b = k\sigma = 10\sqrt{5} \simeq 23.4$. Hence, $P(16.6 \leq X \leq 63.6) \geq 0.95$.

5.19. Let X be a random variable with mean $\mu = 80$ and unknown standard deviation σ. Use Chebyshev's inequality to find a value of σ for which $P(75 \leq X \leq 85) \geq 0.9$.

First solve $1 - \dfrac{1}{k^2} = 0.9$ for k as follows:

$$0.1 = \frac{1}{k^2} \quad \text{or} \quad k^2 = \frac{1}{0.1} = 10 \quad \text{or} \quad k = \sqrt{10}$$

Now, since 75 is 5 units to the left of $\mu = 80$ and 85 is 5 units to the right of μ, we can solve either $\mu - k\sigma = 75$ or $\mu + k\sigma = 85$ for σ. From the latter equation, we get

$$80 + \sqrt{10}\sigma = 85 \quad \text{or} \quad \sigma = \frac{5}{\sqrt{10}} \approx 1.58$$

MISCELLANEOUS PROBLEMS

5.20. Let X be a continuous random variable with distribution:

$$f(x) = \begin{cases} \frac{1}{6}x + k & \text{if } 0 \leq x \leq 3 \\ 0 & \text{elsewhere} \end{cases}$$

(*a*) Evaluate k. (*b*) Find $P(1 \leq X \leq 2)$.

(*a*) The graph of f is drawn in Fig. 5-21(*a*). Since f is a continuous probability function, the shaded region A must have area 1. Note that A forms a trapezoid with parallel bases of lengths k and $k + \frac{1}{2}$, and altitude 3. Setting the area of A equal to 1 yields:

$$\frac{1}{2}\left(k + k + \frac{1}{2}\right)(3) = 1 \quad \text{or} \quad k = \frac{1}{12}$$

Thus $f(x) = x/6 + 1/12$ for $0 \leq x \leq 3$.

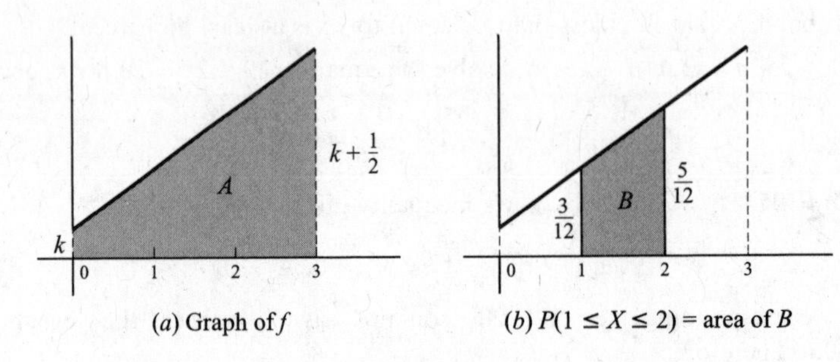

(*a*) Graph of f (*b*) $P(1 \leq X \leq 2)$ = area of B

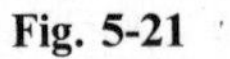

Fig. 5-21

(*b*) $P(1 \leq X \leq 2)$ is equal to the area of B, which is under the graph of f between $x = 1$ and $x = 2$ as shown in Fig. 5-21(*b*). Using $f(x) = x/6 + 1/12$ for $x = 1$ and $x = 2$, we get

$$f(1) = \frac{1}{6} + \frac{1}{12} = \frac{3}{12} \quad \text{and} \quad f(2) = \frac{1}{8} + \frac{1}{12} = \frac{5}{12}$$

Hence

$$P(1 \leq X \leq 2) = \text{area of } B = \frac{1}{2}\left(\frac{3}{12} + \frac{5}{12}\right)(1) = \frac{1}{3}$$

5.21. Let X be a continuous random variable whose distribution f is constant on an interval, say $I = \{a \leq x \leq b\}$, and 0 elsewhere; namely,

$$f(x) = \begin{cases} k & \text{if } a \leq x \leq b \\ 0 & \text{elsewhere} \end{cases}$$

(Such a random variable is said to be *uniformly distributed* on I.) (*a*) Determine k. (*b*) Find the mean μ of X. (*c*) Determine the cumulative distribution function F of X.

(*a*) The graph of f appears in Fig. 5-22(*a*). The region shaded A must have area 1; hence

$$k(b-a) = 1 \qquad \text{or} \qquad k = \frac{1}{b-a}$$

(*b*) If we view probability as weight or mass, and the mean as the center of gravity, then it is intuitively clear that

$$\mu = \frac{a+b}{2}$$

the point midway between a and b. We verify this mathematically using calculus:

$$\mu = E(X) = \int_{\mathbf{R}} xf(x)\,dx = \int_a^b \frac{x}{b-a}\,dx = \left[\frac{x^2}{2(b-a)}\right]_a^b$$

$$= \frac{b^2}{2(b-a)} - \frac{a^2}{2(b-a)} = \frac{a+b}{2}$$

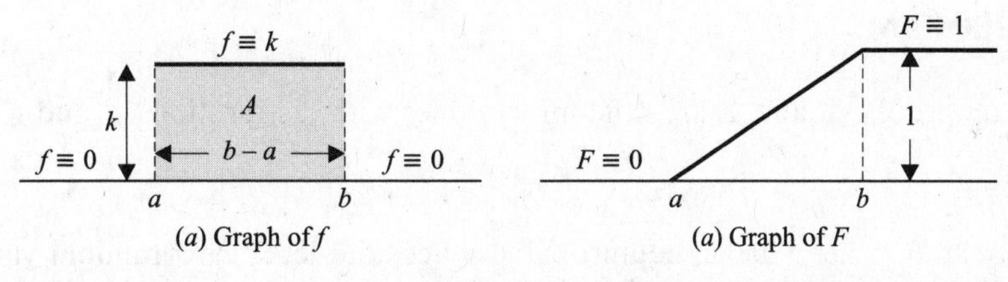

(*a*) Graph of f (*a*) Graph of F

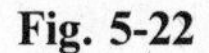

Fig. 5-22

(*c*) Recall that the cumulative distribution function F is defined by $F(k) = P(X \le k)$. Hence $F(k)$ gives the area under the graph of f to the left of $x = k$. Since X is uniformly distributed on the interval $I = \{a \le x \le b\}$, it is intuitive that the graph of F should be as shown in Fig. 5-22(*b*) i.e. $F \equiv 0$ before the point a, $F \equiv 1$ after the point b, and F is linear between a and b. We verify this mathematically using calculus:

(i) For $x < a$,

$$F(x) = \int_{-\infty}^{x} f(t)\,dt = \int_{-\infty}^{x} 0\,dt = 0$$

(ii) For $a \le x \le b$,

$$F(x) = \int_{-\infty}^{x} f(t)\,dt = \int_a^x \frac{1}{b-a}\,dt = \left[\frac{t}{b-a}\right]_a^x = \frac{x-a}{b-a}$$

(iii) For $x > b$,

$$F(x) = P(X \le x) \ge P(X \le b) = F(b) = 1 \qquad \text{and also} \qquad 1 \ge P(X \le x) = F(x)$$

Thus $F(x) \ge 1$ and $F(x) \le 1$, and hence $F(x) = 1$.

5.22. Let X be a random variable with distribution f. The rth *moment* M_r of X is defined by

$$M_r = E(X^r) = \sum x_i^r f(x_i)$$

Find the first five moments of X if X has the following distribution:

x	-2	1	3
$f(x)$	$\frac{1}{2}$	$\frac{1}{4}$	$\frac{1}{4}$

(Note that M_1 is the mean of X, and M_2 is used in computing the variance and standard deviation of X.)

Use the formula for M_r to obtain:

$$M_1 = \sum x_i f(x_i) = -2\left(\frac{1}{2}\right) + 1\left(\frac{1}{4}\right) + 3\left(\frac{1}{4}\right) = 0$$

$$M_2 = \sum x_i^2 f(x_i) = 4\left(\frac{1}{2}\right) + 1\left(\frac{1}{4}\right) + 9\left(\frac{1}{4}\right) = 4.5$$

$$M_3 = \sum x_i^3 f(x_i) = -8\left(\frac{1}{2}\right) + 1\left(\frac{1}{4}\right) + 27\left(\frac{1}{4}\right) = 3$$

$$M_4 = \sum x_i^4 f(x_i) = 16\left(\frac{1}{2}\right) + 1\left(\frac{1}{4}\right) + 81\left(\frac{1}{4}\right) = 28.5$$

$$M_5 = \sum x_i^5 f(x_i) = -32\left(\frac{1}{2}\right) + 1\left(\frac{1}{4}\right) + 243\left(\frac{1}{4}\right) = 45$$

PROOFS OF THEOREMS

Remark: In all proofs, X and Y are random variables with distributions f and g respectively and joint distribution h.

5.23. Prove Theorem 5.1: Let S be an equiprobable space, and let X be a random variable on S with range space $R_X = \{x_i, x_2, \ldots, x_t\}$. Then

$$p_i = f(x_i) = \frac{\text{number of points in } S \text{ whose image is } x_i}{\text{number of points in } S}$$

Let S have n points and let $s_1, s_2, \ldots, s_r$ be the points in S with image x_i. We wish to show that $p_i = f(x_i) = r/n$. By definition,

$$\begin{aligned} p_i = f(x_i) &= \text{sum of the probabilities of the points in } S \text{ whose image is } x_i \\ &= P(s_1) + P(s_2) + \cdots + P(s_r) \end{aligned}$$

Since S is an equiprobable space, each of the n points in S has probability $1/n$. Hence

$$p_i = f(x_i) = \overbrace{\frac{1}{n} + \frac{1}{n} + \cdots + \frac{1}{n}}^{r \text{ times}} = \frac{r}{n}$$

5.24. Show that $f(x_i) = \sum_j h(x_i, y_j)$ and $g(y_j) = \sum_i h(x_i, y_j)$, i.e. that the marginal distributions are the (individual) distributions of X and Y.

Let $A_i = \{X = x_i\}$ and $B_j = \{Y = y_j\}$; that is, let $A_i = X^{-1}(x_i)$ and $B_j = Y^{-1}(y_j)$. Thus the B_j are disjoint and $S = \bigcup_j B_j$. Hence

$$A_i = A_i \cap S = A_i \cap (\textstyle\bigcup_j b_j) = \bigcup_j (A_i \cap B_j)$$

where the $A_i \cap B_j$ are also disjoint. Accordingly,

$$\begin{aligned} f(x_i) = P(X = x_i) = P(A_i) &= \sum_j P(A_i \cap B_j) \\ &= \sum_j P(X = x_i, Y = y_j) = \sum_j h(x_i, y_j) \end{aligned}$$

The proof for g is similar.

5.25. Prove Theorem 5.10: Let X and Y be random variables on S with $Y = \Phi(X)$. Then $E(Y) = \sum_i \Phi(x_i) f(x_i)$, where f is the distribution of X.

(Proof is given for the case X is discrete and finite.)

Suppose X takes on the values $x_i, \ldots, x_n$ and that $\Phi(x_i)$ takes on the values $y_1, \ldots, y_m$ as i runs from 1 to n. Then clearly the possible values of $Y = \Phi(X)$ are $y_1, \ldots, y_m$ and the distribution g of Y is given by

$$g(y_j) = \sum_{\{i:\Phi(x_i)=y_j\}} f(x_i)$$

Therefore

$$E(Y) = \sum_{j=1}^{m} y_j g(y_j) = \sum_{j=1}^{m} y_j \sum_{\{i:0(x_i)=y_j\}} f(x_i)$$

$$= \sum_{i=1}^{n} f(x_i) \sum_{\{j:\Phi(x_i)=y_j\}} y_j = \sum_{i=1}^{n} f(x_i)\Phi(x_i)$$

which proves the theorem.

5.26. Prove Theorem 5.2: Let X be a random variable and let k be a real number. Then: (i) $E(kX) = kE(X)$, (ii) $E(X + k) = E(X) + k$.

(Proof is given for the general discrete case and the assumption that $E(X)$ exists.)

(i) Now $kX = \Phi(X)$ where $\Phi(x) = kx$. Therefore, by Theorem 5.10 (Problem 5.25)

$$E(kX) = \sum_i kx_i f(x_i) = k \sum_i x_i f(x_i) = kE(X)$$

(ii) Here $X + k = \Phi(X)$ where $\Phi(x) = x + k$. Therefore, using $\sum_i f(x_i) = 1$,

$$E(X + k) = \sum_i (x_i + k) f(x_i) = \sum_i x_i f(x_i) + \sum_i kf(x_i) = E(X) + k$$

5.27. Prove Theorem 5.3: Let X and Y be random variables on S. Then $E(X + Y) = E(X) + E(Y)$.

(Proof is given for the general discrete case and the assumption that $E(X)$ and $E(Y)$ both exist.)

Now $X + Y = \Phi(X, Y)$ where $\Phi(x, y) = x + y$. Therefore, by Theorem 5.10 (Problem 5.25),

$$E(X + Y) = \sum_i \sum_j (x_i + y_j) h(x_i, y_j) = \sum_i \sum_j x_i h(x_i, y_j) + \sum_i \sum_j y_j h(x_i, y_j)$$

Applying Problem 5.24 that $f(x_i) = \sum_j h(x_i, y_j)$ and $g(y_j) = \sum_i h(x_i, y_j)$, we get

$$E(X + Y) = \sum_i x_i f(x_i) + \sum_j y_j g(y_j) = E(X) + E(Y)$$

5.28. Prove Corollary 5.4: Let $X_1, X_2, \ldots, X_n$ be random variables on S. Then

$$E(X_1 + X_2 + \cdots + X_n) = E(X_1) + E(X_2) + \cdots E(X_n)$$

(Proof is given for the general discrete case and the assumption that $E(X_1), \ldots, E(X_n)$ all exist.)

The proof is by induction on n. The case $n = 1$ is trivial and the case $n = 2$ is Theorem 5.3 (Problem 5.27). For $n > 2$, we apply the case $n = 2$ to get

$$E(X_1 + \cdots + X_{n-1} + X_n) = E(X_1 + \cdots + X_{n-1}) + E(X_n)$$

By the inductive hypothesis, this becomes $E(X_1) + \cdots + E(X_{n-1}) + E(X_n)$.

5.29. Prove Theorem 5.6: $\text{Var}(aX+b) = a^2\,\text{Var}(X)$.

We prove separately that (i) $\text{Var}(X+k) = \text{Var}(X)$ and (ii) $\text{Var}(kX) = k^2\,\text{Var}(x)$, from which the theorem follows. By Theorem 5.2, $\mu_{X+k} = \mu_X + k$ and $\mu_{kX} = k\mu_X$. Also $\sum x_i f(x_i) = \mu_X$ and $\sum f(x_i) = 1$. Hence

$$\begin{aligned}\text{Var}(X+k) &= \sum(x_i+k)^2 f(x_i) - \mu_X^2 + k\\ &= \sum x_i^2 f(x_i) + 2k\sum x_i f(x_i) + k^2\sum f(x_i) - (\mu_X+k)^2\\ &= \sum x_i^2 f(x_i) + 2k\mu_X + k^2 - (\mu_X^2 + 2k\mu_X + k^2)\\ &= \sum x_i f(x_i) - \mu_X^2 = \text{Var}(X)\end{aligned}$$

and

$$\begin{aligned}\text{Var}(kX) &= \sum(kx_i)^2 f(x_i) - \mu_{kX}^2 = k^2\sum x_i^2 f(x_i) - (k\mu_X)^2\\ &= k^2\sum x_i^2 f(x_i) - k^2\mu_X^2 = k^2\left(\sum x_i^2 f(x_i) - \mu_X^2\right) = k^2\,\text{Var}(X)\end{aligned}$$

5.30. Show that:

$$\text{Cov}(X,Y) = \sum_{i,j}(x_i-\mu_X)(y_j-\mu_Y)h(x_i,y_j) = \sum_{i,j} x_i y_j h(x_i,y_j) - \mu_X\mu_Y$$

(Proof is given for the case that X and Y are discrete and finite.)
We have:

$$\sum_{i,j} y_j h(x_i,y_j) = \sum_j y_j g(y_j) = \mu_Y, \qquad \sum_{i,j} x_i h(x_i,y_j) = \sum_i x_i f(x_i) = \mu_X, \qquad \sum_{i,j} h(x_i,y_j) = 1$$

Therefore:

$$\begin{aligned}&\sum_{i,j}(x_i-\mu_X)(y_j-\mu_Y)h(x_i,y_j)\\ &\quad= \sum_{i,j}(x_i y_j - \mu_X y_j - \mu_Y x_i + \mu_X\mu_Y)h(x_i,y_j)\\ &\quad= \sum_{i,j} x_i y_j h(x_i,y_j) - \mu_X\sum_{i,j} y_j h(x_i,y_j) - \mu_Y\sum_{i,j} x_i h(x_i,y_j) + \mu_X\mu_Y\sum_{i,j} h(x_i,y_j)\\ &\quad= \sum_{i,j} x_i y_j h(x_i,y_j) - \mu_X\mu_Y - \mu_X\mu_Y + \mu_X\mu_Y\\ &\quad= \sum_{i,j} x_i y_j h(x_i,y_j) - \mu_X\mu_Y\end{aligned}$$

5.31. Prove Theorem 5.7: The standardized random variable Z has mean $\mu_Z = 0$ and standard deviation $\sigma_Z = 1$.

By definition $Z = \dfrac{X-\mu}{\sigma}$ where X has mean μ and standard deviation $\sigma > 0$. Using $E(X) = \mu$ and Theorem 5.2, we get

$$\mu_Z = E\left(\frac{X-\mu}{\sigma}\right) = E\left(\frac{X}{\sigma} - \frac{\mu}{\sigma}\right) = \frac{1}{\sigma}E(X) - \frac{\mu}{\sigma} = \frac{\mu}{\sigma} - \frac{\mu}{\sigma} = 0$$

Also, using Theorem 5.6, we get

$$\text{Var}(Z) = \text{Var}\left(\frac{X-\mu}{\sigma}\right) = \text{Var}\left(\frac{X}{\sigma} - \frac{\mu}{\sigma}\right) = \frac{1}{\sigma^2}\text{Var}(X) = \frac{\sigma^2}{\sigma^2} = 1$$

Therefore, $\sigma_Z = \sqrt{\text{Var}(Z)} = \sqrt{1} = 1$.

5.32. Prove Theorem 5.8: Let X and Y be independent random variables on S. Then: (i) $E(XY) = E(X)E(Y)$, (ii) $\text{Var}(X+Y) = \text{Var}(X) + \text{Var}(Y)$, (iii) $\text{Cov}(X, Y) = 0$.

(Proof is given for the case when X and Y are discrete and finite.)
Since X and Y are independent, $h(x_i, y_j) = f(x_i)g(y_j)$. Thus

$$E(XY) = \sum_{i,j} x_i y_j h(x_i, y_j) = \sum_{i,j} x_i y_j f(x_i)g(y_j)$$
$$= \sum_i x_i f(x_i) \sum_j y_j g(y_j) = E(X)E(Y)$$

and
$$\text{Cov}(X, Y) = E(XY) - \mu_X \mu_Y = E(X)E(Y) - \mu_X \mu_Y = 0$$

In order to prove (ii) we also need

$$\mu_{X+Y} = \mu_X + \mu_Y, \qquad \sum_{i,j} x_i^2 h(x_i, y_j) = \sum_i x_i^2 f(x_i), \qquad \sum_{i,j} y_j^2 h(x_i, y_j) = \sum_j y_j^2 g(y_j)$$

Hence

$$\text{Var}(X+Y) = \sum_{i,j} (x_i + y_j)^2 h(x_i, y_j) - \mu_{X+Y}^2$$
$$= \sum_{i,j} x_i^2 h(x_i, y_j) + 2\sum_{i,j} x_i y_j h(x_i, y_j) + \sum_{i,j} y_j^2 h(x_i, y_j) - (\mu_X + \mu_Y)^2$$
$$= \sum_i x_i^2 f(x_i) + 2\sum_i x_i f(x_i) \sum_j y_j g(y_j) + \sum_j y_j^2 g(y_j) - \mu_X^2 - 2\mu_X\mu_Y - \mu_Y^2$$
$$= \sum_i x_i^2 f(x_i) - \mu_X^2 + \sum_j y_j^2 g(y_j) - \mu_Y^2 = \text{Var}(X) + \text{Var}(Y)$$

5.33. Prove Theorem 5.9: Let $X_1, X_2, \ldots, X_n$ be independent random variables on S. Then

$$\text{Var}(X_1 + X_2 + \cdots + X_n) = \text{Var}(X_1) + \text{Var}(X_2) + \cdots + \text{Var}(X_n)$$

(Proof is given for the case when $X_i, \ldots, X_n$ are all discrete and finite.)
We take for granted the analogs of Problem 5.32 and Theorem 5.11 for n random variables. Then:

$$\text{Var}(X_1 + \cdots + X_n) = E((X_1 + \cdots + X_n - \mu_{X_1+\cdots+X_n})^2)$$
$$= \sum (x_1 + \cdots + x_n - \mu_{X_1+\cdots+X_n})^2 h(x_1, \ldots, x_n)$$
$$= \sum (x_1 + \cdots + x_n - \mu_{X_1} - \cdots - \mu_{X_n})^2 h(x_1, \ldots, x_n)$$
$$= \sum \left\{ \sum_i \sum_j x_i x_j + \sum_i \sum_j \mu_{X_i}\mu_{X_j} - 2\sum_i \sum_j \mu_{X_i} x_j \right\} h(x_1, \ldots, x_n)$$

where h is the joint distribution of $X_1, \ldots, X_n$, and $\mu_{X_1+\cdots+X_n} = \mu_{X_1} + \cdots + \mu_{X_n}$. Since the X_i are pairwise independent, $\sum x_i x_j h(x_1, \ldots, x_n) = \mu_{X_i}\mu_{X_j}$ for $i \neq j$. Hence

$$\text{Var}(X_1 + \cdots + X_n) = \sum_{i \neq j} \mu_{X_i}\mu_{X_j} + \sum_{i=1}^n E(X_i^2) + \sum_i \sum_j \mu_{X_i}\mu_{X_j} - 2\sum_i \sum_j \mu_{X_i}\mu_{X_j}$$
$$= \sum_{i=1}^n E(X_i^2) - \sum_{i=1}^n (\mu_{X_i})^2 = \sum_{i=1}^n \text{Var}(X_i)$$

as required.

5.34. Prove Theorem 5.12 (Chebyshev's inequality): For any $k > 0$,

$$P(\mu - k\sigma \le X \le \mu + k\sigma) \ge 1 - \frac{1}{k^2}$$

By definition

$$\sigma^2 = \text{Var}(X) = \sum (x_i - \mu)^2 p_i$$

Delete all terms from the summation for which x_i is in the interval $[\mu - k\sigma, \mu + k\sigma]$; that is, delete all terms for which $|x_i - \mu| \leq k\sigma$. Denote the summation of the remaining terms by $\sum^*(x_i - \mu)^2 p_i$. Then

$$\begin{aligned}\sigma^2 \geq \sum^*(x_i - \mu)^2 p_i \geq \sum^* k^2\sigma^2 p_i = k^2\sigma^2 \sum^* p_i = k^2\sigma^2 P(|X - \mu| > k\sigma) \\ = k^2\sigma^2[1 - P(|X - \mu| \leq k\sigma)] = k^2\sigma^2[1 - P(\mu - k\sigma \leq X \leq \mu + k\sigma)]\end{aligned}$$

If $\sigma > 0$, then dividing by $k^2\sigma^2$ gives

$$\frac{1}{k^2} \geq 1 - P(\mu - k\sigma \leq X \leq \mu + k\sigma) \qquad \text{or} \qquad P(\mu - k\sigma \leq X \leq \mu + k\sigma) \geq 1 - \frac{1}{k^2}$$

which proves Chebyshev's inequality for $\sigma > 0$. If $\sigma = 0$, then $x_i = \mu$ for all $p_i > 0$, and

$$P(\mu - k \cdot 0 \leq X \leq \mu + k \cdot 0) = P(X = \mu) = 1 > 1 - \frac{1}{k^2}$$

which completes the proof.

5.35. Let $X_1, X_2, \ldots, X_n$ be n independent and identically distributed random variables, each with mean μ and variance σ^2, and let $\bar{X}_n$ be the sample mean, that is,

$$\bar{X}_n = \frac{X_1 + X_2 + \cdots + X_n}{n}$$

(*a*) Prove the mean of $\bar{X}_n$ is μ and the variance is σ^2/n.

(*b*) Prove Theorem 5.13 (weak law of large numbers): For any $\alpha > 0$,

$$P(\mu - \alpha \leq \bar{X}_n \leq \mu + \alpha) \to 1 \qquad \text{as} \qquad n \to \infty$$

(*a*) Using Theorems 5.2 and 5.3, we get

$$\begin{aligned}\mu_{\bar{X}_n} = E(\bar{X}_n) = E\left(\frac{X_1 + X_2 + \cdots + X_n)}{n}\right) = \frac{1}{n} E(X_1 + X_2 + \cdots + X_n) \\ = \frac{1}{n}[E(X_1) + E(X_2) + \cdots + E(X_n)] = \frac{n\mu}{n} = \mu\end{aligned}$$

Now using Theorems 5.3 and 5.9, we get

$$\begin{aligned}\text{Var}(\bar{X}_n) = \text{Var}\left(\frac{X_1 + X_2 + \cdots + X_n}{n}\right) = \frac{1}{n^2}\text{Var}(X_1 + X_2 + \cdots + X_n) \\ = \frac{1}{n^2}[\text{Var}(X_1) + \text{Var}(X_2) + \cdots + \text{Var}(X_n)] = \frac{n\sigma^2}{n^2} = \frac{\sigma^2}{n}\end{aligned}$$

(*b*) The proof is based on an application of Chebyshev's inequality to the random variable $\bar{X}_n$. First note that by making the substitution $k\sigma = \alpha$, Chebyshev's inequality can be written as

$$P(\mu - \alpha \leq X \leq \mu + \alpha) \geq 1 - \frac{\sigma^2}{\alpha^2}$$

Applying Chebyshev's inequality in the form above, we get

$$P(\mu - \alpha \leq \bar{X}_n \leq \mu + \alpha) \geq 1 - \frac{\sigma^2}{n\alpha^2}$$

from which the desired result follows.

Supplementary Problems

RANDOM VARIABLES AND EXPECTED VALUE

5.36. Suppose a random variable X takes on the values $-3, 2, 4, 7$ with respective probabilities

$$\frac{k+1}{10}, \quad \frac{2k-2}{10}, \quad \frac{3k-5}{10}, \quad \frac{k+2}{10}$$

Find the distribution and expected value of X.

5.37. A pair of dice is thrown. Let X denote the minimum of the two numbers which occur. Find the distribution and expectation of X.

5.38. A fair coin is tossed four times. Let Y denote the longest string of heads. Find the distribution and expectation of Y. Also draw a probability bar chart and histogram of the distribution. (Compare with the random variables X in Problem 5.2.)

5.39. A coin, weighted so that $P(H) = \frac{3}{4}$ and $P(T) = \frac{1}{4}$, is tossed three times. Let X denote the number of heads that appear. (*a*) Find the distribution of X. (*b*) Find $E(X)$.

5.40. A coin, weighted so that $P(\mathrm{H}) = \frac{1}{3}$ and $P(\mathrm{T}) = \frac{2}{3}$, is tossed until a head or five tails occur. Find the expected number E of tosses of the coin.

5.41. The probability of team A winning any game is $\frac{1}{2}$. Suppose A plays B in a tournament (and there are no ties). The first team to win two games in a row or three games wins the tournament. Find the expected number E of games in the tournament.

5.42. A box contains 10 transistors of which two are defective. A transistor is selected from the box and tested until a nondefective one is chosen. Find the expected number E of transistors to be chosen.

5.43. Solve the preceding Problem 5.42 in the case that three of the 10 items are defective.

5.44. Five cards are numbered 1 to 5. Two cards are drawn at random (without replacement). Let X denote the sum of the numbers drawn. (*a*) Find the distribution of X. (*b*) Find $E(X)$.

5.45. A lottery with 500 tickets gives one prize of \$100, three prizes of \$50 each, and five prizes of \$25 each. (*a*) Find the expected winnings of a ticket. (*b*) If a ticket costs \$1, what is the expected value of the game?

5.46. A player tosses three fair coins. He wins \$5 if 3 heads occur, \$3 if two heads occur, and \$1 if only one 1 head occurs. On the other hand, he loses \$15 if three tails occur. Find the value of the game to the player.

5.47. A player tosses two fair coins. The player wins \$3 if 2 heads occur, and \$1 if 1 heads occurs. For the game to be fair how much should the player lose if no heads occur?

5.48. A coin is weighted so that $P(\mathrm{H}) = p$ and hence $P(\mathrm{T}) = q = 1 - p$. The coin is tossed until a head appears. Let E denote the expected number of tosses. Prove $E = 1/p$. (This is an example of an infinite discrete random variable, and some knowledge of series is required.)

MEAN, VARIANCE, AND STANDARD DEVIATION

5.49. Find the mean μ, variance σ^2, and standard deviation σ of each distribution:

(*a*)

x	2	3	8
$f(x)$	$\frac{1}{4}$	$\frac{1}{2}$	$\frac{1}{4}$

(*b*)

x	-2	-1	7
$f(x)$	$\frac{1}{3}$	$\frac{1}{2}$	$\frac{1}{6}$

5.50. Find the mean μ, variance σ^2, and standard deviation σ of each distribution:

(*a*)

x	-1	0	1	2	3
$f(x)$	0.3	0.1	0.1	0.3	0.2

(*b*)

x	1	2	3	6	7
$f(x)$	0.2	0.1	0.3	0.1	0.3

5.51. Let X be a random variable with the following distribution:

x	1	3	4	5
$f(x)$	0.4	0.1	0.2	0.3

(*a*) Find the mean, variance, and standard deviation of X.
(*b*) Find the distribution, mean, variance, and standard deviation of $Y = X^2 + 2$.

5.52. Find the mean μ, variance σ^2, and standard deviation σ of following two-point distribution where $p + q = 1$:

x	a	b
$f(x)$	p	q

5.53. Let X be a random variable with the following distribution:

x	-1	1	2
$f(x)$	0.2	0.5	0.3

(*a*) Find the mean, variance, and standard deviation of X.
(*b*) Find the distribution, mean, variance, and standard deviation of the random variable $Y = \Phi(X)$, where: (i) $\phi(x) = x^4$, (ii) $\phi(x) = 3^x$, (iii) $\phi(x) = 2^{x+1}$.

5.54. Two cards are selected from a box which contains five cards numbered 1, 1, 2, 2, and 3. Let X denote the sum and Y the maximum of the two numbers drawn. Find the distribution, mean, variance, and standard deviation of the random variables: (*a*) X, (*b*) Y, (*c*) $Z = X + Y$, (*d*) $W = XY$.

JOINT DISTRIBUTIONS, INDEPENDENT RANDOM VARIABLES

5.55. Consider the joint distribution of X and Y in Fig. 5-23(*a*). Find: (*a*) $E(X)$ and $E(Y)$, (*b*) $\mathrm{Cov}(X, Y)$, (*c*) σ_X, σ_Y, and $\rho(X, Y)$.

5.56. Consider the joint distribution of X and Y in Fig. 5-23(*b*). Find: (*a*) $E(X)$ and $E(Y)$, (*b*) $\mathrm{Cov}(X, Y)$, (*c*) σ_X, σ_Y, and $\rho(X, Y)$.

X \ Y	−4	2	7	Sum
1	$\frac{1}{8}$	$\frac{1}{4}$	$\frac{1}{8}$	$\frac{1}{2}$
5	$\frac{1}{4}$	$\frac{1}{8}$	$\frac{1}{8}$	$\frac{1}{2}$
Sum	$\frac{3}{8}$	$\frac{3}{8}$	$\frac{1}{4}$	

(*a*)

X \ Y	−2	−1	4	5	Sum
1	0.1	0.2	0	0.3	0.6
2	0.2	0.1	0.1	0	0.4
Sum	0.3	0.3	0.1	0.3	

(*b*)

Fig. 5-23

5.57. Suppose X and Y are independent random variables with the following respective distributions:

x	1	2
$f(x)$	0.7	0.3

y	−2	5	8
$g(y)$	0.3	0.5	0.2

Find the joint distribution of X and Y, and verify that $\text{Cov}(X, Y) = 0$.

5.58. Consider the joint distribution of X and Y in Fig. 5-24(*a*).
(*a*) Find $E(X)$ and $E(Y)$. (*b*) Determine whether X and Y are independent. (*c*) Find $\text{Cov}(X, Y)$.

5.59. Consider the joint distribution of X and Y in Fig. 5-24(*b*). (*a*) Find $E(X)$ and $E(Y)$. (*b*) Determine whether X and Y are independent. (*c*) Find the distribution, mean, and standard deviation of the random variable $Z = X + Y$.

X \ Y	2	3	4	Sum
1	0.06	0.15	0.09	0.30
2	0.14	0.35	0.21	0.70
Sum	0.20	0.50	0.30	

(*a*)

X \ Y	−2	−1	0	1	2	3	Sum
0	0.05	0.05	0.10	0	0.05	0.05	0.30
1	0.10	0.05	0.05	0.10	0	0.05	0.35
2	0.03	0.12	0.07	0.06	0.03	0.04	0.35
Sum	0.18	0.22	0.22	0.16	0.08	0.14	

(*b*)

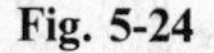

Fig. 5-24

5.60. A fair coin is tossed four times. Let X denote the number of heads occurring, and let Y denote the longest string of heads occurring. (See Problems 5.2 and 5.38.)

(*a*) Determine the joint distribution of X and Y.

(*b*) Find $\text{Cov}(X, Y)$ and $\rho(X, Y)$.

5.61. Two cards are selected at random from a box which contains five cards numbered 1, 1, 2, 2, and 3. Let X denote the sum and Y the maximum of the two numbers drawn. (See Problem 5.54.) (*a*) Determine the joint distribution of X and Y. (*b*) Find $\text{Cov}(X, Y)$ and $\rho(X, Y)$.

5.62. A random sample with replacement of size $n = 2$ is chosen from the set $\{1, 2, 3, 4, 5\}$. Let $X = 0$ if the first number is even, and $X = 1$ otherwise; and let $Y = 1$ if the second number is odd, and $Y = 0$ otherwise. (*a*) Show that the distributions for X and Y are identical. (*b*) Find the joint distribution of X and Y. (*c*) Are X and Y independent?

Remark: It is always possible to find the distributions of X and Y from the joint distribution of X and Y; but, in general, it is not possible to find the joint distribution from the individual distributions of X and Y. Some other information, such as knowing that X and Y are independent, is needed to obtain the joint distribution from the individual distributions.

CHEBYSHEV'S INEQUALITY

5.63. Let X be a random variable with mean μ and standard deviation σ. Use Chebyshev's inequality to estimate $P(\mu - 3\sigma \leq X \leq \mu + 3\sigma)$.

5.64. If Z is the standard normal random variable with mean 0 and standard deviation 1, use Chebyshev's inequality to find a value b for which $P(-b \leq X \leq b) \geq 0.9$.

5.65. Let X be a random variable with mean 0 and standard deviation 1.5. Use Chebyshev's inequality to estimate $P(-3 \leq X \leq 3)$.

5.66. X is a random variable with mean $\mu = 70$. For what value of σ will Chebyshev's inequality give $P(65 \leq X \leq 75) \geq 0.95$?

5.67. X is a random variable with mean $\mu = 100$ and standard deviation $\sigma = 10$. Use Chebyshev's inequality to estimate (*a*) $P(X \geq 120)$, (*b*) $P(X \leq 75)$.

MISCELLANEOUS PROBLEMS

5.68. Let X be a continuous random variable with the following distribution:

$$f(x) = \begin{cases} \frac{1}{8} & \text{if } 0 \leq x \leq 8 \\ 0 & \text{elsewhere} \end{cases}$$

(*a*) Find: (i) $P(2 \leq X \leq 5)$, (ii) $P(3 \leq X \leq 7)$, (iii) $P(X \geq 6)$.

(*b*) Determine and plot the graph of the cumulative distribution function F of X.

5.69. Let X be a continuous random variable with the following distribution:

$$f(x) = \begin{cases} kx & \text{if } 0 \leq x \leq 5 \\ 0 & \text{elsewhere} \end{cases}$$

(*a*) Evaluate k. (*b*) Find: (i) $P(1 \leq X \leq 3)$, (ii) $P(2 \leq X \leq 4)$, (iii) $P(X \leq 3)$.

5.70. Plot the graph of the cumulative distribution function F of the random variable X with the distribution:

x	-3	2	6
$f(x)$	$\frac{1}{4}$	$\frac{1}{2}$	$\frac{1}{4}$

5.71. Find the distribution function $f(x)$ of the continuous random variable X with the cumulative distribution function:

$$(a)\quad F(x) = \begin{cases} 0 & \text{if } x < 0 \\ x^3 & \text{if } 0 \leq x \leq 1 \\ 1 & \text{if } x > 1 \end{cases} \qquad (b)\quad F(x) = \begin{cases} 0 & \text{if } x < 0 \\ \sin x & \text{if } 0 \leq x \leq \pi/2 \\ 1 & \text{if } x > \pi/2 \end{cases}$$

(Hint: $f(x) = F'(x)$, the derivative of $F(x)$, wherever it exists.)

5.72. Show that $\sigma_X = 0$ if and only if X is a *constant function*, that is, $X(s) = k$ for every $s \in S$, or simply $X = k$.

5.73. Suppose $\sigma_X \neq 0$. Show that $\rho(X, X) = 1$ and $\rho(X, -X) = -1$.

5.74. Prove Theorem 5.11: Let X, Y, Z be random variables on S with $Z = \Phi(X, Y)$. Then

$$E(Z) = \sum_{i,j} \Phi(x_i, y_j) h(x_i, y_j)$$

where h is the joint distribution of X and Y.

Answers to Supplementary Problems

The following notation will be used:

$$[x_1, \ldots, x_n;\ f(x_1), \ldots, f(x_n)] \text{ for the distribution } f = \{(x_i, f(x_i))\}$$

$$[x_i;\ y_j;\ \text{row by row}] \text{ for the joint distribution } h = \{[(x_i, y_j), h(x_i, y_j)]\}$$

5.36. $k = 2$; $[-3, 2, 4, 7;\ 0.3, 0.2, 0.1, 0.4]$, $E(X) = 2.7$

5.37. $[1, 2, 3, 4, 5, 6;\ \frac{11}{36}, \frac{9}{36}, \frac{7}{36}, \frac{5}{36}, \frac{3}{36}, \frac{1}{36}]$, $E(X) = 91/36 \approx 2.5$

5.38. $[0, 1, 2, 3, 4;\ \frac{1}{16}, \frac{7}{16}, \frac{5}{16}, \frac{2}{16}, \frac{1}{16}]$, $E(X) = 27/16 \approx 1.7$

5.39. (*a*) $[0, 1, 2, 3;\ \frac{1}{64}, \frac{9}{64}, \frac{27}{64}, \frac{27}{64}]$, (*b*) $E(x) = 2.25$

5.40. $E = 211/81 \approx 2.6$

5.41. $E = 23/8 \approx 2.9$

5.42. $E = 11/9 \approx 1.2$

5.43. $E = 11/8 \approx 1.4$

5.44. (*a*) $[3, 4, 5, \ldots, 9;\ 0.1, 0.1, 0.2, 0.2, 0.2, 0.1, 0.1]$, (*b*) $E(X) = 6$

5.45. (*a*) 0.75, (*b*) -0.25

5.46. 0.25

5.47. \$5

5.48. Hint: Let $y = \sum q^n = 1/(1-q)$, so $dy/dq = \sum nq^{n-1} = 1/(1-q)^2$

5.49. (*a*) $\mu = 4$, $\sigma^2 = 5.5$, $\sigma = 2.3$ (*b*) $\mu = 0$, $\sigma^2 = 10$, $\sigma = 3.2$

5.50. (*a*) $\mu = 1$, $\sigma^2 = 2.4$, $\sigma = 1.5$ (*b*) $\mu = 4.0$, $\sigma^2 = 5.6$, $\sigma = 2.37$

5.51. (*a*) $\mu_X = 3$, $\sigma_X^2 = 3$, $\sigma_X = \sqrt{3} \approx 1.7$
(*b*) $[3, 11, 18, 22;\ 0.4, 0.1, 0.2, 0.3]$, $\mu_Y = 12.5$, $\sigma_Y^2 = 69.5$, $\sigma_Y \approx 8.3$

5.52. $\mu = ap + bq, \quad \sigma^2 = pq(a-b)^2, \quad \sigma = |a-b|\sqrt{pq}$

5.53. $\mu_X = 0.9, \quad \sigma_X^2 = 1.09, \quad \sigma_X = 1.04$

(*a*) $[1, 1, 16; 0.2, 0.5, 0.3]$, $\mu_Y = 5.5$, $\sigma_Y^2 = 47.25$, $\sigma_Y = 6.87$

(*b*) $[\frac{1}{3}, 3, 9; 0.2, 0.5, 0.3]$, $\mu_Y = 4.67$, $\sigma_Y^2 = 5.21$, $\sigma_Y = 2.28$

(*c*) $[1, 2, 8; 0.2, 0.5, 0.3]$, $\mu_Y = 3.6$, $\sigma_Y^2 = 8.44$, $\sigma_Y = 2.91$

5.54. (*a*) $[2, 3, 4, 5; 0.1, 0.4, 0.3, 0.2]$, $\mu_X = 3.6$, $\sigma_X^2 = 0.84$, $\sigma_X = 0.9$

(*b*) $[1, 2, 3; 0.1, 0.5, 0.4]$, $\mu_Y = 2.3$, $\sigma_Y^2 = 0.41$, $\sigma_Y = 0.64$

(*c*) $[3, 5, 6, 7, 8; 0.1, 0.4, 0.1, 0.2, 0.2]$, $\mu_Z = 5.9$, $\sigma_Z^2 = 2.3$, $\sigma_Z = 1.5$

(*d*) $[2, 6, 8, 12, 15; 0.1, 0.4, 0.1, 0.2, 0.2]$, $\mu_W = 8.8$, $\sigma_W^2 = 17.6$, $\sigma_W = 4.2$

5.55. (*a*) $E(X) = 3$, $E(Y) = 1$, (*b*) $\text{Cov}(X, Y) = 1.5$, (*c*) $\sigma_X = 2$, $\sigma_Y = 4.3$, $\rho(X, Y) = 0.17$

5.56. (*a*) $E(X) = 1.4$, $E(Y) = 1$, (*b*) $\text{Cov}(X, Y) = -0.5$, (*c*) $\sigma_X = 0.49$, $\sigma_Y = 3.1$, $\rho(X, Y) = -0.3$

5.57. $[1, 2; -2, 5, 8; 0.21, 0.35, 0.14; 0.09, 0.15, 0.06]$

5.58. (*a*) $E(X) = 1.7$, $E(Y) = 3.1$; (*b*) Yes; (*c*) Must equal 0 since X and Y are independent

5.59. (*a*) $E(X) = 1.05$, $E(Y) = 0.16$; (*b*) No;
(*c*) $[-2, -1, 0, 1, 2, 3, 4, 5; 0.05, 0.15, 0.18, 0.17, 0.22, 0.11, 0.08, 0.041]$,

$$\mu_Z = 1.21, \qquad \sigma_Z = \sqrt{3.21} \approx 1.79$$

5.60. (*a*) $[0, 1, 2, 3, 4; 0, 1, 2, 3, 4; \frac{1}{16}, 0, 0, 0, 0; 0, \frac{4}{16}, 0, 0, 0; 0, \frac{3}{16}, \frac{3}{16}, 0, 0; 0, 0, \frac{2}{16}, \frac{2}{16}; 0, 0, 0, 0, \frac{1}{16}]$

(*b*) $\text{Cov}(X, Y) = 0.85$, $\rho(X, Y) = 0.89$

5.61. (*a*) $[2, 3, 4, 5; 1, 2, 3; 0.1, 0, 0; 0, 0.4, 0; 0, 0.1, 0.2; 0, 0, 0.21]$

(*b*) $\text{Cov}(X, Y) = 0.52$, $\rho(X, Y) = 0.9$

5.62. (*a*) $[0, 1; \frac{10}{25}, \frac{15}{25}]$ (*b*) $[0, 1; 0, 1; \frac{4}{25}, \frac{6}{25}, \frac{6}{25}, \frac{9}{25}]$ (*c*) Yes

5.63. $P \ge 1 - \dfrac{1}{32} \approx 0.89$

5.64. $b = \sqrt{10} \approx 3.16$

5.65. $P \ge 0.75$

5.66. $\sigma = 5/\sqrt{20} \approx 1.12$

5.67. (*a*) $P \ge 0.75$, (*b*) $P \ge 0.84$

5.68. (*a*) $\frac{3}{8}, \frac{1}{2}, \frac{1}{4}$, (*b*) $F(x) = \begin{cases} 0 & \text{if } x < 0 \\ x/8 & \text{if } 0 \le x \le 8 \\ 1 & \text{if } x > 8 \end{cases}$. See Fig. 5-25(*a*)

5.69. (*a*) $k = \frac{2}{25}$, (*b*) (i) $\frac{8}{25}$, (ii) $\frac{12}{25}$, (iii) $\frac{9}{25}$

5.70. See Fig. 5-25(*b*)

5.71. (*a*) $f(x) = 3x^2$ if $0 \leq x \leq 1$, 0 elsewhere

(*b*) $f(x) = \cos x$ if $0 \leq x \leq \pi/2$, 0 elsewhere

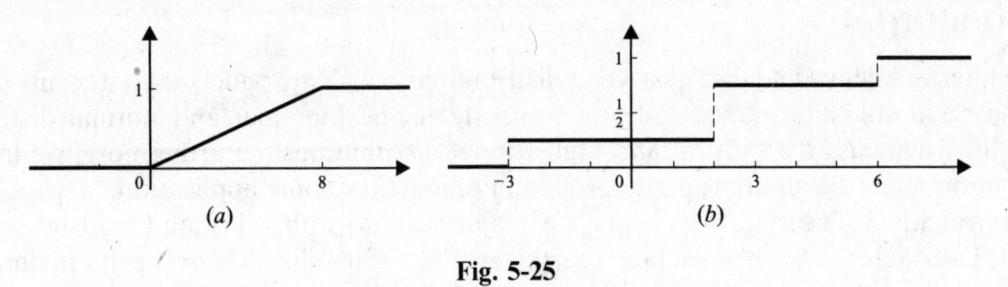

Fig. 5-25

Chapter 6

Binomial and Normal Distributions

6.1 INTRODUCTION

This chapter will define and discuss several distributions which are widely used in many applications of probability and statistics. Specifically, we investigate the binomial and normal distributions in depth, and briefly discuss the Poisson and multinomial distributions. Furthermore, we indicate how each distribution might be an appropriate probability model for some application.

The Central Limit Theorem, which plays a major role in probability and statistics, will also be discussed in this chapter. We will see how this theorem is a generalization of the approximation of the discrete binomial distribution by the continuous normal distribution.

6.2 BERNOULLI TRIALS, BINOMIAL DISTRIBUTION

Consider an experiment with only two outcomes, one called success (S) and the other called failure (F). Independent repeated trials of such an experiment are called Bernoulli trials, named after the Swiss mathematician Jakob Bernoulli (1654–1705). (We emphasize that the term "independent trials" means that the outcome of any trial does not depend on the previous outcomes, such as tossing a coin.)

Let p denote the probability of success in a Bernoulli trial, and so $q = 1 - p$ is the probability of failure. A *binomial experiment* consists of a fixed number of Bernoulli trials. The notation

$$B(n, p)$$

will be used to denote a binomial experiment with n trials and probability p of success.

Frequently, we are interested in the number of successes in a binomial experiment and not in the order in which they occur. The following theorem (proved in Problem 6.8) applies.

Theorem 6.1: The probability of exactly k success in a binomial experiment $B(n, p)$ is given by

$$P(k) = P(k \text{ successes}) = \binom{n}{k} p^k q^{n-k}$$

The probability of one or more successes is $1 - q^n$.

Here $\binom{n}{k}$ is the binomial coefficient, which is defined and discussed in Chapter 2.

Observe that the probability of getting at least k successes, that is, k or more successes, is given by

$$P(k) + P(k+1) + P(k+2) + \cdots + P(n)$$

This follows from the fact that the events of getting k and k' successes are disjoint for $k \neq k'$.

EXAMPLE 6.1

(*a*) A fair coin is tossed 6 times; call heads a success. This is a binomial experiment with $n = 6$ and $p = q = \frac{1}{2}$.

(i) The probability that exactly two heads occurs (i.e. $k = 2$) is:

$$P(2) = \binom{6}{2}\left(\frac{1}{2}\right)^2\left(\frac{1}{2}\right)^4 = \frac{15}{64} \approx 0.23$$

(ii) The probability of getting at least four heads (i.e. $k = 4$, 5, or 6) is:

$$P(4) + P(5) + P(6) = \binom{6}{4}\left(\frac{1}{2}\right)^4\left(\frac{1}{2}\right)^2 + \binom{6}{5}\left(\frac{1}{2}\right)^5\left(\frac{1}{2}\right) + \binom{6}{6}\left(\frac{1}{2}\right)^6$$

$$= \frac{15}{64} + \frac{6}{64} + \frac{1}{64} = \frac{11}{32} \approx 0.34$$

(iii) The probability of getting no heads (i.e. all failures) is $q^6 = (\frac{1}{2})^6 = \frac{1}{64}$, so the probability of one or more heads is $1 - q^n = 1 - \frac{1}{64} = \frac{63}{64} \approx 0.98$.

(*b*) The probability that Ann hits a target is $p = \frac{1}{3}$; hence she misses with probability $q = 1 - p = \frac{2}{3}$. She fires seven times. Find the probability that she hits the target: (i) exactly 3 times, (ii) at least one time.

(i) Here $k = 3$; hence the probability that she hits the target three times is:

$$P(3) = \binom{7}{3}\left(\frac{1}{3}\right)^3\left(\frac{2}{3}\right)^4 = \frac{560}{2187} \approx 0.26$$

(ii) The probability that she never hits the target, that is, all failures, is $q^7 = (\frac{2}{3})^7 = 128/2187 \approx 0.06$. Thus the probability that she hits the target at least once is $1 - q^7 = 2059/2187 \approx 0.94 = 94$ percent.

Binomial Distribution

Consider a binomial experiment $B(n, p)$. That is, $B(n, p)$ consists of n independent repeated trials with two outcomes, success or failure, and p is the probability of success and $q = 1 - p$ is the probability of failure. The number X of k successes is a random variable with the following distribution:

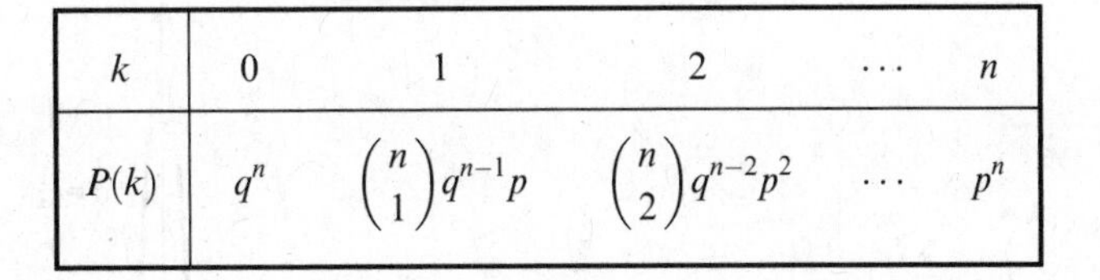

k	0	1	2	$\cdots$	n
$P(k)$	q^n	$\binom{n}{1}q^{n-1}p$	$\binom{n}{2}q^{n-2}p^2$	$\cdots$	p^n

This distribution for a binomial experiment $B(n, p)$ is called the *binomial distribution* since it corresponds to the successive terms of the binomial expansion:

$$(q + p)^n = q^n + \binom{n}{1}q^{n-1}p + \binom{n}{2}q^{n-2}p^2 + \cdots + p^n$$

Thus $B(n, p)$ will also be used to denote the binomial distribution.

Properties of this distribution follow:

Theorem 6.2:

Binomial distribution $B(n, p)$	
Mean or expected number of successes	$\mu = np$
Variance	$\sigma^2 = npq$
Standard deviation	$\sigma = \sqrt{npq}$

EXAMPLE 6.2

(*a*) The probability that Bill hits a target is $p = \frac{1}{5}$. He fires 100 times. Find the expected number μ of times he will hit the target and the standard deviation σ.

Here $p = \frac{1}{5}$ and so $q = \frac{4}{5}$. Hence

$$\mu = np = 100 \cdot \frac{1}{5} = 20 \quad \text{and} \quad \sigma = \sqrt{npq} = \sqrt{100 \cdot \frac{1}{5} \cdot \frac{4}{5}} = 4$$

(*b*) A fair die is tossed 180 times. Find the expected number μ of times a six appears and the standard deviation σ.
Here $p = \frac{1}{6}$ and so $q = \frac{5}{6}$. Hence

$$\mu = np = 180(\tfrac{1}{6}) = 30 \quad \text{and} \quad \sigma = \sqrt{npq} = \sqrt{180(\tfrac{1}{6})(\tfrac{5}{6})} = 5$$

6.3 NORMAL DISTRIBUTION

The most important example of a continuous random variable X is the *normal* random variable, whose density function has a bell-shaped graph. More precisely, there is a normal random variable X for each pair of parameters $\sigma > 0$ and μ, where the corresponding density function is

$$f(x) = \frac{1}{\sqrt{2\pi}\sigma} \exp\left[-\frac{1}{2}\left[\frac{x-\mu}{\sigma}\right]^2\right]$$

Such a normal distribution with parameters μ and σ will be denoted by

$$N(\mu, \sigma^2)$$

If X is such a continuous random variable, then we say X is *normally distributed* or that X is $N(\mu, \sigma^2)$.

Figure 6.1(*a*) shows how the bell-shaped normal curves change as μ varies and σ remains fixed; and Figure 6-1(*b*) shows how the curves change as σ varies and μ remains fixed. Note that each curve reaches its highest point at $x = \mu$ and is symmetric about μ. The inflection points, where the direction of the bend of the curve changes, occur for $x = \mu + \sigma$ and $x = \mu - \sigma$.

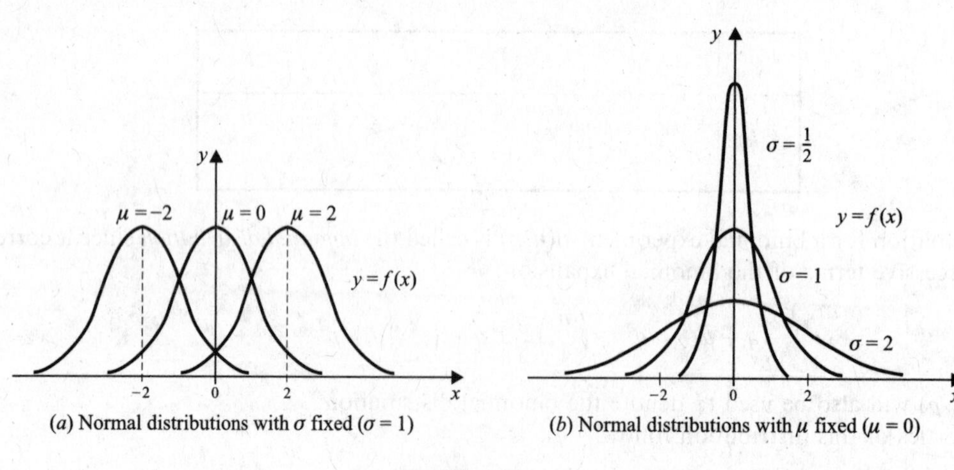

(*a*) Normal distributions with σ fixed ($\sigma = 1$)

(*b*) Normal distributions with μ fixed ($\mu = 0$)

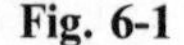
Fig. 6-1

Properties of the normal distribution follow:

Theorem 6.3:

Normal distribution $N(\mu, \sigma^2)$	
Mean or expected value	μ
Variance	σ^2
Standard deviation	σ

That is, the mean, variance, and standard deviation of the normal distribution $N(\mu, \sigma^2)$ are μ, σ^2, and σ, respectively. This is the reason that μ and σ are used as the parameters in the definition of the above density function $f(x)$.

Standardized Normal Distribution

Suppose X is any normal distribution $N(\mu, \sigma^2)$. Recall that the standardized random variable corresponding to X is

$$Z = \frac{X - \mu}{\sigma}$$

In this case Z is also a normal distribution and $\mu = 0$ and $\sigma = 1$, that is, Z is $N(0, 1)$. The density function for Z is

$$\phi(z) = \frac{1}{\sqrt{2\pi}} e^{-z^2/2}$$

whose graph is shown in Fig. 6-2.

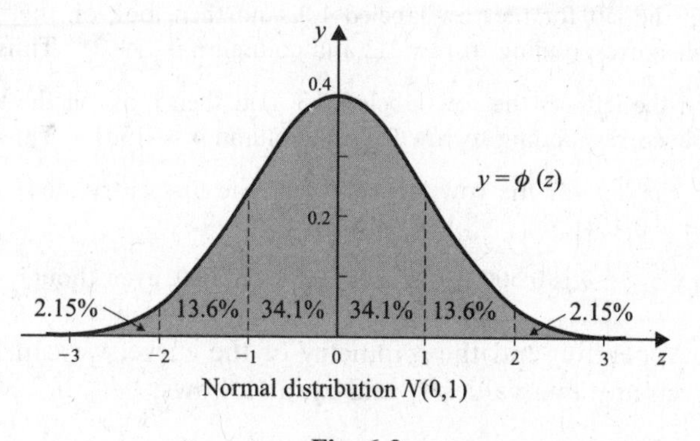

Fig. 6-2

Figure 6-2 also tells us that the percentage of the area under the standardized normal curve $\phi(z)$ and hence also under the corresponding density curve for the normal distribution X is as follows:

68.2 percent	for	$-1 \le z \le 1$	and for	$\mu - \sigma \le x \le \mu + \sigma$
95.4 percent	for	$-2 \le z \le 2$	and for	$\mu - 2\sigma \le x \le \mu + 2\sigma$
99.7 percent	for	$-3 \le z \le 3$	and for	$\mu - 3\sigma \le x \le \mu + 3\sigma$

This gives rise to the so-called:

68–95–99.7 Rule

This rule says that in a normally distributed population, 68 percent (approximately) of the population falls within one standard deviation of the mean, 95 percent falls within two standard deviations of the mean, and 99.7 percent falls within three standard deviations of the mean.

6.4 EVALUATING NORMAL PROBABILITIES

Consider any continuous random variable X on a sample space S with density function $f(x)$. Recall that $\{a \le X \le b\}$ is an event in S and that the probability $P(a \le X \le b)$ is equal to the area under the curve f between $x = a$ and $x = b$. In the language of calculus,

$$P(a \le X \le b) = \int_a^b f(x)\, dx$$

However, if X is a normal distribution, then we are able to evaluate such areas without calculus. We show how to do this in this section in two steps: first with the standard normal distribution Z, and then with any normal distribution X.

Evaluating Standard Normal Probabilities

Table A-1 (see Appendix) gives the area under the standard normal curve ϕ between 0 and z, where $0 \le z < 4$ and z is given in steps of 0.01. This area is denoted by $\Phi(z)$, as indicated by the picture in the table.

EXAMPLE 6.3 Find: (*a*) $\Phi(1.26)$, (*b*) $\Phi(0.34)$, (*c*) $\Phi(1.8)$, (*d*) $\Phi(4.2)$.

(*a*) To find $\Phi(1.26)$, look on the left for the row labeled 1.2, and then look on the top for the column labeled 6. The entry in the table corresponding to row 1.2 and column 6 is .3962. Thus $\Phi(1.26) = 0.3962$.

(*b*) To find $\Phi(0.34)$, look on the left for the row labeled 0.3, and then look on the top for the column labeled 4. The entry in the table corresponding to row 0.3 and column 4 is .1331. Thus $\Phi(0.34) = 0.1331$.

(*c*) To find $\Phi(1.8)$, look on the left for the row labeled 1.8. The first entry .4641 in the row corresponds to $1.8 = 1.80$. Thus $\Phi(1.8) = 0.4641$.

(*d*) The value of $\Phi(z)$ for any $z \ge 3.9$ is 0.5000. Thus $\Phi(4.2) = 0.5000$, even though 4.2 is not in the table.

Using Table A-1 (see Appendix) and the symmetry of the curve, we can find $P(z_1 \le Z \le z_2)$, the area under the curve between any two values z_1 and z_2, as follows:

$$P(z_1 \le Z \le z_2) = \begin{cases} \Phi(z_2) + \Phi(|z_1|) & \text{if } z_1 \le 0 \le z_2 \\ \Phi(z_2) - \Phi(z_1) & \text{if } 0 \le z_1 \le z_2 \\ \Phi(|z_1|) - \Phi(|z_2|) & \text{if } z_1 \le z_2 \le 0 \end{cases}$$

These cases are pictured in Fig. 6-3.

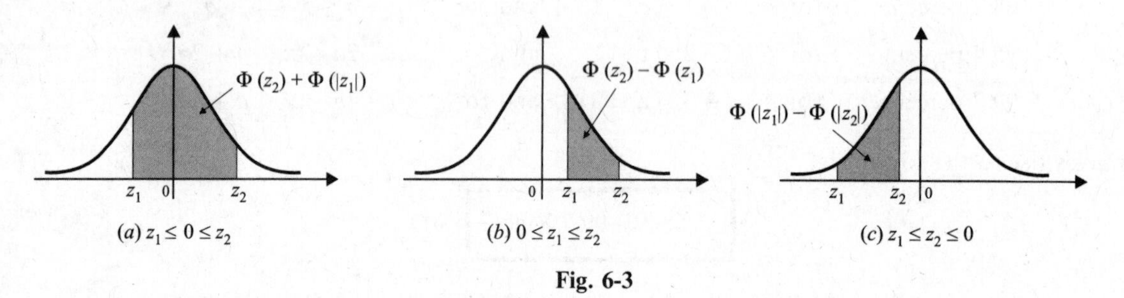

Fig. 6-3

Furthermore, using the fact that the total area under the normal curve is 1 and hence half the area is $\frac{1}{2}$, we can also find the "tail end" of a one-sided probability as follows:

$$P(Z \le z_1) = \begin{cases} 0.5000 + \Phi(z_1) & \text{if } 0 \le z_1 \\ 0.5000 - \Phi(|z_1|) & \text{if } z_1 \le 0 \end{cases}$$

These two cases are pictured in Fig. 6-4(*a*). The complements of these cases give the other one-sided probability, pictured in Fig. 6-4(*b*). Namely,

$$P(Z \ge z_1) = \begin{cases} 0.5000 - \Phi(z_1) & \text{if } 0 \le z_1 \\ 0.5000 + \Phi(|z_1|) & \text{if } z_1 \le 0 \end{cases}$$

The above cover all one-sided probabilities.

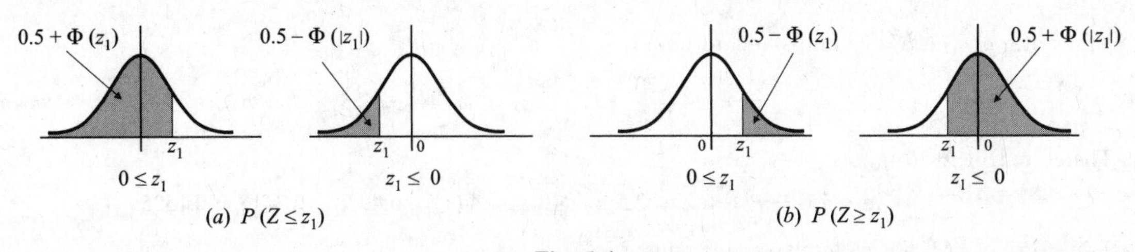

Fig. 6-4

EXAMPLE 6.4 Find the following probabilities for the standard normal distribution Z:

(a) $P(-0.5 \le Z \le 1.1)$ (c) $P(0.2 \le Z \le 1.4)$ (e) $P(Z \ge 1.6)$
(b) $P(-0.38 \le Z \le 1.72)$ (d) $P(-1.5 \le Z \le -0.7)$ (f) $P(Z \le -1.8)$

(a) Referring to Fig. 6-3(a),

$$P(-0.5 \le Z \le 1.1) = \Phi(1.1) + \Phi(0.5) = 0.3643 + 0.1915 = 0.5558$$

(b) Referring to Fig. 6-3(a),

$$P(-0.38 \le Z \le 1.72) = \Phi(1.72) + \Phi(0.38) = 0.4573 + 0.1480 = 0.6053$$

(c) Referring to Fig. 6-3(b),

$$P(0.2 \le Z \le 1.4) = \Phi(1.4) - \Phi(0.2) = 0.4192 - 0.0793 = 0.3399$$

(d) Referring to Fig. 6-3(c),

$$P(-1.5 \le Z \le -0.7) = \Phi(1.5) - \Phi(0.7) = 0.4332 - 0.2580 = 0.1752$$

(e) Referring to Fig. 6-4(b),

$$P(Z \ge 1.6) = 0.5 - \Phi(1.6) = 0.5000 - 0.4452 = 0.0548$$

(f) Referring to Fig. 6-4(a),

$$P(Z \le -1.8) = 0.5 - \Phi(1.8) = 0.5000 - 0.4641 = 0.0359$$

Evaluating Arbitrary Normal Probabilities

Suppose X is a normal distribution, say X is $N(\mu, \sigma^2)$. To evaluate $P(a \le X \le b)$, we usually change a and b into the standard units as follows:

$$z_1 = \frac{a - \mu}{\sigma} \quad \text{and} \quad z_2 = \frac{b - \mu}{\sigma}$$

Then

$$P(a \le X \le b) = P(z_1 \le Z \le z_2)$$

which is the area under the standard normal curve between z_1 and z_2.

EXAMPLE 6.5 Suppose X is the normal distribution $N(70, 4)$. Thus X has mean $\mu = 70$ and standard deviation $\sigma = \sqrt{4} = 2$. Find: (a) $P(68 \le X \le 74)$, (b) $P(72 \le X \le 75)$, (c) $P(63 \le X \le 68)$, (d) $P(X \ge 73)$.

With reference to Figs. 6-3 and 6-4, we make the following computations.

(a) Transform $a = 68$, $b = 74$ into standard units as follows:

$$z_1 = \frac{68 - \mu}{\sigma} = \frac{68 - 70}{2} = -1, \qquad z_1 = \frac{74 - \mu}{\sigma} = \frac{74 - 70}{2} = 2$$

Therefore (Fig. 6-3(a)),

$$P(68 \le X \le 74) = P(-1 \le Z \le 2) = \Phi(2) + \Phi(1) = 0.4772 + 0.3413 = 0.8184$$

(*b*) Transform $a = 72$, $b = 75$ into standard units:

$$z_1 = \frac{72 - 70}{2} = 1, \qquad z_2 = \frac{75 - 70}{2} = 2.5$$

Therefore (Fig. 6-3(*b*)),

$$P(72 \le X \le 75) = P(1 \le Z \le 2.5) = \Phi(2.5) - \Phi(1) = 0.4938 - 0.3413 = 0.1525$$

(*c*) Transform $a = 63$, $b = 68$ into standard units:

$$z_1 = \frac{63 - 70}{2} = -3.5, \qquad z_2 = \frac{68 - 70}{2} = -1$$

Accordingly (Fig. 6-3(*c*)),

$$P(63 \le X \le 68) = P(-3.5 \le Z \le -1) = \Phi(3.5) - \Phi(1) = 0.4998 - 0.3413 = 0.1585$$

(*d*) Transform $a = 73$ into the standard unit $z = (73 - 70)/2 = 1.5$. Thus (Fig. 6.4(*b*)),

$$P(X \ge 73) = P(Z \ge 1.5) = 0.5 - \Phi(1.5) = 0.5000 - 0.4332 = 0.0668$$

Remark: Any continuous random variable X, including the normal random variable, has the property that

$$P(X = a) \equiv P(a \le X \le a) = 0$$

Accordingly, for continuous data, such as heights, weights, and temperatures (whose measurements are really approximations), we usually do not ask for the probability that X is "exactly a" but ask for the probability that X lies in some interval $[a, b]$ or some interval $[a - \epsilon, a + \epsilon]$ centered at a. This is illustrated in the next example.

EXAMPLE 6.6 Suppose the heights of American men are (approximately) normally distributed with mean $\mu = 68$ and standard deviation $\sigma = 2.5$. Find the percentage of American men who are:

(*a*) between $a = 66$ and $b = 71$ inches tall,
(*b*) between $a = 69.5$ and $b = 70.5$ inches tall (that is, "approximately 70 inches" tall),
(*c*) at least 6 feet (72 inches) tall.

(*a*) Transform a and b into standard units, obtaining:

$$z_1 = \frac{66 - 68}{2.5} = -0.80 \quad \text{and} \quad z_2 = \frac{71 - 68}{2.5} = 1.20$$

Here $z_1 < 0 < z_2$. Hence

$$P(66 \le X \le 71) = P(-0.8 \le Z \le 1.2) = \Phi(1.2) + \Phi(0.8)$$
$$= 0.3849 + 0.2881 = 0.6730$$

That is, approximately 67.3 percent of American men are between 66 and 71 inches tall.

(*b*) Transform a and b into standard units, obtaining:

$$z_1 = \frac{69.5 - 68}{2.5} = 0.6 \quad \text{and} \quad z_2 = \frac{70.5 - 68}{2.5} = 1$$

Here $0 < z_1 < z_2$. Therefore,

$$P(69.5 \le X \le 70.5) = P(0.6 \le Z \le 1) = \Phi(1) - \Phi(0.6)$$
$$= 0.3413 - 0.2258 = 0.1155$$

That is, approximately 11.6 percent of American men are between 69.5 and 70.5 inches tall.

(*c*) Transform $a = 72$ into standard units, obtaining $z_1 = (72 - 68)/2.5 = 1.6$. Here $0 < z_1$. Therefore,

$$P(X \ge 72) = P(Z \ge 1.6) = 0.5 - \Phi(1.6) = 0.5 - 0.4452 = 0.0548$$

That is, approximately 5.5 percent of American men are at least 6 feet tall.

6.5 NORMAL APPROXIMATION OF THE BINOMIAL DISTRIBUTION

The binomial probabilities $P(k) = \binom{n}{k} p^k q^{n-k}$ become increasingly difficult to compute as n gets larger. However, there is a way to approximate $P(k)$ by means of a normal distribution when an exact computation is impractical. This is the topic of this section.

Probability Histogram for $B(n, p)$

The probability histograms for $B(10, 0.1)$, $B(10, 0.5)$, $B(10, 0.7)$ are pictured in Fig. 6-5. (Rectangles whose heights are less than 0.01 have been omitted.) Generally speaking, the histogram of a binomial distribution $B(n,p)$ rises as k approaches the mean $\mu = np$ and falls off as k moves away from μ. Furthermore:

(1) For $p = 0.5$, the histogram is symmetric about the mean μ, as in Fig. 6-5(*b*).
(2) For $p < 0.5$, the graph is skewed to the right, as in Fig. 6-5(*a*).
(3) For $p > 0.5$, the graph is skewed to the left, as in Fig. 6-5(*c*).

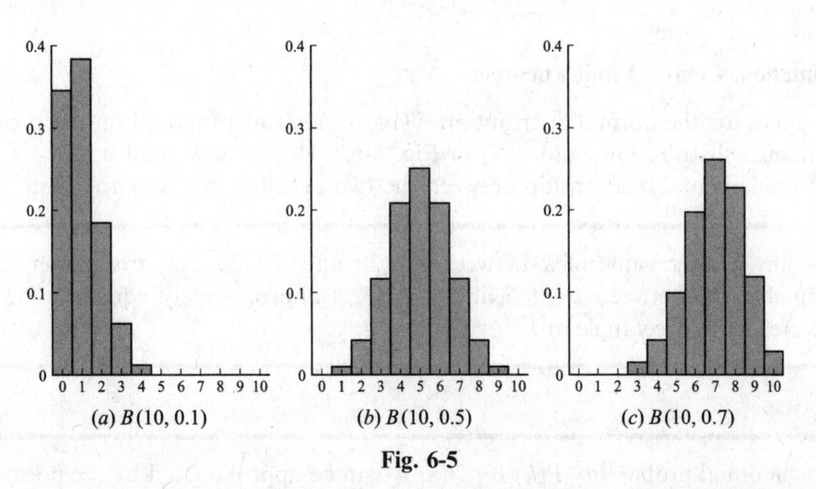

Fig. 6-5

Consider now the following distribution for $B(20, 0.7)$ where an asterisk (*) indicates that $P(k)$ is less than 0.01:

k	0	1	$\cdots$	8	9	10	11	12	13	14	15	16	17	18	19	20
$P(k)$	*	*	$\cdots$	*	0.01	0.03	0.07	0.11	0.16	0.19	0.18	0.13	0.07	0.03	0.01	*

The probability histogram for $B(20, 0.7)$ appears in Fig. 6-6.

Although $p \neq 0.5$, observe that the histogram for $B(20, 0.7)$ is still nearly symmetric about $\mu = 20(0.7) = 14$ for k between 8 and 20, and for k outside that range, $P(k)$ is practically 0. Furthermore, the standard deviation for $B(20, 0.7)$ is approximately $\sigma = 2$, and hence the interval $[8, 20] = [\mu - 3\sigma, \mu + 3\sigma]$. These results are typical for binomial distributions $B(n,p)$ in which both np and nq are at least 5. We state these results more formally:

Basic Property of the Binomial Probability Histogram

For $np \geq 5$ and $nq \geq 5$, the probability histogram for $B(n,p)$ is nearly symmetric about $\mu = np$ over the interval $[\mu - 3\sigma, \mu + 3\sigma]$, where $\sigma = \sqrt{npq}$, and outside this interval $P(k) \simeq 0$.

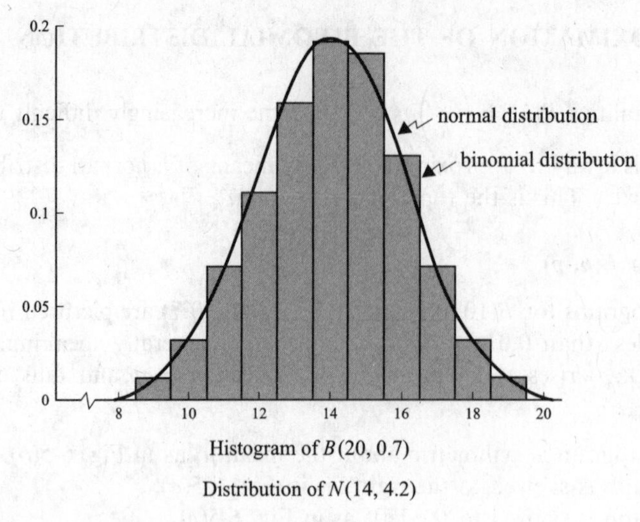

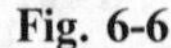
Fig. 6-6

Normal Approximation, Central Limit Theorem

The density curve for the normal distribution $N(14, 4.2)$ is superimposed on the probability histogram for the binomial distribution $B(20, 0.7)$ in Fig. 6-6. Here $\mu = 14$ and $\sigma = \sqrt{4.2}$, for both distributions. The fundamental relationship between the two distributions is as follows:

> For any integer value of k between $\mu - 3\sigma$ and $\mu + 3\sigma$, the area under the normal curve between $k - 0.5$ and $k + 0.5$ is approximately equal to $P(k)$, the area of the rectangle at k.

In other words:

> The binomial probability $P(k)$ for $B(n, p)$ can be approximated by the normal probability $P(k - 0.5 \le X \le k + 0.5)$ for $N(np, npq)$, provided $np \ge 5$ and $nq \ge 5$.

A theoretical justification for the approximation of $B(n,p)$ by $N(np, npq)$ is the fundamental Central Limit Theorem which follows:

Central Limit Theorem 6.4: Let $X_1, X_2, X_3, \ldots$ be a sequence of independent random variables with the same distribution and with mean μ and variance σ^2. Let

$$Z_n = \frac{\bar{X}_n - \mu}{\sigma/\sqrt{n}}$$

where $\bar{X}_n = (X_1 + X_2 + \cdots + X_n)/n$. Then for large n and any interval $\{a \le x \le b\}$,

$$P(a \le Z_n \le b) \approx P(a \le \phi \le b)$$

where ϕ is the standard normal distribution.

Recall that $\bar{X}_n$ was called the sample mean of the random variables $X_1, \ldots, X_n$. Thus Z_n in the above theorem is the standardized sample mean. Roughly speaking, the Central Limit Theorem says that in any sequence of repeated trials the distribution of the standardized sample mean approaches the standard normal distribution as the number of trials increases. Other statements of the Central Limit Theorem are given in Chapter 7.

Calculations Using the Normal Approximation

Let BP denote the binomial probability for $B(n, p)$ and let NP denote the normal probability for $N(np, npq)$, where $np \geq 5$ and $nq \geq 5$. As noted above, for any integer value of k between $\mu - 3\sigma$ and $\mu + 3\sigma$, we have:

$$BP(k) \approx NP(k - 0.5 \leq X \leq k + 0.5)$$

Accordingly, for nonnegative integers n_1 and n_2,

$$BP(n_1 \leq k \leq n_2) \approx NP(n_1 - 0.5 \leq X \leq n_2 + 0.5)$$

Analogous formulas are used for one-sided probabilities. That is,

$$BP(k \leq n_1) \approx NP(X \leq n_1 + 0.5) \qquad \text{and} \qquad BP(k \geq n_1) \approx NP(X \geq n_1 - 0.5)$$

Remark: For the binomial distribution $B(n, p)$, the binomial variable k lies between 0 and n. Thus we should actually replace $BP(k \leq n_1)$ and $BP(k \geq n_1)$ by $BP(0 \leq k \leq n_1)$ and $BP(n_1 \leq k \leq n)$, respectively, which yields the approximations

$$BP(0 \leq k \leq n_1) \approx NP(-0.5 \leq X \leq n_1 + 0.5) = NP(X \leq n_1 + 0.5) - NP(X \leq -0.5)$$

and

$$BP(n_1 \leq k \leq n) \approx NP(n_1 - 0.5 \leq X \leq n + 0.5) = NP(X \geq n_1 - 0.5) - NP(X \geq n + 0.5)$$

However, $NP(X \leq -0.5)$ and $NP(X \geq n + 0.5)$ are very, very small and can be neglected. This is the reason for the above one-sided approximations.

EXAMPLE 6.7 A fair coin is tossed 100 times. Find the probability P that heads occurs: (*a*) exactly 60 times, (*b*) between 48 and 53 times inclusive, (*c*) less than 45 times.

This is a binomial experiment $B(n, p)$ with $n = 100$, $p = 0.5$, and $q = 1 - p = 0.5$. First we find

$$\mu = np = 100(0.5) = 50, \qquad \sigma^2 = npq = 100(0.5)(0.5) = 25, \qquad \text{so} \qquad \sigma = 5$$

(*a*) We can use the normal distribution to approximate the binomial probability $P(60)$ since $np = 50 > 5$ and $nq = 50 > 5$. We have

$$BP(60) \simeq NP(59.5 \leq X \leq 60.5)$$

Transform $a = 59.5$ and $b = 60.5$ into standard units as follows:

$$z_1 = \frac{59.5 - 50}{5} = 1.9 \qquad \text{and} \qquad z_2 = \frac{60.5 - 50}{5} = 2.1$$

Here $0 < z_1 < z_2$. Therefore (Fig. 6-3(*b*)),

$$P = BP(60) \simeq NP(59.5 \leq X \leq 60.5) = NP(1.9 \leq Z \leq 2.1)$$
$$= \Phi(2.1) - \Phi(1.9) = 0.4821 - 0.4713 = 0.0108$$

Remark: This result agrees with the exact value of $BP(60)$ to four decimal places. That is, to four decimal places,

$$BP(60) = \binom{100}{60}(0.5)^{60}(0.5)^{40} = 0.0108$$

(*b*) We seek $BP(48 \leq k \leq 53)$ or, assuming the data is continuous, $NP(47.5 \leq X \leq 53.5)$. Transforming $a = 47.5$ and $b = 53.5$ into standard units yields:

$$z_1 = \frac{47.5 - 50}{5} = -0.5 \qquad \text{and} \qquad z_2 = \frac{53.5 - 50}{5} = 0.7$$

Here, $z_1 < 0 < z_2$. Accordingly (Fig. 6-3(*a*)),

$$P = BP(48 \le k \le 53) \approx NP(47.5 \le X \le 53.5) = NP(-0.5 \le Z \le 0.7)$$
$$= \Phi(0.7) + \Phi(0.5) = 0.2580 + 0.1915 = 0.4495$$

(*c*) We seek $BP(k < 45) = BP(k \le 44)$ or, approximately, $NP(X \le 44.5)$. Transforming $a = 44.5$ into standard units yields

$$z_1 = (44.5 - 50)/5 = -1.1$$

Here $z_1 < 0$. Accordingly (Fig. 6-4(*a*)),

$$P = BP(k \le 44) \approx NP(X \le 44.5) = NP(Z \le -1.1)$$
$$= 0.5 - \Phi(1.1) = 0.5 - 0.3643 = 0.1357$$

6.6 POISSON DISTRIBUTION

A discrete random variable X is said to have the *Poisson distribution* with parameter $\lambda > 0$ if X takes on nonnegative integer values $k = 0, 1, 2, \ldots$ with respective probabilities

$$P(k) = f(k; \lambda) = \frac{\lambda^k e^{-\lambda}}{k!}$$

Such a distribution will be denoted by POI(λ). (This distribution is named after Siméon Poisson (1781–1840) who discovered it in the early part of the 19th century.)

The values of $f(k; \lambda)$ can be obtained using Table 6-1, which gives values of $e^{-\lambda}$ for various values of λ, or by using logarithms.

Table 6-1

Values of $e^{-\lambda}$

λ	0.0	0.1	0.2	0.3	0.4	0.5	0.6	0.7	0.8	0.9
$e^{-\lambda}$	1.000	.905	.819	.741	.670	.607	.549	.497	.449	.407
λ	1	2	3	4	5	6	7	8	9	10
$e^{-\lambda}$	.368	.135	.0498	.0183	.006 74	.002 48	.000 912	.000 335	.000 123	.000 045

The Poisson distribution appears in many natural phenomena, such as the number of telephone calls per minute at some switchboard, the number of misprints per page in a large text, and the number of α particles emitted by a radioactive substance. Bar charts of the Poisson distribution for various values of λ appear in Fig. 6-7.

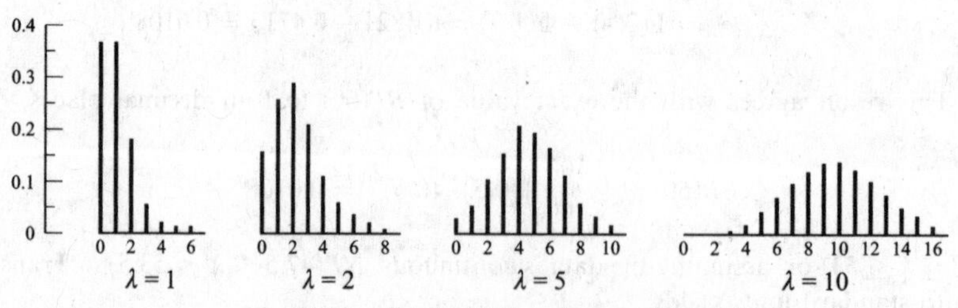

Poisson distribution for selected values of λ

Fig. 6-7

Properties of the Poisson distribution follow:

Theorem 6.5:

Poisson distribution with parameter λ	
Mean or expected value	$\mu = \lambda$
Variance	$\sigma^2 = \lambda$
Standard deviation	$\sigma = \sqrt{\lambda}$

Although the Poisson distribution is of independent interest, it also provides us with a close approximation of the binomial distribution for small k provided p is small and $\lambda = np$; more specifically, if $n \geq 50$ and $np < 5$ (Problem 6.34). This property is indicated in Table 6-2, which compares the binomial and Poisson distributions for small values of k with $n = 100$, $p = 1/100$, and $\lambda = np = 1$.

Table 6-2 Comparison of binomial and Poisson distributions with $n = 100$, $p = 1/100$, and $\lambda = np = 1$.

k	0	1	2	3	4	5
Binomial	0.366	0.370	0.185	0.0610	0.0149	0.0029
Poisson	0.368	0.368	0.184	0.0613	0.0153	0.003 07

EXAMPLE 6.8 Suppose 2 percent of the items made by a factory are defective. Find the probability P that there are 3 defective items in a sample of 100 items.

The binomial distribution with $n = 100$ and $p = 0.2$ applies. However, since p is small, we can use the Poisson approximation with $\lambda = np = 2$. Thus

$$P = f(3, 2) = \frac{2^3 e^{-2}}{3!} = 8(0.135)/6 = 0.180$$

6.7 MULTINOMIAL DISTRIBUTION

The binomial distribution is generalized as follows. Suppose the sample space S of an experiment $\mathscr{E}$ is partitioned into, say, s mutually exclusive events $A_1, A_2, \ldots, A_s$ with respective probabilities $p_1, p_2, \ldots, p_s$. (Hence $p_1 + p_2 + \cdots + p_s = 1$.) Then:

Theorem 6.6: In n repeated trials, the probability that A_1 occurs k_1 times, A_2 occurs k_2 times, ..., and A_s occurs k_s times is equal to

$$\frac{n!}{k_1!k_2!\cdots k_s!} p_1^{k_1} p_2^{k_2} \cdots p_s^{k_s}$$

where $k_1 + k_2 + \cdots + k_s = n$.

The above numbers form the so-called *multinomial distribution*, since they are precisely the terms in the expansion of $(p_1 + p_2 + \cdots + p_s)^n$. Observe that if $s = 2$ then we obtain the binomial distribution, discussed at the beginning of the chapter.

We note that implicitly there are s random variables $X_1, X_2, \ldots, X_s$ connected with the repeated trials of the above experiment $\mathscr{E}$. Specifically, for $i = 1, 2, \ldots, s$, we define X_i to be the number of times A_i occurs when $\mathscr{E}$ is repeated n times. (Observe that the random variables are not independent, since knowledge of any $s - 1$ of them gives the remaining one.)

EXAMPLE 6.9 A fair die is tossed 8 times. Find the probability p of obtaining 5 and 6 exactly twice and the other numbers exactly once.

Here we use the multinomial distribution to obtain:

$$p = \frac{8!}{2!2!1!1!1!1!}\left(\frac{1}{6}\right)^2\left(\frac{1}{6}\right)^2\left(\frac{1}{6}\right)\left(\frac{1}{6}\right)\left(\frac{1}{6}\right)\left(\frac{1}{6}\right) = \frac{35}{5832} \approx 0.006$$

Solved Problems

BINOMIAL DISTRIBUTION

6.1. Compute $P(k)$ for the binomial distribution $B(n, p)$ where:

(*a*) $n = 5,\ p = \frac{1}{4},\ k = 2$ (*b*) $n = 10,\ p = \frac{1}{2},\ k = 7$ (*c*) $n = 8,\ p = \frac{2}{3},\ k = 5$

Use Theorem 6.1, that $P(k) = \binom{n}{k} p^k q^{n-k}$ where $q = 1 - p$.

(*a*) Here $q = \frac{3}{4}$, so $P(2) = \binom{5}{2}\left(\frac{1}{4}\right)^2\left(\frac{3}{4}\right)^3 = 10\left(\frac{1}{16}\right)\left(\frac{27}{64}\right) \approx 0.264.$

(*b*) Here $q = \frac{1}{2}$, so $P(7) = \binom{10}{7}\left(\frac{1}{2}\right)^7\left(\frac{1}{2}\right)^3 = 120\left(\frac{1}{128}\right)\left(\frac{1}{8}\right) \approx 0.117.$

(*c*) Here $q = \frac{1}{3}$, so $P(5) = \binom{8}{5}\left(\frac{2}{3}\right)^5\left(\frac{1}{3}\right)^3 = 56\left(\frac{32}{243}\right)\left(\frac{1}{27}\right) \approx 0.273.$

6.2. The probability that John hits a target is $p = \frac{1}{4}$. He fires $n = 6$ times. Find the probability that he hits the target: (*a*) exactly 2 times, (*b*) more than 4 times, (*c*) at least once.

This is a binomial experiment with $n = 6$, $p = \frac{1}{4}$, and $q = 1 - p = \frac{3}{4}$; hence use Theorem 6.1.

(*a*) $P(2) = \binom{6}{2}(1/4)^2(3/4)^4 = 15(3^4)/(4^6) = 1215/4096 \approx 0.297$

(*b*) John hits the target more than 4 times if he hits it 5 or 6 times. Hence

$$P(X > 4) = P(5) + P(6) = \binom{6}{5}(1/4)^5(3/4)^1 + (1/4)^6$$

$$= 18/4^6 + 1/4^6 = 19/4^6 = 19/4096 \approx 0.0046$$

(*c*) Here $q^6 = (3/4)^6 = 729/4096$ is the probability that John misses all six times; hence

$$P(\text{one or more}) = 1 - 729/4096 = 3367/4096 \approx 0.82$$

6.3. Suppose 20 percent of the items produced by a factory are defective. Suppose 4 items are chosen at random. Find the probability that: (*a*) 2 are defective, (*b*) 3 are defective, (*c*) none are defective.

This is a binomial experiment with $n = 4$, $p = 0.2$ and $q = 1 - p = 0.8$; that is, $B(4,\ 0.2)$. Hence use Theorem 6.1.

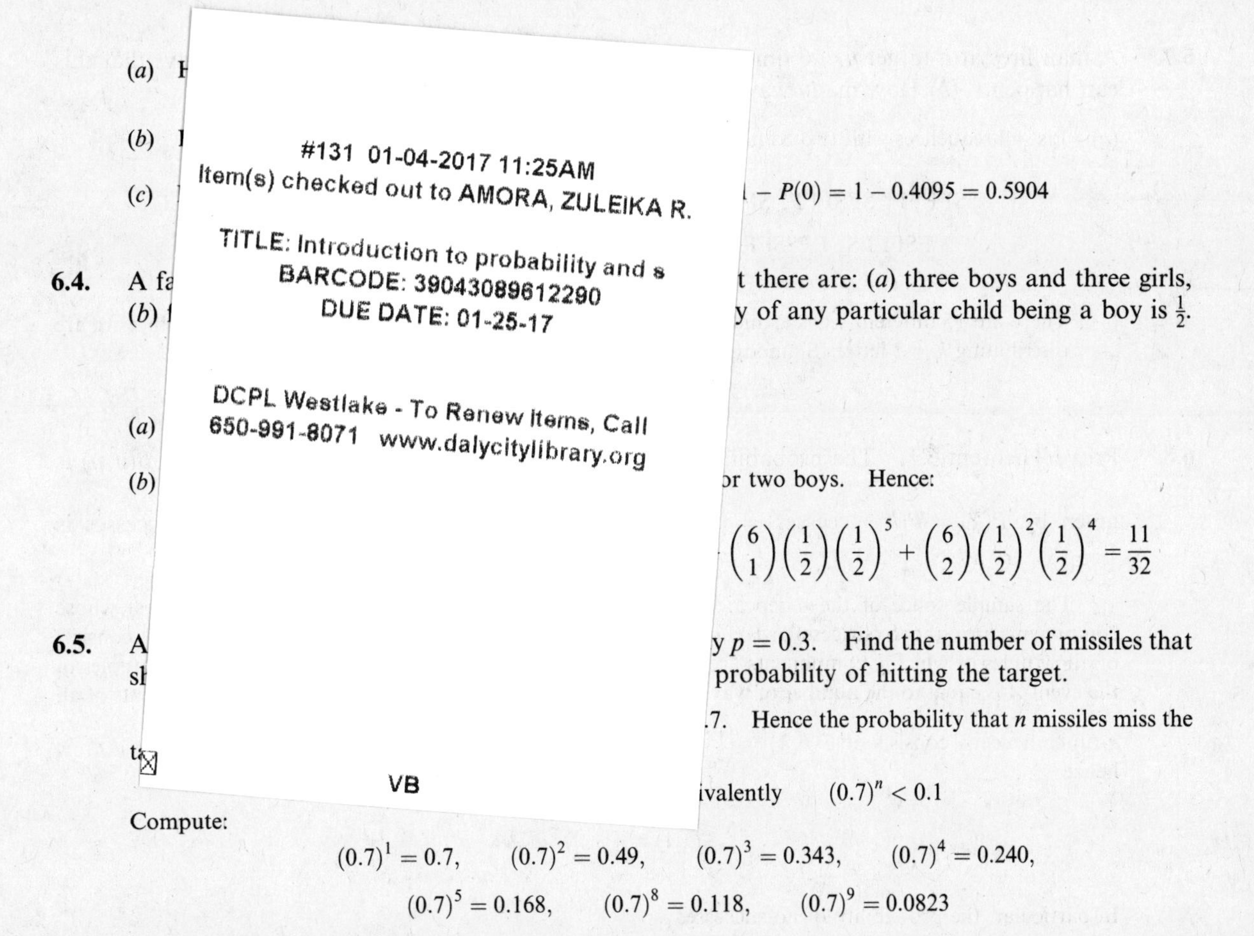

(*a*) F

(*b*) I

(*c*) I ... $1 - P(0) = 1 - 0.4095 = 0.5904$

6.4. A fa... t there are: (*a*) three boys and three girls, (*b*) f... y of any particular child being a boy is $\frac{1}{2}$.

(*a*)

(*b*) ... or two boys. Hence:

$$\binom{6}{1}\left(\frac{1}{2}\right)\left(\frac{1}{2}\right)^5 + \binom{6}{2}\left(\frac{1}{2}\right)^2\left(\frac{1}{2}\right)^4 = \frac{11}{32}$$

6.5. A ... y $p = 0.3$. Find the number of missiles that sh... probability of hitting the target.

... .7. Hence the probability that n missiles miss the ta...

... ivalently $(0.7)^n < 0.1$

Compute:

$$(0.7)^1 = 0.7, \quad (0.7)^2 = 0.49, \quad (0.7)^3 = 0.343, \quad (0.7)^4 = 0.240,$$

$$(0.7)^5 = 0.168, \quad (0.7)^8 = 0.118, \quad (0.7)^9 = 0.0823$$

Thus at least nine missiles should be fired.

6.6. The mathematics department has eight graduate assistants who are assigned the same office. Each assistant is just as likely to study at home as in the office. Find the minimum number m of desks that should be put in the office so that each assistant has a desk at least 90 percent of the time.

This problem can be modeled as a binomial experiment where:

$$n = 8 = \text{number of assistants assigned to the office}$$
$$p = \tfrac{1}{2} = \text{probability that an assistant will study in the office}$$
$$X = \text{number of assistants studying in the office}$$

Suppose there are k desks in the office, where $k \leq 8$. Then a graduate assistant will not have a desk if $X > k$. Note that

$$P(X > k) = P(k+1) + P(k+2) + \cdots + P(8)$$

We want the smallest value of k for which $P(X > k) < 0.10$.

Compute $P(8)$, $P(7)$, $P(6)$, ... until the sum exceeds 10 percent. Using Theorem 6.1, with $n = 8$ and $p = q = \frac{1}{2}$, we obtain:

$$P(8) = (1/2)^8 = 1/256$$
$$P(7) = 8(1/2)^7(1/2) = 8/256$$
$$P(6) = 28(1/2)^6(1/2)^2 = 28/256$$

Now $P(8) + P(7) + P(6) = 37/256 > 10\%$ but $P(7) + P(8) < 10\%$. Thus $m = 6$ desks are needed.

6.7. A man fires at a target $n = 6$ times and hits it $k = 2$ times. (*a*) List the different ways that this can happen. (*b*) How many ways are there?

(*a*) List all sequences with two Ss (successes) and four Fs (failures):

SSFFFF, SFSFFF, SFFSFF, SFFFSF, SFFFFS, FSSFFF, FSFSFF, FSFFSF,

FSFFFS, FFSSFF, FFSFSF, FFSFFS, FFFSSF, FFFSFS, FFFFSS

(*b*) There are 15 different ways as indicated by the list. Observe that this is equal to $\binom{6}{2}$, since we are distributing $k = 2$ letters S among the $n = 6$ positions in the sequence.

6.8. Prove Theorem 6.1. The probability of exactly k successes in a binomial experiment $B(n, p)$ is given by $P(k) = P(k \text{ successes}) = \binom{n}{k} p^k q^{n-k}$. The probability of one or more successes is $1 - q^n$.

The sample space of the n repeated trials consists of all n-tuples (i.e. n-element sequences) whose components are either S (success) or F (failure). Let A be the event of exactly k successes. Then A consists of all n-tuples of which k components are S and $n - k$ components are F. The number of such n-tuples in the event A is equal to the number of ways that k letters S can be distributed among the n components of an n-tuple; hence A consists of $C(n, k) = \binom{n}{k}$ sample points. The probability of each point in A is $p^k q^{n-k}$; hence

$$P(A) = \binom{n}{k} p^k q^{n-k}$$

In particular, the probability of no successes is

$$P(0) = \binom{n}{0} p^0 q^n = q^n$$

Thus the probability of one or more successes is $1 - q^n$.

EXPECTED VALUE AND STANDARD DEVIATION

6.9. Four fair coins are tossed. Let X denote the number of heads occurring. Calculate the expected value of X directly, and compare with Theorem 6.2.

X is binomially distributed with $n = 4$ and $p = q = \frac{1}{2}$. We have:

$$P(0) = \frac{1}{16}, \quad P(1) = \frac{4}{16}, \quad P(2) = \frac{6}{16}, \quad P(3) = \frac{4}{16}, \quad P(4) = \frac{1}{16}$$

Thus the expected value is:

$$E(X) = 0\left(\frac{1}{16}\right) + 1\left(\frac{4}{16}\right) + 2\left(\frac{6}{16}\right) + 3\left(\frac{4}{16}\right) + 4\left(\frac{1}{16}\right) = \frac{32}{16} = 2$$

This agrees with Theorem 6.2, which states that $E(X) = np = 4\left(\frac{1}{2}\right) = 2$.

6.10. A family has eight children. (*a*) Determine the expected number of girls if male and female children are equally probable. (*b*) Find the probability P that the expected number of girls does occur.

(*a*) The number of girls is binomially distributed with $n = 8$ and $p = q = 0.5$. By Theorem 6.2,

$$\mu = np = 8(0.5) = 4$$

(*b*) We seek the probability of 4 girls. By Theorem 6.1, with $k = 4$,

$$P = P(4 \text{ girls}) = \binom{8}{4}(0.5)^4(0.5)^4 \approx 0.27 = 27\%$$

6.11. The probability that a man hits a target is $p = 0.1$. He fires $n = 100$ times. Find the expected number E of times he will hit the target, and the standard deviation σ.

This is a binomial experiment $B(n, p)$ where $n = 100$, $p = 0.1$, and $q = 1 - p = 0.9$. Thus apply Theorem 6.2 to obtain

$$E = np = 100(0.1) = 10 \quad \text{and} \quad \sigma = \sqrt{npq} = \sqrt{100(0.1)(0.9)} = 3$$

6.12. A fair die is tossed 300 times. Find the expected number E and the standard deviation σ of the number of 2's.

The number of 2's is binomially distributed with $n = 300$ and $p = \frac{1}{6}$. Hence $q = 1 - p = \frac{5}{6}$. By Theorem 6.2,

$$E = np = 300\left(\frac{1}{6}\right) = 50 \quad \text{and} \quad \sigma = \sqrt{npq} = \sqrt{300\left(\frac{1}{6}\right)\left(\frac{5}{6}\right)} = \sqrt{41.67} \approx 6.45$$

6.13. A student takes an 18 question multiple-choice exam, with four choices per question. Suppose one of the choices is obviously incorrect, and the student makes an "educated" guess of the remaining choices. Find the expected number E of correct answers, and the standard deviation σ.

This is a binomial experiment $B(n, p)$ where $n = 18$, $p = \frac{1}{3}$, and $q = 1 - p = \frac{2}{3}$. Hence

$$E = np = 18\left(\frac{1}{3}\right) = 6 \quad \text{and} \quad \sigma = \sqrt{npq} = \sqrt{18\left(\frac{1}{3}\right)\left(\frac{2}{3}\right)} = 2$$

6.14. Prove Theorem 6.2: Let X be the binomial random variable $B(n, p)$. Then: (i) $\mu = E(X) = np$, (ii) $\text{Var}(X) = npq$.

On the sample space of n Bernoulli trials, let X_i (for $i = 1, 2, \ldots, n$) be the random variable which has the value 1 or 0 according as the ith trial is a success or a failure. Then each X_i has the distribution

x	0	1
$P(x)$	q	p

and the total number of successes is $X = X_1 + X_2 + \cdots + X_n$.

(i) For each i, we have

$$E(X_i) = 0(q) + 1(p) = p$$

Using the linearity property of E (Theorem 5.4 and Corollary 5.5), we have

$$\begin{aligned} E(X) &= E(X_1 + X_2 + \cdots + X_n) \\ &= E(X_1) + E(X_2) + \cdots + E(X_n) \\ &= p + p + \cdots + p = np \end{aligned}$$

(ii) For each i, we have

$$E(X_i^2) = 0^2(q) + 1^2(p) = p$$

and

$$\text{Var}(X_i) = E(X_i^2) - [E(X)_i]^2 = p - p^2 = p(1-p) = pq$$

The n random variables X_i are independent. Therefore, by Theorem 5.9,

$$\begin{aligned} \text{Var}(X) &= \text{Var}(X_1 + X_2 + \cdots + X_n) \\ &= \text{Var}(X_1) + \text{Var}(X_2) + \cdots + \text{Var}(X_n) \\ &= pq + pq + \cdots + pq = npq \end{aligned}$$

6.15. Give a direct proof of Theorem 6.2: Let X be the binomial random variable $B(n, p)$. Then: (i) $\mu = E(X) = np$, (ii) $\text{Var}(X) = npq$.

(i) Using the notation $b(k; n, p) = P(k) = \binom{n}{k} p^k q^{n-k}$, we obtain:

$$\begin{aligned} E(X) &= \sum_{k=0}^{n} k \cdot b(k; n, p) = \sum_{k=0}^{n} k \frac{n!}{k!(n-k)!} p^k q^{n-k} \\ &= np \sum_{k=1}^{n} \frac{(n-1)!}{(k-1)!(n-k)!} p^{k-1} q^{n-k} \end{aligned}$$

(we drop the term $k = 0$ since its value is zero, and we factor out np from each term). We let $s = k - 1$ in the above sum. As k runs through the values 1 to n, s runs through the values 0 to $n - 1$. Thus

$$E(X) = np \sum_{s=0}^{n-1} \frac{(n-1)!}{s!(n-1-s)!} p^s q^{n-1-s} = np \sum_{s=0}^{n-1} b(s; n-1, p) = np$$

since, by the binomial theorem,

$$\sum_{s=0}^{n-1} b(s; n-1, p) = (p+q)^{n-1} = 1^{n-1} = 1$$

(ii) We first compute $E(X^2)$ as follows:

$$\begin{aligned} E(X^2) &= \sum_{k=0}^{n} k^2 b(k; n, p) = \sum_{k=0}^{n} k^2 \frac{n!}{k!(n-k)!} p^k q^{n-k} \\ &= np \sum_{k=1}^{n} k \frac{(n-1)!}{(k-1)!(n-k)!} p^{k-1} q^{n-k} \end{aligned}$$

Again we let $s = k - 1$ and obtain

$$E(X^2) = np \sum_{s=0}^{n-1} (s+1) \frac{(n-1)!}{s!(n-1-s)!} p^s q^{n-1-s} = np \sum_{s=0}^{n-1} (s+1) b(s; n-1, p)$$

But

$$\begin{aligned} \sum_{s=0}^{n-1} (s+1) b(s; n-1, p) &= \sum_{s=0}^{n-1} s \cdot b(s; n-1, p) + \sum_{s=0}^{n-1} b(s; n-1, p) \\ &= (n-1)p + 1 = np + 1 - p = np + q \end{aligned}$$

where we use (i) to obtain $(n-1)p$. Accordingly,

$$E(X^2) = np(np+q) = (np)^2 + npq$$

and
$$\text{Var}(X) = E(X^2) - \mu_x^2 = (np)^2 + npq - (np)^2 = npq$$

Thus the theorem is proved.

NORMAL DISTRIBUTION

6.16. The mean and standard deviation on an examination are $\mu = 74$ and $\sigma = 12$, respectively. Find the scores in standard units of students receiving: (*a*) 65, (*b*) 74, (*c*) 86, (*d*) 92.

(*a*) $z = \dfrac{x-\mu}{\sigma} = \dfrac{65-74}{12} = -0.75$ (*c*) $z = \dfrac{x-\mu}{\sigma} = \dfrac{86-74}{12} = 1.0$

(*b*) $z = \dfrac{x-\mu}{\sigma} = \dfrac{74-74}{12} = 0$ (*d*) $z = \dfrac{x-\mu}{\sigma} = \dfrac{92-74}{12} = 1.5$

6.17. The mean and standard deviation on an examination are $\mu = 74$ and $\sigma = 12$, respectively. Find the grades corresponding to standard scores: (*a*) -1, (*b*) 0.5, (*c*) 1.25, (*d*) 1.75.

Solving $z = \dfrac{x-\mu}{\sigma}$ for x yields $x = \sigma z + \mu$. Thus:

(*a*) $x = \sigma z + \mu = (12)(-1) + 74 = 62$ (*c*) $x = \sigma z + \mu = (12)(1.25) + 74 = 89$

(*b*) $x = \sigma z + \mu = (12)(0.5) + 74 = 80$ (*d*) $x = \sigma z + \mu = (12)(1.75) + 74 = 95$

6.18. Table A-1 (see Appendix) uses $\Phi(z)$ to denote the area under the standard normal curve ϕ between 0 and z. Find: (*a*) $\Phi(1.47)$, (*b*) $\Phi(0.52)$, (*c*) $\Phi(1.1)$, (*d*) $\Phi(4.1)$.

Use Table A-1 as follows:

(*a*) To find $\Phi(1.47)$, look on the left for the row labeled 1.4, and then look on the top for the column labeled 7. The entry in the table corresponding to row 1.4 and column 7 is 0.4292. Hence $\Phi(1.47) = 0.4292$.

(*b*) To find $\Phi(0.52)$, look on the left for the row labeled 0.5, and then look on the top for the column labeled 2. The entry in the table corresponding to row 0.5 and column 2 is 0.1985. Hence $\Phi(0.52) = 0.1985$.

(*c*) To find $\Phi(1.1)$, look on the left for the row labeled 1.1. The first entry in this row is 0.3643 which corresponds to $1.1 = 1.10$. Hence $\Phi(1.1) = 0.3643$.

(*d*) The value of $\Phi(z)$ for any $z \geq 3.9$ is 0.5000. Thus $\Phi(4.1) = 0.5000$ even though 4.1 is not in the table.

6.19. Let Z be the random variable with standard normal distribution ϕ. Determine the value of z if: (*a*) $P(0 \leq Z \leq z) = 0.4236$, (*b*) $P(Z \leq z) = 0.7967$, (*c*) $P(z \leq Z \leq 2) = 0.1000$.

(*a*) Here $z > 0$. Thus draw a picture of z and $P(0 \leq Z \leq z)$ as in Fig. 6-8(*a*). Here Table 6-1 can be used directly. The entry 0.4236 appears to the right of row 1.4 and under column 3. Thus $z = 1.43$.

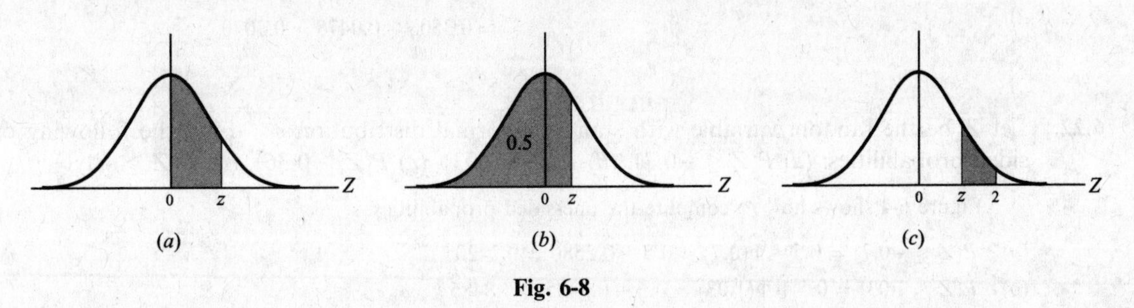

Fig. 6-8

(*b*) Note z must be positive since the probability is greater than 0.5. Thus draw z and $P(Z \le z)$ as in Fig. 6-8(*b*). We have

$$\Phi(z) = P(0 \le Z \le z) = P(Z \le z) - 0.5 = 0.7967 - 0.5000 = 0.2967$$

Since 0.2967 appears in row 0.8 and column 3 in Table 6-1, we have $z = 0.83$.

(*c*) Since $\Phi(2) = 0.4772 > 0.1000$, z must lie between 0 and 2. Thus draw z and $P(z \le Z \le 2)$ as in Fig. 6-8(*c*). We have

$$\Phi(z) = \Phi(2) - P(z \le Z \le 2) = 0.4772 - 0.1000 = 0.3772$$

From Table 6-1, we get $z = 1.16$.

6.20. Let Z be the random variable with standard normal distribution ϕ. Find: (*a*) $P(0 \le Z \le 1.28)$, (*b*) $P(-0.73 \le Z \le 0)$, (*c*) $P(Z = 1.1)$.

(*a*) By definition $\Phi(z)$ is the area under the curve ϕ between 0 and z. Therefore, using Table A-1,

$$P(0 \le Z \le 1.28) = \Phi(1.28) = 0.3997$$

(*b*) By symmetry and Table A-1,

$$P(-0.73 \le Z \le 0) = P(0 \le Z \le 0.73) = \Phi(0.73) = 0.2673$$

(*c*) The area under a single point $a = 1.1$ is 0; hence $P(Z = 1.1) = 0$.

6.21. Let Z be the random variable with standard normal distribution ϕ. Find: (*a*) $P(-1.37 \le Z \le 0.82)$, (*b*) $P(0.65 \le Z \le 1.26)$, (*c*) $P(-1.04 \le Z \le -0.12)$.

Use the following formula (pictured in Fig. 6-3):

$$P(z_1 \le Z \le z_2) = \begin{cases} \Phi(z_2) + \Phi(|z_1|) & \text{if } z_1 \le 0 \le z_2 \\ \Phi(z_2) - \Phi(z_1) & \text{if } 0 \le z_1 \le z_2 \\ \Phi(|z_1|) - \Phi(|z_2|) & \text{if } z_1 \le z_2 \le 0 \end{cases}$$

(*a*) Since $-1.37 < 0 < 0.82$,

$$\begin{aligned} P(-1.37 \le Z \le 0.82) &= \Phi(0.82) + \Phi(1.37) \\ &= 0.2939 + 0.4147 = 0.7086 \end{aligned}$$

(*b*) Since $0 < 0.65 < 1.26$,

$$\begin{aligned} P(0.65 \le Z \le 1.26) &= \Phi(1.26) - \Phi(0.65) \\ &= 0.3962 - 0.2422 = 0.1540 \end{aligned}$$

(*c*) Since $-1.04 < -0.12 < 0$,

$$\begin{aligned} P(-1.04 \le Z \le -0.12) &= \Phi(1.04) - \Phi(0.12) \\ &= 0.3508 - 0.0478 = 0.3030 \end{aligned}$$

6.22. Let Z be the random variable with standard normal distribution ϕ. Find the following one-sided probabilities: (*a*) $P(Z \le -0.7)$, (*b*) $P(Z \le 1.03)$, (*c*) $P(Z \ge 0.36)$, (*d*) $P(Z \ge -1.1)$.

Figure 6-4 shows how to compute the one-sided probabilities.

(*a*) $P(Z \le -0.7) = 0.5 - \Phi(0.7) = 0.5 - 0.2580 = 0.2420$

(*b*) $P(Z \le 1.03) = 0.5 + \Phi(1.03) = 0.5 + 0.3485 = 0.8485$

(*c*) $P(Z \geq 0.36) = 0.5 - \Phi(0.36) = 0.5 - 0.1406 = 0.3594$

(*d*) $P(Z \geq -1.1) = 0.5 + \Phi(-1.1) = 0.5 + 0.3643 = 0.8643$

6.23. Suppose that the student IQ scores form a normal distribution with mean $\mu = 100$ and standard deviation $\sigma = 20$. Find the percentage of students whose scores fall between: (*a*) 80 and 120, (*b*) 60 and 140, (*c*) 40 and 160, (*d*) 100 and 120, (*e*) over 160.

All the scores are units of the standard deviation $\sigma = 20$ from the mean $\mu = 100$; hence we can use the 68–95–99.7 rule or Fig. 6.2 to obtain:

(*a*) 68 percent, (*b*) 95 percent, (*c*) 97.7 percent

(*d*) $\frac{1}{2}$(68 percent) = 34 percent, (*e*) $\frac{1}{2}$(0.3 percent) = 0.15 percent

6.24. Suppose the temperature T during May is normally distributed with mean $\mu = 68°$ and standard deviation $\sigma = 6°$. Find the probability p that the temperature during May is: (*a*) between 70° and 80°, (*b*) less than 60°.

First convert the T values into Z values in standard units, then use Table A-1 (see Appendix).

(*a*) We have:

$$70° \text{ in standard units} = (70 - 68)/6 = 0.33$$
$$80° \text{ in standard units} = (80 - 68)/6 = 2.00$$

Here $0 < 0.33 < 2.00$. Therefore (Fig. 6-9(*a*)),

$$p = P(70 \leq T \leq 80) = P(0.33 \leq Z \leq 2.00)$$
$$= \Phi(2.00) - \Phi(0.33) = 0.4772 - 0.1293 = 0.3479$$

(*b*) We have:

$$60° \text{ in standard units} = (60 - 68)/6 = -1.33$$

This is a one-sided probability with $-1.33 < 0$. Using Fig. 6-9(*b*), symmetry, and that half the area under the curve is 0.5000, we obtain

$$p = P(T \leq 60) = P(Z \leq -1.33) = P(Z \geq 1.33)$$
$$= 0.5 - \Phi(1.33) = 0.5000 - 0.4082 = 0.0918$$

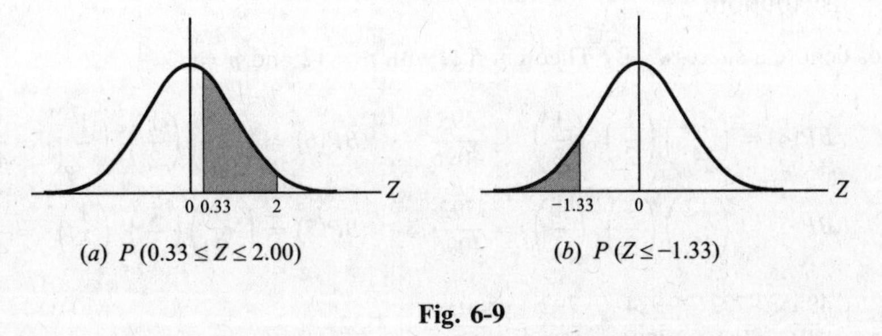

Fig. 6-9

6.25. Suppose the weights W of 800 male students are normally distributed with mean $\mu = 140$ pounds and standard deviation $\sigma = 10$ pounds. Find the number N of students with weights: (*a*) between 138 and 148 pounds, (*b*) more than 152 pounds.

First convert the W values into Z values in standard units, then use Table A-1 (see Appendix).

(*a*) We have:

$$138 \text{ in standard units} = (138 - 140)/10 = -0.2$$
$$148 \text{ in standard units} = (148 - 140)/10 = 0.8$$

Here $-0.2 < 0 < 0.8$. Therefore (Fig. 6-10(*a*)),

$$P(138 \le W \le 148) = P(-0.2 \le Z \le 0.8)$$
$$= \Phi(0.8) + \Phi(-0.2) = 0.2881 + 0.0793 = 0.3674$$

Thus $N = 800(0.3674) \approx 294$.

(*b*) We have:

$$152 \text{ in standard units} = (152 - 140)/10 = 1.20$$

This is a one-sided probability with $0 < 1.20$. Using Fig. 6-10(*b*) and that half the area under the curve is 0.5000, we get

$$P(W \ge 152) = P(Z \ge 1.2) = 0.5 - \Phi(1.2) = 0.5000 - 0.3849 = 0.1151$$

Thus $N = 800(0.1151) \approx 92$.

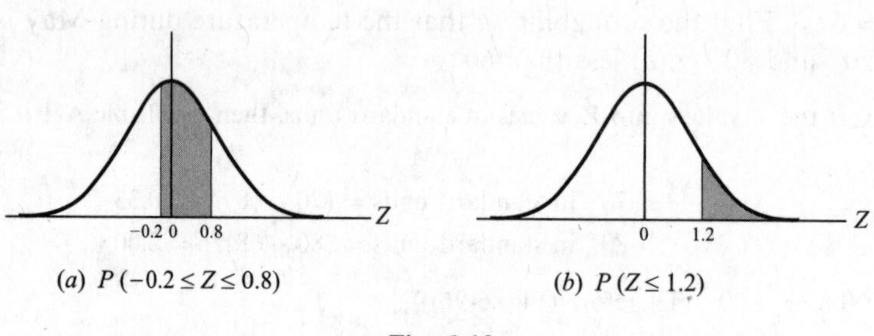

(*a*) $P(-0.2 \le Z \le 0.8)$ (*b*) $P(Z \le 1.2)$

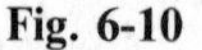

Fig. 6-10

NORMAL APPROXIMATION TO THE BINOMIAL DISTRIBUTION

This section of problems uses *BP* to denote the binomial probability and *NP* to denote the normal probability.

6.26. A fair coin is tossed 12 times. Determine the probability *P* that the number of heads occurring is between 4 and 7 inclusive by using: (*a*) the binomial distribution, (*b*) the normal approximation to the binomial distribution.

(*a*) Let heads denote a success. By Theorem 6.1, with $n = 12$ and $p = q = \frac{1}{2}$:

$$BP(4) = \binom{12}{4}\left(\frac{1}{2}\right)^4\left(\frac{1}{2}\right)^8 = \frac{495}{4096}, \qquad BP(6) = \binom{12}{6}\left(\frac{1}{2}\right)^6\left(\frac{1}{2}\right)^6 = \frac{924}{4096}$$

$$BP(5) = \binom{12}{5}\left(\frac{1}{2}\right)^5\left(\frac{1}{2}\right)^7 = \frac{792}{4096}, \qquad BP(7) = \binom{12}{7}\left(\frac{1}{2}\right)^7\left(\frac{1}{2}\right)^5 = \frac{792}{4096}$$

Hence $P = \dfrac{495}{4096} + \dfrac{792}{4096} + \dfrac{924}{4096} + \dfrac{792}{4096} = \dfrac{3003}{4096} = 0.7332$.

(*b*) Here $\mu = np = 12\left(\frac{1}{2}\right) = 6$ and $\sigma = \sqrt{npq} = \sqrt{12\left(\frac{1}{2}\right)\left(\frac{1}{2}\right)} = 1.73$. Let *X* denote the number of heads occurring. We seek $BP(4 \le X \le 7)$, which corresponds to the shaded area in Fig. 6-11(*a*). On the other hand, if we assume that the data is continuous, in order to apply the binomial approximation, we must find $NP(3.5 \le X \le 7.5)$, as indicated in Fig. 6-11(*a*). We have:

$$3.5 \text{ in standard units} = (3.5 - 6)/1.73 = -1.45$$
$$7.5 \text{ in standard units} = (7.5 - 6)/1.73 = 0.87$$

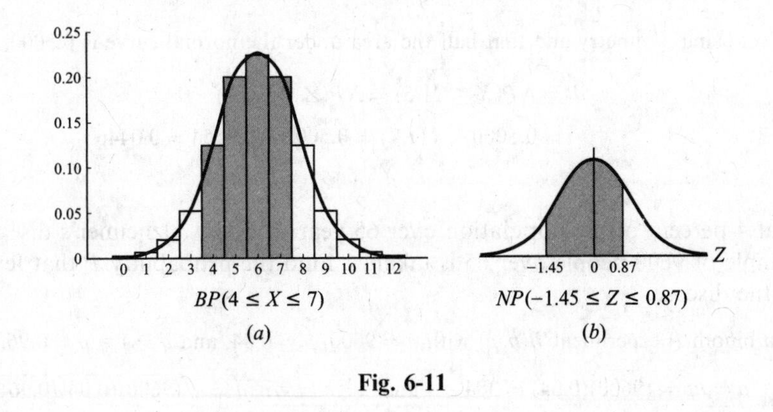

Fig. 6-11

Then, as indicated by Fig. 6-11(*b*),

$$P = NP(3.5 \leq X \leq 7.5) = NP(-1.45 \leq Z \leq 0.87)$$
$$= \phi(0.87) + \Phi(1.45) = 0.3087 + 0.4265 = 0.7343$$

(Note that the relative error $e = |(0.7332 - 0.7343)/0.7332| = 0.0015$ is less than 0.2 percent.)

6.27. A fair die is tossed 180 times. Determine the probability P that the face 6 will appear: (*a*) between 29 and 32 times inclusive, (*b*) between 31 and 35 times inclusive, (*c*) less than 22 times.

This is a binomial experiment $B(n, p)$ with $n = 180$, $p = \frac{1}{6}$ and $q = 1 - p = \frac{5}{6}$. Then

$$\mu = np = 180\left(\frac{1}{6}\right) = 30 \qquad \text{and} \qquad \sigma = \sqrt{npq} = \sqrt{180\left(\frac{1}{6}\right)\left(\frac{5}{6}\right)} = 5$$

Let X denote the number of times the face 6 appears.

(*a*) We seek $BP(29 \leq X \leq 32)$ or, assuming the data is continuous, $NP(28.5 \leq X \leq 32.5)$. We have:

$$28.5 \text{ in standard units} = (28.5 - 30)/5 = -0.3$$
$$32.5 \text{ in standard units} = (32.5 - 30)/5 = \ \ 0.5$$

(This is the case $z_1 \leq 0 \leq z_2$.) Therefore (Fig. 6-3(*a*)),

$$P = NP(28.5 \leq X \leq 32.5) = NP(-0.3 \leq Z \leq 0.5)$$
$$= \Phi(0.5) + \Phi(0.3) = 0.1915 + 0.1179 = 0.3094$$

(*b*) We seek $BP(31 \leq X \leq 35)$ or, assuming the data is continuous, $NP(30.5 \leq X \leq 35.5)$. We have:

$$30.5 \text{ in standard units} = (30.5 - 30)/5 = 0.1$$
$$35.5 \text{ in standard units} = (35.5 - 30)/5 = 1.1$$

(This is the case $0 \leq z_1 \leq z_2$.) Therefore (Fig. 6-3(*b*)),

$$P = NP(30.5 \leq X \leq 35.5) = NP(0.1 \leq Z \leq 1.1)$$
$$= \Phi(1.1) - \Phi(0.1) = 0.3643 - 0.0398 = 0.3245$$

(*c*) We seek $BP(X < 22)$ or, approximately, $NP(X \leq 21.5)$. (See remark in Section 6.5 on the one-sided normal approximation.) We have:

$$21.5 \text{ in standard units} = (21.5 - 30)/5 = -1.7$$

Therefore, using symmetry and that half the area under the normal curve is 0.5000, we get

$$P = NP(X \leq 21.5) = NP(Z \leq -1.7)$$
$$= 0.5000 - \Phi(1.7) = 0.5000 - 0.4554 = 0.0446$$

6.28. Assume that 4 percent of the population over 65 years old has Alzheimer's disease. Suppose a random sample of 9600 people over 65 is taken. Find the probability P that fewer than 400 of them have the disease.

This is a binomial experiment $B(n, p)$ with $n = 9600$, $p = 0.04$, and $q = 1 - p = 0.96$. Then

$$\mu = np = (9600)(0.04) = 384 \quad \text{and} \quad \sigma = \sqrt{npq} = \sqrt{(9600)(0.04)(0.96)} = 19.2$$

Let X denote the number of people with Alzheimer's disease.

We seek $BP(X < 400)$ or, approximately, $NP(X \leq 399.5)$. (See remark in Section 6.5 on the one-sided normal approximation.) We have:

$$399.5 \text{ in standard units} = (399.5 - 384)/19.2 = 0.81$$

Therefore,

$$P = NP(X \leq 399.5) = NP(Z \leq 0.81)$$
$$= 0.5000 + \Phi(0.81) = 0.5000 + 0.2897 = 0.7897$$

POISSON DISTRIBUTION

6.29. Find: (*a*) $e^{-1.3}$, (*b*) $e^{-2.5}$.

Use Table 6-1 and the law of exponents.

(*a*) $e^{-1.3} = (e^{-1})(e^{-0.3}) = (0.368)(0.741) = 0.273$.

(*b*) $e^{-2.5} = (e^{-2})(e^{-0.5}) = (0.135)(0.607) = 0.0819$.

6.30. For the Poisson distribution $f(k; \lambda) = \dfrac{\lambda^k e^{-\lambda}}{k!}$, find: (*a*) $f(2; 1)$, (*b*) $f(3; \frac{1}{2})$, (*c*) $f(2; 0.7)$.

Use Table 6-1 to obtain $e^{-\lambda}$.

(*a*) $f(2; 1) = \dfrac{1^2 e^{-1}}{2!} = \dfrac{e^{-1}}{2} = \dfrac{0.368}{2} = 0.184$.

(*b*) $f(3; \frac{1}{2}) = \dfrac{(\frac{1}{2})^3 e^{-0.5}}{3!} = \dfrac{e^{-0.5}}{48} = \dfrac{0.607}{48} = 0.013$.

(*c*) $f(2; 0.7) = \dfrac{(0.7)^2 e^{-0.7}}{2!} = \dfrac{(0.49)(0.497)}{2} = 0.12$.

6.31. Suppose 300 misprints are distributed randomly throughout a book of 500 pages. Find the probability P that a given page contains (*a*) exactly 2 misprints, (*b*) 2 or more misprints.

We view the number of misprints on one page as the number of successes in a sequence of Bernoulli trials. Here $n = 300$ since there are 300 misprints, and $P = 1/500$, the probability that a misprint appears on the given page. Since p is small, we use the Poisson approximation to the binomial distribution with $\lambda = np = 0.6$.

(*a*) $P = f(2; 0.6) = \dfrac{(0.6)^2 e^{-0.6}}{2!} = (0.36)(0.549)/2 = 0.0988 \approx 0.1$.

(*b*) We have:

$$P(0 \text{ misprints}) = \frac{(0.6)^0 e^{-0.6}}{0!} = e^{-0.6} = 0.549$$

$$P(1 \text{ misprint}) = \frac{(0.6)e^{-0.6}}{1!} = (0.6)(0.549) = 0.329$$

Then $P = 1 - P(0 \text{ or } 1 \text{ misprint}) = 1 - (0.549 + 0.329) = 0.122$.

6.32. Show that the Poisson distribution $f(k; \lambda)$ is a probability distribution, that is,

$$\sum_{k=0}^{\infty} f(k; \lambda) = 1$$

By known results of analysis, $e^{\lambda} = \sum_{k=0}^{\infty} \lambda^k / k!$. Hence

$$\sum_{k=0}^{\infty} f(k; \lambda) = \sum_{k=0}^{\infty} \frac{\lambda^k e^{-\lambda}}{k!} = e^{-\lambda} \sum_{k=0}^{\infty} \lambda^k / k! = e^{-\lambda} e^{\lambda} = 1$$

6.33. Prove Theorem 6.5: Let X be a random variable with the Poisson distribution $f(k; \lambda)$. Then: (i) $E(X) = \lambda$, (ii) $\text{Var}(X) = \lambda$. Hence $\sigma_X = \sqrt{\lambda}$.

(i) Using $f(k; \lambda) = \lambda^k e^{-\lambda}/k!$, we obtain

$$E(X) = \sum_{k=0}^{\infty} k \cdot f(k; \lambda) = \sum_{k=0}^{\infty} k \frac{\lambda^k e^{-\lambda}}{k!} = \lambda \sum_{k=1}^{\infty} \frac{\lambda^{k-1} e^{-\lambda}}{(k-1)!}$$

(we drop the term $k = 0$ since its value is zero, and we factor out λ from each term). Let $s = k - 1$ in the above sum. As k runs through the values 1 to ∞, s runs through the values 0 to ∞. Thus

$$E(X) = \lambda \sum_{k=0}^{\infty} \frac{\lambda^s e^{-\lambda}}{s!} = \lambda \sum_{k=0}^{\infty} f(s; \lambda) = \lambda$$

since $\sum_{k=0}^{\infty} f(s; \lambda) = 1$, by Problem 6.36.

(ii) We first compute $E(X^2)$. We have

$$E(X^2) = \sum_{k=0}^{\infty} k^2 f(k; \lambda) = \sum_{k=0}^{\infty} k^2 \frac{\lambda^k e^{-\lambda}}{k!} = \lambda \sum_{k=1}^{\infty} k \frac{\lambda^{k-1} e^{-\lambda}}{(k-1)!}$$

Again we let $s = k - 1$ and obtain

$$E(X^2) = \lambda \sum_{s=0}^{\infty} (s+1) \frac{\lambda^s e^{-\lambda}}{s!} = \lambda \sum_{s=0}^{\infty} (s+1) f(s; \lambda)$$

But

$$\sum_{s=0}^{\infty} (s+1) f(s; \lambda) = \sum_{s=0}^{\infty} s f(s; \lambda) = \sum_{s=0}^{\infty} f(s; \lambda) = \lambda + 1$$

where we use (i) to obtain λ and Problem 6.36 to obtain 1. Accordingly,

$$E(X^2) = \lambda(\lambda + 1) = \lambda^2 + \lambda$$

and

$$\text{Var}(X) = E(X^2) - \mu_X^2 = \lambda^2 + \lambda - \lambda^2 = \lambda$$

Thus the theorem is proved.

6.34. Show that if p is small and n is large, then the binomial distribution $B(n, p)$ is approximated by the Poisson distribution $POI(\lambda)$ where $\lambda = np$, that is, using

$$BP(k) = \binom{n}{k} p^k q^{n-k} \quad \text{and} \quad f(k; \lambda) = \lambda^k e^{-\lambda}/k!$$

then $BP(k) \approx f(k; \lambda)$ where $np = \lambda$.

We have $BP(0) = (1-p)^n = (1 - \lambda/n)^n$. Taking the natural logarithm of both sides,

$$\ln BP(0) = n \ln(1 - \lambda/n)$$

The Taylor expansion of the natural logarithm is

$$\ln(1 + x) = x - \frac{x^2}{2} + \frac{x^3}{3} - \cdots$$

so

$$\ln\left(1 - \frac{\lambda}{n}\right) = -\frac{\lambda}{n} - \frac{\lambda^2}{2n^2} - \frac{\lambda^3}{3n^3} - \cdots$$

Therefore, if n is large

$$\ln BP(0) = n \ln\left(1 - \frac{\lambda}{n}\right) = -\lambda - \frac{\lambda^2}{2n} - \frac{\lambda^3}{3n^2} \approx -\lambda$$

and hence $BP(0) \approx e^{-\lambda}$.

Furthermore, if p is very small and hence $q \approx 1$, we have

$$\frac{BP(k)}{BP(k-1)} = \frac{(n-k+1)p}{kq} = \frac{\lambda - (k-1)p}{kq} \approx \frac{\lambda}{k}$$

That is, $BP(k) \approx \frac{\lambda}{k} BP(k-1)$. Thus, using $BP(0) \approx e^{-\lambda}$, we obtain $BP(1) \approx \lambda e^{-\lambda}$, $BP(2) \approx \lambda^2 e^{-\lambda}/2$ and, by induction, $BP(k) \approx \lambda^k e^{-\lambda}/k! = f(k; \lambda)$.

MISCELLANEOUS PROBLEMS

6.35. The painted light bulbs produced by a company are 50 percent red, 30 percent green, and 20 percent blue. In a sample of 5 bulbs, find the probability P that 2 are red, 1 is green, and 2 are blue.

By Theorem 6.6 on the multinomial distribution,

$$P = \frac{5!}{2!1!2!}(0.5)^2(0.3)(0.2)^2 = 0.09$$

6.36. Show that the normal distribution

$$f(x) = \frac{1}{\sigma\sqrt{2\pi}} e^{-1/2(x-\mu)^2/\sigma^2}$$

is a continuous probability distribution, i.e. $\int_{-\infty}^{\infty} f(x)\, dx = 1$.

Substituting $t = (x - \mu)/\sigma$ in $\int_{-\infty}^{\infty} f(x)\, dx$, we obtain the integral

$$I = \frac{1}{\sqrt{2\pi}} \int_{-\infty}^{\infty} e^{-t^2/2}\, dt$$

It suffices to show that $I^2 = 1$. We have

$$I^2 = \frac{1}{2\pi}\int_{-\infty}^{\infty} \mathrm{e}^{-t^2/2}\,\mathrm{d}t \int_{-\infty}^{\infty} \mathrm{e}^{-s^2/2}\,\mathrm{d}s = \frac{1}{2\pi}\int_{-\infty}^{\infty}\int_{-\infty}^{\infty} \mathrm{e}^{-(s^2-t^2)/2}\,\mathrm{d}s\,\mathrm{d}t$$

We introduce polar coordinates in the above double integral. Let $s = r\cos\theta$ and $t = r\sin\theta$. Then $\mathrm{d}s\,\mathrm{d}t = r\,\mathrm{d}r\,\mathrm{d}\theta$, $0 \le \theta \le 2\pi$, and $0 \le r \le \infty$. That is,

$$I^2 = \frac{1}{2\pi}\int_0^{2\pi}\int_0^{\infty} r\mathrm{e}^{-r^2/2}\,\mathrm{d}r\,\mathrm{d}\theta$$

But
$$\int_0^{\infty} r\mathrm{e}^{-r^2/2}\,\mathrm{d}r = \left[-\mathrm{e}^{-r^2/2}\right]_0^{\infty} = 1$$

Hence $I^2 = \dfrac{1}{2\pi}\displaystyle\int_0^{2\pi} \mathrm{d}\theta = 1$ and the theorem is proved.

6.37. Prove Theorem 6.3: Let X be a random variable with the normal distribution

$$f(x) = \frac{1}{\sigma\sqrt{2\pi}}\,\mathrm{e}^{-1/2(x-\mu)^2/\sigma^2}$$

Then (i) $E(X) = \mu$ and (ii) $\mathrm{Var}(X) = \sigma^2$. Hence $\sigma_X = \sigma$.

(i) By definition, $E(X) = \dfrac{1}{\sigma\sqrt{2\pi}}\displaystyle\int_{-\infty}^{\infty} x\,\mathrm{e}^{-1/2(x-\mu)^2/\sigma^2}\,\mathrm{d}x$. Setting $t = (x-\mu)/\sigma$, we obtain

$$E(X) = \frac{1}{\sqrt{2\pi}}\int_{-\infty}^{\infty} (\sigma t + \mu)\,\mathrm{e}^{-t^2/2}\,\mathrm{d}t = \frac{\sigma}{\sqrt{2\pi}}\int_{-\infty}^{\infty} t\,\mathrm{e}^{-t^2/2}\,\mathrm{d}t + \mu\frac{1}{\sqrt{2\pi}}\int_{-\infty}^{\infty} \mathrm{e}^{-t^2/2}\,\mathrm{d}t$$

But $g(t) = t\,\mathrm{e}^{-t^2/2}$ is an odd function, i.e. $g(-t) = -g(t)$; hence $\displaystyle\int_{-\infty}^{\infty} t\,\mathrm{e}^{-t^2/2}\,\mathrm{d}t = 0$. Furthermore, $\dfrac{1}{\sqrt{2\pi}}\displaystyle\int_{-\infty}^{\infty} \mathrm{e}^{-t^2/2}\,\mathrm{d}t = 1$, by the preceding problem. Accordingly, $E(X) = \dfrac{\sigma}{\sqrt{2\pi}}\cdot 0 + \mu\cdot 1 = \mu$ as claimed.

(ii) By definition, $E(X^2) = \dfrac{1}{\sigma\sqrt{2\pi}}\displaystyle\int_{-\infty}^{\infty} x^2\,\mathrm{e}^{-1/2(x-\mu)^2/\sigma^2}\,\mathrm{d}x$. Again setting $t = (x-\mu)/\sigma$, we obtain

$$\begin{aligned} E(X^2) &= \frac{1}{\sqrt{2\pi}}\int_{-\infty}^{\infty} (\sigma t + \mu)^2\,\mathrm{e}^{-t^2/2}\,\mathrm{d}t \\ &= \sigma^2\frac{1}{\sqrt{2\pi}}\int_{-\infty}^{\infty} t^2\,\mathrm{e}^{-t^2/2}\,\mathrm{d}t + 2\mu\sigma\frac{1}{\sqrt{2\pi}}\int_{-\infty}^{\infty} t\,\mathrm{e}^{-t^2/2}\,\mathrm{d}t + \mu^2\frac{1}{\sqrt{2\pi}}\int_{-\infty}^{\infty} \mathrm{e}^{-t^2/2}\,\mathrm{d}t \end{aligned}$$

which reduces as above to $E(X^2) = \sigma^2\dfrac{1}{\sqrt{2\pi}}\displaystyle\int_{-\infty}^{\infty} t^2\,\mathrm{e}^{-t^2/2}\,\mathrm{d}t + \mu^2$.

We integrate the above integral by parts. Let $u = t$ and $\mathrm{d}v = t\mathrm{e}^{-t^2/2}\,\mathrm{d}t$. Then $v = -\mathrm{e}^{-t^2/2}$ and $\mathrm{d}u = \mathrm{d}t$. Thus

$$\frac{1}{\sqrt{2\pi}}\int_{-\infty}^{\infty} t^2\,\mathrm{e}^{-t^2/2}\,\mathrm{d}t = \frac{1}{\sqrt{2\pi}}\left[-t\,\mathrm{e}^{-t^2/2}\right]_{-\infty}^{\infty} + \frac{1}{\sqrt{2\pi}}\int_{-\infty}^{\infty} \mathrm{e}^{-t^2/2}\,\mathrm{d}t = 0 + 1 = 1$$

Consequently, $E(X^2) = \sigma^2\cdot 1 + \mu^2 = \sigma^2 + \mu^2$ and

$$\mathrm{Var}(X) = E(X^2) - \mu_X^2 = \sigma^2 + \mu^2 - \mu^2 = \sigma^2$$

Thus the theorem is proved.

Supplementary Problems

BINOMIAL DISTRIBUTION

6.38. Find $P(k)$ for the binomial distribution $B(n, p)$, where:

(*a*) $n = 5, p = \frac{1}{3}, k = 1$; (*b*) $n = 7, p = \frac{1}{2}, k = 2$; (*c*) $n = 4, p = \frac{1}{4}, k = 2$

6.39. A card is drawn and replaced three times from an ordinary 52-card deck. Find the probability that: (*a*) two hearts were drawn, (*b*) three hearts were drawn, (*c*) at least one heart was drawn.

6.40. A box contains three red marbles and two white marbles. A marble is drawn and replaced three times from the box. Find the probability that:
(*a*) one red marble was drawn, (*b*) two red marbles were drawn, (*c*) at least one red marble was drawn.

6.41. A baseball player's batting average is .300. (That is, the probability that he gets a hit is 0.300.) He comes to bat four times. Find the probability that he will get: (*a*) two hits, (*b*) at least one hit.

6.42. A geology quiz consists of 10 multiple-choice questions, there being four choices for each question. Bob is unprepared and decides to guess the answer to every question. Assuming 70 percent is a passing grade, find the probability that Bob will pass the quiz.

6.43. Team A has probability 0.4 of winning whenever it plays (and there are no ties). Suppose A plays four games. Find the probability that A wins:
(*a*) two games, (*b*) at least one game, (*c*) more than half of the games.

6.44. The probability of Ann hitting a target is $\frac{1}{8}$. (*a*) If she fires five times, what is the probability that she hits the target at least twice? (*b*) How many times must she fire so that the probability of hitting the target at least once is more than 90 percent?

6.45. A card is drawn and replaced in an ordinary 52-card deck. Find the number of times a card must be drawn so that: (*a*) there is an even chance of drawing a heart, (*b*) the probability of drawing a heart exceeds 0.75.

EXPECTED VALUE AND STANDARD DEVIATION

6.46. Team A has probability 0.4 of winning whenever it plays (and there are no ties). Let X denote the number of times A wins in four games. (*a*) Find the distribution of X. (*b*) Find the mean, variance and standard deviation of X.

6.47. Suppose 2 percent of the bolts produced by a factory are defective. In a shipment of 3600 bolts from the factory, find the expected number E of defective bolts and the standard deviation σ.

6.48. A fair die is tossed 1620 times. Find the expected number E of times the face 6 occurs and the standard deviation σ.

6.49. Let X be a binomially distributed random variable with $E(X) = 2$ and $\operatorname{Var}(X) = \frac{4}{3}$. Find n and p.

6.50. Consider the binomial distribution $B(n, p)$. Show that:

(a) $\dfrac{P(k)}{P(k-1)} = \dfrac{(n-k+1)p}{kq}$

(b) $P(k-1) < P(k)$ for $k < (n+1)p$;
$P(k-1) > P(k)$ for $k > (n+1)p$

NORMAL DISTRIBUTION

6.51. Let Z be the standard normal random variable. Find:

(a) $P(-0.81 \le Z \le 1.13)$, (b) $P(0.53 \le Z \le 2.03)$
(c) $P(Z \le 0.73)$, (d) $P(|Z| \le 0.25)$

6.52. Let X be normally distributed with mean $\mu = 8$ and standard deviation $\sigma = 4$. Find:
(a) $P(5 \le X \le 10)$, (b) $P(10 \le X \le 15)$, (c) $P(X \ge 15)$, (d) $P(X \le 5)$.

6.53. Suppose the weights of 2000 male students are normally distributed with mean $\mu = 155$ pounds and standard deviation $\sigma = 20$ pounds. Find the number of students with weights:
(a) less than or equal to 100 pounds, (c) between 150 and 175 pounds inclusive,
(b) between 120 and 130 pounds inclusive, (d) greater than or equal to 200 pounds.

6.54. Suppose the diameters d of bolts manufactured by a company are normally distributed with mean $\mu = 0.25$ inches and standard deviation $\sigma = 0.02$ inches. A bolt is considered defective if $d \le 0.20$ inches or $d \ge 0.28$ inches. Find the percentage of defective bolts manufactured by the company.

6.55. Suppose the scores on an examination are normally distributed with mean $\mu = 76$ and standard deviation $\sigma = 15$. The top 15 percent of the students receive As and the bottom 10 percent receive Fs. Find: (a) the minimum score to receive an A, (b) the minimum score to pass (not to receive an F).

6.56. A fair coin is tossed 10 times. Find the probability of obtaining between 4 and 7 heads inclusive by using: (a) the binomial distribution, (b) the normal approximation to the binomial distribution.

6.57. A fair coin is tossed 400 times. Find the probability that the number of heads which occur differs from 200 by: (a) more than 10, (b) more than 25 times.

6.58. A fair die is tossed 720 times. Find the probability that the face 6 will occur:
(a) between 100 and 125 times inclusive, (b) more than 150 times.

6.59. Among 625 random digits, find the probability that the digit 7 appears:
(a) between 50 and 60 times inclusive, (b) between 60 and 70 times inclusive.

POISSON DISTRIBUTION

6.60. Find: (a) $e^{-1.6}$, (b) $e^{-2.3}$

6.61. For the Poisson distribution $f(k, \lambda) = \lambda^k e^{-\lambda}/k!$, find: (a) $f(2; 1.5)$, (b) $f(3; 1)$, (c) $f(2; 0.6)$.

6.62. Suppose 220 misprints are distributed randomly throughout a book of 200 pages. Find the probability that a given page contains: (*a*) no misprints, (*b*) 1 misprint, (*c*) 2 misprints, (*d*) 2 or more misprints.

6.63. Suppose 1 percent of the items made by a machine are defective. Find the probability that 3 or more items are defective in a sample of 100 items.

6.64. Suppose 2 percent of people on the average are left-handed. Find the probability of finding 3 or more left-handed among 100 people.

6.65. Suppose there is an average of 2 suicides per year per 50,000 population. In a city of 100,000, find the probability that in a given year there are: (*a*) 0, (*b*) 1, (*c*) 2, (*d*) 2 or more suicides.

MULTINOMIAL DISTRIBUTION

6.66. A die is loaded so that the faces 1, 2, 3, 4, 5, 6 occur with respective probabilities 0.1, 0.15, 0.15, 0.15, 0.15, 0.3. The die is tossed 6 times. Find the probability that:
(*a*) each face appears once, (*b*) the faces 4, 5, 6 each appear twice.

6.67. A box contains 5 red, 3 white, and 2 blue marbles. A sample of six marbles is drawn with replacement; that is, each marble is replaced before the next marble is drawn. Find the probability that:
(*a*) 3 are red, 2 are white, 1 is blue, (*b*) 2 are red, 3 are white, 1 is blue, (*c*) 2 of each color appear.

Answers to Supplementary Problems

6.38. (*a*) 80/243, (*b*) 21/128, (*c*) 27/128

6.39. (*a*) 9/64, (*b*) 1/64, (*c*) 37/64

6.40. (*a*) 36/125, (*b*) 54/125, (*c*) 117/125

6.41. (*a*) 0.2646, (*b*) 0.7599

6.42. 0.0035

6.43. (*a*) 216/625, (*b*) 544/625, (*c*) 112/625

6.44. (*a*) 131/243, (*b*) 6

6.45. (*a*) 3, (*b*) 5

6.46. (*a*) [0, 1, 2, 3, 4; 0.1296, 0.3456, 0.3456, 0.1536, 0.0256]
(*b*) $\mu = 1.6$, $\sigma^2 = 0.96$, $\sigma = 0.9798$

6.47. $\mu = 72$, $\sigma = 8.4$

6.48. $\mu = 270$, $\sigma = 15$

6.49. $n = 6$, $p = 1/3$

6.51. (*a*) $0.2910 + 0.3708 = 0.6618$, (*b*) $0.4788 - 0.2019 = 0.2769$,
(*c*) $0.5000 + 0.2673 = 0.7673$, (*d*) $2(0.0987) = 0.1974$

6.52. (*a*) 0.4649, (*b*) 0.2684, (*c*) 0.0401, (*d*) 0.2266

6.53. (*a*) 6, (*b*) 131, (*c*) 880, (*d*) 24

6.54. 7.3 percent

6.55. (*a*) 92, (*b*) 57

6.56. (*a*) 0.7734, (*b*) 0.7718

6.57. (*a*) 0.2938, (*b*) 0.0108

6.58. (*a*) 0.6886, (*b*) 0.0011

6.59. (*a*) 0.3518, (*b*) 0.5131

6.60. (*a*) 0.202, (*b*) 0.100

6.61. (*a*) 0.251, (*b*) 0.0613, (*c*) 0.0988

6.62. (*a*) 0.333, (*b*) 0.366, (*c*) 0.201, (*d*) 0.301

6.63. 0.080

6.64. 0.325

6.65. (*a*) 0.0183, (*b*) 0.0732, (*c*) 0.1464, (*d*) 0.909

6.66. (*a*) 0.0109, (*b*) 0.00103

6.67. (*a*) 0.135, (*b*) 0.0810, (*c*) 0.0810

PART II: **Inferential Statistics**

Chapter 7

Sampling Distributions

7.1 INTRODUCTION: SAMPLING WITH AND WITHOUT REPLACEMENT

Inferential statistics is used to draw conclusions about a population, based on a probability model of random samples of the population. For example, a pollster may want to estimate the proportion of all eligible voters favoring a particular presidential candidate by polling a random sample of eligible voters. Or, a statistician may want to use the mean starting income of a random sample of recent college graduates to estimate the mean starting income of all college graduates. *Since different random samples will most likely give different estimates, some knowledge of the variability of all possible estimates derived from random samples is needed to arrive at reasonable conclusions.* Before investigating this variability, some technical terminology is needed.

In general, a *population* is any finite set of objects being investigated. A sample of objects drawn from a population is a *random sample* if it is selected by a process in which every member of the population has essentially the same chance of being chosen. We consider two types of random sample: those drawn *with replacement* and those drawn *without replacement*. The probability distribution of a random variable defined on a space of random samples is called a *sampling distribution*. Sampling distributions are discussed in this chapter and their application to inferential statistics in the following chapters.

EXAMPLE 7.1 Suppose it is desired to determine the average age of students graduating from colleges in the U.S. in a given year. Here the population is the set of all college graduates in the U.S. for the given year. The age X of each graduate is a random variable defined on the population. The average age $\bar{X}$ of the students in a random sample of n graduates is a random variable defined on the space of all random samples of n graduates. The probability distribution of $\bar{X}$ is a sampling distribution.

Sampling With Replacement

In sampling with replacement, each object chosen is returned to the population before the next object is drawn. We define a random sample of size n, drawn with replacement, as an *ordered n-tuple* of objects from the population, repetitions allowed.

EXAMPLE 7.2 A population consists of the set $S = \{4, 7, 10\}$. The space of all random samples of size 2, drawn with replacement, consists of all ordered pairs (a, b), including repetitions, of numbers in S. There are nine such pairs, which are

$$(4,4), \quad (4,7), \quad (4,10), \quad (7,4), \quad (7,7), \quad (7,10), \quad (10,4), \quad (10,7), \quad (10,10)$$

The Space of Random Samples Drawn With Replacement

In general, if samples of size n are drawn with replacement from a population of size N, then the fundamental principle of counting says there are

$$N \cdot N \cdot \ldots \cdot N = N^n$$

such samples. In any survey involving samples of size n, each of these should have the same probability of being chosen. This is equivalent to making the collection of all N^n samples a probability space in which each sample has probability $\frac{1}{N^n}$ of being chosen. Hence, in Example 7.2, there are $3^2 = 9$ random samples of size 2, and each of the nine random samples has probability $\frac{1}{9}$ of being chosen.

Sampling Without Replacement

In sampling without replacement, an object chosen is not returned to the population before the next object is drawn. We define a random sample of size n, drawn without replacement, as an *unordered subset* of n objects from the population.

EXAMPLE 7.3 When sampling without replacement from the population $S = \{4, 7, 10\}$, there are only three random samples of size 2, which are the three subsets of S containing two elements, namely

$$\{4,7\}, \quad \{4,10\}, \quad \{7,10\}.$$

The Space of Random Samples Drawn Without Replacement

If samples of size n are drawn without replacement from a population of size N, then there are

$$\binom{N}{n} = \frac{N!}{n!(N-n)!}$$

such samples, which is the number of subsets of the population containing n elements. For instance, in Example 7.3, there are

$$\binom{3}{2} = \frac{3!}{2! \cdot 1!} = 3$$

random samples of size 2. As in the case of sampling with replacement, the collection of all random samples drawn without replacement can be made into a probability space in which any two samples have the same chance of being selected. In Example 7.3, each of the three random samples has probability $\frac{1}{3}$ of being chosen.

EXAMPLE 7.4 Suppose 75 out of the 100 seniors in a high-school senior class prefer candidate A over candidate B for class president. If 20 different seniors, chosen randomly, are polled about their preference, what is the probability that exactly 15 of them favor candidate A? To answer this question, first note that the 20 different seniors can be interpreted as a sample of size 20, drawn without replacement, from a population of size 100. There are $\binom{100}{20}$ such samples. The number of these samples in which 15 seniors favor candidate A is $\binom{75}{15}\binom{25}{5}$, where

$\binom{75}{15}$ = the number of ways 15 seniors can be chosen from the 75 that favor A, and

$\binom{25}{5}$ = the number of ways the remaining 5 seniors of the sample can be chosen from the 25 that do not favor candidate A

Therefore, the probability that exactly 15 seniors in the sample with favor A is

$$P(15) = \frac{\binom{75}{15}\binom{25}{5}}{\binom{100}{20}} \approx 0.226$$

Comparing Sampling With and Without Replacement

We saw in Example 7.4 that if 20 seniors, chosen without replacement, were polled, then the probability that exactly 15 of them would favor candidate A is approximately 0.226. If the 20 seniors were chosen with replacement, then their selection would be a binomial experiment, $b(20, \frac{75}{100})$, and the probability of exactly 15 in the sample favoring candidate A is

$$P(15) = \binom{20}{15}\left(\frac{75}{100}\right)^{15}\left(\frac{25}{100}\right)^{5} \approx 0.202$$

Figure 7-1 shows the complete probability histograms for the number of seniors favoring candidate A when samples of size 20 are drawn with or without replacement, respectively. For $k = 0, 1, 2, \ldots, 9$, and 20, the probability that k seniors favor candidate A is 0 to two decimal places in both types of sampling.

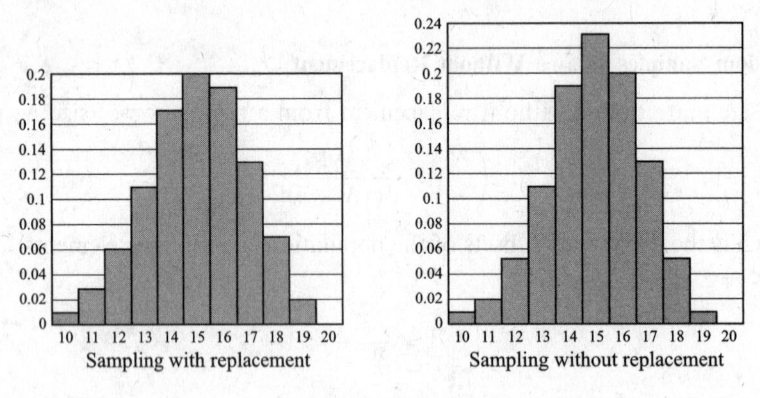

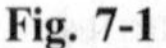
Fig. 7-1

The main difference between the two types of sampling is that when sampling with replacement, the individual outcomes in each sample are *independent*, whereas when sampling without replacement, the outcomes are not independent. For example, if two coins are drawn at random from three dimes and two quarters, then the probability of getting two quarters is $\frac{2}{5} \cdot \frac{2}{5} = 0.16$ if the coins are drawn with replacement, and $\frac{2}{5} \cdot \frac{1}{4} = 0.1$ without replacement. However, when the population is large in comparison with the sample size, results obtained by sampling are very similar whether the sampling is with or without replacement. Therefore, *when the population size is much larger than the sample size, a probability model that assumes the individual outcomes in each sample are independent can be applied to the sampling process regardless of whether the samples are obtained with or without replacement.*

EXAMPLE 7.5 Suppose that 55 percent of all eligible voters in a state favor candidate B for governor. If a random sample of 1000 eligible voters is chosen, find the probability that between 52 percent and 58 percent of the voters in the sample will favor candidate B.

A sample of 1000 voters is small in comparison with the number of eligible voters in any state, so we may use sampling with replacement as a probability model. The sample selection is then a binomial experiment $b(n, p)$, where $n = 1000$ and $p = 0.55$. The probability that between 52 percent and 58 percent of the voters sampled will favor candidate B is the probability that there will be between 520 and 580 successes in the experiment. This probability is equal to

$$\sum_{r=520}^{580} \binom{1000}{r} (0.55)^r (0.45)^{1000-r} \approx 0.95$$

The result was determined by using the normal approximation of the binomial (see Problem 7.8). Hence, approximately 95 percent of the time the random sample will be within 3 percentage points of the true percentage of the population favoring candidate B. Also, the result does not depend on the actual size of the total voting population, only that the sample is small by comparison.

7.2 SAMPLE MEAN

Sampling With Replacement

Suppose X is a random variable with mean μ and standard deviation σ, defined on some population. A random sample of size n, drawn with replacement, yields n values, $x_1, x_2, \ldots, x_n$, for X. Since the sample is drawn with replacement, these values are independent of one another. They can therefore be considered to be values of n independent random variables $X_1, X_2, \ldots, X_n$, each with mean μ and standard deviation σ. For example, if X is the age of college graduates in a given year, then X_i would be the age of the ith graduate ($i = 1, 2, \ldots, n$) in a random sample from this population. The random variable

$$\bar{X} = \frac{X_1 + X_2 + \cdots + X_n}{n}$$

is called the *sample mean* of $X_1, X_2, \ldots, X_n$. As a random variable, $\bar{X}$ also has a mean, $\mu_{\bar{X}}$, and a standard deviation, $\sigma_{\bar{X}}$. It can be shown that these sample parameters are related to the corresponding population parameters μ and σ, as stated in Theorem 7.1 below.

Theorem 7.1 (Mean and Standard Deviation of $\bar{X}$: Sampling With Replacement): Suppose a population random variable X has mean μ and standard deviation σ. Then the sample mean $\bar{X}$, for random samples of size n drawn with replacement, has mean $\mu_{\bar{X}}$ and standard deviation $\sigma_{\bar{X}}$ given by

$$\mu_{\bar{X}} = \mu \qquad \text{and} \qquad \sigma_{\bar{X}} = \frac{\sigma}{\sqrt{n}}$$

Furthermore, if X is approximately normally distributed, then so is $\bar{X}$.

EXAMPLE 7.6 A population consists of the set $S = \{4, 7, 10\}$ as an equiprobable space. Random samples of size 2 are drawn with replacement.

(*a*) Compute the population mean, μ, and standard deviation, σ.
(*b*) Find the sampling distribution (probability distribution) for the sample mean, $\bar{X}$.
(*c*) Compute the mean, $\mu_{\bar{X}}$, and standard deviation, $\sigma_{\bar{X}}$, of $\bar{X}$, and compare with μ and σ.

(*a*) Since the population is an equiprobable space, the probability of each number in S occurring is $\frac{1}{3}$. Therefore, the mean and standard deviation of the population are

$$\mu = \sum xP(x) = 4 \cdot \frac{1}{3} + 7 \cdot \frac{1}{3} + 10 \cdot \frac{1}{3} = \frac{21}{3} = 7$$

and
$$\sigma = \sqrt{\sum (x - \mu)^2 P(x)} = \sqrt{(4-7)^2 \cdot \frac{1}{3} + (7-7)^2 \cdot \frac{1}{3} + (10-7)^2 \cdot \frac{1}{3}} = \sqrt{\frac{18}{3}} = \sqrt{6}$$

(*b*) Table 7-1 lists the mean value $\frac{(a+b)}{2}$ for every possible sample pair, and Table 7-2 gives the sampling distribution for the sample mean, $\bar{X}$.

(*c*) From Table 7-2,

$$\mu_{\bar{X}} = E(\bar{X}) = 4 \cdot \frac{1}{9} + 5.5 \cdot \frac{2}{9} + 7 \cdot \frac{3}{9} + 8.5 \cdot \frac{2}{9} + 10 \cdot \frac{1}{9} = \frac{63}{9} = 7$$

Table 7-1. Samples of size 2, sampling with replacement.

(a, b)	$\bar{x}$
(4, 4)	4
(4, 7)	5.5
(4, 10)	7
(7, 4)	5.5
(7, 7)	7
(7, 10)	8.5
(10, 4)	7
(10, 7)	8.5
(10, 10)	10

Table 7-2. Sampling distribution for $\bar{X}$, sampling with replacement.

$\bar{x}$	$P(\bar{x})$
4	$\frac{1}{9}$
5.5	$\frac{2}{9}$
7	$\frac{3}{9}$
8.5	$\frac{2}{9}$
10	$\frac{1}{9}$

and

$$\begin{aligned}\sigma_{\bar{X}} &= \sqrt{\sum(\bar{x}-\mu_{\bar{X}})^2 P(\bar{x})} \\ &= \sqrt{(4-7)^2\cdot\frac{1}{9}+(5.5-7)^2\cdot\frac{2}{9}+(7-7)^2\cdot\frac{3}{9}+(8.5-7)^2\cdot\frac{2}{9}+(10-7)^2\cdot\frac{1}{9}} \\ &= \sqrt{\frac{27}{9}} = \sqrt{3}\end{aligned}$$

Therefore, $\mu_{\bar{X}} = 7 = \mu$, and $\sigma_{\bar{X}} = \sqrt{3} = \frac{\sqrt{6}}{\sqrt{2}} = \frac{\sigma}{\sqrt{2}}$, which agree with the formulas of Theorem 7.1, where $n = 2$.

Sampling Without Replacement

If samples are drawn without replacement, then the sample values, $x_1, x_2, \ldots, x_n$, of a random variable X are not independent. Nevertheless, the average of the values, namely

$$\frac{x_1 + x_2 + \cdots + x_n}{n}$$

defines a sample random variable which will also be denoted by $\bar{X}$ and will also be called the *sample mean*. In this case, the mean and standard deviation of $\bar{X}$ are given by Theorem 7.2 below.

Theorem 7.2 (Mean and Standard Deviation of $\bar{X}$: Sampling Without Replacement): Suppose a population random variable X has mean μ and standard deviation σ. Then the sample mean $\bar{X}$, for random sample of size n drawn without replacement, has mean $\mu_{\bar{X}}$ and standard deviation $\sigma_{\bar{X}}$ given by

$$\mu_{\bar{X}} = \mu \qquad \text{and} \qquad \sigma_{\bar{X}} = \frac{\sigma}{\sqrt{n}}\sqrt{\frac{N-n}{N-1}}$$

where N is the size of the population and $n < N$. Furthermore, if X is approximately normally distributed, so is $\bar{X}$.

EXAMPLE 7.7 Assume that random samples of size 2 are drawn without replacement from the population $S = \{4, 7, 10\}$ as an equiprobable space.

(*a*) Find the sampling distribution for the sample mean, $\bar{X}$.

(*b*) Compute the mean $\mu_{\bar{X}}$ and standard deviation $\sigma_{\bar{X}}$ of $\bar{X}$, and compare with the population mean μ and standard deviation σ.

(*a*) Table 7-3 lists the mean value $\frac{(a+b)}{2}$ for every possible sample pair, and Table 7-4 gives the sampling distribution for the sample mean, $\bar{X}$.

Table 7-3. Samples of size 2, sampling without replacement.

$\{a, b\}$	$\bar{x}$
{4, 7}	5.5
{4, 10}	7
{7, 10}	8.5

Table 7-4. Sampling distribution for $\bar{X}$, sampling without replacement.

$\bar{x}$	$P(\bar{x})$
5.5	$\frac{1}{3}$
7	$\frac{1}{3}$
8.5	$\frac{1}{3}$

(*b*) From Table 7-4,

$$\mu_{\bar{X}} = E(\bar{X}) = 5.5 \cdot \frac{1}{3} + 7 \cdot \frac{1}{3} + 8.5 \cdot \frac{1}{3} = \frac{21}{3} = 7,$$

and

$$\sigma_{\bar{X}} = \sqrt{\sum(\bar{x} - \mu_{\bar{X}})^2 P(\bar{x})} = \sqrt{(5.5-7)^2 \cdot \frac{1}{3} + (7-7)^2 \cdot \frac{1}{3} + (8.5-7)^2 \cdot \frac{1}{3}}$$

$$= \sqrt{\frac{4.5}{3}} = \sqrt{1.5}$$

From Example 7.5, the population mean and standard deviation are $\mu = 7$ and $\sigma = \sqrt{6}$. Hence, $\mu_{\bar{X}} = 7 = \mu$; also, $\sigma_{\bar{X}} = \sqrt{1.5}$ and $\frac{\sigma}{\sqrt{2}} \cdot \sqrt{\frac{N-n}{N-1}} = \frac{\sqrt{6}}{\sqrt{2}} \cdot \sqrt{\frac{3-2}{3-1}} = \sqrt{3} \cdot \sqrt{\frac{1}{2}} = \sqrt{1.5}$. These equations agree with the formulas of Theorem 7.2.

The Sampling Distribution of $\bar{X}$

The second parts of Theorems 7.1 and 7.2 say that if X is approximately normally distributed, then $\bar{X}$ is also approximately normally distributed. Since we are assuming that the population is finite, X cannot be exactly normal, but many random variables for large populations can, for most practical purposes, be considered to be normally distributed. For example, the national SAT scores in a given year are approximately normally distributed with mean 500 and standard deviation 100. The mean SAT scores for random samples of size n will also be approximatey normal with mean 500 and standard deviation $\frac{100}{\sqrt{n}}$. (Since the population size, N, of students taking the SAT is large in comparison to a typical sample size, n, we may assume $\sqrt{\frac{N-n}{N-1}} \approx 1$; equivalently, we may assume that the scores in each sample are independent.) The following remarkable theorem says that if the sample size is large, then the sample mean $\bar{X}$ is approximately normally distributed regardless of the distribution of X.

Theorem 7.3 (Central Limit Theorem): Supose X is a random variable with mean μ and standard deviation $\sigma > 0$, defined on some population. If n is large, then the sample mean $\bar{X}$ is approximately normally distributed.

As a rule of thumb, the central limit theorem applies when $n \geq 30$. Note that Theorems 7.1 and 7.2 still apply when n is large. That is, $\bar{X}$ has mean μ and standard deviation $\frac{\sigma}{\sqrt{n}}$ if the samples are drawn with replacement, whereas $\bar{X}$ has mean μ and standard deviation $\frac{\sigma}{\sqrt{n}}\sqrt{\frac{N-n}{N-1}}$ if the samples are drawn without replacement and $n < N$.

Sampling From a Large Population

As noted earlier, when a random sample is drawn from a large population, it can be assumed that the values $x_1, x_2, \ldots, x_n$ of the sample are independent. The assumption of independence is key to much of the probability theory used as a model for statistical inference. In the following sections, phrases such as "the population is much larger than the sample size" or "the population is large in comparison to the sample size" are meant to convey that the x values obtained in samples are essentially independent. In practice, if the assumption $\sqrt{\frac{N-n}{N-1}} \approx 1$ is reasonable in a given context, then independence may be assumed. Hence the Central Limit Theorem can be rephrased as follows.

Theorem 7.3′ (Central Limit Theorem): Suppose X is a random variable with mean μ and standard deviation $\sigma > 0$, defined on some population. If n is large ($n \geq 30$) and the population size is large in comparison to n, then the sample mean $\bar{X}$ is approximately normally distributed with mean $\mu_{\bar{X}} = \mu$ and standard deviation $\sigma_{\bar{X}} = \frac{\sigma}{\sqrt{n}}$.

EXAMPLE 7.8 Suppose that the number of customers entering Dee's Grocery each day over a five-year period is a random variable with mean 100 and standard deviation 10. Then the average number of customers computed over randomly selected 30-day periods can be modeled as a normal random variable with mean 100 and standard deviation $\frac{10}{\sqrt{30}} \approx 1.8$. To see this, first note that the sample size is 30, which is large enough to assume that the sample average is a normal random variable. Also, the population size is the number of days in a 5-year period, which is at least 1826 and sufficiently large compared with the sample size to enable us to assume that the numbers of customers in the days of a sample are independent; equivalently, $\sqrt{\frac{N-n}{N-1}} \geq \sqrt{\frac{1826-30}{1826-1}} \approx 0.9920 \approx 1$.

EXAMPLE 7.9 With reference to Example 7.8, what is the probability that the average number of customers entering Dee's Grocery daily over a 30-day period is between 95 and 105?

As indicated in Example 7.8, the average number of customers, or sample mean $\bar{X}$, can be modeled as a normal random variable with mean 100 and standard deviation 1.8. Then

$$Z = \frac{\bar{X} - 100}{1.8}$$

is a normal random variable with mean 0 and standard deviation 1, that is, a standard normal random variable. Using the standard normal table,

$$P(95 \leq \bar{X} \leq 105) = P\left(\frac{95-100}{1.8} \leq \frac{\bar{X}-100}{1.8} \leq \frac{105-100}{1.8}\right) = P(-2.78 \leq Z \leq 2.78) = 0.9946$$

Hence, it is almost certain that the average number of customers entering the store daily over a 30-day period is between 95 and 105.

7.3 SAMPLE PROPORTION

Suppose a proportion p of a population favor candidate A for president. In a random sample of size n drawn from the population, a certain proportion $\hat{p}$ of the sample will favor candidate A, and the collection of all such proportions defines a random variable $\hat{P}$, called the *sample proportion*. The mean and standard deviation of $\hat{P}$ are given in the next two theorems.

Theorem 7.4 (Mean and Standard Deviation of $\hat{P}$: Sampling With Replacement): Suppose the population proportion is p, and random samples of size n are drawn with replacement. Then the sample proportion $\hat{P}$ has mean p and standard deviation $\sqrt{\dfrac{p(1-p)}{n}}$.

Theorem 7.5 (Mean and Standard Deviation of $\hat{P}$: Sampling Without Replacement): Suppose the population size is N, the population proportion is p, and random samples of size n are drawn without replacement. Then the sample proportion $\hat{P}$ has mean p and standard deviation $\sqrt{\dfrac{p(1-p)}{n}} \cdot \sqrt{\dfrac{N-n}{N-1}}$.

If the population is much larger than the sample size, then $\sqrt{\dfrac{N-n}{N-1}} \approx 1$, and if the sample size itself is also large ($n \geq 30$), then the central limit theorem (Theorem 7.3′) can be used to obtain the following result.

Theorem 7.6 (Central Limit Theorem for Sample Proportions): Suppose the sample size n is large ($n \geq 30$), and the population size is large in comparison to n. Then the sample proportion $\hat{P}$ is approximately normally distributed with mean p and standard deviation $\sqrt{\dfrac{p(1-p)}{n}}$.

Theorems 7.4 and 7.5 follow from Theorems 7.1 and 7.2, respectively, and Theorem 7.6 follows from theorem 7.3′ (see Problems 7.63–7.65).

EXAMPLE 7.10 Suppose 25 percent of all U.S. workers belong to a labor union. What is the probability that in a random sample of 100 U.S. workers, at least 20 percent will belong to a labor union?

The sample size, $n = 100$, is greater than 30, and the total number of U.S. workers is much larger than 100. Therefore, the sample proportion $\hat{P}$ of workers that belong to a labor union can be modeled as a normal random variable with mean $p = 0.25$ and standard deviation $\sqrt{\dfrac{p(1-p)}{n}} = \sqrt{\dfrac{0.25 \times 0.75}{100}} \approx 0.0433$. Then

$$Z = \frac{\hat{P} - 0.25}{0.0433}$$

is a standard normal random variable. Using the standard normal table,

$$\begin{aligned} P(\hat{P} \geq 0.2) &= P\left(\frac{\hat{P} - 0.25}{0.0433} \geq \frac{0.2 - 0.25}{0.0433}\right) \\ &\approx P(Z \geq -1.15) \\ &= P(Z \leq 1.15) \\ &= 0.8749 \end{aligned}$$

Hence, it is very likely that there will be at least 20 percent union workers in a random sample of 100 workers.

7.4 SAMPLE VARIANCE

Let X be a population random variable with mean μ and standard deviation σ. We assume that random samples of size n are taken with replacement, or if they are taken without replacement, we assume that the population size is much larger than n. Then the values $x_1, x_2, \ldots, x_n$ of X in a random sample are, in effect, values of n independent random variables $X_1, X_2, \ldots, X_n$, each with mean μ and standard deviation σ. The random variable

$$S^2 = \frac{(X_1 - \bar{X})^2 + (X_2 - \bar{X})^2 + \cdots + (X_n - \bar{X})^2}{n-1}$$

where $\bar{X}$ is the sample mean, is called the *sample variance*.

Since S^2 is intended to be an average of the square deviations from $\bar{X}$, it may seem more natural to divide by n rather than $n-1$. In fact, some statisticians do define S^2 with n as the denominator, and there are pros and cons for each choice. A reason in favor of dividing by $n-1$, as above, is that the expected value of S^2 is then equal to σ^2, the variance of X (see Problem 7.31). In technical terms, the above S^2 is an *unbiased estimator* of σ^2. Before discussing a sampling distribution related to S^2, we must introduce the chi-square random variable.

The Chi-Square Distribution

Because of the central limit theorem, the normal distribution plays a major role in inferential statistics. Another continuous probability distribution that plays an important role in inferential statistics is the chi-square distribution, which can be defined as follows.

Definition: Let $Z_1, Z_2, \ldots, Z_k$ be k independent normal random variables, each with mean 0 and standard deviation 1. Then, the random variable

$$\chi^2 = Z_1^2 + Z_2^2 + \cdots + Z_k^2$$

is called a *chi-square random variable with k degrees of freedom.*

Properties of the Chi-Square Distribution

A chi-square random variable χ^2 with k degrees of freedom is often denoted by $\chi^2(k)$ to emphasize its dependence on the parameter k, which can be any positive integer, including 1. There is a density curve for each value of k, several of which are illustrated in Fig. 7-2. Note that $\chi^2(k)$ assumes only non-negative values (since it is a sum of squares). Also, as k increases, the corresponding density curve becomes less skewed to the right and more symmetric about the mode, which is $k-2$; $\chi^2(k)$ has mean k and standard deviation $\sqrt{2k}$.

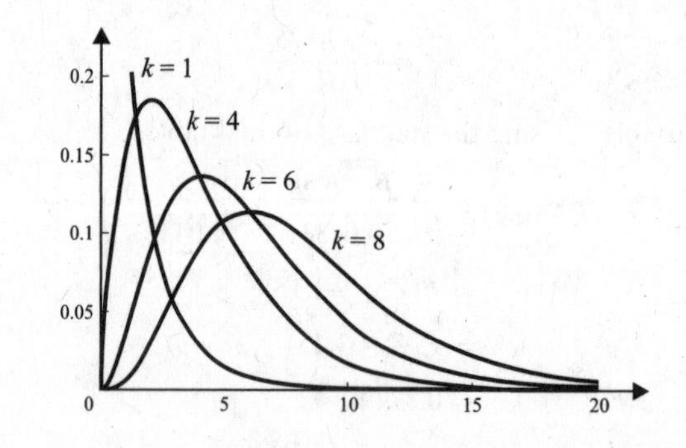

Fig. 7-2 Chi-square distribution for k degrees of freedom

EXAMPLE 7.11 Suppose X_1, X_2, and X_3 are independent normal random variables, each with mean 100 and standard deviation 15, and let $Z_i = (X_i - 100)/15$ for $i = 1, 2, 3$. Then Z_1, Z_2, and Z_3 are independent normal random variables, each with mean 0 and standard deviation 1. Therefore, Z_1^2, Z_2^2, and Z_3^2 are each $\chi^2(1)$, with mean 1 and standard deviation $\sqrt{2}$; and $Z_1^2 + Z_2^2 + Z_3^2$ is $\chi^2(3)$, with mean 3 and standard deviation $\sqrt{6}$.

The Sampling Distribution of $(n-1)S^2/\sigma^2$

We are now in a position to determine a sampling distribution related to the sample variance S^2. Note that if $X_1, X_2, \ldots, X_n$ are n random variables, each with mean μ and standard deviation $\sigma > 0$, then (see Problem 7.30)

$$\sum(X_i - \bar{X})^2 = \sum(X_i - \mu)^2 - n(\bar{X} - \mu)^2$$

Dividing both sides by σ^2 gives

$$\sum \frac{(X_i - \bar{X})^2}{\sigma^2} = \sum \frac{(X_i - \mu)^2}{\sigma^2} - \frac{(\bar{X} - \mu)^2}{\sigma^2/n}$$

If the X_is are independent random variables, the left side of the above equation is $(n-1)S^2/\sigma^2$; and if the X_is are also normally distributed, the right side is the difference of a $\chi^2(n)$ random variable (by definition) and a $\chi^2(1)$ random variable (by the Central Limit Theorem 7.3′). The following result can then be established.

Theorem 7.7: Suppose random samples of size n corresponding to some random variable X are drawn from a population whose size is much larger than n. Suppose also that X is (approximately) a normal random variable with mean μ and standard deviation $\sigma > 0$. Then $(n-1)S^2/\sigma^2$ is (approximately) a chi-square random variable with $n-1$ degrees of freedom.

Mean and Standard Deviation of S^2

As stated above, the expected value of S^2 is σ^2, the variance of X. That is, the mean of S^2 is σ^2. Also, since $(n-1)S^2/\sigma^2$ is a chi-square random variable with $n-1$ degrees of freedom, the standard deviation of $(n-1)S^2/\sigma^2$ is $\sqrt{2(n-1)}$. Therefore, the standard deviation of S^2 is $[\sqrt{2(n-1)}]\sigma^2/(n-1)$, which is equal to $[\sqrt{2/(n-1)}]\sigma^2$.

EXAMPLE 7.12 The annual college SAT scores are (approximately) normally distributed with mean $\mu = 500$ and standard deviation $\sigma = 100$. If S^2 is the sample variance on the space of all random samples of 50 SAT scores, then $49S^2/\sigma^2$ is (approximately) a $\chi^2(49)$ random variable, which has mean 49 and standard deviation $\sqrt{2.49} = 7\sqrt{2} \approx 9.9$. S^2 itself has mean $\mu_{S^2} = \sigma^2 = 10{,}000$ and standard deviation $\sigma_{S^2} = [\sqrt{2/49}] \cdot 100^2 \approx 2020$.

BASIC ASSUMPTION REGARDING FUTURE SAMPLING

In Chapters 8, 9, 10, and 11, unless otherwise stated, we will assume, for simplicity, that either sampling is done with replacement or that the population size N is much larger than the sample size n. This will ensure that the individual outcomes of a random sample are essentially independent, and make the correction factor $\sqrt{\dfrac{N-n}{N-1}}$ for the sample variance unnecessary.

Solved Problems

SAMPLING WITH AND WITHOUT REPLACEMENT

7.1. Let $S = \{1, 5, 6, 8\}$.

(*a*) List all samples of size 3, with replacement.

(*b*) How many samples, with replacement, are there of size 4, size 5, size n?

(*a*) A sample with replacement is a 3-tuple of numbers from S, repetitions allowed. By the fundamental counting principle, there are $4 \times 4 \times 4 = 64$ such samples:

(1, 1, 1), (1, 1, 5), (1, 1, 6), (1, 1, 8) (1, 5, 1), (1, 5, 5,), (1, 5, 6), (1, 5, 8)
(1, 6, 1), (1, 6, 5), (1, 6, 6), (1, 6, 8), (1, 8, 1), (1, 8, 5), (1, 8, 6), (1, 8, 8)
(5, 1, 1), (5, 1, 5), (5, 1, 6), (5, 1, 8), (5, 5, 1), (5, 5, 5), (5, 5, 6), (5, 5, 8)
(5, 6, 1), (5, 6, 5), (5, 6, 6), (5, 6, 8), (5, 8, 1), (5, 8, 5), (5, 8, 6), (5, 8, 8)
(6, 1, 1), (6, 1, 5), (6, 1, 6), (6, 1, 8), (6, 5, 1), (6, 5, 5), (6, 5, 6), (6, 5, 8)
(6, 6, 1), (6, 6, 5), (6, 6, 6), (6, 6, 8), (6, 8, 1), (6, 8, 5), (6, 8, 6), (6, 8, 8)
(8, 1, 1), (8, 1, 5), (8, 1, 6), (8, 1, 8), (8, 5, 1), (8, 5, 5), (8, 5, 6), (8, 5, 8)
(8, 6, 1), (8, 6, 5), (8, 6, 6), (8, 6, 8), (8, 8, 1), (8, 8, 5), (8, 8, 6), (8, 8, 8)

(*b*) There are $4^4 = 256$ samples of size 4, $4^5 = 1024$ samples of size 5, and, in general, 4^n samples of size n for any positive integer n.

7.2. Let $S = \{1, 5, 6, 8\}$.

(*a*) List all samples of size 3, without replacement.

(*b*) How many samples, without replacement, are there of size 4, size n?

(*a*) A sample of size 3, without replacement, is a subset of S containing 3 elements. There are $\binom{4}{3} = 4$ subsets: {1, 5, 6}, {1, 5, 8}, {1, 6, 8}, {5, 6, 8}.

(*b*) For $n = 1, 2, 3, 4$, there are $\binom{4}{n}$ samples of size n; for $n > 4$, there are no samples of size n.

7.3. Five different banks draw a name at random from the same list of 100 names to send a credit-card application. How many random samples of five applications, one application for each bank, are possible? How many of the samples contain the same name more than once?

Let the banks be denoted by A, B, C, D, E. Each sample of five applications can be considered as a 5-tuple of names, where the first name is chosen by bank A, the second by bank B, and so on. Since repetitions are allowed, the sampling is with replacement, and there are $100^5 = 10{,}000{,}000{,}000$, or 10 billion, possible samples. By the fundamental counting principle, the number of samples with five different names is $100 \times 99 \times 98 \times 97 \times 96 = 9{,}034{,}502{,}400$. Subtracting this number from 10 billion, we find that there are 965,497,600 applications with the same name appearing more than once.

7.4. How many committees of 5 people can be randomly selected from a group of 10 women and 15 men. How many of the committees will have all men? How many will have all women? How many have three women and two men?

The number of 5-person committees is the number of ways that 5 people can be chosen from a group of 25 people, or the number of samples of size 5 that can be chosen, without replacement, from a population of size 25, which is $\binom{25}{5} = 53{,}130$. The number that have all men is $\binom{15}{5} = 3003$, and the number that have all women is $\binom{10}{5} = 252$. The number that have three women and two men is $\binom{10}{3}\binom{15}{2} = 12{,}600$.

7.5. What is the most likely breakdown of men and women in a committee of five randomly chosen from 15 men and 10 women?

Since the ratio of 15 men to 10 women is 3 to 2, it seems reasonable that a committee of 3 men and 2 women would be the most likely to occur at random. This expectation can be checked by simply counting the number of each type of committee. From Problem 7.4 we have the following counts.

5 men	3003
5 women:	252
3 women, 2 men:	12,600

Similarly, we get the following counts.

$$\text{1 man, 4 women:} \quad \binom{15}{1}\binom{10}{4} = 3150$$

$$\text{3 men, 2 women:} \quad \binom{15}{3}\binom{10}{2} = 20{,}475$$

$$\text{4 men, 1 woman:} \quad \binom{15}{4}\binom{10}{1} = 13{,}650$$

As expected, a committee with 3 men and 2 women is the most likely to occur.

7.6. A professor asks her class to determine the number of random samples of size 3 that can be selected, without replacement, from a population of three Democrats and two Republicans. James answers that there are three random samples, one consisting of 3 Democrats, one consisting of 2 Democrats and 1 Republican, and 1 consisting of 1 Democrat and 2 Republicans. Is James right? If not, how many are there?

All random samples of size 3 should have the same chance of occurring. However, since there are more Democrats than Republicans, a Democrat is more likely to be selected than a Republican. Therefore, a sample consisting of 2 Democrats and 1 Republican is more likely than a sample consisting of 1 Democrat and 2 Republicans. So James's answer is incorrect. To arrive at the correct answer, label the Democrats as D_1, D_2, D_3 and the Republicans as R_1, R_2. Then, there are $\binom{5}{3} = 10$ random samples of size 3, without replacement, namely:

$$\{D_1, D_2, D_3\}, \quad \{D_1, D_2, R_1\}, \quad \{D_1, D_2, R_2\}, \quad \{D_1, D_3, R_1\}, \quad \{D_1, D_3, R_2\},$$
$$\{D_2, D_3, R_1\}, \quad \{D_2, D_3, R_2\}, \quad \{D_1, R_1, R_2\}, \quad \{D_2, R_1, R_2\}, \quad \{D_3, R_1, R_2\}$$

Note that the probability that a random sample of size 3 will have 2 Democrats and 1 Republican is $\frac{6}{10}$, whereas the probability that the sample will have 1 Democrat and 2 Republicans is only $\frac{3}{10}$.

7.7. How many random samples of size 3, with replacement, are there for the population in Problem 7.6? How many are there in each of the categories: 3 Democrats; 2 Democrats, 1 Republican; 1 Democrat, 2 Republicans; 3 Republicans?

There are $5^3 = 125$ such random samples, broken down as follows.

3 Democrats: 27 random samples. They are: (D_1, D_1, D_1), (D_2, D_2, D_2), (D_3, D_3, D_3); 3 permutations each of (D_1, D_1, D_2), (D_1, D_1, D_3), (D_2, D_2, D_1), (D_2, D_2, D_3), (D_3, D_3, D_1), (D_3, D_3, D_2); and 6 permutations of (D_1, D_2, D_3).

2 Democrats, 1 Republican: 54 random samples. They are: 3 permutations each of (D_1, D_1, R_1), (D_1, D_1, R_2), (D_2, D_2, R_1), (D_2, D_2, R_2), (D_3, D_3, R_1), (D_3, D_3, R_2); and 6 permutations each of (D_1, D_2, R_1), (D_1, D_2, R_2), (D_1, D_3, R_1), (D_1, D_3, R_2), (D_2, D_3, R_1), (D_2, D_3, R_2).

1 Democrat, 2 Republicans: 36 random samples. They are: 6 permutations each of (D_1, R_1, R_2), (D_2, R_1, R_2), (D_3, R_1, R_2); and 3 permutations each of (D_1, R_1, R_1), (D_1, R_2, R_2), (D_2, R_1, R_1), (D_2, R_2, R_2), (D_3, R_1, R_1), (D_3, R_2, R_2).

3 Republicans: 8 random samples. They are: (R_1, R_1, R_1), (R_2, R_2, R_2); and 3 permutations each of (R_1, R_1, R_2), (R_1, R_2, R_2).

7.8. In Example 7.5 it was stated that $\sum_{r=520}^{580} \binom{1000}{r}(0.55)^r(0.45)^{1000-r} \approx 0.95$. Show that this is true.

The result says that $P(520 \leq X \leq 580) \approx 0.95$, where X is a binomial random variable with mean $np = 1000(0.55) = 550$, and standard deviation $\sqrt{np(1-p)} = \sqrt{1000(0.55)(0.45)} \approx 15.73$. By approximating X by a normal random variable with the same mean and standard deviation, and using the continuity correction, we get

$$P(520 \leq X \leq 1000) = P\left(\frac{519.5 - 550}{15.73} \leq \frac{X - 550}{15.73} \leq \frac{580.5 - 550}{15.73}\right)$$
$$\approx P(-1.94 \leq Z \leq 1.94)$$

where Z is the standard normal random variable. Then, from the standard normal table,

$$P(-1.94 \leq Z \leq 1.94) = 2P(0 \leq Z \leq 1.94)$$
$$\approx 2(0.4738)$$
$$\approx 0.95$$

SAMPLE MEAN

7.9. A population random variable X has mean 100 and standard deviation 16. What are the mean and standard deviation of the sample mean $\bar{X}$ for random samples of size 4 drawn with replacement?

For the population, $\mu = 100$ and $\sigma = 16$. By Theorem 7.1 the mean $\mu_{\bar{X}}$ and standard deviation $\sigma_{\bar{X}}$ of $\bar{X}$ are:

$$\mu_{\bar{X}} = \mu = 100 \quad \text{and} \quad \sigma_{\bar{X}} = \frac{\sigma}{\sqrt{n}} = \frac{16}{\sqrt{4}} = 8$$

7.10. With reference to Problem 7.9, what are the mean and standard deviation of $\bar{X}$ if the population size is 250, and the samples of size 4 are drawn without replacement?

By Theorem 7.2, where $N = 250$ and $n = 4$,

$$\mu_{\bar{X}} = \mu = 100 \quad \text{and} \quad \sigma_{\bar{X}} = \frac{\sigma}{\sqrt{n}}\sqrt{\frac{N-n}{N-1}} = \frac{16}{\sqrt{4}}\sqrt{\frac{246}{249}} \approx 7.95$$

7.11. Suppose the random variable X in Problem 7.9 is approximately normally distributed. What is $P(95 \leq \bar{X} \leq 105)$ for samples of size 4 drawn with replacement?

By Problem 7.9, the mean and standard deviation of $\bar{X}$ are $\mu_{\bar{X}} = 100$ and $\sigma_{\bar{X}} = 8$. By Theorem 7.1, $\bar{X}$ is approximately normally distributed. Therefore,

$$P(95 \leq \bar{X} \leq 105) = P\left(\frac{95 - 100}{8} \leq \frac{\bar{X} - 100}{8} \leq \frac{105 - 100}{8}\right)$$
$$= P(-0.625 \leq Z \leq 0.625),$$

where Z is the standard normal random variable. Using a standard normal table,

$$P(-0.625 \leq Z \leq 0.625) = 2P(0 \leq Z \leq 0.625)$$
$$\approx 2(0.2324)$$
$$\approx 0.46$$

7.12. Suppose the random variable X in Problem 7.9 is approximately normally distributed. What is $P(95 \le \bar{X} \le 105)$ for samples of size 4 drawn without replacement?

By Problem 7.10, the mean and standard deviation of $\bar{X}$ are $\mu_{\bar{X}} = 100$ and $\sigma_{\bar{X}} \approx 7.95$. By Theorem 7.2, $\bar{X}$ is approximately normally distributed. Therefore,

$$P(95 \le \bar{X} \le 105) = P\left(\frac{95-100}{7.95} \le \frac{\bar{X}-100}{7.95} \le \frac{105-100}{7.95}\right)$$
$$\approx P(-0.63 \le Z \le 0.63)$$

where Z is the standard normal random variable. Using a standard normal table,

$$P(-0.63 \le Z \le 0.63) = 2P(0 \le Z \le 0.63)$$
$$\approx 2(0.2357)$$
$$\approx 0.47$$

7.13. Let $S = \{1, 5, 6, 8\}$. Find the probability distribution of the sample mean $\bar{X}$ for random samples of size 2 drawn with replacement.

Since S has 4 elements, there are $4^2 = 16$ random samples of size 2 drawn with replacement. These pairs and their average values are given in the following table.

Sample	$\bar{x}$	Sample	$\bar{x}$	Sample	$\bar{x}$	Sample	$\bar{x}$
(1, 1)	1	(1, 5)	3	(1, 6)	3.5	(1, 8)	4.5
(5, 1)	3	(5, 5)	5	(5, 6)	5.5	(5, 8)	6.5
(6, 1)	3.5	(6, 5)	5.5	(6, 6)	6	(6, 8)	7
(8, 1)	4.5	(8, 5)	6.5	(8, 6)	7	(8, 8)	8

The probability distribution of $\bar{X}$ is given in the following table:

$\bar{x}$	1	3	3.5	4.5	5	5.5	6	6.5	7	8
$p(\bar{x})$	$\frac{1}{16}$	$\frac{2}{16}$	$\frac{2}{16}$	$\frac{2}{16}$	$\frac{1}{16}$	$\frac{2}{16}$	$\frac{1}{16}$	$\frac{2}{16}$	$\frac{2}{16}$	$\frac{1}{16}$

7.14. Let $S = \{1, 5, 6, 8\}$. Find the probability distribution of the sample mean $\bar{X}$ for random samples of size 2 drawn without replacement.

Since S has 4 elements, there are $\binom{4}{2} = 6$ random samples of size 2 drawn without replacement. These, their average value, and the probability distribution of $\bar{X}$ are given in the following two tables:

Sample	$\bar{x}$
{1, 5}	3
{1, 6}	3.5
{1, 8}	4.5
{5, 6}	5.5
{5, 8}	6.5
{6, 8}	7

$\bar{x}$	$p(\bar{x})$
3	$\frac{1}{6}$
3.5	$\frac{1}{6}$
4.5	$\frac{1}{6}$
5.5	$\frac{1}{6}$
6.5	$\frac{1}{6}$
7	$\frac{1}{6}$

7.15. Let $S = \{1, 5, 6, 8\}$. Compute the population mean μ and standard deviation σ. Also, compute the mean $\mu_{\bar{X}}$ and standard deviation $\sigma_{\bar{X}}$ of the sample mean $\bar{X}$ for random samples of size 2 drawn with replacement. Verify that $\mu_{\bar{X}} = \mu$ and $\sigma_{\bar{X}} = \sigma/\sqrt{2}$, as stated in Theorem 7.1.

The population, taken as an equiprobable space, has mean $\mu = \dfrac{1+5+5+8}{4} = 5$, and standard deviation $\sigma = \sqrt{\dfrac{(1-5)^2+(5-5)^2+(6-5)^2+(8-5)^2}{4}} = \sqrt{\dfrac{26}{4}} = \dfrac{\sqrt{26}}{2}$. Using the probability distribution table in Problem 7.13, the mean of $\bar{X}$ is

$$\begin{aligned}\mu_{\bar{X}} &= 1\times\frac{1}{16}+3\times\frac{2}{16}+3.5\times\frac{2}{16}+4.5\times\frac{2}{16}+5\times\frac{1}{16}+5.5\times\frac{2}{16}+6\times\frac{1}{16}\\ &\quad+6.5\times\frac{2}{16}+7\times\frac{2}{16}+8\times\frac{1}{16}\\ &= \frac{80}{16} = 5\end{aligned}$$

which is the same as the population mean. The variance of $\bar{X}$ is

$$\begin{aligned}\sigma_{\bar{X}}^2 &= (1-5)^2\frac{1}{16}+(3-5)^2\frac{2}{16}+(3.5-5)^2\frac{2}{16}+(4.5-5)^2\frac{2}{16}+(5-5)^2\frac{1}{16}\\ &\quad+(5.5-5)^2\frac{2}{16}+(6-5)^2\frac{1}{16}+(6.5-5)^2\frac{2}{16}+(7-5)^2\frac{2}{16}+(8-5)^2\frac{1}{16}\\ &= \frac{1}{16}(16+8+4.5+0.5+0+0.5+1+4.5+8+9)\\ &= \frac{52}{16} = \frac{13}{4}\end{aligned}$$

Therefore, the standard deviation of $\bar{X}$ is $\sigma_{\bar{X}} = \dfrac{\sqrt{13}}{2}$. Since $\sigma = \dfrac{\sqrt{26}}{2}$ and $n = 2$, it follows that $\dfrac{\sigma}{\sqrt{n}} = \dfrac{\sqrt{26}}{2\sqrt{2}} = \dfrac{\sqrt{13}}{2} = \sigma_{\bar{X}}$, as stated in Theorem 7.1.

7.16. Let $S = \{1, 5, 6, 8\}$. Compute the mean $\mu_{\bar{X}}$ and standard deviation $\sigma_{\bar{X}}$ of $\bar{X}$ for random samples of size 2 drawn without replacement. Verify that $\mu_{\bar{X}} = \mu$ and $\sigma_{\bar{X}} = \dfrac{\sigma}{\sqrt{n}}\sqrt{\dfrac{N-n}{N-1}}$, as stated in Theorem 7.2.

As computed in Problem 7.15, the population mean and standard deviation are $\mu = 5$ and $\sigma = \dfrac{\sqrt{26}}{2}$. Using the probability distribution table in Problem 7.14, the mean of $\bar{X}$ is

$$\begin{aligned}\mu_{\bar{X}} &= 3\times\frac{1}{6}+3.5\times\frac{1}{6}+4.5\times\frac{1}{6}+5.5\times\frac{1}{6}+6.5\times\frac{1}{6}+7\times\frac{1}{6}\\ &= \frac{30}{6} = 5\end{aligned}$$

as stated in Theorem 7.2. The standard deviation of $\bar{X}$ is

$$\begin{aligned}\sigma_{\bar{X}} &= \sqrt{(3-5)^2\frac{1}{6}+(3.5-5)^2\frac{1}{6}+(4.5-5)^2\frac{1}{6}+(5.5-5)^2\frac{1}{6}+(6.5-5)^2\frac{1}{6}+(7-5)^2\frac{1}{6}}\\ &= \sqrt{\frac{13}{6}}\end{aligned}$$

Since $N = 4$ and $n = 2$, we have $\dfrac{\sigma}{\sqrt{n}}\sqrt{\dfrac{N-n}{N-1}} = \dfrac{\sqrt{26}}{2\sqrt{2}}\sqrt{\dfrac{2}{3}} = \dfrac{\sqrt{13}}{2}\dfrac{\sqrt{2}}{\sqrt{3}} = \dfrac{\sqrt{13}}{\sqrt{2}\sqrt{3}} = \sqrt{\dfrac{13}{6}}$, which is equal to $\sigma_{\bar{X}}$, as stated in Theorem 7.2.

7.17. Let $S = \{1, 5, 6, 8\}$. Find the probability distribution of the sample mean $\bar{X}$ for random samples of size 3 drawn (*a*) with replacement, (*b*) without replacement.

(*a*) There are $4^3 = 64$ random samples of size 3 drawn with replacement. These are shown in Problem 7.1. By finding the average of the three entries in each triple, we arrive at the following probability distribution.

$\bar{x}$	$p(\bar{x})$	$\bar{x}$	$p(\bar{x})$	$\bar{x}$	$p(\bar{x})$
1	1/64	13/3	3/64	19/3	6/64
7/3	3/64	14/3	6/64	20/3	3/64
8/3	3/64	5	7/64	7	3/64
10/3	3/64	16/3	3/64	22/3	3/64
11/3	3/64	17/3	6/64	24/3	1/64
4	6/64	6	4/64		

(*b*) There are $\binom{4}{3} = 4$ random samples of size 4 drawn without replacement. They are $\{1, 5, 6\}$, $\{1, 5, 8\}$, $\{1, 6, 8\}$, $\{5, 6, 8\}$. Computing the average of the entries in each of these, we arrive at the following probability distribution table.

$\bar{x}$	4	14/3	5	19/3
$p(\bar{x})$	1/4	1/4	1/4	1/4

7.18. Find $P(4 \le \bar{X} \le 6)$, where $\bar{X}$ is the sample mean for random samples of size 3 drawn with replacement from the population $\{1, 5, 6, 8\}$.

Using the probability distribution table in Problem 7.17(*a*), we find that

$$P(4 \le \bar{X} \le 6) = p(4) + p\left(\frac{13}{3}\right) + p\left(\frac{14}{3}\right) + p(5) + p\left(\frac{16}{3}\right) + p\left(\frac{17}{3}\right) + p(6)$$

$$= \frac{6}{64} + \frac{3}{64} + \frac{6}{64} + \frac{7}{64} + \frac{3}{64} + \frac{6}{64} + \frac{4}{64}$$

$$= \frac{35}{64} \approx 0.55$$

7.19. Find $P(4 \le \bar{X} \le 6)$, where $\bar{X}$ is the sample mean for random samples of size 3 drawn without replacement from the population $\{1, 5, 6, 8\}$.

Using the probability distribution table in Problem 7.17(*b*), we find that

$$P(4 \le \bar{X} \le 6) = p(4) + p\left(\frac{14}{3}\right) + p(5)$$

$$= \frac{1}{4} + \frac{1}{4} + \frac{1}{4} = \frac{3}{4}$$

7.20. Does the Central Limit Theorem (Theorem 7.3) apply to the sample mean $\bar{X}$ for random samples of size 36 drawn with replacement from the population $\{1, 5, 6, 8\}$? If so, use the theorem to compute $P(4 \le \bar{X} \le 6)$.

Since the sample size 36 is larger than 30, Theorem 7.3 does apply. Hence, we may assume that $\bar{X}$ is approximately normally distributed. Also, by Theorem 7.1, $\bar{X}$ has mean $\mu_{\bar{X}} = 5$ and standard deviation

$\sigma_{\bar{X}} = \dfrac{\sigma}{\sqrt{36}}$, where $\sigma = \dfrac{\sqrt{26}}{2}$ is the population standard deviation (as computed in Problem 7.15).

Therefore, $\sigma_{\bar{X}} = \dfrac{\sqrt{26}}{2\sqrt{36}} = \dfrac{\sqrt{26}}{12} \approx 0.4249$. Then

$$\begin{aligned} P(4 \le \bar{X} \le 6) &= P\left(\frac{4-5}{0.4249} \le \frac{\bar{X}-5}{0.4249} \le \frac{6-5}{0.4249}\right) \\ &\approx P(-2.35 \le Z \le 2.35) \end{aligned}$$

where Z is the standard normal random variable. Using a standard normal table, we find that $P(-2.35 \le Z \le 2.35) = 2P(0 \le Z \le 2.35) \approx 2(0.4906) \approx 0.98$.

7.21. Does the Central Limit Theorem (Theorem 7.3′) apply to the sample mean $\bar{X}$ for random samples drawn without replacement from the population {1, 5, 6, 8}?

No, only samples of size 4 or less can be drawn without replacement. Furthermore, the population size, 4, can never be much larger than the sample size.

SAMPLE PROPORTION

7.22. The proportion of unmarried men between ages 21 and 30 years in a town is $\frac{2}{3}$. Suppose random samples of size 16 are drawn with replacement from all men in the town between ages 21 and 30. What are the mean and standard deviation of the proportion $\hat{P}$ for all such samples?

By Theorem 7.4, the mean of $\hat{P}$ is $\frac{2}{3}$, and the standard deviation of $\hat{P}$ is

$$\sqrt{\frac{\frac{2}{3}(1-\frac{2}{3})}{16}} = \sqrt{\frac{\frac{2}{3}\times\frac{1}{3}}{16}} = \frac{\sqrt{2}}{12} \approx 0.1179$$

7.23. Suppose the town in Problem 7.22 has 225 men between ages 21 and 30 years, and the sampling is without replacement. Then what are the mean and standard deviation of $\hat{P}$?

By Theorem 7.5, the mean of $\hat{P}$ is still $\frac{2}{3}$; but the standard deviation is the standard deviation without replacement, 0.1179, multiplied by

$$\sqrt{\frac{225-16}{225-1}} = \sqrt{\frac{209}{224}} \approx 0.9659$$

Hence, the new standard deviation is approximately $0.1179 \times 0.9659 \approx 0.1139$.

7.24. The proportion of Democrats in a population consisting of three Democrats, D_1, D_2, D_3, and two Republicans, R_1, R_2 is $p = \frac{3}{5}$. There are 125 random samples of size $n = 3$ that can be drawn with replacement from the population. Find the probability distribution for the sample proportion $\hat{P}$ of Democrats defined by the collection of all 125 such samples (see also Problem 7.60).

Let $\hat{p}$ denote the proportion of Democrats in a given sample; $\hat{p}$ can assume the values: 1 (three Democrats), $\frac{2}{3}$ (two Democrats, one Republican), $\frac{1}{3}$ (one Democrat, two Republicans), 0 (three Republicans). Problem 7.7 gives the breakdown of the samples into these categories, which results in the following probability distribution table.

Category	$\hat{p}$	Frequency	$P(\hat{p})$
3 Democrats	1	27	$\frac{27}{125}$
2 Democrats, 1 Republican	$\frac{2}{3}$	54	$\frac{54}{125}$
1 Democrat, 2 Republicans	$\frac{1}{3}$	36	$\frac{36}{125}$
3 Republicans	0	8	$\frac{8}{125}$

7.25. Verify that the sample proportion $\hat{P}$ in Problem 7.24 has mean $p = \frac{3}{5}$ and standard deviation $\sqrt{\frac{p(1-p)}{n}}$, as stated in Theorem 7.4.

From the probability distribution table in Problem 7.24, the mean of $\hat{P}$ is

$$\sum \hat{p}P(\hat{p}) = 1 \times \frac{27}{125} + \frac{2}{3} \times \frac{54}{125} + \frac{1}{3} \times \frac{36}{125} + 0 \times \frac{8}{125} = \frac{75}{125} = \frac{3}{5}$$

The variance of $\hat{P}$ is

$$\begin{aligned}
\sum (\hat{p} - p)^2 P(\hat{p}) &= \left(1 - \frac{3}{5}\right)^2 \frac{27}{125} + \left(\frac{2}{3} - \frac{3}{5}\right)^2 \frac{54}{125} + \left(\frac{1}{3} - \frac{3}{5}\right)^2 \frac{36}{125} + \left(0 - \frac{3}{5}\right)^2 \frac{8}{125} \\
&= \frac{4}{25} \cdot \frac{27}{125} + \frac{1}{225} \cdot \frac{54}{125} + \frac{16}{225} \cdot \frac{36}{125} + \frac{9}{25} \cdot \frac{8}{125} \\
&= \frac{4}{25} \cdot \frac{27}{125} + \frac{1}{25} \cdot \frac{6}{125} + \frac{16}{25} \cdot \frac{4}{125} + \frac{9}{25} \cdot \frac{8}{125} \\
&= \frac{250}{25 \times 125} \\
&= \frac{2}{25}
\end{aligned}$$

Therefore, the standard deviation of $\hat{P}$ is $\frac{\sqrt{2}}{5}$. Also,

$$\sqrt{\frac{p(1-p)}{n}} = \sqrt{\frac{\frac{3}{5}(1 - \frac{3}{5})}{3}} = \frac{\sqrt{2}}{5}$$

7.26. There are only $\binom{5}{3} = 10$ random samples of size $n = 3$ that can be drawn without replacement from the population D_1, D_2, D_3, R_1, R_2. Find the probability distribution for the sample proportion $\hat{P}$ of Democrats defined by the collection of all 10 random samples.

The ten random samples of size $n = 3$, drawn without replacement are:

$$\{D_1, D_2, D_3\},\ \{D_1, D_2, R_1\},\ \{D_1, D_2, R_2\},\ \{D_1, D_3, R_1\},\ \{D_1, D_3, R_2\},$$
$$\{D_2, D_3, R_1\},\ \{D_2, D_3, R_2\},\ \{D_1, R_1, R_3\},\ \{D_2, R_1, R_2\},\ \{D_3, R_1, R_3\}$$

Let $\hat{p}$ denote the proportion of Democrats in a given sample; $\hat{p}$ can assume the values: 1 (three Democrats), $\frac{2}{3}$ (two Democrats, one Republican), or $\frac{1}{3}$ (one Democrat, two Republicans). We obtain the following probability distribution table.

Category	$\hat{p}$	Frequency	$P(\hat{p})$
3 Democrats	1	1	$\frac{1}{10}$
2 Democrats, 1 Republican	$\frac{2}{3}$	6	$\frac{6}{10}$
1 Democrat, 2 Republicans	$\frac{1}{3}$	3	$\frac{3}{10}$

7.27. Verify that the sample proportion $\hat{P}$ in Problem 7.26 has mean $p = \frac{3}{5}$ and standard deviation $\sqrt{\frac{p(1-p)}{n}} \cdot \sqrt{\frac{N-n}{N-1}}$, as stated in Theorem 7.5.

From the probability distribution table in Problem 7.26, the mean of $\hat{P}$ is

$$\sum \hat{p}P(\hat{p}) = 1 \times \frac{1}{10} + \frac{2}{3} \times \frac{6}{20} + \frac{1}{3} \times \frac{3}{10} = \frac{6}{10} = \frac{3}{5}$$

The standard deviation is

$$\sqrt{\left(1-\frac{3}{5}\right)^2 \frac{1}{10} + \left(\frac{2}{3}-\frac{3}{5}\right)^2 \frac{6}{10} + \left(\frac{1}{3}-\frac{3}{5}\right)^2 \frac{3}{10}} = \sqrt{\left(\frac{2}{5}\right)^2 \frac{1}{10} + \left(\frac{1}{15}\right)^2 \frac{6}{10} + \left(\frac{4}{15}\right)^2 \frac{3}{10}}$$

$$= \sqrt{\frac{90}{2250}} = \sqrt{\frac{1}{15}} = \frac{1}{5}$$

Furthermore,

$$\sqrt{\frac{p(1-p)}{n}} \cdot \sqrt{\frac{N-n}{n-1}} = \sqrt{\frac{\frac{3}{5}(1-\frac{3}{5})}{3}} \cdot \sqrt{\frac{5-3}{5-1}}$$

$$= \frac{\sqrt{2}}{5} \cdot \sqrt{\frac{1}{2}} = \frac{1}{5}$$

SAMPLE VARIANCE

7.28. Suppose Z_1, Z_2, Z_3 are three independent standard normal random variables. Use these to generate three chi-square random variables, each with 2 degrees of freedom. What are the mean and variance of each of the three chi-square random variables?

$Z_1^2 + Z_2^2$, $Z_1^2 + Z_3^2$, and $Z_2^2 + Z_3^2$ are each chi-square random variables with $k = 2$ degrees of freedom. Each one has mean 2 and variance $2k = 4$.

7.29. Suppose Z is a standard normal random variable. Is $Z^2 + Z^2$ a chi-square random variable with 2 degrees of freedom?

No. If $Z^2 + Z^2$ were $\chi^2(2)$, it would have variance 4, as in Problem 7.28, but $Z^2 + Z^2 = 2Z^2$, and $\text{Var}(2Z^2) = 2^2\,\text{Var}(Z^2) = 4 \times 2 = 8$.

7.30. Let X_1, X_2 be two random variables, each with mean μ. Show that $\sum_{i=1}^{2}(X_i - \bar{X})^2 = \sum_{i=1}^{2}(X_i - \mu)^2 - 2(\bar{X} - \mu)^2$, where $\bar{X}$ is the sample mean.

$$\begin{aligned}
\sum_{i=1}^{2}(X_i - \bar{X})^2 &= (X_1 - \bar{X})^2 + (X_2 - \bar{X})^2 \\
&= [(X_1 - \mu) + (\mu - \bar{X})]^2 + [(X_2 - \mu) + (\mu - \bar{X})]^2 \\
&= (X_1 - \mu)^2 + 2(X_1 - \mu)(\mu - \bar{X}) + (\mu - \bar{X})^2 \\
&\quad + (X_2 - \mu)^2 + 2(X_2 - \mu)(\mu - \bar{X}) + (\mu - \bar{X})^2 \\
&= \sum_{i=1}^{2}(X_i - \mu)^2 + (\mu - \bar{X})(2X_1 - 2\mu + \mu - \bar{X} + 2X_2 - 2\mu + \mu - \bar{X}) \\
&= \sum_{i=1}^{2}(X_i - \mu)^2 + (\mu - \bar{X})(2\bar{X} - 2\mu) \\
&= \sum_{i=1}^{2}(X_i - \mu)^2 - 2(\bar{X} - \mu)^2
\end{aligned}$$

The same procedure can be used to show that, for any positive integer n, $\sum_{i=1}^{n}(X_i - \bar{X})^2 = \sum_{i=1}^{n}(X_i - \mu)^2 - n(\bar{X} - \mu)^2$.

7.31. Let $X_1, X_2, \ldots, X_n$ be n independent random variables, each with mean μ and standard deviation σ. Show that the expected value of the sample variance,

$$S^2 = \frac{(X_1 - \bar{X})^2 + (X_2 - \bar{X})^2 + \cdots + (X_n - \bar{X})^2}{n-1}$$

is equal to σ^2

First note that $\sigma^2_{X_i} = E(X_i - \mu)^2 = \sigma^2$ and $\sigma^2_{\bar{X}} = E(\bar{X} - \mu)^2 = \sigma^2/n$ by Theorem 7.1. Then

$$\begin{aligned} E(S^2) = E\left(\frac{\sum(X_i - \bar{X})^2}{n-1}\right) &= \frac{1}{n-1}E\left(\sum(X_i - \mu)^2 - n(\bar{X} - \mu)^2\right) \\ &= \frac{1}{n-1}\sum E(X_i - \mu)^2 - \frac{n}{n-1}E(\bar{X} - \mu)^2 \\ &= \frac{1}{n-1}\sum \sigma^2 - \frac{n}{n-1}\cdot\frac{\sigma^2}{n} \\ &= \frac{n\sigma^2}{n-1} - \frac{\sigma^2}{n-1} \\ &= \sigma^2 \end{aligned}$$

7.32. Let $S = \{1, 5, 6, 8\}$. Find the probability distribution of the sample variance S^2 for random samples of size 3 drawn without replacement.

There are four random samples of size 3 drawn without replacement: {1, 5, 6}, {1, 5, 8}, {1, 6, 8}, {5, 6, 8}. There are four corresponding values of

$$S^2 = \frac{(X_1 - \bar{X})^2 + (X_2 - \bar{X})^2 + (X_3 - \bar{X})^2}{2}$$

and each has probability $\frac{1}{4}$, as indicated in the following table.

Sample	X_1	X_2	X_3	$\bar{X}$	S^2	p
{1, 5, 6}	1	5	6	4	7	$\frac{1}{4}$
{1, 5, 8}	1	5	8	$\frac{14}{3}$	$\frac{37}{3}$	$\frac{1}{4}$
{1, 6, 8}	1	6	8	5	13	$\frac{1}{4}$
{5, 6, 8}	5	6	8	$\frac{19}{3}$	$\frac{7}{3}$	$\frac{1}{4}$

7.33. Use the probability distribution determined in Problem 7.32 to compute the mean μ_{S^2} and the standard deviation σ_{S^2} of the sample variance S^2 for random samples of size 3 drawn without replacement from the population {1, 5, 6, 8}.

The mean of S^2 is $\mu_{S^2} = 7 \times \frac{1}{4} + \frac{37}{3} \times \frac{1}{4} + 13 \times \frac{1}{4} + \frac{7}{3} \times \frac{1}{4} = \frac{104}{12} = \frac{26}{3}$; and the variance of S^2 is

$$\begin{aligned} \sigma^2_{S^2} &= \left(7 - \frac{26}{3}\right)^2 \times \frac{1}{4} + \left(\frac{37}{3} - \frac{26}{3}\right)^2 \times \frac{1}{4} + \left(13 - \frac{26}{3}\right)^2 \times \frac{1}{4} + \left(\frac{7}{3} - \frac{26}{3}\right)^2 \times \frac{1}{4} \\ &= \frac{676}{36} \\ &= \frac{169}{9} \end{aligned}$$

Therefore, the standard deviation of S^2 is $\sigma_{S^2} = \sqrt{\frac{169}{9}} = \frac{13}{3}$.

7.34. It can be shown that when sampling without replacement, the mean of the corresponding sample variance is $\mu_{S^2} = \frac{N}{N-1}\sigma^2$, where σ^2 is the population variance, taken as an equiprobable space, and N is the population size. Use Problem 7.33 to verify this result for the population $\{1, 5, 6, 8\}$, when samples of size 3 are drawn without replacement.

In Problem 7.15 it was determined that the population variance for $\{1, 5, 6, 8\}$, taken as an equiprobable space, is $\sigma^2 = \frac{26}{4}$. From Problem 7.33, we have $\mu_{S^2} = \frac{26}{3}$. Since $N = 4$, we get $\frac{N}{N-1}\sigma^2 = \frac{4}{3} \cdot \frac{26}{4} = \frac{26}{3}$, as was to be shown.

7.35. Suppose samples of size 10 corresponding to a population random variable X are drawn without replacement. Suppose also that X is normal, with mean 75 and standard deviation 5. What are the mean μ_{S^2} and standard deviation σ_{S^2} of the sample variance S^2?

The mean μ_{S^2} of S^2 is equal to the variance of X regardless of the sample size. Therefore, $\mu_{S^2} = 5^2 = 25$. Also, by Theorem 7.7, $9s^2/25$ is chi-square with 9 degrees of freedom. Therefore, the standard deviation of $9s^2/25$ is $\sqrt{2 \times 9} = 3\sqrt{2}$; and the standard deviation of S^2 is $\sigma_{S^2} = \frac{25}{9} \times 3\sqrt{2} = \frac{25\sqrt{2}}{3}$.

Supplementary Problems

INTRODUCTION: SAMPLING WITH AND WITHOUT REPLACEMENT

7.36. How many samples of size 3 can be drawn from $S = \{2, 4, 8, 10, 12\}$, (*a*) with replacement, (*b*) without replacement?

7.37. In Problem 7.36, how many of the samples drawn with replacement have three different numbers?

7.38. If a population has size 10, what is the sample size n for which there are the most samples drawn (*a*) with replacement, (*b*) without replacement?

7.39. Repeat Problem 7.38 for a population of size N.

7.40. If a student guesses each answer in a 5-question True–False test, what is the most likely number of correct answers the student will get? What is the least likely number of correct answers the student will get?

7.41. Suppose there are 20 business majors in a statistics class of 32 students. If a random sample of 4 students is chosen without replacement, what is the probability that at least 2 of them will be business majors?

7.42. Repeat Problem 7.41 if the samples are chosen with replacement.

SAMPLE MEAN

7.43. A population random variable X has mean 75 and standard deviation 8. Find the mean and standard deviation of $\bar{X}$, based on random samples of size 25 taken with replacement.

7.44. Repeat Problem 7.43 if the random samples are taken without replacement.

7.45. Suppose the random variable X in Problem 7.43 is approximately normally distributed. Find $P(72 \leq \bar{X} \leq 78)$.

7.46. Repeat Problem 7.45 if the samples are taken without replacement, and the population has size 400.

7.47. SAT scores are approximately normally distributed with mean 500 and standard deviation 100. If a random sample of size 50 is taken, what is $P(\bar{X} \geq 525)$?

7.48. With reference to Problem 7.47, how large must the sample size be so that $P(475 \leq \bar{X} \leq 525) = 0.95$?

7.49. A population random variable X has mean 250 and standard deviation 75. Suppose $\bar{X}$ has standard deviation 13.5, based on random samples of size 25 taken without replacement. How large is the population?

7.50. Let $S = \{2, 4, 8, 16, 32\}$. Find the probability distribution of the sample mean $\bar{X}$ for samples of size 2 drawn without replacement.

7.51. Find the mean and standard deviation of $\bar{X}$ in Problem 7.50.

7.52. Repeat Problem 7.51 if the samples are drawn with replacement.

7.53. A population random variable X has mean 25 and standard deviation 5. Samples of size 40 are drawn with replacement. Find $P(24 \leq \bar{X} \leq 26)$.

7.54. Suppose the waiting time for a bus is a random variable with mean 8 minutes and standard deviation 4 minutes. In a given month, what is the probability that the average waiting time is less than 6 minutes?

7.55. Let X be a 4-place decimal number drawn at random from the interval [0, 10]. X has mean 5 and standard deviation 2.89. Suppose 100 numbers are drawn at random from the interval. What is the probability that the average of the numbers is between 4.8 and 5.2?

SAMPLE PROPORTION

7.56. Thirty-three percent of the first-year students at an urban university live in university housing. What are the mean and standard deviation of the proportion $\hat{P}$ of first-year students in university housing for all samples of size 50, drawn with replacement, from the population of first-year students?

7.57. With reference to Problem 7.56, suppose there are a total of 3970 first-year students. What are the mean and standard deviation of the proportion $\hat{P}$ if the samples are drawn without replacement?

7.58. With reference to Problem 7.56, what is the probability that between 15 and 18 of first-year students in a random sample of 50 live in university housing?

7.59. Show that if the random variable $\hat{P}$ is the sample proportion, with mean p and variance $\dfrac{p(1-p)}{n}$, corresponding to random samples of size n, then $n\hat{P}$ is a binomial random variable with mean np and variance $np(1-p)$.

7.60. In Problem 7.24, the probability distribution of the sample proportion $\hat{P}$ of Democrats in random samples of size 3, drawn with replacement from a population of three Democrats and two Republicans, was obtained by listing the frequency of all possible proportions in samples of size 3. Use the fact that $n\hat{P}$ is a binomial random variable (see Problem 7.59) to obtain the probability distribution $\hat{P}$ without listing all possible frequencies.

7.61. A population is broken down into two categories, A and B. Suppose the proportion of the population in category A is 0.7, and let $\hat{P}$ be the proportion in category A in random samples of size 5 drawn with replacement from the population. Use the fact that $n\hat{P}$ is a binomial random variable (Problem 7.59) to find the probability distribution of $\hat{P}$.

7.62. A population is broken down into two categories A and B, and p is the proportion in category A. Selecting a single individual from the population can be modeled as a Bernoulli random variable X, where $X = 1$ if the individual is in category A, and $X = 0$ if the individual is in category B. Show that the sample mean $\bar{X}$, corresponding to random samples of size n, is the proportion $\hat{P}$ of individuals in the sample that are in category A.

7.63. Since the random variable X in Problem 7.62 is a Bernoulli random variable, X has mean $\mu = p$ and standard deviation $\sigma = \sqrt{p(1-p)}$. Use these equations and the fact that $\bar{X} = \hat{P}$ to show that Theorem 7.4 follows from Theorem 7.1.

7.64. Use the equations in Problem 7.63 and the fact that $\bar{X} = \hat{P}$ (Problem 7.62) to show that Theorem 7.5 follows from Theorem 7.2.

7.65. Use the results of the previous two problems to show that Theorem 7.6 follows from Theorem 7.3′.

SAMPLE VARIANCE

7.66. Let $X_1, X_2, \ldots, X_n$ be n independent normal random variables, each with mean 20 and variance 4. Explain why $\frac{(X_1-20)^2}{4} + \frac{(X_2-20)^2}{4} + \cdots + \frac{(X_n-20)^2}{4}$ is a chi-square random variable with n degrees of freedom.

7.67. Let X_1, X_2, X_3 be three random variables, each with mean 25 and variance 7, and let $\bar{X}$ be the sample mean. Show that

$$\sum \frac{(X_i - \bar{X})^2}{7} = \sum \frac{(X_i - 25)^2}{7} - \frac{(\bar{X} - 25)^2}{7/3}$$

7.68. As stated in Example 7.12, the annual SAT scores are approximately normally distributed with mean $\mu = 500$ and standard deviation $\sigma = 100$. Let S^2 be the sample variance defined for random samples of 25 SAT scores. Find the mean and standard deviation of S^2.

7.69. With reference to Problem 7.68, for what value $\hat{S}^2$ of S^2 is $P(S^2 \leq \hat{S}^2) = 0.95$?

7.70. With reference to Problem 7.68, for what value $\hat{S}^2$ of S^2 is $P(S^2 \geq \hat{S}^2) = 0.95$?

Answers to Supplementary Problems

7.36. (*a*) $5^3 = 125$; (*b*) $\binom{5}{3} = 10$

7.37. $5 \cdot 4 \cdot 3 = 60$

7.38. (*a*) There is no such n since the number of samples drawn with replacement, which is 10^n, increases as n increases. (*b*) $\binom{10}{5} = 252$.

7.39. (*a*) There is no sample size n which gives the most samples drawn with replacement; the number of such samples, N^n, increases as n increases.
(*b*) If N is even, then the maximum number of samples of size n, drawn without replacement, occurs when $n = N/2$. If N is odd, then the maximum number occurs when $n = (N-1)/2$ and when $n = (N+1)/2$.

7.40. The probability of exactly n correct answers is $P(n) = \binom{5}{n} \times (0.5)^5$, which is a maximum when n is either 2 or 3, and is a minimum when n is either 0 or 5. Hence the most likely number of correct answers is 2 or 3, and the least likely is 0 or 5.

7.41. $1 - [P(0) + P(1)] = 1 - \left[\binom{20}{0}\binom{12}{4}\bigg/\binom{32}{4} + \binom{20}{1}\binom{12}{3}\bigg/\binom{32}{4}\right] \approx 0.86$

7.42. $1 - [P(0) + P(1)] = 1 - \left[\left(\frac{12}{32}\right)^4 + 4 \times \frac{20}{32} \times \left(\frac{12}{32}\right)^3\right] \approx 0.85$

7.43. $\mu_{\bar{X}} = 75$; $\sigma_{\bar{X}} = 8/\sqrt{25} = 1.6$

7.44. $\mu_{\bar{X}} = 75$; $\sigma_{\bar{X}} = \frac{8}{\sqrt{25}}\sqrt{\frac{N-25}{N-1}} = 1.6\sqrt{\frac{N-25}{N-1}}$, where N is the size of the population.

7.45. $P(72 \le \bar{X} \le 78) = P\left(\frac{72-75}{1.6} \le \frac{\bar{X}-75}{1.6} \le \frac{78-75}{1.6}\right) \approx P(-1.875 \le Z \le 1.875) \approx 0.94$

7.46. $P(72 \le \bar{X} \le 78) = P\left(\frac{72-75}{1.55} \le \frac{\bar{X}-75}{1.55} \le \frac{78-75}{1.55}\right) \approx P(-1.935 \le Z \le 1.935) \approx 0.95$

7.47. $P(\bar{X} \ge 525) = P\left(\frac{\bar{X}-500}{100/\sqrt{50}} \ge \frac{525-500}{100/\sqrt{50}}\right) \approx P(Z \ge 1.77) \approx 0.04$

7.48. $P(475 \le \bar{X} \le 525) = P\left(\frac{475-500}{100/\sqrt{n}} \le \frac{\bar{X}-500}{100/\sqrt{n}} \le \frac{525-500}{100/\sqrt{n}}\right) \approx P\left(-\frac{\sqrt{n}}{4} \le Z \le \frac{\sqrt{n}}{4}\right) = 0.95$ for $\frac{\sqrt{n}}{4} = 1.96$, or $n = 61.5$; round up to $n = 62$.

7.49. $\frac{75}{\sqrt{25}}\sqrt{\frac{N-25}{N-1}} = 13.5$; $\sqrt{\frac{N-25}{N-1}} = \frac{13.5}{15} = 0.9$; $N = 128$

7.50.

$\bar{x}$	3	5	6	9	10	12	17	18	20	24
$P(\bar{x})$	0.1	0.1	0.1	0.1	0.1	0.1	0.1	0.1	0.1	0.1

7.51. $\mu_{\bar{X}} = 12.4$; $\sigma_{\bar{X}} = 6.68$

7.52. $\mu_{\bar{X}} = 12.4$; $\sigma_{\bar{X}} = 7.715$

7.53. $P(24 \le \bar{X} \le 26) = P\left(\frac{24-25}{5/\sqrt{40}} \le \frac{\bar{X}-25}{5/\sqrt{40}} \le \frac{26-25}{5/\sqrt{40}}\right) \approx P(-1.26 \le Z \le 1.26) = 0.79$

7.54. For a 30 day month, $P(\bar{X} < 6) = P\left(\frac{\bar{X}-8}{4/\sqrt{30}} < \frac{6-8}{4/\sqrt{30}}\right) \approx P(Z < -2.74) = 0.003$

7.55. $P(4.8 \le \bar{X} \le 5.2) = P\left(\frac{4.8-5}{2.89/\sqrt{100}} \le \frac{\bar{X}-5}{2.89/\sqrt{100}} \le \frac{5.2-5}{2.89/\sqrt{100}}\right) \approx P(-0.69) \le Z \le 0.69) = 0.51$

7.56. $\mu_{\hat{P}} = 0.33$; $\sigma_{\hat{P}} = \sqrt{\frac{0.33(1-0.33)}{50}} = 0.0665$

7.57. $\mu_{\hat{P}} = 0.33$; $\sigma_{\hat{P}} = \sqrt{\frac{0.33(1-0.33)}{50}} \cdot \sqrt{\frac{3970-50}{3970-1}} = 0.0661$

7.58. $P\left(\frac{15}{50} \le \hat{P} \le \frac{18}{50}\right) = P\left(\frac{0.3-0.33}{0.0665} \le \frac{\hat{P}-0.33}{0.0665} \le \frac{0.36-0.33}{0.0665}\right) \approx P(-0.45 \le Z \le 0.45) = 0.35$

7.59. $n\hat{P}$ is the number of "successes" in n trials, where p is the probability of success, due to sampling with replacement, in each trial.

7.60. $p = \frac{3}{5} = 0.6$; $P(\hat{P} = 0) = P(3\hat{P} = 0) = (0.4)^3 = 0.064$

$P\left(\hat{P} = \frac{1}{3}\right) = P(3\hat{P} = 1) = 3 \times 0.6 \times (0.4)^2 = 0.288$

$P\left(\hat{P} = \frac{2}{3}\right) = P(3\hat{P} = 2) = 3 \times (0.6)^2 \times 0.4 = 0.432$

$P(\hat{P} = 1) = P(3\hat{P} = 3) = (0.6)^3 = 0.216$

7.61. $P(\hat{P} = 0) = P(5\hat{P} = 0) = (0.3)^5 = 0.00243$

$P\left(\hat{P} = \frac{1}{5}\right) = P(5\hat{P} = 1) = 5 \times 0.7 \times (0.3)^4 = 0.02835$

$P\left(\hat{P} = \frac{2}{5}\right) = P(5\hat{P} = 2) = 10 \times (0.7)^2 \times (0.3)^3 = 0.1323$

$P\left(\hat{P} = \frac{3}{5}\right) = P(5\hat{P} = 3) = 10 \times (0.7)^3 \times (0.3)^2 = 0.3087$

$P\left(\hat{P} = \frac{4}{5}\right) = P(5\hat{P} = 4) = 5 \times (0.7)^4 \times 0.3 = 0.36015$

$P(\hat{P} = 1) = P(5\hat{P} = 5) = (0.7)^5 = 0.16807$

7.62. $\bar{x} = \dfrac{x_1 + x_2 + \cdots + x_n}{n} = \dfrac{\text{number of } x_i\text{s equal to 1}}{n} = \hat{p}$

7.63. $\mu_{\hat{P}} = \mu_{\bar{X}} = p$ by Theorem 7.1, and $\sigma_{\hat{P}} = \sigma_{\bar{X}} = \dfrac{\sigma}{\sqrt{n}} = \sqrt{\dfrac{p(1-p)}{n}}$, also by Theorem 7.1.

7.64. $\mu_{\hat{P}} - \mu_{\bar{X}} = p$ by Theorem 7.2, and $\sigma_{\hat{P}} = \sigma_{\bar{X}} = \dfrac{\sigma}{\sqrt{n}} \cdot \sqrt{\dfrac{N-n}{N-1}} = \sqrt{\dfrac{p(1-p)}{n}} \cdot \sqrt{\dfrac{N-n}{N-1}}$, also by Theorem 7.2.

7.65. $\hat{P} = \bar{X}$, and $\bar{X}$ is approximately normally distributed with mean p and standard deviation $\sigma_{\bar{X}} = \dfrac{\sigma}{\sqrt{n}} = \sqrt{\dfrac{p(1-p)}{n}} = \sigma_{\hat{P}}$ by Theorem 7.6.

7.66. $\dfrac{(X_1 - 20)^2}{4}, \dfrac{(X_2 - 20)^2}{4}, \ldots, \dfrac{(X_n - 20)^2}{4}$ are n independent normal random variables, each with mean 0 and standard deviation 1. By definition, their sum is a chi-square random variable with n degrees of freedom.

7.67.

$$\begin{aligned}
\sum \frac{(X_i - 25)^2}{7} &= \sum \frac{(X_i - \bar{X} + \bar{X} - 25)^2}{7} \\
&= \sum \left[\frac{(X_i - \bar{X})^2}{7} + \frac{2(X_i - \bar{X})(\bar{X} - 25)}{7} + \frac{(\bar{X} - 25)^2}{7} \right] \\
&= \sum \frac{(X_i - \bar{X})^2}{7} + \frac{2(\bar{X} - 25)}{7} \sum (X_i - \bar{X})^2 + \sum \frac{(\bar{X} - 25)^2}{7} \\
&= \sum \frac{(X_i - \bar{X})^2}{7} + 0 + 3 \sum \frac{(\bar{X} - 25)^2}{7}
\end{aligned}$$

7.68. $\mu_{S^2} = \sigma^2 = 10{,}000$; $\sigma_{S^2} = [\sqrt{2/(n-1)}]\sigma^2 = \sqrt{\dfrac{2}{24}} \times 10{,}000 = \dfrac{5000}{\sqrt{3}}$

7.69. $P(S^2 \le \hat{S}^2) = P\left(\dfrac{24S^2}{10{,}000} \le \dfrac{24\hat{S}^2}{10{,}000}\right) \approx P(\chi^2(24) \le 0.0024\hat{S}^2) = 0.95$ for $0.0024\hat{S}^2 = 36.4$, or $\hat{S}^2 = 15{,}166.67$

7.70. $P(S^2 \ge \hat{S}^2) = P\left(\dfrac{24S^2}{10{,}000} \ge \dfrac{24\hat{S}^2}{10{,}000}\right) \approx P(\chi^2(24) \ge 0.0024\hat{S}^2) = 0.95$

$P(\chi^2(24) \le 0.0024\hat{S}^2) = 0.05$ for $0.0024\hat{S}^2 = 13.8$, or $\hat{S}^2 = 5750$

Chapter 8

Confidence Intervals for a Single Population

8.1 PARAMETERS AND STATISTICS

The mean μ and standard deviation σ of a population random variable X are called *parameters*; and the mean $\bar{x}$ and standard deviation s of a random sample are called *statistics*. In general, any numerical characteristic of a population is called a *parameter*, and any quantity computed from a random sample is called a *statistic*. Statistics are used to estimate parameters.

EXAMPLE 8.1 In 1994, the median income for all four-person families in the state of Pennsylvania was $49,120. A random sample of 25 four-person families in Pennsylvania had a median income of $48,500. The value $49,120 is a parameter, and $48,500 is a statistic.

A random variable defined for random samples is called a *statistic* if it does not explicitly depend on any unknown population parameters. For instance, the sample mean $\bar{X}$ and the sample variance S^2 are statistics. If the value of the parameter σ is known, then S^2/σ^2 is a statistic; but if the value of σ is not known, then S^2/σ^2 is not a statistic. Because their values are used to estimate parameters, random-variable statistics should not depend on unknown population parameters.

Note that the word "statistic" can refer to a numerical value, as in $\bar{x}$, and also to a random variable, as in $\bar{X}$. We will know from context which meaning to attach to the word.

Random-Variable Samples

As stated at the end of Chapter 7, we will assume from here on that either samples are chosen with replacement or that the population is large in comparison with the size of random samples, so that, in effect, the values in a random sample are independent. If $x_1, x_2, \ldots, x_n$ are a random sample of values of a random variable X, it is often convenient to consider them to be values of n independent random variables $X_1, X_2, \ldots, X_n$, each with the same probability distribution as X. Then the collection $X_1, X_2, \ldots, X_n$ is called a *random-variable sample corresponding to* X or, simply, a *random sample*. Hence, just as the term "statistic" can refer to either a numerical value or a random variable, the expression "random sample" can refer to a collection of numerical values or a collection of independent random variables. We will know from context what meaning to attach to these terms.

EXAMPLE 8.2 Suppose X is a normal random variable with mean 100 and standard deviation 8. Then a collection of three independent normal random variables X_1, X_2, X_3, each with mean 100 and standard deviation 8, is a random-variable sample corresponding to X. If the values $x_1 = 104, x_2 = 92, x_3 = 100$ of X are obtained in a random sample, then 104, 92, and 110 can be considered to be sample values of X_1, X_2, and X_3, respectively.

Point Estimates

A value of a statistic used to estimate a population parameter is called a *point estimate* of the parameter. In Example 8.1, the median income $48,500 for a sample of 25 four-person families is a point estimate of the median income $49,120 for all four-person families in Pennsylvania. A different sample would most likely yield a different point estimate of the median income for the population.

Unbiased and Biased Estimators

A random-variable statistic is called an *unbiased estimator* of a population parameter if the expected value of the statistic is equal to the parameter. The sample mean $\bar{X}$ and sample variance S^2 are unbiased estimators of the corresponding population mean μ and variance σ^2, respectively. That is,

$$E(\bar{X}) = \mu \qquad \text{and} \qquad E(S^2) = \sigma^2$$

(see Problems 8.5 and 7.31). Also, suppose a population is divided into two groups, and the members of one group are designated as "successors". Let p be the proportion of successes in the population. The sample proportion $\hat{P}$, whose value on a random sample of size n is the proportion of successes in the sample (see Section 7.3), has mean p and variance $\dfrac{p(1-p)}{n}$. $\hat{P}$ is an unbiased estimator of p, and $\dfrac{1}{n-1}\hat{P}(1-\hat{P})$ is an unbiased estimator of $\dfrac{p(1-p)}{n}$. That is,

$$E(\hat{P}) = p \qquad \text{and} \qquad E\left(\frac{1}{n-1}\hat{P}(1-\hat{P})\right) = \frac{p(1-p)}{n}$$

(see Problems 8.6 and 8.7).

In general, the value of an unbiased estimator obtained from a numerical random sample is called an *unbiased point estimate* of the corresponding population parameter.

EXAMPLE 8.3 With reference to Example 8.2, find unbiased point estimates of the mean μ and variance σ^2 of X, based on the sample values $x_1 = 104, x_2 = 92, x_3 = 110$.

$\bar{x} = \dfrac{1}{n}\sum x_i = \dfrac{104 + 92 + 110}{3} = \dfrac{306}{3} = 102$ is an unbiased point estimate of μ, which is equal to 100; and

$$s^2 = \frac{1}{n-1}\sum (x_i - \bar{x})^2 = \frac{(104-102)^2 + (92-102)^2 + (110-102)^2}{2} = \frac{4 + 100 + 64}{2} = 84$$

is an unbiased point estimate of σ^2, which is equal to 64.

A random-variable statistic used to estimate a population parameter is called a *biased estimator* if the expected value of the statistic is not equal to the parameter. For example, the statistic

$$\tilde{S}^2 = \frac{(X_1 - \bar{X})^2 + (X_2 - \bar{X})^2 + \cdots + (X_n - \bar{X})^2}{n}$$

is a biased estimator of the population variance σ^2 since

$$\tilde{S}^2 = \frac{n-1}{n}S^2 \qquad \text{and} \qquad E(\tilde{S}^2) = \frac{n-1}{n}E(S^2) = \frac{n-1}{n}\sigma^2 \neq \sigma^2$$

It can also be shown that the sample standard deviation $S = \sqrt{S^2}$ is a biased estimator of the population standard deviation σ. That is, $E(S) \neq \sigma$. Finally, $\dfrac{1}{n}\hat{P}(1-\hat{P})$ is a biased estimator of $\dfrac{p(1-p)}{n}$.

In general, the value of a biased estimator obtained from a numerical random sample is called a *biased point estimate* of the corresponding population parameter.

8.2 THE NOTION OF A CONFIDENCE INTERVAL

In addition to point estimates of a parameter, there are *interval estimates* which stipulate, with a certain degree of confidence, that the parameter lies between two values of the estimating statistic. Suppose a population random variable X has mean μ whose value is unknown. From a random sample of size n, the value $\bar{x}$ of the sample mean $\bar{X}$ can be used to estimate μ at the 95 percent confidence level as follows. First, determine the value E (see Example 8.4) for which

$$P(\mu - E \leq \bar{X} \leq \mu + E) = 0.95$$

which is equivalent to the equation

$$P(\bar{X} - E \leq \mu \leq \bar{X} + E) = 0.95$$

(see Problem 8.10). Then $[\bar{X} - E, \bar{X} + E]$ is called a *random 95 percent confidence interval* for μ, and E is called the *margin of error*. This means that the probability is 0.95 that a random sample will result in a value $\bar{x}$ of $\bar{X}$ for which the numerical interval $[\bar{x} - E, \bar{x} + E]$ contains μ. In other words, as $\bar{x}$ ranges through all possible values of $\bar{X}$, 95 percent of the intervals $[\bar{x} - E, \bar{x} + E]$ will contain μ (see Fig. 8-1). Each interval $[\bar{x} - E, \bar{x} + E]$ is called a *95 percent confidence interval* for μ.

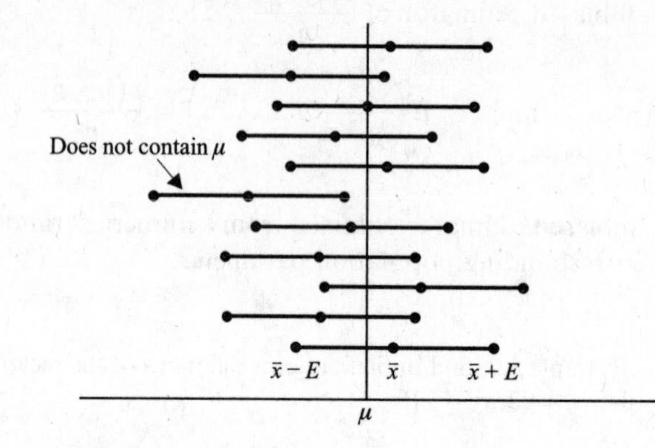

Fig. 8-1 Ninety-five percent of all intervals $[\bar{x} - E,\ \bar{x} + x]$ contain μ.

EXAMPLE 8.4 Suppose X is a normal random variable with mean μ, which is unknown, and standard deviation σ, which is known to be 2. A random sample (with replacement) of 25 values of X results in a sample mean $\bar{x} = 10$. Determine the margin of error E for a 95 percent confidence interval for μ and find the corresponding confidence interval. Give an interpretation of the result.

To determine E, we first convert to standard units. The random variable

$$Z = \frac{\bar{X} - \mu}{\sigma/\sqrt{n}}$$

has mean 0 and standard deviation 1. Also, since X is normal, so is Z. Therefore, Z is a standard normal random variable. Now $\dfrac{\sigma}{\sqrt{n}} = \dfrac{2}{5} = 0.4$, so $Z = \dfrac{\bar{X} - \mu}{0.4}$. The margin of error E satisfies the equation

$$P(\mu - E \leq \bar{X} \leq \mu + E) = 0.95,$$

which, in standard units, is equivalent to

$$P\left(-\frac{E}{0.4} \leq \frac{\bar{X} - \mu}{0.4} \leq \frac{E}{0.4}\right) = 0.95, \qquad \text{or} \qquad P\left(-\frac{E}{0.4} \leq Z \leq \frac{E}{0.4}\right) = 0.95$$

From Table A-1 in the Appendix, we find that

$$P(-1.96 \leq Z \leq 1.96) = 0.95; \qquad \text{equivalently,} \qquad P(0 \leq Z \leq 1.96) = \frac{0.95}{2} = 0.475$$

That is, 1.96 is the critical value of Z corresponding to probability 0.95. Therefore,

$$\frac{E}{0.4} = 1.96 \qquad \text{or} \qquad E = 0.4 \times 1.96 = 0.784$$

The corresponding 95 percent confidence interval is

$$[\bar{x} - E, \bar{x} + E] = [10 - 0.784, 10 + 0.784] = [9.216, 10.784]$$

We are 95 percent confident that the mean μ of X is some value in this interval, which means that, as $\bar{x}$ ranges through all possible values of $\bar{X}$, 95 percent of all the intervals $[\bar{x} - 0.784, \bar{x} + 0.784]$ will contain μ. Note that although different random samples of size 25 can give different values of $\bar{x}$, the value of E is the same for each sample.

Confidence Level

The confidence level 0.95 is the probability that μ will lie in the random interval $[\bar{X} - E, \bar{X} + E]$, and $1 - 0.95 = 0.05$ is the probability that μ will *not* lie in the random interval $[\bar{X} - E, \bar{X} + E]$. That is, 5 percent of the intervals in Figure 8-1 will not contain μ. In general, if 0.95 is replaced by γ, where $0 < \gamma < 1$, and

$$P(\bar{X} - E \leq \mu \leq \bar{X} + E) = \gamma$$

then $[\bar{X} - E, \bar{X} + E]$ is called a random 100γ percent confidence interval for μ; γ is called the *confidence level* and is equal to the probability that $[\bar{X} - E, \bar{X} + E]$ will contain μ; $1 - \gamma$ is equal to the probability that $[\bar{X} - E, \bar{X} + E]$ will *not* contain μ. For a given value $\bar{x}$ of $\bar{X}$, the numerical interval $[\bar{x} - E, \bar{x} + E]$ is called a 100γ percent confidence interval.

EXAMPLE 8.5 In Example 8.3, the confidence level is $\gamma = 0.95$, and the margin of error is $E = 0.784$. The probability that a random interval $[\bar{X} - 0.784, \bar{X} + 0.784]$ will contain μ is 0.95, and the probability that $[\bar{X} - 0.784, \bar{X} + 0.784]$ will not contain μ is 0.05.

Comment on Terminology and Notation

The confidence level γ is also called the *confidence coefficient*, and instead of the Greek letter γ (gamma), the notation $1 - \alpha$, where α is the Greek letter alpha, is often used to denote the confidence level. For simplicity, we will use the single letter γ for the confidence level.

Finding Margin of Error

As illustrated in Example 8.4, when the standard deviation σ of X is known, the margin of error E is given by

$$E = \frac{z^*\sigma}{\sqrt{n}}$$

where z^* is the value of the standard normal random variable Z satisfying

$$P(-z^* \leq Z \leq z^*) = \gamma; \qquad \text{equivalently,} \qquad P(0 \leq Z \leq z^*) = \frac{\gamma}{2}$$

When σ is not known, it will be replaced by the sample variance, and Z will be replaced by a t random variable, which will be defined in Section 8.3.

Sample Size

As illustrated in the following example, the formula $E = \frac{z^*\sigma}{\sqrt{n}}$ can also be used to determine the sample size needed to obtain a desired margin of error at a given confidence level.

EXAMPLE 8.6 Suppose X is a normal random variable with mean μ and standard deviation 2, and we wish to obtain a 95 percent confidence interval for μ with a margin of error no larger than 0.5. How large must the sample size be?

We can solve the equation $E = \frac{z^*\sigma}{\sqrt{n}}$ for $\sqrt{n}$ in terms of E, z^*, and σ, obtaining

$$\sqrt{n} = \frac{z^*\sigma}{E}$$

As E decreases, $\sqrt{n}$ will increase. We have $\sigma = 2, E \leq 0.05$, and we saw in Example 8.3 that the critical Z value for a 95 percent confidence interval is $z^* = 1.96$. Therefore,

$$\sqrt{n} \geq \frac{1.96 \times 2}{0.5} = 7.84; \qquad \text{equivalently,} \qquad n \geq (7.84)^2 \approx 61.5$$

Since n must be a positive integer, a sample size of 62 or larger is needed for a margin of error of 0.5 or less.

In general, the larger the sample size for a given level of confidence, the smaller the margin of error. On the other hand, if in Example 8.6, a value of n as large as 62 is impractical, then a margin of error of 0.5 or less can only be obtained by decreasing the confidence level. For instance, suppose the maximum possible sample size is 36, but we still want a margin of error of 0.5 or less. We then solve the equation $E = \frac{z^*\sigma}{\sqrt{n}}$ for z^* in terms of E, σ, and $\sqrt{n}$, obtaining

$$z^* = \frac{E\sqrt{n}}{\sigma}$$

As E or $\sqrt{n}$ decreases, so does z^*. We have $E \leq 0.5, n \leq 36$, and $\sigma = 2$. Therefore,

$$z^* \leq \frac{0.5 \times 6}{2} = 1.5$$

From the standard normal table,

$$P(0 \leq Z \leq 1.5) = 0.4332; \qquad \text{equivalently,} \qquad P(-1.5 \leq Z \leq 1.5) = 0.8664$$

Therefore, the confidence level is 86.64 percent with a margin of error of 0.5 and a sample size of 36.

Ideally, we want a small margin of error and a high confidence level. The price for obtaining both of these is a large sample size. If a sample size large enough to achieve both objectives is impractical, then we must settle for either a higher margin of error or a lower confidence level.

Models and Reality

The probability theory for deriving confidence intervals for μ requires that the sample mean $\bar{X}$ be normally distributed. For large samples, the Central Limit Theorem is used to conclude that $\bar{X}$ is approximately normally distributed, and for small samples we will require that X itself is approximately normally distributed, which then implies that $\bar{X}$ is also approximately normal. In applications, X will always be a random variable defined on a finite population, so that X and therefore $\bar{X}$ can never be exactly normal. Hence, the confidence intervals obtained in applications are only approximate. The theory provides a model for dealing with real populations, but our conclusions will be valid only to the extent that the real $\bar{X}$ approximates a normal distribution.

8.3 CONFIDENCE INTERVALS FOR MEANS

Let X be a random variable defined on some population, and suppose the mean μ of X is unknown. Suppose also that $\bar{x}$ is the value of the sample mean obtained in a random sample of size n. Then a confidence interval for μ is

$$[\bar{x} - E, \bar{x} + E]$$

where E is the margin of error. We know how to find E for a given confidence level when the standard deviation σ of X is known by using the formula

$$E = \frac{z^*\sigma}{\sqrt{n}}$$

developed in Section 8.2. We will also show how to find E when σ is unknown.

The confidence intervals prescribed for μ require that the sample mean $\bar{X}$ be approximately normally distributed. This condition can be met for small samples ($n < 30$) if X itself is approximately normally distributed. For large samples ($n \geq 30$), the Central Limit Theorem enables us to assume that $\bar{X}$ is approximately normally distributed regardless of the distribution of X.

PRESCRIPTION 8.1 (Confidence interval for μ when σ is known)

Requirements: X has known standard deviation σ, and $\bar{X}$ is approximately normally distributed.

Let γ be the specified confidence level, and suppose that a value $\bar{x}$ of the sample mean $\bar{X}$ is obtained in a random sample of size n. Complete the following steps.

(1) *Find Critical Z Value*: Using a standard normal table (or computer software), find the value z^* of the standard normal random variable Z for which $P(-z^* \leq Z \leq z^*) = \gamma$ (equivalently, $P(0 \leq Z \leq z^*) = \gamma/2$) (see Fig. 8-2).

(2) *Compute Margin of Error*: Compute $E = \dfrac{z^*\sigma}{\sqrt{n}}$.

(3) *Determine Confidence Interval*: An approximate 100γ percent confidence interval for the mean μ of X is $[\bar{x} - E, \bar{x} + E]$.

The value z^* is the *critical value* of Z corresponding to the confidence level γ (see Fig. 8-2).

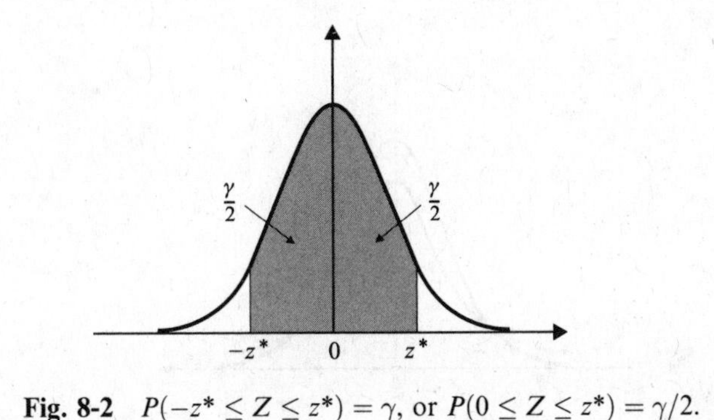

Fig. 8-2 $P(-z^* \leq Z \leq z^*) = \gamma$, or $P(0 \leq Z \leq z^*) = \gamma/2$.

EXAMPLE 8.7 A population random variable X has unknown mean μ and standard deviation $\sigma = 20$. A random sample of size 100 results in a sample mean $\bar{x} = 250$. Find the corresponding 90 percent confidence interval for μ.

Since the sample size is greater than 30, we may assume that $\bar{X}$ is approximately normally distributed. The confidence level is 0.90. From Table A-1 in the Appendix, the critical value z^* satisfying $P(-z^* \leq Z \leq z^*) = 0.9$ (equivalently, $P(0 \leq Z \leq z^*) = 0.45$) is $z^* = 1.65$. Therefore, by Prescription 8.1, the margin of error is

$$E = \frac{1.65 \times 20}{\sqrt{100}} = 3.3$$

The approximate 90 percent confidence interval for μ is $[250 - 3.3, 250 + 3.3]$, or $[246.7, 253.3]$.

Confidence Intervals for μ When σ is Unknown

When the standard deviation, σ, of X is not known, values of the sample standard deviation

$$S = \sqrt{\frac{1}{n-1}\sum (X_i - \bar{X})^2}$$

are used in place of σ. Recall that $(n-1)S^2/\sigma^2$ is a chi-square random variable with $n-1$ degrees of freedom (see Section 7.4). However, we must first introduce the t random variable.

The t Distribution

The standard normal and chi-square distributions combine to produce the t distribution, which is defined as follows.

Definition: Let Z be the standard normal random variable, and χ^2 the chi-square random variable with k degrees of freedom. Suppose Z and χ^2 are independent. Then the random variable

$$t = \frac{Z}{\sqrt{\chi^2/k}}$$

is called the *t random variable with k degrees of freedom.*

Properties of the t distribution

The random variable t is also denoted by $t(k)$ to emphasize its dependence on the parameter k. The density curve of $t(k)$ is a bell-shaped curve, as illustrated in Fig. 8-3. The curve is similar to a normal density curve, but flatter and with thicker tails; $t(k)$ has mean 0 for $k \geq 2$ and standard deviation $\sqrt{k/(k-2)}$ for $k \geq 3$. For $k = 1$, the mean is not defined, and for $k = 1, 2$, the standard deviation is not defined. For large values of k ($k \geq 30$), the t distribution closely approximates a standard normal distribution.

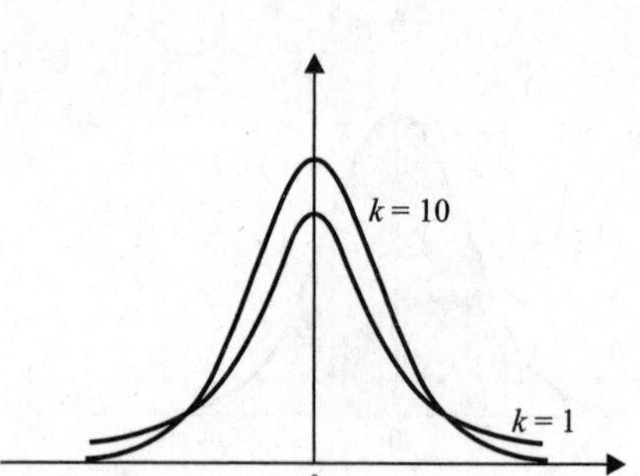

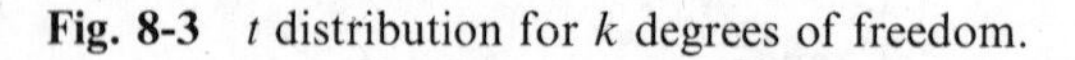

Fig. 8-3 t distribution for k degrees of freedom.

EXAMPLE 8.8 Let X be a normal random variable with mean $\mu = 10$ and standard deviation $\sigma = 2$. Suppose $\bar{X}$ is the sample mean for random samples of size $n = 25$, and S^2 is the corresponding sample variance. Then $Z = \dfrac{\bar{X} - 10}{2/5}$ is a standard normal random variable, and $\chi^2 = \dfrac{24S^2}{4}$ is a chi-square random variable with $k = 24$ degrees of freedom (see Section 7.4). Furthermore, these two random variables are independent. Dividing χ^2 by k and taking the square root of the result gives $\sqrt{\chi^2/k} = \dfrac{S}{2}$. Finally, when Z is divided by $\sqrt{\chi^2/k}$ the result is $t = \dfrac{\bar{X} - 10}{S/5}$ which, by definition, is a t random variable with 24 degrees of freedom.

Example 8.8 can be generalized to obtain the following result (see Problem 8.26).

Theorem 8.1: Suppose a random variable X has mean μ. Let $\bar{X}$ be the sample mean corresponding to random samples of size n, and let S be the corresponding sample standard deviation. If $\bar{X}$ is normally distributed, then the random variable

$$t = \frac{\bar{X} - \mu}{S/\sqrt{n}}$$

has a t distribution with $n - 1$ degrees of freedom.

We can now state a prescription for finding a confidence interval for μ when σ is unknown. In it, the t random variable takes the place of Z, and the statistic s takes the place of the unknown parameter σ.

PRESCRIPTION 8.2 (Confidence interval for μ when σ is unknown)

Requirement: $\bar{X}$ is approximately normally distributed.

Let γ be the specified confidence level. Suppose that values $x_1, x_2, \ldots, x_n$ of X are obtained in a random sample of size n. First compute the sample statistics $\bar{x} = (x_1 + x_2 + \cdots + x_n)/n$ and $s = \sqrt{\frac{1}{n-1}\sum(x_i - \bar{x})^2}$. Then complete the following steps.

(1) *Find Critical t Value*: Using a t table (or computer software), find the value t^* of the t random variable with $n-1$ degrees of freedom that satisfies $P(-t^* \le t \le t^*) = \gamma$ (equivalently, $P(0 \le t \le t^*) = \gamma/2$) (see Fig. 8-4).

(2) *Compute Margin of Error*: Compute $E = \frac{t^* s}{\sqrt{n}}$.

(3) *Determine Confidence Interval*: An approximate 100γ percent confidence interval for the mean μ of X is $[\bar{x} - E, \bar{x} + E]$.

The value t^* is called the *critical value* of t corresponding to the confidence level γ (see Fig. 8-4).

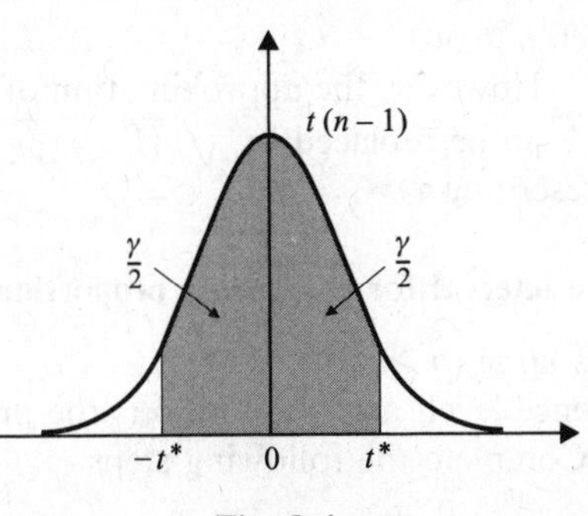

Fig. 8-4

EXAMPLE 8.9 The average of a random sample of 10 scores on a college placement exam is 75, and the sample standard deviation is 8.4. Assuming that the collection of all scores is approximately normally distributed, find a 95 percent confidence interval for the mean score.

Using Table A-2 in the Appendix, with $10 - 1 = 9$ degrees of freedom, we find that $P(-t^* \le t \le t^*) = 0.95$ (equivalently, $P(0 \le t \le t^*) = 0.475$) for $t^* = 2.26$. Therefore, the margin of error is

$$E = \frac{t^* s}{\sqrt{n}} = \frac{2.26 \times 8.4}{\sqrt{10}} \approx 6.00$$

and an approximate 95 percent confidence interval for the mean score is $[75 - 6, 75 + 6] = [69, 81]$.

Small Samples

As illustrated in Example 8.9, the t distribution can be used when σ is unknown, provided $\bar{X}$ is approximately normally distributed. If the sample size is 30 or greater, then we may assume, on the basis of the Central Limit Theorem, that $\bar{X}$ is approximately normally distributed. In fact, for large samples, the t distribution is very close to the normal distribution, and both have essentially the same critical values, as is illustrated in Example 8.10. The t distribution is needed mainly for small samples, but in this case, X itself must be approximately normal in order to guarantee that $\bar{X}$ is.

EXAMPLE 8.10 The critical Z value at the 95 percent confidence level is 1.96. The critical t values at the 95 percent confidence level are: $t^* = 2.04$ for 30 degrees of freedom, $t^* = 2.02$ for 40 degrees of freedom, $t^* = 2.00$ for 60 degrees of freedom, and $t^* = 1.98$ for 120 degrees of freedom.

8.4 CONFIDENCE INTERVALS FOR PROPORTIONS

Suppose a population is broken up into two groups. The members of one of these groups will be referred to as "successes". Let p be the (unknown) proportion of successes in the population. A numerical confidence interval for p is of the form

$$[\hat{p} - E, \hat{p} + E]$$

where $\hat{p}$ is the proportion of successes obtained in a random sample, and E is the margin of error. We give a prescription for finding E when the sample size is large. As defined in Sections 7.3 and 8.1, the sample proportion $\hat{P}$ is the random variable whose value on a random sample size n is the proportion $\hat{p}$ of successes in the sample. $\hat{P}$ has mean

$$\mu_{\hat{P}} = p$$

and standard deviation

$$\sigma_{\hat{P}} = \sqrt{\frac{p(1-p)}{n}}$$

and is approximately normal when $n \geq 30$.

Since p is unknown, so is $\sigma_{\hat{P}}$. However, the approximation of $\hat{P}$ as normally distributed is usually sufficiently robust for $\sqrt{p(1-p)/n}$ to be replaced by $\sqrt{\hat{p}(1-\hat{p})/n}$ in a confidence interval for p. We can then deduce the following prescription.

PRESCRIPTION 8.3 (Confidence interval for population proportion p)

Requirement: The sample size n is large ($n \geq 30$).

Let γ be the specified confidence level, and suppose $\hat{p}$ is the proportion of successes obtained in a random sample of size $n \geq 30$. Complete the following steps.

(1) *Find Critical Z Value*: Using a standard normal table (or computer software), find the value z^* of the standard normal random variable Z for which $P(-z^* \leq Z \leq z^*) = \gamma$ (equivalently, $P(0 \leq Z \leq z^*) = \gamma/2$) (see Fig. 8-2).

(2) *Compute Margin of Error*: Compute $E = z^* \sqrt{\frac{\hat{p}(1-\hat{p})}{n}}$.

(3) *Determine Confidence Interval*: An approximate 100γ percent confidence interval for the proportion p of successes in the population is $[\hat{p} - E, \hat{p} + E]$.

EXAMPLE 8.11 In a random sample of 900 registered voters, 55 percent favored the Democratic candidate for President. Find an approximate confidence interval for the proportion of all registered voters that favor the Democratic candidate at confidence level (*a*) 90 percent, (*b*) 99 percent.

(*a*) We have $\hat{p} = 0.55$, and from Table A-1 in the Appendix, we find that $P(-z^* \leq Z \leq z^*) = 0.9$ (equivalently, $P(0 \leq Z \leq 0.45)$ for $z^* = 1.65$). Therefore, the margin of error is

$$E = z^* \sqrt{\frac{\hat{p}(1-\hat{p})}{n}} = 1.65 \sqrt{\frac{0.55 \times 0.45}{900}} \approx 0.03$$

and the corresponding 90 percent confidence interval is $[0.55 - 0.03, 0.55 + 0.03] = [0.52, 0.58]$.

(*b*) The critical Z value at the 99 percent level is $z^* = 2.58$, the margin of error is

$$E = 2.58 \sqrt{\frac{0.55 \times 0.45}{900}} \approx 0.04$$

and the corresponding 99 percent confidence interval is [0.51, 0.59].

Sample Size

From the formula $E = z^* \sqrt{\frac{\hat{p}(1-\hat{p})}{n}}$, we see that as the sample size n increases, the margin of error E for a given confidence level decreases. We may want to know, before sampling, how large a sample must be to guarantee that the margin of error does not exceed some value for a given confidence level. For example, suppose we want the margin of error to be at most 0.03 at a 99 percent confidence level. From Example 8.11, we know that $z^* = 2.58$ for a confidence level of 99 percent. That means n must be chosen so that

$$2.58\sqrt{\frac{\hat{p}(1-\hat{p})}{n}} \le 0.03$$

However, we won't know what $\hat{p}$ is until we do the sampling. We can get around this problem by considering a worst-case scenario. It can be shown that the product $\hat{p}(1-\hat{p})$ is at most 0.25, which occurs when $\hat{p} = 0.5$. Therefore, $\sqrt{\hat{p}(1-\hat{p})}$ is at most $\sqrt{0.25} = 0.5$. If we choose n so that

$$\frac{2.58 \times 0.5}{\sqrt{n}} \le 0.03$$

then E will be at most 0.03 regardless of the sample value $\hat{p}$ obtained. The above inequality is equivalent to

$$\sqrt{n} \ge \frac{2.58 \times 0.5}{0.03} = 43, \qquad \text{or} \qquad n \ge (43)^2 = 1849$$

Hence, a sample size of 1849 or higher will result in a margin of error of 0.03 or less. Often when making such a calculation for n, the result turns out not to be a whole number. In this case, we round up to the next integer. For example, if n came out to be 1426.2, we would round up to 1427.

8.5 CONFIDENCE INTERVALS FOR VARIANCES

Suppose a random variable X is approximately normally distributed with mean μ and unknown variance σ^2. A prescription will be given here for finding confidence intervals for σ^2 when μ is unknown (see Problems 8.38 and 8.64 for the case in which μ is known). The confidence intervals depend on the chi-square distribution (see Section 7.4), which is not symmetric, so they do not take the usual form $s^2 \pm E$, where s^2 is a value of the sample variance S^2 obtained in a random sample. Prescription 8.4 gives a method for finding the endpoints of the intervals.

The confidence intervals for means in Section 8.3 require that the sample mean $\bar{X}$ be approximately normally distributed. Here we require that X itself be approximately normally distributed.

Confidence Intervals for σ^2 When μ is Unknown

Suppose $X_1, X_2, \ldots, X_n$ is a random-variable sample corresponding to a normal random variable X (Section 8.1). $\bar{X} = (X_1 + X_2 + \cdots + X_n)/n$ is the sample mean, and $S^2 = \sum(X_i - \bar{X})^2/(n-1)$ is the sample variance. By Theorem 7.7, the random variable

$$\frac{(n-1)S^2}{\sigma^2}$$

is a chi-square random variable with $n-1$ degrees of freedom, denoted by $\chi^2(n-1)$, or simply χ^2 if $n-1$ understood from context (Section 7.4).

Let us first consider the case of a 95 percent confidence interval for σ^2. With reference to Table A-3 in the Appendix, choose the constants a and b, corresponding to $n-1$ degrees of freedom, to satisfy

$$P(\chi^2 \le a) = \frac{1-0.95}{2} = 0.025 \qquad \text{and} \qquad P(\chi^2 \le b) = \frac{1+0.95}{2} = 0.975$$

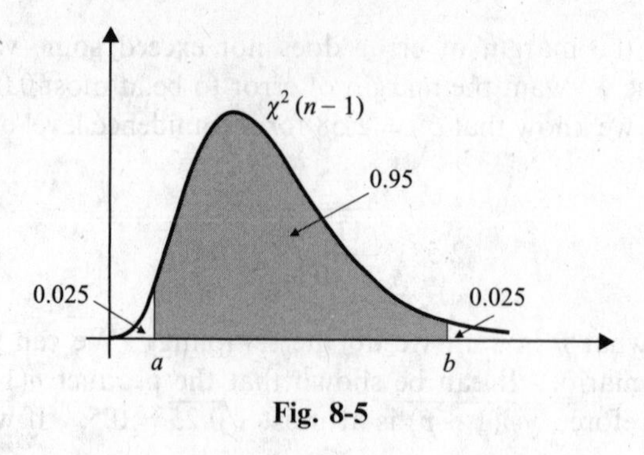

Fig. 8-5

(Fig. 8-5). Then

$$P(a \leq \chi^2 \leq b) = P(\chi^2 \leq b) - P(\chi^2 \leq a)$$
$$= 0.975 - 0.025$$
$$= 0.95$$

Replacing χ^2 by $\dfrac{(n-1)S^2}{\sigma^2}$, we get

$$P\left(a \leq \frac{(n-1)S^2}{\sigma^2} \leq b\right) = 0.95$$

which is equivalent to

$$P\left(\frac{(n-1)S^2}{b} \leq \sigma^2 \leq \frac{(n-1)S^2}{a}\right) = 0.95$$

(see Problem 8.31). Therefore, $\left[\dfrac{(n-1)S^2}{b}, \dfrac{(n-1)S^2}{a}\right]$ is a random 95 percent confidence interval for σ^2.

In general, if a level of confidence γ is specified, and the constants a and b are chosen for $n-1$ degrees of freedom to satisfy

$$P(\chi^2 \leq a) = \frac{1-\gamma}{2} \quad \text{and} \quad P(\chi^2 \leq b) = \frac{1+\gamma}{2}$$

then

$$\left[\frac{(n-1)S^2}{b}, \frac{(n-1)S^2}{a}\right]$$

is a random 100γ percent confidence interval for σ^2. We therefore arrive at the following prescription.

PRESCRIPTION 8.4 (Confidence interval for σ^2 when μ is unknown)

Requirement: X is approximately normally distributed.

Let γ be the specified confidence level. Suppose the values $x_1, x_2, \ldots, x_n$ of X are obtained in a random sample of size n. First compute the sample values $\bar{x} = (x_1 + x_2 + \cdots + x_n)/n$ and $s^2 = \sum (x_i - \bar{x})^2/(n-1)$. Then complete the following steps.

(1) *Find Critical χ^2 Values*: Using a chi-square table (or computer software), find values a and b of the chi-square random variable with $n-1$ degrees of freedom that satisfy $P(\chi^2 \leq a) = \dfrac{1-\gamma}{2}$, $P(\chi^2 \leq b) = \dfrac{1+\gamma}{2}$.

(2) *Determine Confidence Interval*: An approximate 100γ percent confidence interval for σ^2 is $\left[\dfrac{(n-1)s^2}{b}, \dfrac{(n-1)s^2}{a}\right]$.

EXAMPLE 8.12 The values of a normal random variable X obtained in random sample are

$$55, \quad 65, \quad 82, \quad 48, \quad 55, \quad 75, \quad 70, \quad 62$$

Find a 90 percent confidence interval for the variance of X.

The value of the sample mean is

$$\bar{x} = \frac{55 + 65 + 82 + 48 + 55 + 75 + 70 + 62}{8} = \frac{512}{8} = 64$$

and the value of the sample variance is

$$s^2 = \frac{(55-64)^2 + (65-64)^2 + (82-64)^2 + (48-64)^2 + (55-64)^2 + (75-64)^2 + (70-64)^2 + (62-64)^2}{7}$$

$$= \frac{904}{7} \approx 129.14$$

From Table A-3 in the Appendix, with 7 degrees of freedom, we find that $P(\chi^2 \le a) = \dfrac{1-0.90}{2} = 0.05$ for $a = 2.17$, and $P(\chi^2 \le b) = \dfrac{1+0.90}{2} = 0.95$ for $b = 14.1$. Therefore, the corresponding 90 percent confidence interval for σ^2 is

$$\left[\frac{7 \times 129.14}{14.1}, \frac{7 \times 129.14}{2.17}\right] = [64.1, 416.6]$$

(See also Problem 8.34.)

Confidence Intervals for the Standard Deviation

Note that the confidence interval obtained for the variance σ^2 in Example 8.12 is quite large, having a left endpoint of 64.1 and a right endpoint of 416.6. The corresponding confidence interval for the standard deviation σ is smaller. By definition, if $\left[\dfrac{(n-1)S^2}{b}, \dfrac{(n-1)S^2}{a}\right]$ is a random 100γ percent confidence interval for σ^2, then $\left[\sqrt{\dfrac{(n-1)S^2}{b}}, \sqrt{\dfrac{(n-1)S^2}{a}}\right]$ is a random 100γ percent confidence interval for σ, meaning that the probability that σ lies in this interval is γ. If s^2 is the value of the sample variance S^2 obtained in a random sample of size n, then the numerical interval $\left[\sqrt{\dfrac{(n-1)s^2}{b}}, \sqrt{\dfrac{(n-1)s^2}{a}}\right]$ is called a 100γ percent confidence interval for σ. In Example 8.12, the corresponding 90 percent confidence interval for σ is $[\sqrt{64.1}, \sqrt{416.6}] = [8.01, 20.41]$.

Comment

The interval given in Prescription 8.4 has the property that the probability in each of the two tails of the chi-square distribution for $n-1$ degrees of freedom is $(1-\gamma)/2$, where γ is the level of confidence. This choice of confidence interval is consistent with the confidence intervals for means and proportions, in which the probability in each of the two tails of the standard normal or t distribution is $(1-\gamma)/2$. In the case of the standard normal and t distributions, these intervals are the smallest possible 100γ percent confidence intervals. Because the chi-square distribution is not symmetric, the confidence interval in Prescription 8.4 may not be the smallest possible.

WARNING

All confidence intervals obtained in this and previous sections are approximate to the extent that X, or $\bar{X}$, is normally distributed. The confidence intervals for means, using the normal or the t distribution, and those for proportions, are *robust* in the sense that they are very close to the true confidence intervals for bell-shaped distributions, even when the distributions are not very close to being normal. (The deviation of a bell-shaped distribution from a normal distribution can be measured by $E[(X-\mu)^4]/\sigma^4$, which is called the *kurtosis* of the distribution; the kurtosis is 3 for a normal distribution, less than 3 for a flatter distribution, and greater than 3 for a steeper bell-shaped distribution.) However, *confidence intervals obtained for the variance are not robust, and can deviate very significantly from true confidence intervals when X is not normally distributed.* Therefore, the practical use of confidence intervals for variances is limited.

Solved Problems

PARAMETERS AND STATISTICS

8.1. Fill in each blank below with the number of each item in the second list that describes the expression to the left of the blank.

(a) μ ______ *(h)* $\dfrac{(x_1-\bar{x})^2+(x_2-\bar{x})^2+\cdots+(x_n-\bar{x})^2}{n-1}$ ______

(b) σ ______ *(i)* $\dfrac{(X_1-\bar{X})^2+(X_2-\bar{X})^2+\cdots+(X_n-\bar{X})^2}{n-1}$ ______

(c) χ^2 ______ *(j)* $\dfrac{(x_1-\bar{x})^2+(x_2-\bar{x})^2+\cdots+(x_n-\bar{x})^2}{n}$ ______

(d) p ______ *(k)* $\dfrac{(X_1-\bar{X})^2+(X_2-\bar{X})^2+\cdots+(X_n-\bar{X})^2}{n}$ ______

(e) $x_1, x_2, \ldots, x_n$ ______ *(l)* $X_1, X_2, \ldots, X_n$ ______

(f) $\dfrac{x_1+x_2+\cdots+x_n}{n-1}$ ______ *(m)* $\dfrac{X_1+X_2+\cdots+X_n}{n-1}$ ______

(g) $\dfrac{x_1+x_2+\cdots+x_n}{n}$ ______ *(n)* $\dfrac{X_1+X_2+\cdots+X_n}{n}$ ______

(1) parameter
(2) random variable
(3) numerical statistic
(4) random-variable statistic
(5) numerical random sample
(6) random-variable sample
(7) unbiased point estimate of μ
(8) unbiased estimator of μ
(9) biased point estimate of μ
(10) biased estimator of μ
(11) unbiased point estimate of σ^2
(12) unbiased estimator of σ^2
(13) biased point estimate of σ^2
(14) biased estimator of σ^2

(a) 1 *(b)* 1 *(c)* 2 *(d)* 1 *(e)* 5 *(f)* 3, 9 *(g)* 3, 7 *(h)* 3, 11
(i) 4, 12 *(j)* 3, 13 *(k)* 4, 14 *(l)* 6 *(m)* 4, 10 *(n)* 4, 8

8.2. The values 2, 7, 3, 8 were obtained as a random sample of a random variable X. Give unbiased point estimates of the mean μ and variance σ^2 of X.

$\bar{x} = \dfrac{2+7+3+8}{4} = \dfrac{20}{4} = 5$ is an unbiased point estimate of μ.

$s^2 = \dfrac{(2-5)^2+(7-5)^2+(3-5)^2+(8-5)^2}{3} = \dfrac{9+4+4+9}{3} = \dfrac{26}{3} \approx 8.67$ is an unbiased point estimate of σ^2.

8.3. Show that a value $s^2 = \frac{1}{n-1}\sum (x - \bar{x})^2$ of the sample variance can be computed by means of the formula $s^2 = \frac{1}{n-1}\sum x^2 - \frac{n}{n-1}\bar{x}^2$.

$$\begin{aligned}
\frac{1}{n-1}\sum (x-\bar{x})^2 &= \frac{1}{n-1}\sum (x^2 - 2x\bar{x} + \bar{x}^2) \\
&= \frac{1}{n-1}\left(\sum x^2 - 2\bar{x}\sum x + \sum \bar{x}^2\right) \\
&= \frac{1}{n-1}\left(\sum x^2 - 2\bar{x}\cdot n\bar{x} + n\bar{x}^2\right) \\
&= \frac{1}{n-1}\left(\sum x^2 - n\bar{x}^2\right) \\
&= \frac{1}{n-1}\sum x^2 - \frac{n}{n-1}\bar{x}^2
\end{aligned}$$

8.4. The formula for the sample variance in Problem 8.3 is convenient for computational purposes, especially for large samples. Use the formula to verify the point estimate of σ^2 obtained in Problem 8.2.

$$\frac{1}{n-1}\sum x^2 - \frac{n}{n-1}\bar{x}^2 = \frac{1}{3}(4 + 49 + 9 + 64) - \frac{4}{3} \times 25 = \frac{126}{3} - \frac{100}{3} = \frac{26}{3}$$

which is the value of s^2 obtained in Problem 8.2.

8.5. Show that the sample mean $\bar{X}$, for random samples of size n, is an unbiased estimator of the population mean μ.

By definition, $\bar{X} = \frac{X_1 + X_2 + \cdots + X_n}{n}$, where $X_1, X_2, \ldots, X_n$ are independent random variables, each with mean μ and standard deviation σ. The expected value of a constant times a random variable is that constant times the expected value of the random variable, and the expected value of a sum of random variables is equal to the sum of the expected values of the random variables. Therefore,

$$\begin{aligned}
E\left(\frac{X_1 + X_2 + \cdots + X_n}{n}\right) &= \frac{1}{n}E(X_1 + X_2 + \cdots + X_n) \\
&= \frac{1}{n}\left(E(X_1) + E(X_2) + \cdots + E(X_n)\right) \\
&= \frac{1}{n}(\mu + \mu + \cdots + \mu) \\
&= \frac{1}{n}(n\mu) \\
&= \mu
\end{aligned}$$

8.6. Show that the sample proportion $\hat{P}$ is an unbiased estimator of the population proportion p.

The parameter p is the proportion of successes in the population. Let X be the random variable whose value is 1 for a success, 0 otherwise. X is a Bernoulli random variable with mean $\mu = p$ and variance $\sigma^2 = p(1-p)$. The sample mean $\bar{X}$, for random samples of size n, is the number of successes in the sample divided by n. Therefore, $\bar{X} = \hat{P}$, the sample proportion. As shown in Problem 8.5, for any random variable X, the expected value of $\bar{X}$ is the same as the expected value of X, in this case p. Therefore, p is the expected value of $\hat{P}$, which makes $\hat{P}$ an unbiased estimator of p.

8.7. Show that $\frac{1}{n-1}\hat{P}(1-\hat{P})$ is an unbiased estimator of $\frac{p(1-p)}{n}$.

The Bernoulli random variable X of Problem 8.6 has mean p and variance $\sigma^2 = p(1-p)$. Also, we know that for any random variable, the sample variance $S^2 = \frac{1}{n-1}\sum (X_i - \bar{X})^2$ is an unbiased estimator of σ^2. Now X_i is the value of X on the ith member of the sample, and since X_i is either 0 or 1, it follows that X_i^2 is equal to X_i, which is the key to the following.

$$\begin{aligned}
S^2 &= \frac{1}{n-1}\sum (X_i - \bar{X})^2 \\
&= \frac{1}{n-1}\sum (X_i^2 - 2X_i\bar{X} + \bar{X}^2) \\
&= \frac{1}{n-1}\left(\sum X_i^2 - 2\bar{X}\sum X_i + \sum \bar{X}^2\right) \\
&= \frac{1}{n-1}\left(\sum X_i - 2\bar{X}\cdot n\bar{X} + n\bar{X}^2\right) \\
&= \frac{1}{n-1}\left(n\bar{X} - n\bar{X}^2\right) \\
&= \frac{n}{n-1}\bar{X}(1-\bar{X}) \\
&= \frac{n}{n-1}\hat{P}(1-\hat{P})
\end{aligned}$$

Now $E(S^2) = \sigma^2$. Therefore, $E\left(\frac{n}{n-1}\hat{P}(1-\hat{P})\right) = p(1-p)$. Therefore,

$$E\left(\frac{1}{n-1}\hat{P}(1-\hat{P})\right) = \frac{p(1-p)}{n}$$

8.8. A random sample of 25 students at Greentree College had 10 males and 15 females. Give unbiased point estimates for the proportion of male students and for the proportion of female students in the college.

$\hat{p} = \frac{10}{25} = \frac{2}{5}$ is an unbiased point estimate for the proportion of male students, and $\hat{q} = 1 - \hat{p} = \frac{15}{25} = \frac{3}{5}$ is an unbiased estimate for the proportion of female students.

8.9. With reference to Problem 8.8, give an unbiased point estimate of the variance of the sample proportion $\hat{P}$ of males for all random samples of size 25. Also give an unbiased point estimate of the variance of the sample proportion $\hat{Q}$ of females.

$\frac{1}{n-1}\hat{p}(1-\hat{p}) = \frac{1}{24}\cdot\frac{2}{5}\cdot\frac{3}{5} = \frac{1}{100} = 0.01$ is an unbiased point estimate of the variance of $\hat{P}$; $\frac{1}{n-1}\hat{q}(1-\hat{q}) = \frac{1}{24}\cdot\frac{3}{5}\cdot\frac{2}{5} = \frac{1}{100} = 0.01$ is an unbiased point estimate of the variance of $\hat{Q}$. In general, $\hat{P}$ and $\hat{Q} = 1 - \hat{P}$ have the same variance.

THE NOTION OF A CONFIDENCE INTERVAL

8.10. Show that, for any value of γ between 0 and 1, $P(\mu - E \leq \bar{X} \leq \mu + E) = \gamma$ is equivalent to $P(\bar{X} - E \leq \mu \leq \bar{X} + E) = \gamma$.

$$\begin{aligned} \mu - E \leq \bar{X} \leq \mu + E &\Leftrightarrow \mu - E \leq \bar{X} \quad \text{and} \quad \bar{X} \leq \mu + E \\ &\Leftrightarrow \mu \leq \bar{X} + E \quad \text{and} \quad \bar{X} - E \leq \mu \\ &\Leftrightarrow \bar{X} - E \leq \mu \leq \bar{X} + E \end{aligned}$$

from which the desired result follows.

8.11. Let X be a normal random variable with mean μ and standard deviation $\sigma = 10$. Find the margin of error for a 90 percent confidence interval for μ corresponding to a sample size of 12.

The formula for the margin of error E is $E = \dfrac{z^*\sigma}{\sqrt{n}}$. We have $\sigma = 10$, $n = 12$, and from Table A-1, we find that $P(-z^* \leq Z \leq z^*) = 0.9$ for $z^* = 1.65$. Therefore, the margin of error is $E = \dfrac{1.65 \times 10}{\sqrt{12}} \approx 4.76$.

8.12. Interpret the result of Problem 8.11.

The probability is 0.9 that the mean μ of X is in the random interval $[\bar{X} - 4.76, \ \bar{X} + 4.76]$, where $\bar{X}$ is the sample mean of X; the probability is 0.1 that μ is not in the random interval $[\bar{X} - 4.76, \ \bar{X} + 4.76]$.

8.13. With reference to Problem 8.11, find an approximate 90 percent confidence interval for μ if the 12 sample values of X are as follows.

95	103	107	98	90	110
92	108	90	94	105	100

The value of the sample mean determined by the sample values is

$$\bar{x} = \frac{95 + 103 + 107 + 98 + 90 + 110 + 92 + 108 + 90 + 94 + 105 + 100}{12} = \frac{1192}{12} \approx 99.33$$

From Problem 8.11, the margin of error is $E \approx 4.76$. The corresponding confidence interval is $[\bar{x} - E, \ \bar{x} + E] = [99.33 - 4.76, \ 99.33 + 4.76] = [94.57, \ 104.09]$.

8.14. Interpret the result of Problem 8.13.

We are 90 percent confident that the interval [94.57, 104.09] contains the mean μ of X, meaning that as $\bar{x}$ ranges through all possible values of $\bar{X}$, approximately 90 percent of the intervals $[\bar{x} - 4.76, \ \bar{x} + 4.76]$ will contain μ. That is, as more and more random samples of 12 values of X are taken, approximately 90 percent of the corresponding confidence intervals $[\bar{x} - 4.76, \ \bar{x} + 4.76]$ will actually contain μ.

8.15. Using the random sample in Problem 8.13, find an approximate 99 percent confidence interval for μ.

We have $\bar{x} \approx 99.33$, but must determine the margin of error corresponding to a 99 percent confidence interval for μ. At a 99 percent level of confidence, the critical Z value z^* satisfies $P(-z^* \leq Z \leq z^*) = 0.99$; equivalently, $P(0 \leq Z \leq z^*) = \dfrac{0.99}{2} = 0.495$. From Table A-1, we find that $z^* \approx 2.58$. Since the random variable X in question has standard deviation 10, and the sample size is 12, it follows that the margin of error is $E = \frac{2.58 \times 10}{\sqrt{12}} \approx 7.45$. Therefore, an approximate 99 percent confidence interval for μ is $[99.33 - 7.45, 99.33 + 7.45] = [91.88, 106.78]$.

8.16. Let X be a normal random variable with unknown mean μ and standard deviation $\sigma = 5$. Find the sample size needed for a 99 percent confidence interval with a margin of error 2.5.

The critical Z value for a 99 percent confidence interval for μ, as determined in Problem 8.15, is $z^* = 2.58$. From the equation for the margin of error, $E = \frac{z^*\sigma}{\sqrt{n}}$, we see that $\sqrt{n} = \frac{z^*\sigma}{E} = \frac{2.58 \times 5}{2.5} = 5.16$. Therefore, n must be at least $(5.16)^2 \approx 26.6$. The desired value of n, which must be an integer, is 27.

8.17. Let X be a normal random variable with unknown mean μ and standard deviation $\sigma = 3$. It is desired to obtain a confidence interval for μ with a margin of error of 1.5, based on a random sample of size 16. What is the corresponding confidence level?

The formula for the margin of error is $E = \frac{z^*\sigma}{\sqrt{n}}$. We therefore have $1.5 = \frac{z^* \times 3}{\sqrt{16}} = \frac{3z^*}{4}$, so $z^* = \frac{4 \times 1.5}{3} = 2$. From Table A-1, we find that $P(-2 \le Z \le 2) = 2P(0 \le Z \le 2) = 2(0.4772) = 0.9544$. Therefore, the confidence level is 0.9544, or 95.44 percent.

CONFIDENCE INTERVALS FOR MEANS

8.18. A random variable X has unknown mean and standard deviation 25. A random sample of 50 values of X has mean $\bar{x} = 112$. Find an approximate 85 percent confidence interval for the mean μ of X.

Since the sample size, 50, is larger than 30, we assume that $\bar{X}$ is approximately normally distributed. We then apply Prescription 8.1. Using Table A-1, we find that the critical Z value satisfying $P(-z^* \le Z \le z^*) = 0.85$ is $z^* = 1.44$. Therefore, the margin of error is $E = \frac{z^*\sigma}{\sqrt{n}} = \frac{1.44 \times 25}{\sqrt{50}} \approx 5.09$. The corresponding approximate 85 percent confidence interval for μ is $[112 - 5.09, 112 + 5.09] = [106.91, 117.09]$.

8.19. With reference to Problem 8.18, how large must the sample size be to obtain an 85 percent confidence interval for μ with a margin of error equal to 2.5?

Substitute $E = 2.5$, $z^* = 1.44$, and $\sigma = 25$ into the formula $E = \frac{z^*\sigma}{\sqrt{n}}$ to obtain $2.5 = \frac{1.44 \times 25}{\sqrt{n}} = \frac{36}{\sqrt{n}}$. Therefore, $\sqrt{n} = \frac{36}{2.5} = 14.4$, and $n = (14.4)^2 = 207.36$. Since n must be an integer, the sample size needed is 208.

8.20. With reference to Problem 8.18, suppose that the sample size can be no larger than 100. What is the smallest possible margin of error?

If $n \le 100$, then $E \ge \frac{1.44 \times 25}{\sqrt{100}} = 3.6$. Therefore, 3.6 is the smallest possible margin of error.

8.21. Find the values t^* in each of the following cases for a t distribution with 10 degrees of freedom.
(a) $P(0 \le t \le t^*) = 0.45$ (b) $P(-t^* \le t \le t^*) = 0.90$ (c) $P(t \le t^*) = 0.95$

As illustrated in Fig 8-6, Table A-2 in the Appendix gives t^* values for various values of $P(0 \le t \le t^*)$ corresponding to different degrees of freedom.

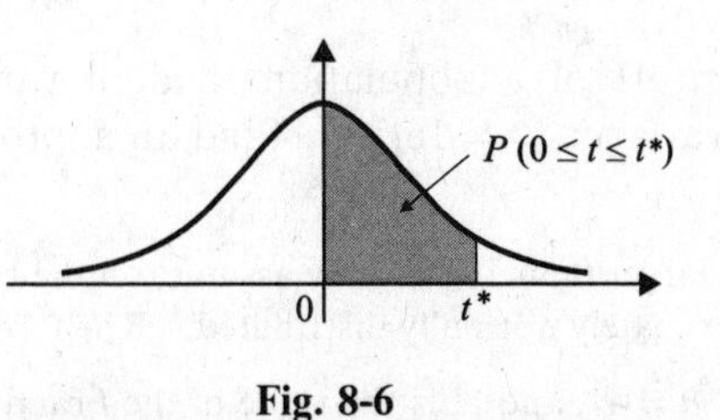

Fig. 8-6

(*a*) From Table A-2, corresponding to 10 degrees of freedom, we find that $P(0 \le t \le t^*) = 0.45$ for $t^* = 1.81$.

(*b*) By the symmetry of the t distribution, $P(-t^* \le t \le t^*) = 2P(0 \le t \le t^*) = 2 \times 0.45 = 0.90$. Therefore, from part (*a*), $t^* = 1.81$.

(*c*) $P(t \le t^*) = 0.95 = 0.5 + 0.45 = P(t \le 0) + P(0 \le t \le t^*)$. Therefore, from part (*a*), $t^* = 1.81$.

8.22. The numbers

24.4, 18.9, 12.8, 20.5, 19.1, 15.2, 21.7, 14.6

form a random sample of values of a normally distributed random variable. Find a 98 percent confidence interval for the mean μ of X.

Since X is normally distributed, so is $\bar{X}$; therefore Prescription 8.2 can be used. The mean of the sample values is

$$\bar{x} = \frac{24.4 + 18.9 + 12.8 + 20.5 + 19.1 + 15.2 + 21.7 + 14.6}{8} = \frac{147.2}{8} = 18.4$$

The value of the sample variance is

$$\begin{aligned} s^2 &= [(24.4 - 18.4)^2 + (18.9 - 18.4)^2 + (12.8 - 18.4)^2 + (20.5 - 18.4)^2 \\ &\quad + (19.1 - 18.4)^2 + (15.2 - 18.4)^2 + (21.7 - 18.4)^2 + (14.6 - 18.4)^2]/7 \\ &= \frac{108.08}{7} \\ &= 15.44 \end{aligned}$$

Therefore the sample value of the standard deviation is $s = \sqrt{15.44} \approx 3.93$. By the symmetry of the t distribution, $P(-t^* \le t \le t^*) = 0.98$ is equivalent to $P(0 \le t \le t^*) = 0.98/2 = 0.49$. Using Table A-2 in the Appendix, with $8 - 1 = 7$ degrees of freedom, we find that $P(0 \le t \le t^*) = 0.49$ for $t^* = 3.00$. By Prescription 8.2, the margin of error is $E = \frac{t^* s}{\sqrt{n}} \approx \frac{3.00 \times 3.93}{\sqrt{8}} \approx 4.17$. The 98 percent confidence interval for the mean of X is $[18.4 - 4.17,\ 18.4 + 4.17] = [14.23,\ 22.57]$.

8.23. A random sample of size 10 from a normal population variable X results in the value $\bar{x} = 124$ for the sample mean and $s^2 = 21$ for the sample variance. Find an approximate 90 percent confidence interval for the mean μ of X.

Since X is normally distributed, so is $\bar{X}$. Therefore, Prescription 8.2 applies. The margin of error is $E = \frac{t^* s}{\sqrt{n}}$, where $s = \sqrt{21}$, $n = 10$, and t^* is the value of the t random variable with 9 degrees of freedom satisfying $P(-t^* \le t \le t^*) = 0.90$; equivalently, $P(0 \le t \le t^*) = 0.45$. From Table A-2 in the Appendix, $t^* = 1.83$. Therefore, $E = \frac{1.83\sqrt{21}}{\sqrt{10}} \approx 2.65$; and the corresponding 90 percent confidence interval for μ is $[124 - 2.65,\ 124 + 2.65] = [121.35\ ,126.65]$.

8.24. A random sample of size 41 of a population random variable X results in a sample mean $\bar{x} = 75.82$ and a sample variance $s^2 = 16.16$. Find an approximate 99 percent confidence interval for the mean μ of X.

Since the sample size is larger than 30, we may assume, on the basis of the Central Limit Theorem, that the sample mean $\bar{X}$ is approximately normally distributed. Then, by Prescription 8.2, the margin of error is $E = \frac{t^* s}{\sqrt{n}}$, where $s = \sqrt{16.16}$, $n = 41$, and t^* is the value of the t random variable with 40 degrees of freedom satisfying $P(-t^* \leq t \leq t^*) = 0.99$, equivalently, $P(0 \leq t \leq t^*) = 0.495$. From Table A-2 in the Appendix, $t^* = 2.70$. Therefore, $E = \frac{2.70\sqrt{16.16}}{\sqrt{41}} \approx 1.70$; and the corresponding 99 percent confidence interval for μ is $[75.82 - 1.70,\ 75.82 + 1.70] = [74.12,\ 77.52]$.

8.25. Suppose the sample size in Problem 8.24 was 200. What would be the corresponding 99 percent confidence interval?

Table A-2 does not have a t^* value for 199 degrees of freedom. The closest value for t^* in the t table is 2.62 for 120 degrees of freedom. Using computer software, we get $t^* = 2.60$. Then $E = \frac{2.60\sqrt{16.16}}{\sqrt{200}} \approx 0.74$ with a corresponding approximate 99 percent confidence interval of $[75.82 - 0.74,\ 75.82 + 0.74] = [75.08,\ 76.56]$. If the value 2.62 were used for t^*, the corresponding margin of error is 0.74 to two places, which gives the same confidence interval. Furthermore, as the number of degrees of freedom increases, the t distribution approaches the standard normal distribution. The critical value of the standard normal Z at the 99 percent confidence level is 2.58, which gives a margin of error of 0.73, and the slightly smaller confidence interval, [75.09, 76.55].

8.26. Prove Theorem 8.1. That is, suppose X has mean μ and standard deviation σ, and the sample mean $\bar{X}$, for random samples of size n, is normally distributed. Let S be the sample standard deviation. Show that the random variable $t = \frac{\bar{X} - \mu}{S/\sqrt{n}}$ has a t distribution with $n - 1$ degrees of freedom.

The random variable $Z = \frac{\bar{X} - \mu}{\sigma/\sqrt{n}}$ has a standard normal distribution (Theorem 7.1), and $\chi^2 = \frac{(n-1)S^2}{\sigma^2}$ has a chi-square distribution with $n - 1$ degrees of freedom (Theorem 7.7). Therefore, by definition, $\frac{Z}{\sqrt{\chi^2/(n-1)}}$ has a t distribution with $n - 1$ degrees of freedom. Now $\sqrt{\chi^2/(n-1)} = S/\sigma$, and when Z is divided by S/σ, the σs cancel, leaving $\frac{\bar{X} - \mu}{S/\sqrt{n}}$, as desired.

CONFIDENCE INTERVALS FOR PROPORTIONS

8.27. In a random sample of 100 transistors, 92 were within the specifications stated by the manufacturer. Find a 99.5 percent confidence interval for the proportion p of all of the manufacturer's transistors that meet the stated specifications.

Since the sample size is larger than 30, Prescription 8.3 can be used. We have $\hat{p} = 92/100 = 0.92$, and from Table A-1, $P(-z^* \leq Z \leq z^*) = 0.995$ (equivalently, $P(0 \leq Z \leq z^*) = 0.995/2 = 0.4975$) for $z^* = 2.81$. The margin of error is $E = z^*\sqrt{\frac{\hat{p}(1-\hat{p})}{n}} = 2.81\sqrt{\frac{0.92 \times 0.08}{100}} \approx 0.08$; and the corresponding 99.5 percent confidence interval for p is $[0.92 - 0.08,\ 0.92 + 0.08] = [0.84,\ 1.00]$.

8.28. Find a 95 percent confidence interval for the proportion p in Problem 8.27.

From Table A-1, $P(-z^* \leq Z \leq z^*) = 0.95$ (equivalently, $P(0 \leq Z \leq z^*) = 0.95/2 = 0.475$) for $z^* = 1.96$. The margin of error is $E = z^*\sqrt{\frac{\hat{p}(1-\hat{p})}{n}} = 1.96\sqrt{\frac{0.92 \times 0.08}{100}} \approx 0.05$, and the corresponding 95 percent confidence interval is $[0.92 - 0.05, 0.92 + 0.05] = [0.87, 0.97]$.

8.29. Suppose a pollster states that [0.52, 0.57] is a 98 percent confidence interval for the proportion of eligible voters favoring candidate A. What percentage of the sample favored candidate A, and what is the margin of error?

The sample proportion $\hat{p}$ favoring candidate A is the center of the interval, which is the average of the two endpoints. Hence, $\hat{p} = \frac{0.52 + 0.57}{2} = \frac{1.09}{2} = 0.545$. Therefore, 54.5 percent of the sample favored candidate A. The interval [0.52, 0.57] is of the form $[\hat{p} - E, \hat{p} + E]$, where $\hat{p}$ is the proportion in the sample favoring A, and E is the margin of error. The length of the interval is $0.57 - 0.52 = 0.05 = \hat{p} + E - (\hat{p} - E) = 2E$. Therefore, $E = 0.05/2 = 0.025$, or 2.5 percent.

8.30. In Problem 8.29, how many eligible voters responded to the pollster?

From Table A-1, we find that critical Z value at the 98 percent level is $z^* = 2.33$; that is, $P(-2.33 \leq Z \leq 2.33) = 0.98$. We then substitute $z^* = 2.33$, $\hat{p} = 0.545$, and $E = 0.025$ into the formula $E = z^*\sqrt{\frac{\hat{p}(1-\hat{p})}{n}}$ to obtain $0.025 = 2.33\sqrt{\frac{0.545 \times 0.455}{n}}$, or $0.025 = \frac{1.16}{\sqrt{n}}$. Therefore, $\sqrt{n} = \frac{1.16}{0.025} = 46.4$. By squaring 46.4 and rounding up to the next integer we find that 2153 eligible voters responded.

8.31. A pollster obtained a confidence interval [0.51, 0.55] for the proportion of eligible voters favoring candidate B based on a sample of 1200 eligible voters. What is the level of confidence of the interval?

Proceeding as in Problem 8.29, we find that the sample proportion favoring B is $\hat{p} = 0.53$, and the margin of error is $E = 0.02$. Substituting these values, along with $n = 1200$, into the formula $E = z^*\sqrt{\frac{\hat{p}(1-\hat{p})}{n}}$, we get $0.02 = z^*\sqrt{\frac{0.53 \times 0.47}{1200}} \approx 0.0144z^*$. Therefore, $z^* = \frac{0.02}{0.0144} \approx 1.39$. From Table A-1, we find that 1.39 is the critical value of Z at the 83.54 percent level. Hence the level of confidence is 83.54 percent.

8.32. Suppose a pollster wants to determine a 95 percent confidence interval for the proportion of citizens that favor a balanced budget, even if some social programs must be cut. A margin of error of no more than two percentage points is desired. How large must the sample size be?

The margin of error is $E = z^*\sqrt{\frac{\hat{p}(1-\hat{p})}{n}}$. Using Table A-1, we find that, for a 95 percent confidence interval, $z^* = 1.96$. As stated at the end of Section 8.4, the most $\sqrt{\hat{p}(1-\hat{p})}$ can be is 0.5. We therefore set $\frac{1.96 \times 0.5}{\sqrt{n}} \leq 0.03$ and solve for n. We get $\sqrt{n} \geq \frac{1.96 \times 0.5}{0.03} = 32.67$. Therefore, $n \geq (32.67)^2 = 1067.33$. Since n must be a whole number, we round up to 1068 as the desired sample size.

CONFIDENCE INTERVALS FOR VARIANCES

8.33. Show that, for any level of confidence γ, $P\left(a \le \frac{(n-1)S^2}{\sigma^2} \le b\right) = \gamma$ is equivalent to $P\left(\frac{(n-1)S^2}{b} \le \sigma^2 \le \frac{(n-1)S^2}{a}\right) = \gamma$.

The inequality $a \le \frac{(n-1)S^2}{\sigma^2}$ is equivalent to $\sigma^2 \le \frac{(n-1)S^2}{a}$, and the inequality $\frac{(n-1)S^2}{\sigma^2} \le b$ is equivalent to $\frac{(n-1)S^2}{b} \le \sigma^2$, from which the desired result follows.

8.34. Suppose that the value $s^2 = 129.14$ obtained in Example 8.12 was based on a random sample of size 101. What would be the corresponding 90 percent confidence interval for σ^2?

With reference to Prescription 8.4, and using Table A-3 in the Appendix, with $101 - 1 = 100$ degrees of freedom, we find that $P(\chi^2 \le a) = \frac{1-0.90}{2} = 0.05$ for $a = 77.9$, and $P(\chi^2 \le b) = \frac{1+0.90}{2} = 0.95$ for $b = 124$. The corresponding 90 percent confidence interval for σ^2 is $\left[\frac{100 \times 129.14}{124}, \frac{100 \times 129.14}{77.9}\right] =$ [104.15, 165.78]. Note that this interval is considerably smaller than the interval [64.1, 416.6] obtained in Example 8.12 for 7 degrees of freedom.

8.35. What is the corresponding 90 percent confidence interval for the standard deviation σ in Problem 8.34?

$$[\sqrt{104.15}, \sqrt{165.78} = [10.21,\ 12.88]$$

8.36. A pollster states that 225 is the left endpoint of a 95 percent confidence interval for the variance of a normally distributed random variable, based on a random sample of size 31. If the interval was determined using Prescription 8.4, what was the value of the sample variance obtained in the sample, and what is the right endpoint of the interval?

According to Prescription 8.4, the confidence interval is of the form $\left[\frac{30s^2}{b}, \frac{30s^2}{a}\right]$, where a satisfies $P(\chi^2 \le a) = \frac{1-0.95}{2} = 0.025$, and $P(\chi^2 \le b) = \frac{1+0.95}{2} = 0.975$. From Table A-3 in the Appendix, we find that $a = 16.8$ and $b = 47.0$. Since the left endpoint is 225, the sample value s^2 of the sample variance must satisfy $\frac{30s^2}{47} = 225$. We can therefore conclude that $s^2 = \frac{225 \times 47}{30} = 352.5$. It then follows that the right endpoint must be $\frac{30s^2}{a} = \frac{30 \times 352.5}{16.8} = 629.46$.

8.37. A random sample of 28 values of a normal random variable X results in a sample standard deviation $s = 6$. Find a 98 percent confidence interval for the standard deviation σ of X.

We first find a 98 percent confidence interval for the variance σ^2 of X, using the value $s^2 = 6^2 = 36$ for the sample variance. Following Prescription 8.4, we find values a and b for the chi-square random variable with 27 degrees of freedom that satisfy $P(\chi^2 \le a) = \frac{1-0.98}{2} = 0.01$ and $P(\chi^2 \le b) = \frac{1+0.98}{2} = 0.99$. Table A-3 in the Appendix gives $a = 12.9$ and $b = 47.0$. The 98 percent confidence interval for σ^2 is $\left[\frac{27 \times 36}{47}, \frac{27 \times 36}{12.9}\right] = [20.68,\ 75.35]$. Therefore, the desired 90 percent confidence interval for σ is $[\sqrt{20.68}, \sqrt{75.35} = [4.55,\ 8.68]$.

8.38. (*Confidence interval for σ^2 when μ is known*) Suppose X is a normal random variable with known mean μ and unknown variance σ^2, and suppose that $X_1, X_2, \ldots, X_n$ is a random-variable sample of size n corresponding to X. Let γ be a specified confidence level. Show that $\left[\frac{\sum(X_i - \mu)^2}{b}, \frac{\sum(X_i - \mu)^2}{a}\right]$ is a 100γ percent confidence interval for σ^2, where the constants a and b are chosen to satisfy, with n degrees of freedom, $P(\chi^2 \le a) = \frac{1-\gamma}{2}$, $P(\chi^2 \le b) = \frac{1+\gamma}{2}$.

Each $(X_i - \mu)/\sigma$ is a normal random variable with mean 0 and standard deviation 1. Therefore, by definition, the random variable $\sum(X_i - \mu)^2/\sigma^2$ is chi-square with n degrees of freedom. The desired result follows from the reasoning leading to Prescription 8.4 with $\sum(x_i - \mu)^2$ in place of $(n-1)s^2$.

Supplementary Problems

PARAMETERS AND STATISTICS

8.39. The median age for the total number of first-year students at a university in 1997 was 17.6, and in a sample of 20 first-year students, the median age was 18.1. Which number is a parameter, and which is a statistic?

8.40. How is a random sample $x_1, x_2, \ldots, x_n$ of values of a random variable X related to a random-variable sample $X_1, X_2, \ldots, X_n$ corresponding to X?

8.41. Suppose $X_1, X_2, \ldots, X_n$ is a random-variable sample corresponding to X, which has unknown mean μ, and let $\bar{X}$ be the corresponding sample mean. Explain why $\sum(X_i - \bar{X})^2$ is a statistic and $\sum(X_i - \mu)^2$ is not.

8.42. Let $x_1 = 12$, $x_2 = 15$, $x_3 = 10$, $x_4 = 11$ be a random sample of values of a random variable X. Find unbiased point estimates of the mean μ and variance σ^2 of X, respectively.

8.43. Suppose that p is the proportion of computer users that are connected to the Internet. In a random sample of 36 computer users, 20 were connected to the Internet. Give an unbiased point estimate for p and also an unbiased estimate for the variance of the proportion of computer users connected to the Internet in all random samples of 36 computer users.

8.44. Explain the difference between an unbiased estimator of a parameter and an unbiased point estimate of the parameter.

THE NOTION OF A CONFIDENCE INTERVAL

8.45. A normal random variable X has unknown mean μ and standard deviation 5. What is the margin of error for a 92 percent confidence interval for μ, based on a random sample of size 25?

8.46. What sample size is needed to have a margin of error equal to one-half the margin of error in Problem 8.45?

8.47. Suppose E is the margin of error in a confidence interval for the mean μ of a normal random variable X, based on a random sample of size n. If the sample size is doubled, what is the new margin of error?

8.48. Suppose 4.5 is the margin of error in a 98 percent confidence interval for the mean μ of a normal random variable X with known standard deviation σ, based on a random sample of size n. What would the margin of error be in a 90 percent confidence interval for μ based on the same random sample?

8.49. Suppose 2.6 is the margin of error in a confidence interval for the mean μ of a normal random variable X with standard deviation 5.4, based on a random sample of size 16. What is the confidence level for the confidence interval?

8.50. Suppose 3.332 is the margin of error in a 95 percent confidence interval for the mean μ of a normal random variable X with standard deviation 8.5. What is the size of the random sample?

CONFIDENCE INTERVALS FOR MEANS

8.51. A random variable X has unknown mean μ and standard deviation 12.5. The sample mean for a random sample of size 50 is $\bar{x} = 72.4$. Find a 95 percent confidence interval for μ.

8.52. A random sample of interest rates charged by area banks for personal loans is: 12.8 percent, 12.2 percent, 13.4 percent, 11.9 percent, 13 percent. Assuming the rates are normally distributed with a standard deviation of 0.9 percent, find a 90 percent confidence interval for the average interest rate.

8.53. [126.4, 132.8] is a 95 percent confidence interval for the mean μ of a normally distributed random variable with known variance. Find a 98 percent confidence interval for μ, based on the same random sample.

8.54. A random sample of 25 grade-point averages at a university has a sample mean $\bar{x} = 2.68$ and a sample standard deviation $s = 0.32$. Assuming that the grade-point averages are approximately normally distributed, find a 95 percent confidence interval for the mean grade-point average.

8.55. A random sample of five gas stations in a certain area gave the following prices in cents for a gallon of regular gasoline: 124.9, 127.9, 130.9, 128.9, 122.9. Assuming the price per gallon is normally distributed, find a 90 percent confidence interval for the average price per gallon.

8.56. [42.7, 49.3] is a 95 percent confidence interval for the mean μ of a normally distributed random variable with unknown variance, based on a random sample of size 16. Find a 90 percent confidence interval for μ.

CONFIDENCE INTERVALS FOR PROPORTIONS

8.57. In a random sample of 200 first-year students at a large urban university, 35 percent said they planned on working from 16 to 20 hours per week to earn money. Give a 95 percent confidence interval for the proportion of all first-year students at the university who plan on working between 16 and 20 hours per week.

8.58. In the same student sample as in Problem 8.57, 9.5 percent of the students said they would be traveling to and from campus by train. Find a 90 percent confidence interval for all first-year students at the university that will be traveling to and from campus by train.

8.59. How large must a random sample be to obtain a 95 percent confidence interval for a population proportion with a margin of error of at most 0.04?

8.60. Suppose [0.46, 0.51] is a 99 percent confidence interval for a population proportion p based on a random sample of size n. Using the same random sample, find a 95 percent confidence interval for p.

CONFIDENCE INTERVALS FOR VARIANCES

8.61. A random sample of 20 values of a normally distributed random variable X results in a sample variance $s^2 = 48.5$. Find a 90 percent confidence interval for the variance σ^2 of X.

8.62. The values of a normally distributed random variable X obtained in a random sample are 25, 28, 26, 25, 22, 30. Find a 95 percent confidence interval for the variance σ^2 of X.

8.63. Suppose 127 is the right endpoint of a 90 percent confidence interval for the variance of a normally distributed random variable, based on a random sample of size 26. If the interval was determined by Prescription 8.4, what is the value of the sample variance obtained in the sample, and what is the left endpoint of the confidence interval?

8.64. Suppose X is a normally distributed random variable with mean $\mu = 50$. The values 46.5, 52.1, 48.6, 50.8 are a random sample of values of X. Use the result of Problem 8.38 to find a 95 percent confidence interval for the standard deviation σ of X.

8.65. Suppose the mean of the random variable X in Problem 8.64 were not known. What would be the corresponding 95 percent confidence interval for σ?

Answers to Supplementary Problems

8.39. 17.6 is a parameter; 18.1 is a statistic.

8.40. x_i is the value of $X_i, i = 1, 2, \ldots, n$, in a random sample of size n.

8.41. $\sum (X_i - \bar{X})^2$ can be computed in terms of sample values x_i of X_i; it does not depend on any unknown population parameters. $\sum (X_i - \mu)^2$ cannot be computed in terms of sample values; it depends on the unknown parameter μ.

8.42. $\bar{x} = \dfrac{12 + 15 + 10 + 11}{4} = 12$ is an unbiased point estimate of μ;

$s^2 = \dfrac{(12 - 12)^2 + (15 - 12)^2 + (10 - 12)^2 + (11 - 12)^2}{3} \approx 4.67$ is an unbiased point estimate of σ^2.

8.43. $\hat{p} = \dfrac{20}{36} \approx 0.56$ is an unbiased point estimate of p; $\dfrac{1}{n-1}\hat{p}(1 - \hat{p}) = \dfrac{1}{35} \times \dfrac{20}{36}\left(1 - \dfrac{20}{36}\right) \approx 0.007$ is an unbiased estimate of $\dfrac{p(1-p)}{36}$.

8.44. An unbiased estimator of a parameter is a random variable whose expected value is equal to the parameter; an unbiased point estimate of a parameter is a numerical value obtained from a random sample of an unbiased estimator of the parameter.

8.45. $E = \dfrac{z^*\sigma}{\sqrt{n}} = \dfrac{1.75 \times 5}{\sqrt{25}} = 1.75$

8.46. $\dfrac{1.75 \times 5}{\sqrt{n}} = \dfrac{1.75}{2}$; $\sqrt{n} = 10$; $n = 100$

8.47. If σ is known, $E_1 = \frac{z^*\sigma}{\sqrt{n}}$; $E_2 = \frac{z^*\sigma}{\sqrt{2n}} = \frac{z^*\sigma}{\sqrt{2}\sqrt{n}} = \frac{1}{\sqrt{2}} E_1$. If σ is unknown,

$$E_1 = \frac{t_1^* s}{\sqrt{n}}; \quad E_2 = \frac{t_2^* s}{\sqrt{2n}} = \frac{t_2^* s}{\sqrt{2}\sqrt{n}} = \frac{t_2^*}{t_1^*} \cdot \frac{E_1}{\sqrt{2}}$$

8.48. $4.5 = \frac{2.33\sigma}{\sqrt{n}}$; $E = \frac{1.65\sigma}{\sqrt{n}} = \frac{1.65 \times 4.5}{2.33} \approx 3.19$

8.49. $2.6 = \frac{z^* \times 5.4}{\sqrt{16}}$; $z^* = \frac{4 \times 2.6}{5.4} \approx 1.93$; confidence level = 94.64 percent.

8.50. $3.332 = \frac{1.96 \times 8.5}{\sqrt{n}}$; $\sqrt{n} = \frac{1.96 \times 8.5}{3.332} = 5$; $n = 25$

8.51. [68.94, 75.87]

8.52. [12.00 percent, 13.32 percent]

8.53. $\bar{x} = 129.6$; $E = 3.8$ (see Problem 8.48); [125.8, 133.4]

8.54. [2.55, 2.81]

8.55. [124.06, 130.14]

8.56. $\bar{x} = 46$; $E = 2.7$ (see Problem 8.48); [43.3, 48.7]

8.57. [0.28, 0.42]

8.58. [0.06, 0.13]

8.59. $1.96\sqrt{\frac{\hat{p}(1-\hat{p})}{n}} \leq \frac{1.96 \times 0.5}{\sqrt{n}} \leq 0.04$; $n \geq 600.25$; round up to 601

8.60. $\hat{p} = 0.485$; $E = 0.019$ (see Problem 8.48); [0.466, 0.504]

8.61. $a = 10.1$, $b = 30.1$; [30.6, 91.2]

8.62. $s^2 = 7.6$, $a = 0.831$, $b = 12.8$; [2.97, 45.7]

8.63. $a = 14.6$, $127 = \frac{25s^2}{14.6}$, $s^2 = 74.2$; $b = 37.7$; left endpoint $= \frac{25 \times 74.2}{37.7} = 49.2$

8.64. $\sum (x_i - \mu)^2 = 19.26$, $a = 0.484$, $b = 11.1$; [1.32, 6.31]

8.65. $s^2 = 6.09$, $a = 0.216$, $b = 9.35$; [1.40, 9.20]

Chapter 9

Hypotheses Tests for a Single Population

9.1 INTRODUCTION: TESTING HYPOTHESES ABOUT PARAMETERS

In Chapter 8, confidence intervals were prescribed for means, proportions, and variances. Here we discuss another type of statistical inference regarding these same parameters. As an illustration, consider the following example.

EXAMPLE 9.1 A bank institutes a new teller procedure designed to shorten the average customer waiting time on busy Friday evenings. The old waiting time was normally distributed with mean $\mu = 12$ minutes and standard deviation $\sigma = 3$ minutes. As a test of the new procedure, a random sample of 36 Friday-evening customers was chosen and found to have an average waiting time of 11 minutes. Determine the probability that such an average waiting time or less would have occurred by chance under the old system, and interpret the result.

Let X denote the random variable representing the old waiting time. Then, for samples of size 36, the sample mean $\bar{X}$ is normally distributed with mean 12 and standard deviation $3/\sqrt{36} = 0.5$. Therefore, the random variable

$$Z = \frac{\bar{X} - 12}{0.5}$$

is the standard normal random variable with mean 0 and standard deviation 1. Using the standard normal table,

$$P(\bar{X} \leq 11) = P\left(\frac{\bar{X} - 12}{0.5} \leq \frac{11 - 12}{0.5}\right) = P(Z \leq -2) = 0.0228$$

Hence, there are only 228 chances in 10,000, or 2.28 out of 100, that a waiting time of 11 minutes or less would have occurred at random under the old system. Such a low theoretical probability for what in fact did occur under the new system is fairly strong evidence that the new procedure really does reduce the average waiting time.

Null Hypothesis and Alternative Hypothesis

Example 9.1 illustrates a typical situation in which some sort of system is modified and it is desired to evaluate the effect of the changes based on sample results of the new system. Specifically, we identify some parameter of a random variable associated with the old system, and ask whether the sample results indicate that a real change has occurred in the value of the parameter, or can the results be merely attributed to chance?

To address this question, we make two opposing hypotheses concerning the parameter, and then use probability to test these hypotheses in light of the sample results. The first hypothesis, called the *null hypothesis*, denoted by H_0, plays devil's advocate and says that the value of the parameter has not really changed; the sample results are simply due to chance. The second hypothesis, called the *alternative hypothesis*, denoted by H_a, maintains that there really has been a change in the value of the parameter; the sample results are not due to chance.

EXAMPLE 9.2 In Example 9.1, the random variable X is the waiting time on Friday evenings, and the parameter is the mean μ of X; that is, the mean waiting time. The null and alternative hypotheses are as follows:

H_0: The waiting time is still normally distributed with mean 12 minutes and standard deviation 3 minutes.
H_a: The average waiting time is less than 12 minutes.

Or, more briefly, letting μ denote the mean waiting time,

$$H_0: \mu = 12$$

$$H_a: \mu < 12$$

Both hypotheses are concerned only with the mean of X. The standard deviation of X is assumed to be unchanged by the new system.

One-Sided and Two-Sided Alternatives

The alternative hypothesis in Example 9.1 could also take on the form $H_a: \mu > 12$ or $H_a: \mu \neq 12$. Each of the hypotheses $H_a: \mu < 12$ and $H_a: \mu > 12$ is called a *one-sided alternative*, while $H_a: \mu \neq 12$ is called a *two-sided alternative*. A one-sided alternative hypothesis says that the mean μ has changed in a specified direction, namely either "the new mean is less than the old mean" or "the new mean is greater than the old mean". A two-sided alternative hypothesis is bi-directional; it says "the new mean is either greater than or less than the old mean". An hypothesis test with a one-sided alternative is called a *one-sided test*. An hypothesis test with a two-sided alternative is called a *two-sided test*.

Simple and Composite Hypotheses

The null hypothesis in Example 9.1, $H_0: \mu = 12$, which says that the mean of X is equal to a specific value, is called a *simple hypothesis*. The alternative hypothesis, $H_a: \mu < 12$, which says that the mean can take on a whole range of values, is called a *composite hypothesis*. A simple hypothesis completely determines the distribution of X, whereas a composite hypothesis does not. For example, the simple null hypothesis $H_0: \mu = 12$ says that the distribution of X is exactly the same as it was before the change, whereas the composite alternative hypothesis $H_a: \mu < 12$ is less specific; it says that the new mean is less than 12, but does not specify what the new mean is. In our examples, the null hypothesis will be simple, and the alternative hypothesis will usually be composite.

Test Statistic and *P*-value of a Test

After the null and alternative hypotheses have been made, a *test statistic* is defined that will enable us to perform the test. Performing the test means determining the likelihood that the sample results would have occurred if the null hypothesis were true. More specifically, the test statistic is a statistic whose value can be computed from the sample results; and the *P-value* of the test is the probability that a value of the test statistic in the direction of the alternative hypothesis and as extreme as the one that actually did occur would have occurred if H_0 were true.

EXAMPLE 9.3 In Example 9.1, the underlying random variable X is the Friday evening waiting time, and the test statistic is obtained by standardizing the sample mean $\bar{X}$. That is, the test statistic is

$$Z = \frac{\bar{X} - 12}{0.5}$$

which, if $H_0: \mu = 12$ is true, is the standard normal random variable. The value of $\bar{X}$ obtained in the sample of 36 customers is $\bar{x} = 11$ minutes. The corresponding z-score of $\bar{x}$, namely

$$z = \frac{11 - 12}{0.5} = -2$$

is the test value of the test statistic. Now the alternative hypothesis says that the new average waiting time is less than 12 minutes. Hence, assuming $H_0: \mu = 12$ is true, then a sample waiting time in the direction of the alternative hypothesis as extreme as 11 minutes is a sample waiting time at least one minute less than the sample mean of 12 minutes, or at least two standard units less than zero. Therefore, the *P*-value of the test is

$$P(Z \leq -2) = 0.0228$$

EXAMPLE 9.4 Suppose the alternative hypothesis in Example 9.1 were $H_a: \mu \neq 12$. Then a sample waiting time as extreme as 11 minutes is a sample waiting time at least one minute less than or one minute greater than 12 minutes; equivalently, at least two standard units less than or greater than zero. The P-value of the test would be

$$P(Z \leq -2) + P(Z \geq 2) = 2P(Z \geq 2) = 2(0.0228) = 0.0456$$

Determining the *P*-value

In general, the P-value of a test depends on the null hypothesis, the test statistic, the test value of the test statistic, and the alternative hypothesis. Suppose the null hypothesis is a simple hypothesis, say $H_0: \mu = \mu_0$, the test statistic, assuming H_0 is true, is the standard normal variable Z, and the test value of Z is z. Then the P-value is determined as follows:

For $H_a: \mu < \mu_0$, the P-value is $P(Z \leq z)$
For $H_a: \mu > \mu_0$, the P-value is $P(Z \geq z)$
For $H_a: \mu \neq \mu_0$, the P-value is $P(Z \leq -|z|) + P(Z \geq |z|)$ [equivalently, $2P(Z \geq |z|)$]

Significance Level and Statistical Significance

A high P-value is evidence in support of the null hypothesis, H_0, and a low P-value provides evidence against H_0. To assess the weight of the evidence, a threshold P-value, called the *significance level of the test*, is often selected before conducting the test. The significance level is usually denoted by α, and values of 0.01 and 0.05 have traditionally been used for α, but other values can be used as well. If the P-value of the test is less than or equal to α, then the corresponding value of the test statistic is said to be *statistically significant at the level* α. If the P-value is greater than α, then the value of the test statistic is *not statistically significant at the level* α.

EXAMPLE 9.5 In Example 9.1, the value of the test statistic $Z = (\bar{X} - 12)/0.5$ obtained in the sample is -2, whose P-value was computed to be 0.0228 in Example 9.3. Since 0.0228 is less than 0.05, the test value -2 of Z is statistically significant at the level 0.05. However, 0.0228 is not less than 0.01, so -2 is not statistically significant at the level 0.01.

Using Significance Level and *P*-value for Decision Making

Suppose now that the sample results of a test are going to be used to decide whether or not to reject the null hypothesis, H_0, as being true. A low P-value obtained in a test says that if H_0 is true, then a rare event has taken place; equivalently, if the event is not so rare, then H_0 must be false, and should be rejected. The significance level α, chosen before the test, is used as the measure of rarity. *If the P-value of the test is less than or equal to* α, *then the null hypothesis is rejected at the* α *level of significance.* If the P-value is greater than α, then we either accept H_0 or hedge a bit and conclude that there is insufficient evidence to reject H_0 at the α level of significance. Simply stated, *if the P-value of the test is greater than* α, *then the null hypothesis is not rejected.*

EXAMPLE 9.6 The P-value for the average waiting time in Example 9.1 was computed to be 0.0228 in Example 9.3. Since 0.0228 is less than 0.05, the null hypothesis would be rejected at the 0.05 level of significance. Here we are inclined to believe that the average waiting time has been reduced from 12 minutes by the new teller procedure rather than maintain that the average waiting time is still 12 minutes and an event whose chances are less than 5 in 20 did occur. On the other hand, since 0.0228 is not less than 0.01, the null hypothesis would not be rejected at the 0.01 level. At this level, to conclude that the average waiting time has been reduced from 12 minutes, we require that the average waiting time for the sample be so rare as to have only 1 chance out of 100, or less, of occurring.

The Critical Region

All values of the test statistic in the direction of the alternative hypothesis with a P-value less than or equal to the significance level α define a set called the *critical region* of the test statistic. By definition, α is the probability that the test statistic will lie in the critical region.

EXAMPLE 9.7 Suppose the test statistic, assuming $H_0: \mu = \mu_0$ is true, is the standard normal random variable Z, and the level of significance is $\alpha = 0.05$. Then, as illustrated in Fig. 9-1, the critical regions for one-sided and two-sided alternative hypotheses are as follows:

For $H_a: \mu < \mu_0$, the critical region is all values $z \leq -1.65$, since $P(Z \leq -1.65) = 0.05$
For $H_a: \mu > \mu_0$, the critical region is all values $z \geq 1.65$, since $P(Z \geq 1.65) = 0.05$
For $H_a: \mu \neq \mu_0$, the critical region is all values $z \leq -1.96$ or $z \geq 1.96$, since $P(Z \leq -1.96) + P(Z \geq 1.96) = 0.05$ [equivalently, $P(Z \geq 1.96) = 0.05/2 = 0.025$]

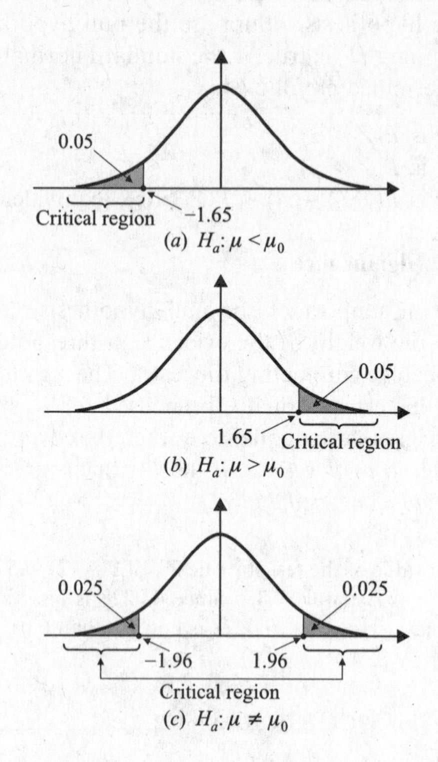

Fig. 9-1 Critical Z regions, $\alpha = 0.05$.

EXAMPLE 9.8 Suppose the test statistic, assuming $H_0: \mu = \mu_0$ is true, is the standard normal random variable Z, and $\alpha = 0.01$. At the 0.01 level of significance, the corresponding critical regions are the following (see Fig. 9-2):

For $H_a: \mu < \mu_0$, the critical region is all values $z \leq -2.33$, since $P(Z \leq -2.33) = 0.01$
For $H_a: \mu > \mu_0$, the critical region is all values $z \geq 2.33$, since $P(Z \geq 2.33) = 0.01$
For $H_a: \mu \neq \mu_0$, the critical region is all values $z \leq -2.58$ or $z \geq 2.58$, since $P(Z \leq -2.58) + P(Z \geq 2.58) = 0.01$ [equivalently, $P(Z \geq 2.58) = 0.01/2 = 0.005$]

Determining the Critical Region

In general, the critical region depends on the null hypothesis, the test statistic, the significance level, and the alternative hypothesis. Suppose the null hypothesis is the simple hypothesis $H_0: \mu = \mu_0$, the test statistic, assuming H_0 is true, is the standard normal random variable Z, and the significance level is α. Then the critical region is determined as follows:

For $H_a: \mu < \mu_0$, the critical region is all values $z \leq z^*$, where $P(Z \leq z^*) = \alpha$
For $H_a: \mu > \mu_0$, the critical region is all values $z \geq z^*$, where $P(Z \geq z^*) = \alpha$
For $H_a: \mu \neq \mu_0$, the critical region is all values $z \leq -z^*$ or $z \geq z^*$, where $P(Z \leq -z^*) + P(Z \geq z^*) = \alpha$ [equivalently, $P(Z \geq z^*) = \alpha/2$]

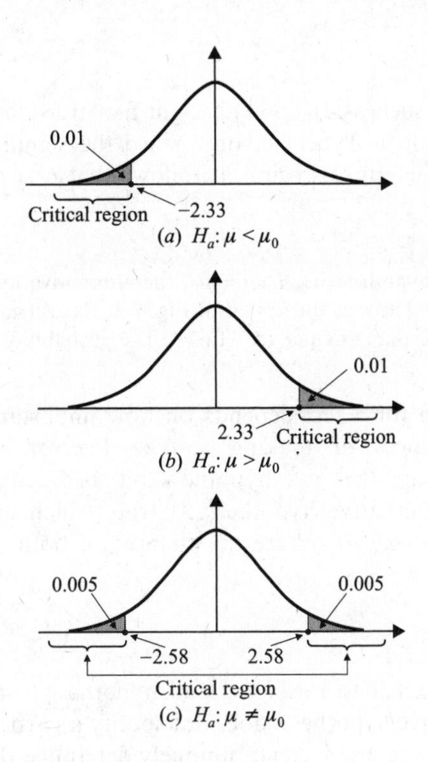

Fig. 9-2 Critical Z regions, $\alpha = 0.01$.

Using Significance Level and Critical Region for Decision Making

The P-value of a test will be less than or equal to the significance level α precisely when the test value of the test statistic lies in the critical region. Hence, *if the test value lies in the critical region, then the null hypothesis is rejected at the α level of significance; if the test value is not in the critical region, then the null hypothesis is not rejected.*

EXAMPLE 9.9 The critical region for the standard normal random variable at the 0.05 level of significance for the alternative hypothesis $H_a: \mu < \mu_0$ is all values $z \leq -1.65$ (Example 9.7). Suppose the test value of the test statistic is -2 (as in Example 9.3). Since -2 is less than -1.65, the null hypothesis is rejected at the 0.05 significance level. However, the critical region at the 0.01 level of significance is all values $z \leq -2.33$ (Example 9.8). Since -2 is not less than -2.33, the null hypothesis is not rejected at the 0.01 significance level.

Type I and Type II Errors

There are two important types of mistake that can be made when reaching a decision on the basis of a hypothesis test: rejecting the null hypothesis when it is true is called a *Type I* error; not rejecting the null hypothesis when the alternative hypothesis is true is called a *Type II* error. Each type of error depends on the specified significance level.

EXAMPLE 9.10 To illustrate a Type I error, suppose the significance level were chosen to be 0.05, and the P-value of the test is 0.0228 (Example 9.3). Since 0.0228 is less than 0.05, H_0 would be rejected at the 0.05 level of significance. However, this would be a mistake if H_0 were in fact true.

EXAMPLE 9.11 Suppose the level of significance were chosen to be 0.01, and the P-value of the test is 0.0228. Since 0.0228 is not less than 0.01, the null hypothesis H_0 would not be rejected at the 0.01 level of significance. This would be a Type II error if the alternative hypothesis H_a were true.

Probability of a Type I Error

If a simple null hypothesis, such as $H_0: \mu = \mu_0$, is in fact true, then all values of the test statistic in the critical region will result in a Type I error. Since the significance level α is the probability that the test statistic will lie in the critical region, it follows that *α is the probability of making a Type I error.*

EXAMPLE 9.12 Suppose the null hypothesis is $H_0: \mu = \mu_0$, the alternative hypothesis is $H_a: \mu < \mu_0$, and the significance level is chosen to be 0.05. Then, as illustrated in Fig. 9-1, the critical region consists of all values of the standard normal random variable less than or equal to -1.65. The probability that a test value will lie to the left of -1.65 is precisely 0.05.

The significance level chosen for a test depends on how important it is to avoid a Type I error. Decreasing α reduces the chances of making a type I error. However, since decreasing α reduces the likelihood of rejecting the null hypothesis, it also reduces the likelihood of rejecting the null hypothesis when the alternative hypothesis is true, which increases the chances of making a Type II error. The only sure way to reduce the chances of both types of error is to increase the sample size.

Probability of a Type II Error

A Type II error results if we fail to reject the null hypothesis when the alternative hypothesis is true. Now a composite alternative hypothesis does not specify a particular value of the parameter that forms the basis of the test, and therefore does not uniquely determine the distribution of the underlying random variable. Hence, we cannot determine the probability of a Type II error simply by assuming the alternative hypothesis is true. We can, however, determine the probability of a Type II error for each specific value of the parameter for which the alternative hypothesis is true (see Examples 9.13 and 9.14). This determination leads to the notion of the power of a test, described below.

Power of a Test

Suppose that the null and alternative hypotheses are hypotheses about a parameter, say the mean μ, of a random variable X, whose value completely determines the distribution of X. The test is to be conducted at a specified level of significance α, which therefore determines a critical region for the test statistic. Let μ_1 be a specific value of the parameter. Then the *power of the test at* μ_1, denoted by $K(\mu_1)$, is defined to be the probability that the null hypothesis $H_0: \mu = \mu_0$ will be rejected when $\mu = \mu_1$. Therefore, $1 - K(\mu_1)$ is the probability of a Type II error, given that $\mu = \mu_1 \neq \mu_0$.

$K(\mu_1)$ is the probability that the test statistic lies in the critical region, given that $\mu = \mu_1$. Recall that the critical region for the test statistic $\dfrac{\bar{X} - \mu_0}{\sigma/\sqrt{n}}$ is determined under the assumption that this statistic is the standard normal random variable; but if $\mu = \mu_1$ and $\mu_1 \neq \mu_0$, then the test statistic is *not* the standard normal random variable, so the probability that it lies in the critical region is not equal to the level of significance α.

EXAMPLE 9.13 With reference to Example 9.1, suppose that the null hypothesis $H_0: \mu = 12$ is to be tested against the alternative hypothesis $H_a: \mu < 12$ at the 0.05 level of significance, based on a random sample of size 36. Suppose also that the new teller policy actually results in the average waiting time being reduced from 12 minutes to 11 minutes, and the standard deviation of the waiting time is still 3 minutes. What is the power $K(11)$ of the test?

The test statistic is $\dfrac{\bar{X} - \mu_0}{\sigma/\sqrt{n}} = \dfrac{\bar{X} - 12}{3/\sqrt{36}} = \dfrac{\bar{X} - 12}{0.5}$. The critical value of the standard normal random variable at the 0.05 percent level for the alternative hypothesis $H_a: \mu < 12$ is $z^* = -1.65$ (see Example 9.7). We want to compute the probability that $\frac{\bar{X}-12}{0.5} \le -1.65$, given that $\frac{\bar{X}-11}{0.5}$ is the standard normal random variable. Now

$$\begin{aligned}
K(11) &= P\left(\frac{\bar{X}-12}{0.5} \le -1.65\right) \\
&= P(\bar{X} \le -1.65 \times 0.5 + 12) \\
&= P\left(\frac{\bar{X}-11}{0.5} \le \frac{-1.65 \times 0.5 + 12 - 11}{0.5}\right) \\
&\approx P(Z \le 0.35) \\
&\approx 0.6368
\end{aligned}$$

Powerful Tests

If the null hypothesis is $H_0\colon \mu = \mu_0$, then $K(\mu_0) = \alpha$, which is the probability of a Type I error; therefore we want $K(\mu_0)$ to be small. We can achieve this by choosing the level of significance α to be small. On the other hand, if μ_1 is a value of μ for which the alternative hypothesis H_a is true, then we want $K(\mu_1)$ to be large. $1 - K(\mu_1)$ is the probability of a Type II error, which will be small when $K(\mu_1)$ is large. Hence, a powerful test is one in which $K(\mu_0)$ is small, and $K(\mu_1)$ is large whenever $\mu_1 \neq \mu_0$. Increasing sample size increases the power of a test.

EXAMPLE 9.14 Suppose the sample size in Example 9.13 is increased to 64. Find the power $K(11)$ of the test and the probability of a Type II error at the 0.05 significance level when $\mu = 11$.

With a sample of size 64, the test statistic is $\frac{\bar{X}-12}{3/\sqrt{64}} = \frac{\bar{X}-12}{0.375}$. Following the method of Example 9.13, we get

$$\begin{aligned}
K(11) &= P\left(\frac{\bar{X}-12}{0.375} \le -1.65\right) \\
&= P(\bar{X} \le -1.65 \times 0.375 + 12) \\
&= P\left(\frac{\bar{X}-11}{0.375} \le \frac{-1.65 \times 0.375 + 12 - 11}{0.375}\right) \\
&\approx P(Z \le 1.02) \\
&\approx 0.8461
\end{aligned}$$

The probability of a Type II error when $\mu = 11$ is $1 - K(11) = 0.1539$.

Using Tables or Computer Software to Find the *P*-value of a Test

The test statistic in hypothesis testing is often the standard normal random variable, a chi-square random variable, a t random variable, or an F random variable (see Section 10.5). Tables in statistics texts are usually adequate to compute the P-value for the test value of a standard normal random variable. However, the t and chi-square random variables require a different table for each degree of freedom, so most statistics texts include test values for a very limited number of P-values in addition to 0.05 and 0.01. The F random variable depends on a pair of degrees of freedom; and textbook tables are often limited to P-values of 0.05 and 0.01 only. More detailed reference tables are available, but the simpler procedure is to use a computer software package to determine the P-value when the test statistic is not a standard normal random variable.

Reasonable Doubt

Hypothesis testing for rejecting or not rejecting the null hypothesis is similar to weighing evidence against the defendant in a criminal or civil trial. If, in the mind of a jury in a criminal trial, there is no reasonable doubt that the defendant has committed the crime, then the jury should find the defendant guilty. In a civil trial, only a "preponderance" of evidence is needed to find the defendant liable. In hypothesis testing, the null hypothesis is an assumption of innocence, and the evidence against the null hypothesis is provided by the sample results. The level of significance α sets the standard for reasonable doubt or preponderance of the evidence. If the P-value of the test is less than or equal to α, then the criterion for eliminating reasonable doubt or for establishing a preponderance of evidence has been achieved, and the null hypothesis is rejected. A very low significance level corresponds to a criminal case that requires a very high degree of certainty of guilt; relatively higher levels of significance correspond to civil cases in which a lighter weight of evidence can be used for a finding of liability.

Absolute certainty is rare in trials and in hypothesis testing. Committing a Type I error in hypothesis testing corresponds to finding an innocent defendant guilty or liable. A Type II error corresponds to not reaching a guilty or liable verdict when the defendant is guilty.

9.2 HYPOTHESES TESTS FOR MEANS

Let X be a random variable with mean μ, which is unknown, and standard deviation σ, defined on some population. We give prescriptions for hypotheses tests regarding μ when σ is known and when σ is unknown. As in the case of confidence intervals for μ, hypotheses tests for μ require that the sample mean, $\bar{X}$, be approximately normally distributed. This condition can be met for small samples ($n < 30$) if X itself is normally distributed. For large samples ($n \geq 30$), the Central Limit Theorem allows us to assume that $\bar{X}$ is approximately normally distributed regardless of the distribution of X. We consider hypotheses tests for μ where the null hypothesis is

$$H_0: \mu = \mu_0$$

and the alternative hypothesis is one of the following:

$$H_a: \mu < \mu_0, \quad H_a: \mu > \mu_0, \quad \text{or} \quad H_a: \mu \neq \mu_0$$

PRESCRIPTION 9.1 (*P*-value hypotheses tests for μ when σ is known)

Requirements: X has known standard deviation σ, and the sample mean $\bar{X}$ is approximately normally distributed.

Let α be the specified level of significance for the test, and suppose that a value $\bar{x}$ of the sample mean $\bar{X}$ is obtained in a random sample of size n. Complete the following steps.

(1) *State Hypotheses*: State null hypothesis $H_0: \mu = \mu_0$ and alternative hypothesis H_a.

(2) *Compute Test Statistic*: The test statistic is the standardized sample mean, namely $Z = \dfrac{\bar{X} - \mu_0}{\sigma/\sqrt{n}}$ which, assuming H_0 is true, is (approximately) the standard normal random variable. Compute the test value of Z, which is the z score of $\bar{x}$: $z = \dfrac{\bar{x} - \mu_0}{\sigma/\sqrt{n}}$.

(3) *Determine P-value*: Using a standard normal table (or computer software), find the P-value of the test corresponding to H_a:

For $H_a: \mu < \mu_0$, the P-value is $P(Z \leq z)$
For $H_a: \mu > \mu_0$, the P-value is $P(Z \geq z)$
For $H_a: \mu \neq \mu_0$, the P-value is $P(Z \leq -|z|) + P(Z \geq |z|)$ [equivalently, $2P(Z \geq |z|)$].

(4) *Draw Conclusion*: If P-value $\leq \alpha$, then both z and $\bar{x}$ are said to be statistically significant at level α, and H_0 is rejected. If P-value $> \alpha$, then z and $\bar{x}$ are not statistically significant at level α, and H_0 is not rejected.

Alternative Version of Prescription 9.1

Instead of computing the P-values in Step 3 of Prescription 9.1, we could determine the critical region for the alternative hypothesis at the specified level of significance, α (see Section 9.1). If the test value z is in the critical region, then the test result is significant at level α, and the null hypothesis would be rejected. If the test value z is not in the critical region, then the test result is not significant at level α, and the null hypothesis would not be rejected. Hence Prescription 9.1 can be replaced by the following.

PRESCRIPTION 9.1*a* (Critical-region hypotheses tests for μ and σ is known)

Requirements: X has known standard deviation σ, and the sample mean $\bar{X}$ is approximately normally distributed.

(1) and 2 Same as in Prescription 9.1.

(3) *Determine Critical Region*: Using a standard normal table (or computer software), find the critical region corresponding to H_a and α:

For $H_a: \mu < \mu_0$, the critical region is all z scores $z \leq z^*$, where z^* is the (negative) value satisfying $P(Z \leq z^*) = \alpha$ (Fig. 9-3(*a*)).

For $H_a: \mu > \mu_0$, the critical region is all z scores $z \geq z^*$, where z^* is the (positive) value satisfying $P(Z \geq z^*) = \alpha$ (Fig. 9-3(*b*)).

For $H_a: \mu \neq \mu_0$, the critical region is all z scores for which $z \leq -z^*$ or $z \geq z^*$, where z^* is the (positive) value satisfying $P(Z \leq -z^*) + P(Z \geq z^*) = \alpha$ [equivalently, $P(Z \geq z^*) = \alpha/2$] (Fig. 9-3(*c*).

(4) *Draw Conclusion*: If the sample value z of the test statistic lies in the critical region, then both z and $\bar{x}$ are said to be statistically significant at level α, and H_0 is rejected. If z does not lie in the critical region, then z and $\bar{x}$ are not statistically significant at level α, and H_0 is not rejected.

Both versions of Prescription 9.1 are applied to the following example.

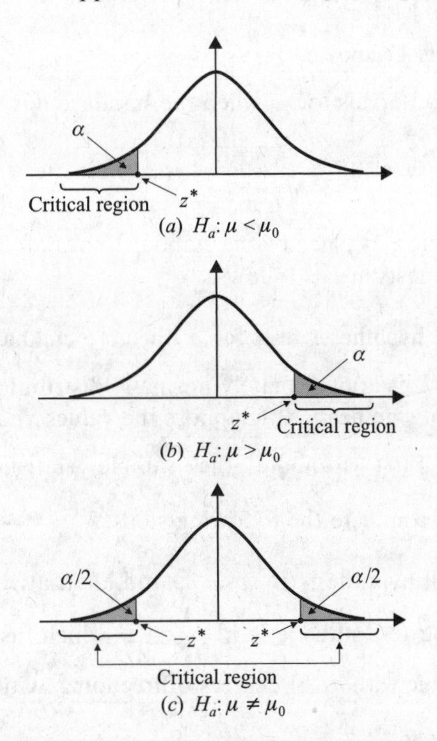

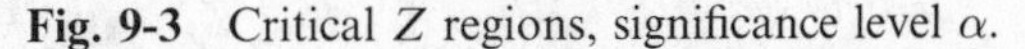

Fig. 9-3 Critical Z regions, significance level α.

EXAMPLE 9.15 A population random variable X is normally distributed with unknown mean μ and with standard deviation $\sigma = 2$. The null hypothesis is $H_0: \mu = 15$. A random sample of size 25, drawn from the population, resu'ts in a sample mean $\bar{x} = 16$. Test the null hypothesis at significance level $\alpha = 0.01$ against each of the following alternative hypothesis:

(*a*) $H_a: \mu < 15$ (*b*) $H_a: \mu > 15$ (*c*) $H_a: \mu \neq 15$

P-value solution: Since X is normally distributed, so is $\bar{X}$. The null and alternative hypotheses have already been stated in each case. The value of the test statistic is $z = \dfrac{16 - 15}{2/5} = 2.5$ in all three cases. We now do each case individually.

(*a*) By Step 3 in Prescription 9.1, the *P*-value is $P(Z \leq 2.5) = 0.9938$, which is certainly not less than 0.01. Therefore, the test result $z = 2.5$ (or $\bar{x} = 16$) is not statistically significant at level 0.01, and the null hypothesis would not be rejected.

(*b*) By Step 3 in Prescription 9.1, the *P*-value is $P(Z \geq 2.5) = 0.0062$, which is less than 0.01. Therefore, the test result $z = 2.5$ (or $\bar{x} = 16$) is statistically significant at level 0.01, and the null hypothesis would be rejected.

(*c*) By Step 3 in Prescription 9.1, the *P*-value is

$$P(Z \leq -2.5) + P(Z \geq 2.5) = 0.0062 + 0.0062 = 0.0124$$

which is not less than 0.01. Therefore, the test result $z = 2.5$ (or $\bar{x} = 16$) is not statistically significant at level 0.01, and the null hypothesis would not be rejected.

Critical-region solution

(*a*) The critical region for the one-sided alternative $H_a: \mu < 15$ at significance level $\alpha = 0.01$ is $z \leq -2.33$ (Fig. 9-2(*a*)). Since the test value $z = 2.5$ is not in this region, the test result is not statistically significant at level 0.01, and the null hypothesis $H_0: \mu = 15$ would not be rejected.

(*b*) The critical region for the one-sided alternative $H_a: \mu > 15$ at significance level $\alpha = 0.01$ is $z \geq 2.33$ (Fig. 9-2(*b*)). Since the test value $z = 2.5$ is in this region, the test result is statistically significant at level 0.01, and the null hypothesis $H_0: \mu = 15$ would be rejected.

(*c*) The critical region for the two-sided alternative $H_a: \mu \neq 15$ at significance level $\alpha = 0.01$ consists of the z scores satisfying $z \leq -2.58$ or $z \geq 2.58$ (Fig. 9-2(*c*)). Since the test value $z = 2.5$ is not in this region, the test result is not statistically significant at level 0.01, and the null hypothesis $H_0: \mu = 15$ would not be rejected.

Hypotheses Tests for μ When σ is Unknown

As in the case of confidence intervals for μ, when the standard deviation σ is not known, we use the sample standard deviation

$$S = \sqrt{\frac{1}{n-1}\sum (X_i - \bar{X})^2}$$

in place of σ, and the t distribution in place of the standard normal distribution. The corresponding prescriptions for the hypotheses tests are as follows.

PRESCRIPTION 9.2 (*P*-value hypotheses tests for μ when σ is unknown)

Requirement: The sample mean $\bar{X}$ is approximately normally distributed.

Let α be the specified level of significance. Suppose the values $x_1, x_2, \ldots, x_n$ of X are obtained in a random sample of size n. First compute the sample statistics $\bar{x} = \dfrac{x_1 + x_2 + \cdots + x_n}{n}$ and $s = \sqrt{\dfrac{1}{n-1}\sum (x_i - \bar{x})^2}$. Then complete the following steps.

(1) *State Hypotheses*: State null hypothesis $H_0: \mu = \mu_0$ and alternative hypothesis H_a.

(2) *Compute Test Statistic*: The test statistic is $t = \dfrac{\bar{X} - \mu_0}{S/\sqrt{n}}$ which, assuming H_0 is true, is (approximately) the t random variable with $n - 1$ degrees of freedom. Compute the test value of t as the t score of the sample mean: $\hat{t} = \dfrac{\bar{x} - \mu_0}{s/\sqrt{n}}$.

(3) *Determine P-value*: Using a t table, if adequate, or computer software for the t random variable with $n - 1$ degrees of freedom, find the P-value of the test corresponding to H_a:

For $H_a: \mu < \mu_0$, the P-value is $P(t < \hat{t})$

For $H_a: \mu > \mu_0$, the P-value is $P(t > \hat{t})$

For $H_a: \mu \neq \mu_0$, the P-value is $P(t < -|\hat{t}|) + P(t > |\hat{t}|)$ [equivalently, $2P(t > |\hat{t}|)$]

(4) *Draw Conclusion*: If P-value $\leq \alpha$, then both the value $\hat{t}$ of the test statistic and the value $\bar{x}$ of the sample mean are said to be statistically significant at level α, and H_0 is rejected at the α level of significance. If P-value $> \alpha$, then $\hat{t}$ and $\bar{x}$ are not statistically significant at level α, and H_0 is not rejected.

EXAMPLE 9.16 A population random variable X is normally distributed with unknown mean and standard deviation. A random sample of size 16 yields a sample mean $\bar{x} = 110$ and sample standard deviation $s = 18.18$. Test the null hypothesis $H_0: \mu = 100$ against the alternative hypothesis $H_a: \mu \neq 100$ at the significance level $\alpha = 0.05$ by computing the P-value of the test.

Since σ is unknown, we apply Prescription 9.2. X is normally distributed, so $\bar{X}$ also is. Step 1 has already been completed since the null and alternative hypotheses are given. In Step 2, the test statistic is the t random variable $t = \dfrac{\bar{X} - 100}{S/\sqrt{16}} = \dfrac{\bar{X} - 100}{S/4}$, with 15 degrees of freedom. The test value is the t score of $\bar{X}: \hat{t} = \dfrac{110 - 100}{18.18/4} = 2.2$. The P-value of the test is $P(t \leq -2.2) + P(t \geq 2.2) = 2P(t \geq 2.2)$. The closest value to 2.2 for 15 degrees of freedom in Table A-2 of the Appendix is 2.13, which corresponds to a P-value of 0.05. The actual P-value is less than 0.05. Using computer software, we find that the P-value is 0.0439. Since 0.0439 is less than 0.05, we reject the null hypothesis.

Alternative Version of Prescription 9.2

Most t tables in textbooks are inadequate to determine the P-value in many cases when the test statistic is the t random variable. The alternative version of Prescription 9.2 uses the critical region corresponding to the level of significance α and the alternative hypothesis H_a, and does not require the determination of the P-value. The critical region can be determined for various levels of significance and degrees of freedom from Table A-2 for the t random variable in the Appendix.

PRESCRIPTION 9.2*a* (Critical-region hypotheses tests for μ when σ is unknown)

Requirement: The sample mean $\bar{X}$ is approximately normally distributed.

(1) and (2) Same as in Prescription 9.2.

(3) *Determine Critical Region*: Using a t table with $n - 1$ degrees of freedom (or computer software), find the critical region corresponding to H_a and α:

For $H_a: \mu < \mu_0$, the critical region is all values $\hat{t} \leq t^*$, where t^* is the (negative) value satisfying $P(t \leq t^*) = \alpha$ (Fig. 9-4(*a*)).

For $H_a: \mu > \mu_0$, the critical region is all values $\hat{t} \geq t^*$, where t^* is the (positive) value satisfying $P(t \geq t^*) = \alpha$ (Fig. 9-4(*b*)).

For $H_a: \mu \neq \mu_0$, the critical region is all values $\hat{t}$ for which $\hat{t} \leq -t^*$ or $\hat{t} \geq t^*$, where t^* is the (positive) value satisfying $P(t \leq -t^*) + P(t \geq t^*) = \alpha$ [equivalently, $P(t \geq t^*) = \alpha/2$] (Fig. 9-4(*c*)).

(4) *Draw Conclusion*: If the sample value, $\hat{t}$, of the test statistic lies in the critical region, then $\hat{t}$ and $\bar{x}$ are statistically significant at level α, and H_0 is rejected. If $\hat{t}$ does not lie in the critical region, then $\hat{t}$ and $\bar{x}$ are not statistically significant at level α, and H_0 is not rejected.

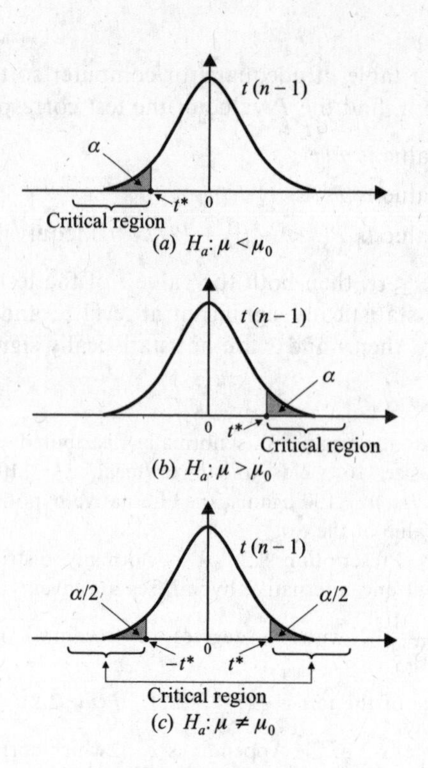

Fig. 9-4 Critical t regions.

EXAMPLE 9.17 Test the null hypothesis $H_0: \mu = 100$ against the alternative hypothesis $H_a: \mu \neq 100$ in Example 9.16 at the significance level $\alpha = 0.05$ by determining the critical region for H_a at level $\alpha = 0.05$.

We follow Prescription 9.2*a*. The critical region for the two-sided alternative $H_a: \mu \neq 100$ at significance level $\alpha = 0.05$ and 15 degrees of freedom consists of the t scores for which $\hat{t} \leq -2.13$ or $\hat{t} \geq 2.13$ (Fig. 9-5). Since the test value $\hat{t} = 2.2$ is in this region, the test result is statistically significant at level 0.05, and the null hypothesis $H_0: \mu = 100$ would be rejected.

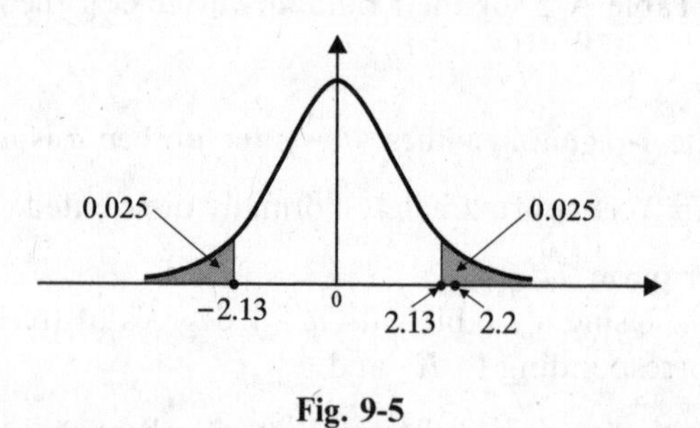

Fig. 9-5

9.3 HYPOTHESES TESTS FOR PROPORTIONS

As in the case of confidence intervals, we assume that a population is broken up into two groups, and the members of one of the groups are referred to as "successes." Let p be the (unknown) proportion of successes in the population, and let $\hat{P}$ be the random variable whose value on a random sample of size n is the proportion $\hat{p}$ of successes in the sample. $\hat{P}$ has mean p and standard deviation $\sqrt{p(1-p)/n}$, and is approximately normal when $n \geq 30$. We then arrive at the following prescription.

PRESCRIPTION 9.3 (*P*-value hypotheses tests for population proportion *p*)

Requirement: The sample size n is large, $n \geq 30$.

Let α be the specified level of significance for the test, and suppose $\hat{p}$ is the proportion of successes obtained in a random sample of size $n \geq 30$. Complete the following steps.

(1) *State Hypotheses*: State null hypothesis $H_0: p = p_0$ and alternative hypothesis H_a.

(2) *Compute Test Statistic*: The test statistic is the standardized sample proportion, namely $Z = \dfrac{\hat{P} - p_0}{\sqrt{p_0(1-p_0)/n}}$ which, assuming H_0 is true, is approximately normally distributed with mean 0 and standard deviation 1. Compute the test value of Z as the z score of $\hat{p}$, namely $z = \dfrac{\hat{p} - p_0}{\sqrt{p_0(1-p_0)/n}}$.

(3) *Determine P-value*: Using a standard normal table (or computer software), find the P-value of the test corresponding to H_a:

For $H_a: p < p_0$, the P-value is $P(Z \leq z)$

For $H_a: p > p_0$, the P-value is $P(Z \geq z)$

For $H_a: p \neq p_0$, the P-value is $P(Z \leq -|z|) + P(Z \geq |z|)$ [equivalently, $2P(Z \geq |z|)$]

(4) *Draw Conclusion*: If P-value $\leq \alpha$, then the z score of the sample proportion $\hat{p}$ is statistically significant at level α, and H_0 is rejected. We would also say that $\hat{p}$ is statistically significant at level α. If P-value $> \alpha$, then z and $\hat{p}$ are not statistically significant at level α, and H_0 is not rejected.

EXAMPLE 9.18 A pharmaceutical company claims that 90 percent of smokers that use their anti-tobacco product, Kickit, break the smoking habit in two months. In a random sample of 100 smokers who used Kickit as prescribed, 84 stopped smoking in two months. Determine the P-value of the test of the null hypothesis $H_0: p = 0.9$ against the alternative hypothesis $H_a: p < 0.9$. Is the sample proportion $\hat{p} = \dfrac{84}{100} = 0.84$ statistically significant at the 0.01 level?

Since the sample size $n = 100$ is greater than 30, we can use Prescription 9.3. The test statistic is $Z = \dfrac{\hat{P} - 0.9}{\sqrt{0.9(1-0.9)/100}} = \dfrac{\hat{P} - 0.9}{0.03}$, and the test value of Z is $\dfrac{0.84 - 0.9}{0.03} = -2$. Therefore the P-value of the test is $P(Z \leq -2)$. Using the standard normal table, we find that $P(Z \leq -2) = 0.0228$. Since 0.0228 is not less than 0.01, the sample proportion $\hat{p} = 0.84$ is not statistically significant at the 0.01 significance level. The test does not provide enough evidence to reject the null hypothesis at the 0.01 level. It does, however, provide evidence to reject the null hypothesis at any significance level greater than or equal to 0.0228.

Alternative Version of Prescription 9.3

Instead of computing the P-values in step 3 of Prescription 9.3, we could determine the critical region for the alternative hypothesis at the specified level of significance, α. If the test value z is in the critical region, then the test result is significant at level α, and the null hypothesis would be rejected. If the test value z is not in the critical region, then the test result is not significant at level α, and the null hypothesis would not be rejected. Hence, Prescription 9.3 can be replaced by the following.

PRESCRIPTION 9.3*a* (Critical-region hypotheses tests for population proportion *p*)

Requirement: The sample size n is large, $n \geq 30$.

(1) and (2) Same as in Prescription 9.3.

(3) *Determine Critical Region*: Using a standard normal table (or computer software), find the critical region corresponding to H_a and α:

For $H_a: \mu < \mu_0$, the critical region is all z scores $z \leq z^*$, where z^* is the (negative) value satisfying $P(Z \leq z^*) = \alpha$ (Fig. 9-3(*a*)).

For $H_a: \mu > \mu_0$, the critical region is all z scores $z \geq z^*$, where z^* is the (positive) value satisfying $P(Z \geq z^*) = \alpha$ (Fig. 9-3(*b*)).

For $H_a: \mu \neq \mu_0$, the critical region is all z scores for which $z \geq z^*$ or $z \leq -z^*$, where z^* is the (positive) value satisfying $P(Z \leq -z^*) + P(Z \geq z^*) = \alpha$ [equivalently, $P(Z \geq z^*) = \alpha/2$] (Fig. 9-3(*c*)).

(4) *Draw Conclusion*: If the z score of the sample proportion $\hat{p}$ lies in the critical region, then z and therefore $\hat{p}$ are statistically significant at level α, and H_0 is rejected. If z does not lie in the critical region, then z and $\hat{p}$ are not statistically significant at level α, and H_0 is not rejected.

EXAMPLE 9.19 The critical region at the 0.01 level for the test in Example 9.18 consists of all z scores less than or equal to z^*, where $P(Z \leq z^*) = 0.01$. From the normal table, we find that $z^* = -2.33$. Therefore, the critical region is all $z \leq -2.33$. For any z score of the sample proportion in this region, the null hypothesis $H_0: p = 0.9$ would be rejected at the 0.01 significance level. Since the z score in Example 9.18 is -2, which is not ≤ -2.33, H_0 is not rejected.

EXAMPLE 9.20 What sample size would be needed for the test value $\hat{p} = 0.84$ of Example 9.18 to be statistically significant at the 0.01 significance level?

From Example 9.19, the z score of the sample proportion 0.84 must be less than or equal to -2.33. The z score of $\hat{p} = 0.84$ is $z = \dfrac{0.84 - 0.9}{\sqrt{0.9(1-0.9)/n}} = \dfrac{-0.06}{\sqrt{0.09}/\sqrt{n}} = -0.2\sqrt{n}$. Setting $-0.2\sqrt{n} = -2.33$, we get $\sqrt{n} = 11.65$. Squaring 11.65 and rounding upward, we find that $n = 136$.

9.4 HYPOTHESIS TESTS FOR VARIANCES

Suppose X is approximately a normally distributed random variable with mean μ and unknown variance σ^2. We consider hypotheses tests for σ^2 when μ is unknown; the case where μ is known is covered in the exercises. The null hypothesis will be

$$H_0: \sigma^2 = \sigma_0^2$$

and the alternative hypothesis will be one of the following:

$$H_a: \sigma^2 < \sigma_0^2, \quad H_a: \sigma^2 > \sigma_0^2, \quad \text{or} \quad H_a: \sigma^2 \neq \sigma_0^2$$

Hypotheses Tests for σ^2 When μ is Unknown

As with confidence intervals, hypotheses tests for σ^2 depend on the chi-square random variable with $n-1$ degrees of freedom,

$$\chi^2 = \frac{(n-1)S^2}{\sigma^2}$$

where $S^2 = \dfrac{1}{n-1}\sum (X_i - \bar{X})^2$ is the sample variance, $X_1, X_2, \ldots, X_n$ being a random-variable sample corresponding to X, and $\bar{X}$ is the sample mean. Proceeding as in the case of confidence intervals, we arrive at the following prescription.

PRESCRIPTION 9.4 (*P*-value hypotheses tests for σ^2 when μ is unknown)

Requirement: X is approximately normally distributed.

Let α be the level of significance for the test. Suppose the values $x_1, x_2, \ldots, x_n$ of X are obtained in a random sample of size n. Compute the corresponding values of the sample mean $\bar{x} = \dfrac{x_1 + x_2 + \cdots + x_n}{n}$, and sample variance $s^2 = \dfrac{1}{n-1}\sum (x_i - \bar{x})^2$. Then complete the following steps.

(1) *State Hypotheses*: State null hypothesis $H_0: \sigma^2 = \sigma_0^2$ and alternative hypothesis H_a.

(2) *Compute Test Statistic*: The test statistic is $\chi^2 = \dfrac{(n-1)S^2}{\sigma_0^2}$ which, assuming H_0 is true, is (approximately) a chi-square random variable with $n - 1$ degrees of freedom. Compute the test value $\hat{\chi}^2 = \dfrac{(n-1)s^2}{\sigma_0^2}$.

(3) *Determine P-value*: Using a table or computer software for the chi-square random variable with $n - 1$ degrees of freedom, find the P-value of the test corresponding to H_a:

For $H_a: \sigma^2 < \sigma_0^2$, the P-value is $P(\chi^2 \leq \hat{\chi}^2)$

For $H_a: \sigma^2 > \sigma_0^2$, the P-value is $P(\chi^2 \geq \hat{\chi}^2)$

For $H_a: \sigma^2 \neq \sigma_0^2$, the P-value is $\begin{cases} 2P(\chi^2 \leq \hat{\chi}^2) & \text{if } s^2 < \sigma_0^2 \\ 2P(\chi^2 \geq \hat{\chi}^2) & \text{if } s^2 > \sigma_0^2 \end{cases}$

(4) *Draw Conclusion*: If P-value $\leq \alpha$, then both $\hat{\chi}^2$ and s^2 are said to be statistically significant at level α, and H_0 is rejected. If P-value $> \alpha$, then $\hat{\chi}^2$ and s^2 are not statistically significant at level α, and H_0 is not rejected.

Alternative Version of Prescription 9.4

Chi-square tables in textbooks are often inadequate for computing P-values due to the need to include data for many different degrees of freedom. An alternative version of Prescription 9.4 replaces P-values by the critical region for the alternative hypothesis at the specified level of significance, α (see Section 9.1). If the test value $\hat{\chi}^2$ is in the critical region, then the test result is significant at level α, and the null hypothesis would be rejected. If $\hat{\chi}^2$ is not in the critical region, then the test result is not significant at level α, and the null hypothesis would not be rejected. Hence Prescription 9.4 can be replaced by the following.

PRESCRIPTION 9.4*a* (Critical-region hypotheses tests for σ^2; μ unknown)

Requirement: X is approximately normally distributed.

(1) and (2) Same as in Prescription 9.4.

(3) *Determine Critical Region*: Using a chi-square table with $n - 1$ degrees of freedom (or computer software), find the critical region corresponding to H_a and α:

For $H_a: \sigma^2 < \sigma_0^2$, the critical region is all values $\hat{\chi}^2 \leq \chi^*$, where χ^* is the value satisfying $P(\chi^2 \leq \chi^*) = \alpha$ (Fig. 9-6(*a*)).

For $H_a: \sigma^2 > \sigma_0^2$, the critical region is all values $\hat{\chi}^2 \geq \chi^*$, where χ^* is the value satisfying $P(\chi^2 \geq \chi^*) = \alpha$ (Fig. 9-6(*b*)).

For $H_a: \sigma^2 \neq \sigma_0^2$, the critical region is all values $\hat{\chi}^2 \leq \chi_1^*$ or $\chi^2 \geq \chi_2^*$, where χ_1^* is the value satisfying $P(\chi^2 \leq \chi_1^*) = \alpha/2$, and χ_2^* is the value satisfying $P(\chi^2 \geq \chi_2^*) = \alpha/2$ (Fig. 9-6(*c*)).

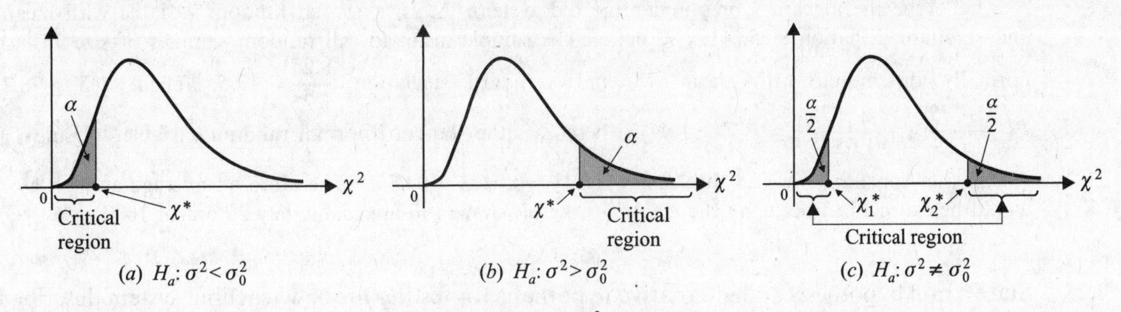

Fig. 9-6 Critical χ^2 regions.

(4) *Draw Conclusion*: If the sample value, $\hat{\chi}^2$, of the test statistic lies in the critical region, then $\hat{\chi}^2$ and s^2 are statistically significant at level α, and H_0 is rejected. If $\hat{\chi}^2$ does not lie in the critical region, then $\hat{\chi}^2$ and s^2 are not statistically significant at level α, and H_0 is not rejected.

EXAMPLE 9.21 Over the years the grades in a mathematics professor's calculus classes have been normally distributed with mean 75 and standard deviation 8. Recently the grades seem to have fallen and show more variation. A sample of 41 recent grades has mean $\bar{x} = 73$ and standard deviation $s = 9.6$. Assuming the grades are still normally distributed, test the null hypothesis $H_0: \sigma^2 = 64$ against the alternative hypothesis $H_a: \sigma^2 > 64$ at the 0.05 significance level.

P-value solution: The grades are a random variable X which we are assuming is normally distributed. However, since the grades seem to have fallen, we will not assume that the mean of X is 75. Therefore, Prescriptions 9.4 and 9.4*a* apply. In either case, the test statistic is $\chi^2 = \frac{(n-1)S^2}{\sigma_0^2} = \frac{40S^2}{64} = 0.625S^2$ which, if H_0 is true, is a chi-square random variable with 40 degrees of freedom. The test value is $\hat{\chi}^2 = 0.625 \times (9.6)^2 = 57.6$. To apply Prescription 9.4, we must compute the *P*-value of the test, which is $P(\chi^2 \geq 57.6)$. Table A-3 in the Appendix shows that $0.025 < P(\chi^2 \geq 57.6) < 0.05$; using computer software, we find that $P(\chi^2 \geq 57.6) \approx 0.035$. Since $0.035 < 0.05$, the test is statistically significant at the 0.05 level; and we reject the null hypothesis that the variance is still 64.

Critical region solution: We now apply Prescription 9.4*a*. The critical region for $\chi^2(40)$ for the alternative hypothesis $H_a: \sigma^2 > 64$ at the 0.05 significance level is all values $\hat{\chi}^2 \geq \chi^*$, where χ^* satisfies $P(\chi^2 \geq \chi^*) = 0.05$. From Table A-3 in the Appendix, with 40 degrees of freedom, we find that $\chi^* = 55.8$. The test value is $\hat{\chi}^2 = 57.6$ (see *P*-value solution), and since $57.6 > 55.8$, the test value is in the critical region, which means that the test is significant at the 0.05 level, so the null hypothesis $H_0: \sigma^2 = 64$ is rejected at this level.

Warning

As with confidence intervals for the variance, hypotheses tests for the variance, based on the chi-square test statistic, are not robust, meaning that decisions made may not be very reliable when X is not close to being normally distributed. Therefore the practical use of hypotheses testing for the variance is limited.

Solved Problems

TESTING HYPOTHESES ABOUT PARAMETERS

9.1. The 9th grade algebra scores in a school district have been normally distributed with a mean of 75 and a standard deviation of 8.25. A new teaching system is introduced to a random sample of 25 students, and in the first year under the new system the average score is 78.2. What is the probability that an average this high would occur for a random sample of 25 students in a given year under the old system?

Let X be the algebra scores under the old system. X is a normal random variable with mean 75 and standard deviation 8.25. Let $\bar{X}$ denote the sample mean for all random samples of size 25. $\bar{X}$ is normally distributed with mean 75 and standard deviation $\frac{8.25}{\sqrt{25}} = 1.65$. Then $P(\bar{X} \geq 78.2) = P\left(\frac{\bar{X} - 75}{1.65} \geq \frac{78.2 - 75}{1.65}\right) = P(Z \geq 1.94)$, where Z is the standard normal random variable. Using Table A-1 in the Appendix, we find that $P(Z \geq 1.94) = 0.0262 \approx 0.026$. Therefore an average as high as 78.2 would be expected to occur by chance under the old system in approximately 26 out of 1000 cases.

9.2. State a null hypothesis and alternative hypothesis for testing the new teaching system described in Problem 9.1.

The null hypothesis states that the average score under the old system has not changed with the new system, that is, $H_0: \mu = 75$. The alternative hypothesis, for one who feels that the new system is better, states that the new mean score has increased, that is, $H_a: \mu > 75$. It is assumed that $\sigma = 8.25$ under both systems.

9.3. What is the test statistic and the P-value of the test in Problems 9.1 and 9.2?

The test statistic is the standardized sample mean, namely $Z = \dfrac{\bar{X} - 75}{1.65}$ which, if H_0 is true, is the standard normal random variable. The P-value of the test corresponding to the alternative hypothesis $H_a: \bar{X} > 75$ is $P(\bar{X} \geq 78.2) = P(Z \geq 1.94) = 0.0262$.

9.4. At which significance levels would the null hypothesis be rejected in Problems 9.1 and 9.2? Specifically, would the null hypothesis be rejected at significance level 0.05, at significance level 0.01?

Since the P-value of the test is 0.0262 (Problem 9.3), the null hypothesis, $H_0: \mu = 75$, would be rejected at any significance level α for which 0.0262 is less than or equal to α, and would not be rejected if 0.0262 is greater than α. Since 0.0262 is less than 0.05, the null hypothesis would be rejected at the 0.05 level of significance; since 0.0262 is greater than 0.01, H_0 would not be rejected at the 0.01 level of significance.

9.5. What is the critical region for the test in Problems 9.1 and 9.2 at the (*a*) 0.05 significance level, (*b*) 0.01 significance level?

(*a*) Since the test statistic, assuming $H_0: \mu = 75$ is true, is the standard normal random variable, and the alternative hypothesis is $H_a: \mu > 75$, it follows from Example 9.7 that the critical region at the 0.05 significance level consists of all z scores greater than or equal to 1.65.

(*b*) From Example 9.8, the critical region at the 0.01 significance level consists of all z scores greater than or equal to 2.33.

9.6. For what values of the sample mean $\bar{X}$ in Problems 9.1 and 9.2 will the test statistic lie in the critical region at the (*a*) 0.05 significance level, (*b*) 0.01 significance level?

(*a*) By Problem 9.5, all z scores of the sample mean that are greater than or equal to 1.65 lie in the critical region at the 0.05 significance level. The inequality $\dfrac{\bar{x} - 75}{1.65} \geq 1.65$ is equivalent to $\bar{x} \geq 1.65 \times 1.65 + 75 \approx 77.72$. Therefore the test statistic will lie in the critical region, and H_0 will be rejected when $\bar{x} \geq 77.72$.

(*b*) By Problem 9.5, all z scores of the sample mean that are greater than or equal to 2.33 lie in the critical region at the 0.01 significance level. The inequality $\dfrac{\bar{x} - 75}{1.65} \geq 2.33$ is equivalent to $\bar{x} \geq 2.33 \times 1.65 + 75 \approx 78.84$. Therefore, the test statistic will lie in the critical region, and H_0 will be rejected, when $\bar{x} \geq 78.84$.

9.7. Suppose the null hypothesis $H_0: \mu = 75$ in Problems 9.1 and 9.2 is true. What values of the sample mean will result in a Type I error at the (*a*) 0.05 significance level, (*b*) 0.01 significance level?

A Type I error occurs if $H_0: \mu = 75$ is rejected when it is true. H_0 will be rejected whenever the z score of the sample mean lies in the critical region. (*a*) By Problem 9.6, part (*a*), a Type I error at the 0.05 level will occur if $\bar{x} \geq 77.72$ and H_0 is true. (*b*) By Problem 9.6, part (*b*), a Type I error at the 0.01 level will occur if $\bar{x} \geq 78.84$ and H_0 is true.

9.8. Suppose the new average score is actually 78 in Problems 9.1 and 9.2, and the standard deviation of the scores is still 8.25. What is the power $K(78)$ of the test at the 0.01 significance level? What is the probability of a Type II error when $\mu = 78$?

$K(78)$ is the probability that the test statistic $\frac{\bar{X} - 75}{1.65}$ lies in the critical region, given that $\frac{\bar{X} - 78}{1.65}$ is the standard normal random variable. At the 0.01 level, the critical region is all z scores greater than or equal to 2.33 (Problem 9.5), and $\frac{\bar{x} - 75}{1.65} \geq 2.33$ is equivalent to $\bar{x} \geq 78.84$ (Problem 9.6, part (*b*)). Therefore, $K(78) = P(\bar{X} \geq 78.84) = P\left(\frac{\bar{X} - 78}{1.65} \geq \frac{78.84 - 78}{1.65}\right) = P(Z \geq 0.51) = 0.305$. The probability that a Type II error will occur when $\mu = 78$ is $1 - K(78) = 1 - 0.305 = 0.695$.

9.9. With reference to Problem 9.8, what is the power $K(78)$ of the test at the 0.01 significance level if the sample size is 100? What is the probability of a Type I error when the sample size is 100? What is the probability of a Type II error when $\mu = 78$ and the sample size is 100?

If the sample size is 100, then the test statistic is $\frac{\bar{X} - 75}{8.25/\sqrt{100}} = \frac{\bar{X} - 75}{0.825}$, and $\frac{\bar{X} - 78}{0.825}$ is the standard normal random variable. Therefore

$$K(78) = P\left(\frac{\bar{X} - 75}{0.825} \geq 2.33\right) = P(\bar{X} \geq 2.33 \times 0.825 + 75) = P(\bar{X} \geq 76.92)$$

$$= P\left(\frac{\bar{X} - 78}{0.825} \geq \frac{76.92 - 78}{0.825}\right) = P(Z \geq -1.31) = 0.9049$$

The probability of a Type I error is equal to the significance level 0.01, regardless of the sample size. The probability of a Type II error, when $\mu = 78$ and the sample size is 100, is $1 - K(98) = 1 - 0.9049 = 0.0951$.

9.10. With reference to Problem 9.8, what sample size is needed to raise the power of the test to 0.98 when $\mu = 78$?

Let n be the sample size. Then the test statistic is $\frac{\bar{X} - 75}{8.25/\sqrt{n}}$, and $\frac{\bar{X} - 78}{8.25/\sqrt{n}}$ is the standard normal random variable. The sample size n must satisfy $P\left(\frac{\bar{X} - 75}{8.25/\sqrt{n}} \geq 2.33\right) = 0.98$, equivalently, $P(\bar{X} \geq 75 + 2.33 \times 8.25/\sqrt{n}) = 0.98$, or $P\left(\frac{\bar{X} - 78}{8.25/\sqrt{n}} \geq \frac{75 - 78}{8.25/\sqrt{n}} + 2.33\right) = P\left(Z \geq \frac{-3\sqrt{n}}{8.25} + 2.33\right) = 0.98$. From the standard normal table, we find that the Z value for which $P(Z \geq z^*) = 0.98$ is $z^* = -2.05$. Solving $\frac{-3\sqrt{n}}{8.25} + 2.33 = -2.05$ for $\sqrt{n}$, we get $\sqrt{n} = \frac{8.25}{3}(2.05 + 2.33) = 12.045$. Squaring 12.045 and rounding upward, we find that a sample size of $n = 146$ is needed for the power $K(78)$ to equal 0.98 at the 0.01 level of significance.

HYPOTHESES TESTS FOR MEANS

9.11. The useful lifetime of Everlast's 1.5 volt battery is a normally distributed random variable with mean 40 hours and standard deviation 4 hours. A new chemical composition is introduced to make the production of the batteries more efficient. The company wants to see if the useful lifetime of the battery has been affected by the new process. Specifically, they wish to test the null hypothesis H_0: $\mu = 40$ against the alternative hypothesis H_a: $\mu \neq 40$. It is assumed that the standard deviation is still 4 hours. A sample of 100 batteries has a useful lifetime of 39.1 hours. Determine the test statistic and P-value of the test.

We follow Prescription 9.1. The test statistic is the standardized sample mean, namely $Z = \dfrac{\bar{X} - 40}{4/\sqrt{100}} = \dfrac{\bar{X} - 40}{0.4}$ whose test value is $z = \dfrac{39.1 - 40}{0.4} = -2.25$. The P-value of the test is computed under the assumption that the null hypothesis is true, that is, that Z is the standard normal random variable. Using the standard normal table, we find that the P-value of the test is $2P(Z \geq |-2.25|) = 2P(Z \geq 2.25) = 2 \times 0.0122 = 0.0244$.

9.12. At which significance levels would the null hypothesis be rejected in Problem 9.11? Specifically, would the null hypothesis be rejected at significance level 0.05, at significance level 0.01?

As determined in Problem 9.11, the P-value of the test is 0.0244. The null hypothesis would be rejected at any significance level α for which $0.0244 \leq \alpha$, and would not be rejected if $0.0244 > \alpha$. Therefore, the null hypothesis would be rejected at the 0.05 significance level but not at the 0.01 significance level.

9.13. What is the critical region for the test in Problem 9.11 at the (*a*) 0.05 significance level, (*b*) 0.01 significance level?

We use Prescription 9.1*a*. The critical region is all z scores greater than or equal to z^* or less than or equal to $-z^*$, where $P(Z \geq z^*) = \alpha/2$, α being the level of significance.

(*a*) Here $\alpha/2 = 0.05/2 = 0.025$. From the standard normal table, we find that $P(Z \geq z^*) = 0.025$ for $z^* = 1.96$. Therefore the critical region is all z scores greater than or equal to 1.96 or less than or equal to -1.96.

(*b*) Here $\alpha/2 = 0.01/2 = 0.005$. From the standard normal table, we find that $P(Z \geq z^*) = 0.005$ for $z^* = 2.58$. Therefore the critical region is all z scores greater than or equal to 2.58 or less than or equal to -2.58.

9.14. Use the critical regions obtained in Problem 9.13 to determine whether to reject or not reject the null hypothesis in Problem 9.11.

We use Prescription 9.1*a*. In Problem 9.11, the test value of the sample mean $\bar{X}$ is 3.91, and the value of the test statistic $Z = \dfrac{\bar{X} - 40}{0.4}$ is $z = \dfrac{39.1 - 40}{0.4} = -2.25$. Since -2.25 is less than -1.96, the test value of Z is the critical region for the 0.05 significance level (Problem 9.13), and the null hypothesis $H_0: \mu = 40$ would be rejected at that level. Since -2.25 is not less than -2.58, the test value of Z is not in the critical region for the 0.01 significance level (Problem 9.13), and the null hypothesis would not be rejected at that level.

9.15. For what values $\bar{x}$ of the sample mean in Problem 9.11 will the z score of $\bar{x}$ lie in the critical region (*a*) at the 0.05 significance level, (*b*) at the 0.01 significance level?

The z score of $\bar{x}$ is $z = \dfrac{\bar{x} - \mu_0}{\sigma/\sqrt{n}} = \dfrac{\bar{x} - 40}{4/\sqrt{100}} = \dfrac{\bar{x} - 40}{0.4}$.

(*a*) At the 0.05 significance level, the critical region consists of all z scores greater than or equal to 1.96 or less than or equal to -1.96 (Problem 9.13). Setting $\dfrac{\bar{x} - 40}{0.4} \geq 1.96$, we get $\bar{x} \geq 1.96 \times 0.4 + 40 \approx 40.78$. Similarly, setting $\dfrac{\bar{x} - 40}{0.4} \leq -1.96$, we get $\bar{x} \leq 39.22$. Hence, the z score of $\bar{x}$ will lie in the critical region at the 0.05 significance level when $\bar{x} \geq 40.78$ or $\bar{x} \leq 39.22$.

(*b*) At the 0.01 significance level, the critical region consists of all z scores greater than or equal to 2.58 or less than or equal to -2.58 (Problem 9.13). Setting $\dfrac{\bar{x} - 40}{0.4} \geq 2.58$, we get $\bar{x} \geq 2.58 \times$

$0.4 + 40 \approx 41.03$. Similarly, setting $\frac{\bar{x} - 40}{0.4} \leq -2.58$, we get $\bar{x} \leq 38.97$. Hence, the z score of $\bar{x}$ will lie in the critical region at the 0.01 significance level when $\bar{x} \geq 41.03$ or $\bar{x} \leq 38.97$.

9.16. Suppose it is decided to reject the null hypothesis $H_0: \mu = 40$ in Problem 9.11 if a random sample of 100 batteries gives an average useful life of less than 39.5 or greater than 40.5. At what significance level is the test being conducted?

The significance level α is determined under the assumption that H_0 is true, in which case the test statistic $Z = \frac{\bar{X} - 40}{0.4}$ is the standard normal random variable; α is the probability that a test value of Z is less than $\frac{39.5 - 40}{0.4} = -1.25$ or greater than $\frac{40.5 - 40}{0.4} = 1.25$, which is $2P(Z \geq 1.25) = 2 \times 0.1056 = 0.2122$.

9.17. What is the power of the test in Problem 9.16 at the value 39 for μ? What is the probability of a Type II error when $\mu = 39$?

The power $K(39)$ of the test in Problem 9.16 is the probability that the sample mean $\bar{X}$ will assume a value greater than or equal to 40.5 or less than or equal to 39.5, assuming $\mu = 39$, that is, assuming that $Z = \frac{\bar{X} - 39}{0.4}$ is the standard normal random variable. Therefore,

$$K(39) = P(\bar{X} \geq 40.5) + P(\bar{X} \leq 39.5) = \left(P\frac{\bar{X} - 39}{0.4} \geq \frac{40.5 - 39}{0.4}\right) + P\left(\frac{\bar{X} - 39}{0.4} \leq \frac{39.5 - 39}{0.4}\right)$$

$$= P(Z \geq 3.75) + P(Z \leq 1.25) = 0.8945$$

The probability of a Type II error when $\mu = 39$ is $1 - K(39) = 1 - 0.8945 = 0.1055$.

9.18. The following cholesterol levels were found in a random sample of 10 women aged 20 to 24 engaged in a low-fat diet program:

176, 180, 175, 186, 182, 188, 180, 186, 168, 184

The null hypothesis is that the average cholesterol level of all women who maintain the diet is normally distributed with mean $\mu = 184$. The alternative hypothesis is $H_a: \mu < 184$. Use the data to determine the P-value of the test.

Since σ is not given, we follow Prescription 9.2. The value of the sample mean is

$$\bar{x} = \frac{176 + 180 + 175 + 186 + 182 + 188 + 180 + 186 + 168 + 184}{10} = \frac{1805}{10} = 180.5$$

and the value of the sample variance (see Problem 8.3) is

$$s^2 = \frac{1}{9}\sum x^2 - \frac{10}{9}\bar{x}^2 = \frac{1}{9}(326,141) = \frac{10}{9}(180.5)^2 = 37.611$$

so the sample standard deviation is $s = \sqrt{s^2} = 6.133$. The test statistic is $t = \frac{\bar{X} - 184}{S/\sqrt{10}}$ whose test value is $\hat{t} = \frac{180.5 - 184}{6.133/\sqrt{10}} \approx -1.80$. The P-value of the test is $P(t \leq -1.80) = 1 - P(t \leq 1.80)$, which is computed under the assumption that t is a t random variable with 9 degrees of freedom. The value closest to 1.80 in the t table for 9 degrees of freedom is 1.83, which gives a P-value of $1 - 0.95 = 0.05$. The P-value for 1.80 is slightly larger than 0.05; using computer software, we find that the P-value for 1.80 is 0.0527.

9.19. Find the critical region for the test in Problem 9.18 at the (*a*) 0.05 significance level, (*b*) 0.01 significance level.

We follow Prescription 9.2*a*. (*a*) The critical region corresponding to the alternative hypothesis $H_a: \mu < 184$ is all values $\hat{t}$ of the test statistic satisfying $\hat{t} \leq t^*$, where $P(t \leq t^*) = 0.05$. Using the t table with 9 degrees of freedom we find that $t^* = -1.83$. (*b*) Here the critical region is all values $\hat{t} \leq t^*$, where $P(t \leq t^*) = 0.01$; the t table gives $t^* = -2.82$.

9.20. For what values $\bar{x}$ of the sample mean, corresponding to other samples of 10 women in Problem 9.18, will the null hypothesis $H_0: \mu = 184$ be rejected (*a*) at the 0.05 significance level, (*b*) at the 0.01 significance level?

The null hypothesis will be rejected when the t score of $\bar{x}$, namely $\dfrac{\bar{x} - 184}{s/\sqrt{10}}$, lies in the critical region. The value s of the sample standard deviation will vary from sample to sample.

(*a*) From Problem 9.19, the critical region for the alternative hypothesis at the 0.05 significance level consists of all t scores less than or equal to -1.83. Setting $\dfrac{\bar{x} - 184}{s/\sqrt{10}} \leq -1.83$, we get $\bar{x} \leq 184 - \dfrac{1.83s}{\sqrt{10}} \approx 184 - 0.5787s$. The result depends on the value s of the sample standard deviation obtained in the sample.

(*b*) From Problem 9.19, the critical region for the alternative hypothesis at the 0.01 significance level consists of all t scores less than or equal to -2.82. Setting $\dfrac{\bar{x} - 184}{s/\sqrt{10}} \leq -2.82$, we get $\bar{x} \leq 184 - \dfrac{2.82s}{\sqrt{10}} \approx 1.84 - 0.8918s$. As in part (*a*), the result depends on the sample value s of the sample standard deviation.

9.21. Suppose the cholesterol levels of a random sample of 10 women in the low-fat diet program of Problem 9.18 have a sample standard deviation value of $s = 5.2$. What values $\bar{x}$ of the sample mean will result in a rejection of the null hypothesis $H_0: \mu = 184$ in favor of the alternative hypothesis $H_a: \mu < 184$ (*a*) at the 0.05 significance level, (*b*) at the 0.01 significance level?

(*a*) According to Problem 9.20, the null hypothesis will be rejected at the 0.05 significance level if $\bar{x} \leq 184 - 0.5787s$. Substituting $s = 5.2$, we find that if $\bar{x} \leq 184 - 0.5787 \times 5.2 \approx 180.99$, then the null hypothesis will be rejected.

(*b*) According to Problem 9.20, the null hypothesis will be rejected at the 0.01 significance level if $\bar{x} \leq 184 - 0.8918s$. Substituting $s = 5.2$, we find that if $\bar{x} \leq 184 - 0.8918 \times 5.2 \approx 179.36$, then the null hypothesis will be rejected.

9.22. The bumpers on a new Saber automobile are supposed to sustain only minor damage in collisions at speeds up to 5 miles per hour. In a test of 5 Sabers, the mean speed for minor damage was 4.8 miles per hour with a sample standard deviation of 0.3 miles per hour. Are the test results statistically significant at the 0.05 level?

We assume that the top speed for minor damage is normally distributed. The test statistic is the t score of the sample mean, and the results are statistically significant if the test statistic lies in the critical region of the t random variable at 4 degrees of freedom. The null hypothesis is $H_0: \mu = 5$, and the alternative hypothesis is $H_a: \mu < 5$. From the t table, the critical t value at the 0.05 level with 4 degrees of freedom is $t^* = -2.13$. Therefore, the critical region, which is in the direction of alternative hypothesis, consists of all t scores less than or equal to -2.13. The t score for the sample mean is $\hat{t} = \dfrac{4.8 - 5}{0.3/\sqrt{5}} = -1.49$. Since

-1.49 is not less than -2.13, the test results are not statistically significant at the 0.05 level. There is not enough evidence to reject the null hypothesis at this level.

9.23. At what levels are the test results in Problem 9.22 statistically significant?

The test results are statistically significant at any level α for which the P-value of the test is less than or equal to α. The P-value of the test is the probability that a sample mean of 4.8 or lower would occur if the actual mean were equal to 5. That is, the P-value is equal to $P(\bar{X} \le 4.8) = P\left(\frac{\bar{X}-5}{0.3/\sqrt{5}} \le \frac{4.8-5}{0.3/\sqrt{5}}\right) = P(t \le -1.49)$, where t is a t random variable with 4 degrees of freedom. The value closest to -1.49 obtainable from Table A-2 in the Appendix is -1.53 which corresponds to $\alpha = 0.1$. Computer software gives $P(t \le -1.49) \approx 0.105$.

9.24. Show that (*a*) $-z^* < \frac{\bar{x}-\mu_0}{\sigma/\sqrt{n}} < z^*$ is equivalent to (*b*) $\bar{x} - \frac{z^*\sigma}{\sqrt{n}} < \mu_0 < \bar{x} + \frac{z^*\sigma}{z\sqrt{n}}$.

The inequality $-z^* < \frac{\bar{x}-\mu_0}{\sigma/\sqrt{n}}$ is equivalent to $-\frac{z^*\sigma}{\sqrt{n}} < \bar{x} - \mu_0$ which in turn is equivalent to $\mu_0 < \bar{x} + \frac{z^*\sigma}{\sqrt{n}}$. That is, the left inequality of (*a*) is equivalent to the right inequality of (*b*). Also, the inequality $\frac{\bar{x}-\mu_0}{\sigma/\sqrt{n}} < z^*$ is equivalent to $\bar{x} - \mu_0 < \frac{z^*\sigma}{\sqrt{n}}$ which in turn is equivalent to $\bar{x} - \frac{z^*\sigma}{\sqrt{n}} < \mu_0$. Therefore, the right inequality of (*a*) is equivalent to the left inequality of (*b*).

HYPOTHESES TESTS FOR PROPORTIONS

9.25. In a random sample of 125 cola drinkers, 68 said they preferred Coke over Pepsi. Let p denote the percentage of all cola drinkers that prefer Coke over Pepsi. Do a P-value test of the null hypothesis $H_0\colon p = 0.5$ against the alternative hypothesis $H_a\colon p > 0.5$ at the 0.05 percent level.

Since the sample size $n = 125$ is greater than 30, we can use Prescription 9.3. Letting $\hat{P}$ denote the sample proportion random variable, the test statistic is $Z = \frac{\hat{P} - 0.5}{\sqrt{0.5(1-0.5)/125}} \approx \frac{\hat{P}-0.5}{0.0447}$. If H_0 is true, then Z is (approximately) the standard normal random variable. The test proportion of those that prefer Coke is $\hat{p} = \frac{68}{125} = 0.544$, and the test value of Z is $z = \frac{0.544 - 0.5}{0.0447} \approx 0.98$. Using a standard normal table, the P-value of the test is $P(Z \ge 0.98) = 0.1635$. Since 0.1635 is not less than 0.05, there is not enough evidence to reject H_0 at the 0.05 significance level.

9.26. What is the critical region for the hypothesis test in Problem 9.25? Is the test value of the test statistic in the critical region? What conclusions can you make regarding the null hypothesis $H_0\colon p = 0.5$?

The critical region is determined under the assumption that the null hypothesis $H_0\colon p = 0.5$ is true, in which case the test statistic $Z = \frac{\hat{P} - 0.5}{\sqrt{0.5(1-0.5)/125}} \approx \frac{\hat{P}-0.5}{0.0447}$ is (approximately) the standard normal random variable. By Prescription 9.3*a*, the critical region for Z corresponding to the alternative hypothesis $H_a\colon p > 0.5$ at the 0.05 significance level is all z scores $z \ge z^*$, where $P(Z \ge z^*) = 0.05$. From the standard normal table, we find that $z^* = 1.65$. The test value of Z is $z = 0.98$ (Problem 9.25), and since 0.98 is less than 1.65, the test value of Z is not in the critical region. We therefore would not reject the null hypothesis.

9.27. In Problem 9.25, how large of a sample is needed for a sample proportion $\hat{p} = 0.544$ to be statistically significant at the 0.05 level? Use the notion of P-value to answer the question.

We choose the sample size n so that the P-value of the test is at most 0.05. The test statistic is $Z = \dfrac{\hat{P} - 0.5}{\sqrt{0.5(1-0.5)/n}} = \dfrac{\hat{P} - 0.5}{0.5/\sqrt{n}} = \dfrac{(\hat{P} - 0.5)\sqrt{n}}{0.5}$, and the test value of Z is $z = \dfrac{(0.544 - 0.5)\sqrt{n}}{0.5} = 0.088\sqrt{n}$. Hence we want n to satisfy $P(Z \geq 0.088\sqrt{n}) \leq 0.05$. From the standard normal table, we find that $P(Z \geq z^*) = 0.05$ for $z^* = 1.65$. Setting $0.088\sqrt{n} = 1.65$, we get $\sqrt{n} = \dfrac{1.65}{0.088} = 18.75$. Squaring and rounding upward, we find that $n = 352$ is the smallest sample size for a test proportion $\hat{p} = 0.544$ to be statistically significant at the 0.05 level.

9.28. Use the critical region to answer the question posed in Problem 9.27.

For samples of size n, the critical region for the test in Problem 9.25 consists of all z scores $z = \dfrac{\hat{p} - 0.5}{0.5/\sqrt{n}}$ of the sample proportion $\hat{p}$ for which $z \geq 1.65$ (Problem 9.26). Substituting $\hat{p} = 0.544$ and setting $z \geq 1.65$, we get $\dfrac{0.544 - 0.5}{0.5/\sqrt{n}} \geq 1.65$, which is equivalent to $0.088\sqrt{n} \geq 1.65$, or $\sqrt{n} \geq \dfrac{1.65}{0.088} \approx 18.75$. Squaring and rounding upward, we find that $n \geq 352$, as in Problem 9.27.

9.29. In Problem 9.25, what test proportion $\hat{p}$ is needed for the test to be statistically significant at the 0.05 level, based on a sample of size $n = 125$? Use the notion of P-value to answer the question.

We choose $\hat{p}$ so that the P-value of the test is at most 0.05, that is, $P\left(Z \geq \dfrac{\hat{p} - 0.5}{0.0447}\right) \leq 0.05$. From Problem 9.27 (or from the standard normal table), we know that $P(Z \geq z^*) = 0.05$ for $z^* = 1.65$. Setting $\dfrac{\hat{p} - 0.5}{0.0447} \geq 1.65$, we get $\hat{p} \geq 1.65 \times 0.0447 + 0.5 \approx 0.5738$. Hence, a test proportion of at least 0.5738 is needed for statistical significance at the 0.05 level.

9.30. Use the critical region to answer the question posed in Problem 9.29.

The test value of the test statistic is $z = \dfrac{\hat{p} - 0.5}{0.0447}$, and we want to find the test proportion $\hat{p}$ for which z is in the critical region $z \geq 1.65$ (Problem 9.26). Setting $\dfrac{\hat{p} - 0.5}{0.0447} \geq 1.65$, we get $\hat{p} \geq 1.65 \times 0.0447 + 0.5 \approx 0.5738$, as in Problem 9.29.

9.31. What is the power of the test in Problem 9.25 at $p = 0.6$?

The power of the test at $p = 0.6$, denoted by $K(0.6)$, is the probability that the null hypothesis $H_0\colon p = 0.5$ will be rejected at the 0.05 significance level when the true proportion of cola drinkers that prefer Coke over Pepsi is $p = 0.6$. H_0 will be rejected in favor of $H_a\colon p > 0.5$ if the test statistic lies in the critical region for H_a at the 0.05 level, which is all z scores $z \geq 1.65$. The test statistic $\dfrac{\hat{P} - 0.5}{0.0447}$ is no longer the standard normal since we are now assuming that $p = 0.6$, not 0.5. Hence, we must compute $P\left(\dfrac{\hat{P} - 0.5}{0.0447} \geq 1.65\right)$, given that $\dfrac{\hat{P} - 0.6}{\sqrt{0.6(1-0.6)/125}}$, which is equal to $\dfrac{\hat{P} - 0.6}{0.0438}$, is the standard normal random variable. We get

$$P\left(\frac{\hat{P}-0.5}{0.0447} \geq 1.65\right) = P(\hat{P} \geq 1.65 \times 0.0447 + 0.5) = P(\hat{P} \geq 0.5738)$$

$$= P\left(\frac{\hat{P}-0.6}{0.0438} \geq \frac{0.5738}{0.0438}\right) = P(Z \geq -0.6),$$

where Z is the standard normal random variable. From the standard normal table, we find that $P(Z \geq -0.6) = 0.7258$. Hence, the power of the test at $p = 0.6$ is 0.7258.

9.32. What sample size is needed in Problem 9.25 for the power of the test at $p = 0.6$ to equal 0.9?

The test statistic is $\dfrac{\hat{P}-0.5}{\sqrt{0.5(1-0.5)/n}} = \dfrac{(\hat{P}-0.5)\sqrt{n}}{0.5}$, and we are assuming that the random variable $Z = \dfrac{\hat{P}-0.6}{\sqrt{0.6(1-0.6)}/\sqrt{n}} = \dfrac{(\hat{P}-0.6)\sqrt{n}}{0.4899}$ is the standard normal. With reference to Problem 9.31, we want to determine the sample size n so that $P\left(\dfrac{(\hat{P}-0.5)\sqrt{n}}{0.5} \geq 1.65\right) = 0.9$. We get

$$P\left(\frac{(\hat{P}-0.5)\sqrt{n}}{0.5} \geq 1.65\right) = P\left(\hat{P} \geq \frac{1.65 \times 0.5}{\sqrt{n}} + 0.5\right)$$

$$= P\left(\frac{(\hat{P}-0.6)\sqrt{n}}{0.4899} \geq \left[\frac{1.65 \times 0.5}{\sqrt{n}} + 0.5 - 0.6\right] \times \frac{\sqrt{n}}{0.4899}\right)$$

$$= P\left(Z \geq \frac{0.825 - 0.1\sqrt{n}}{0.4899}\right) = 0.9$$

where Z is the standard normal random variable. From the standard normal table, we find that $P(Z \geq -1.28) = 0.9$. Setting $\dfrac{0.825 - 0.1\sqrt{n}}{0.4899} = -1.28$, we get $\sqrt{n} = \dfrac{-1.28 \times 0.4899 - 0.825}{-0.1} \approx 14.52$. Squaring and rounding upward, we get $n = 211$.

HYPOTHESES TESTS FOR VARIANCES

9.33. Find the critical region at the 0.01 significance level for the test in Example 9.21, and determine whether the null hypothesis would be rejected at this level.

We apply Prescription 9.4*a*. The critical region for $\chi^2(40)$ for the alternative hypothesis $H_a\colon \chi^2 > 64$ at the 0.01 significance level is all values $\hat{\chi}^2 \geq \chi^*$, where χ^* satisfies $P(\chi^2 \geq \chi^*) = 0.01$. From Table A-3 in the Appendix, with 40 degrees of freedom, we find that $\chi^* = 63.7$. The test value obtained in Example 9.21 is $\hat{\chi}^2 = 57.6$, and since $57.6 < 63.7$, the test value is not in the critical region, which means that the null hypothesis $H_0\colon \sigma^2 = 64$ is not rejected at this level.

9.34. The amount of soda in 96 oz bottles of Andy's Root Beer is normally distributed with mean $\mu = 96$ and standard deviation $\sigma = 12$ oz. A new bottling procedure is designed to decrease the variability of the amount of soda in the bottles. A sample of 101 bottles has a standard deviation of 0.98 oz. Test the null hypothesis $H_0\colon \sigma^2 = 1.44$ against the alternative hypothesis $H_a\colon \sigma^2 < 1.44$ at the 0.025 level.

We apply Prescription 9.4*a*. The critical region for $\chi^2(100)$ for the alternative hypothesis $H_a\colon \sigma^2 < 1.44$ at the 0.025 significance level is all values $\hat{\chi}^2 \leq \chi^*$, where χ^* satisfies $P(\chi^2 \leq \chi^*) = 0.025$. From Table A-3 in the Appendix, with 100 degrees of freedom, we find that

$\chi^* = 74.2$. The test value of the test statistic is $\hat{\chi}^2 = \dfrac{(n-1)s^2}{\sigma_0^2} = \dfrac{100 \times (0.98)^2}{1.44} \approx 66.69$; and since $66.69 < 74.2$, the test value is in the critical region, which means that the null hypothesis $H_0: \sigma^2 = 1.44$ is rejected at the 0.025 significance level.

9.35. The number of hours spent sleeping by an undergraduate college student is a normal random variable with mean $\mu = 7.5$ and variance $\sigma^2 = 1.25$. In graduate school the student's sleep pattern changes. A sample of 15 days gives an average of $\bar{x} = 6.25$ hours and $s^2 = 1.5$. Assuming that the sleeping hours are normally distributed, test the null hypothesis $H_0: \sigma^2 = 1.25$ against the alternative hypothesis $H_a: \sigma^2 \neq 1.25$ at the 0.05 significance level.

We apply Prescription 9.4*a*. The critical region for $\chi^2(14)$ for the alternative hypothesis $H_a: \sigma^2 \neq 1.25$ at the 0.05 significance level is all values $\hat{\chi}^2 \leq \chi_1^*$ or $\hat{\chi}^2 \geq \chi_2^*$, where χ_1^* satisfies $P(\chi^2 \leq \chi_1^*) = 0.05/2 = 0.025$, and χ_2^* satisfies $P(\chi^2 \geq \chi_2^*) = 0.025$. From Table A-3 in the Appendix, with 14 degrees of freedom, we find that $\chi_1^* = 5.63$ and $\chi_2^* = 26.1$. The test value of the test statistic is $\hat{\chi}^2 = \dfrac{(n-1)s^2}{\sigma_0^2} = \dfrac{14 \times 1.5}{1.25} = 16.8$, which is not in the critical region. Therefore, the null hypothesis $H_0: \sigma^2 = 1.25$ is not rejected at the 0.05 significance level.

9.36. Suppose X is a normal random variable with mean μ and variance σ^2, and $X_1, X_2, \ldots, X_n$ is a random sample of size n corresponding to X. Show that $\chi^2 = \sum \dfrac{(X_i - \mu)^2}{\sigma^2}$ is a chi-square random variable with n degrees of freedom.

For $i = 1, 2, \ldots, n$, the random variable $Z_i = \dfrac{X - \mu}{\sigma}$ is standard normal, and the Z_is are independent. Therefore, by definition (see Section 7.4), $\chi^2 = \sum \dfrac{(X_i - \mu)^2}{\sigma^2}$ is a chi-square random variable with n degrees of freedom.

9.37. The weight of a 16 oz bag of C&P Potato Chips is a random variable X with mean $\mu = 16$ oz and standard deviation $\sigma = 0.5$ oz. A new quality control procedure is introduced to reduce the variability of X. The weights of a random sample of 25 bags are as follows:

15.8	15.4	15.9	16.5	16.3
15.9	16.0	15.9	16.6	15.5
16.4	15.2	16.6	16.2	15.8
16.6	15.7	15.4	15.9	16.1
15.5	16.4	15.4	15.5	16.4

Assuming that the mean of all bags produced under the new system is still 16 oz, test the null hypothesis $H_0: \sigma^2 = 0.25$ against the alternative hypothesis $H_a: \sigma^2 < 0.25$ at the 0.01 level. Use $\chi^2 = \sum \dfrac{(X_i - \mu)^2}{\sigma^2}$ as the test statistic.

By Problem 9.36, χ^2 is a chi-square random variable with 25 degrees of freedom. From the chi-square table, the critical region for $H_a: \sigma^2 < 0.25$ at the 0.01 level is all test values of χ^2 less than or equal to 11.5. Computing with the help of a calculator, we find that the test value is $\hat{\chi}^2 = \sum \dfrac{(x_i - 16)^2}{0.25} = 18.52$. The test value is not in the critical region, so the null hypothesis is not rejected at the 0.01 level of significance. The sample does not supply enough evidence at this level to conclude that the new quality control procedure has actually decreased the variability of X.

9.38. At what significance level would the null hypothesis in Problem 9.37 be rejected, based on the sample data?

From the chi-square table, with 25 degrees of freedom, we find that the critical region for the alternative hypothesis $H_a: \sigma^2 < 0.25$ at the 0.1 level is all test values less than or equal to 16.5, and the critical region for the alternative hypothesis at the 0.25 level is all test values less than or equal to 19.9. Since the test value obtained in Problem 9.37 is 18.52, we can conclude that the smallest significance level that the null hypothesis $H_0: \sigma^2 = 0.25$ would be rejected is between 0.1 and 0.25. Using computer software, we find that the P-value of the test is 0.18. Therefore, H_0 will be rejected at any level greater than or equal to 0.18.

Supplementary Problems

INTRODUCTION: TESTING HYPOTHESIS ABOUT PARAMETERS

9.39. The tread life of Goodwear's all-weather tire is normally distributed with mean $\mu = 39{,}000$ miles and standard deviation $\sigma = 3000$ miles. A test of 16 new model all-weather tires results in an average tread life of 40,500 miles. What is the probability that an average tread life of 40,500 miles or greater would occur with the previous model all-weather tires?

9.40. Identify the null and alternative hypotheses in Problem 9.39, and classify each as either simple or composite.

9.41. What is the P-value of the test in Problem 9.39?

9.42. Would the null hypothesis in Problem 9.39 be rejected at the 0.01 significance level? What about the 0.05 significance level?

9.43. Find the critical region for the test in Problem 9.39 at the 0.05 significance level, and find the power of the test at (*a*) $\mu = 39{,}000$ miles, (*b*) $\mu = 40{,}500$ miles, (*c*) $\mu = 42{,}000$ miles.

9.44. Repeat part (*b*) of Problem 9.43 under the assumption that the test result in Problem 9.39 was obtained for a sample of 36 tires.

9.45. Suppose the null hypothesis is $H_0: \mu = \mu_0$ and the test value $\bar{x}$ of the sample mean of a normal random variable is greater than μ_0. Under which of the following alternative hypotheses is H_0 more likely to be rejected?

(*a*) $H_a: \mu > \mu_0$, (*b*) $H_a: \mu \neq \mu_0$.

HYPOTHESES TESTS FOR MEANS

9.46. The sample mean of a random sample of 50 values of a random variable X is $\bar{x} = 72.4$. Assuming that X has standard deviation $\sigma = 9$, test the null hypothesis $H_0: \mu = 70$ against the alternative hypothesis $H_a: \mu > 70$ at the 0.05 significance level by computing the P-value of the test.

9.47. Perform the test in Problem 9.46 by using the critical region for the test.

9.48. Repeat Problems 9.46 and 9.47 for the significance level 0.01.

9.49. Determine the power of the test in Problem 9.46 at $\mu = 71.5$.

9.50. The numbers 125.5, 130.2, 112.8, 120.2, 111.3 form a random sample of five values of a normal random variable X. Test the null hypothesis $H_0: \mu = 110$ against the alternative hypothesis $H_a: \mu > 110$ at the 0.05 significance level.

9.51. Repeat Problem 9.50 at the 0.01 significance level.

9.52. Repeat Problems 9.50 and 9.51 for the alternative hypothesis $H_a: \mu \neq 110$.

9.53. Using Table A-2 in the Appendix, estimate the P-value of the test in Problem 9.50. If appropriate computer software is available, find the P-value.

9.54. Suppose a test of the null hypothesis $H_0: \mu = \mu_0$ against the alternative hypothesis $H_a: \mu > \mu_0$ is rejected at the 0.05 significance level but not at the 0.01 significance level. If possible, determine what the decision would be if H_0 were tested against $H_a: \mu \neq \mu_0$ at each of the levels 0.05 and 0.01. Assume that X is normally distributed and σ is known.

HYPOTHESIS TESTS FOR PROPORTIONS

9.55. In a random sample of 25 students at a private liberal arts college, 17 were receiving some sort of financial aid. Letting p denote the proportion of all students at the college receiving financial aid, test the hypothesis $H_0: p = 0.5$ against the alternative hypothesis $H_a: p > 0.5$ at the 0.05 significance level by computing the P-value of the test.

9.56. Perform the test in Problem 9.55 by using the critical region for the test.

9.57. Repeat Problems 9.55 and 9.56 at the 0.01 significance level.

9.58. How large of a sample is needed in Problem 9.55 for the test value $\hat{p} = 0.68$ to be statistically significant at the 0.01 level?

9.59. Determine the power of the test in Problem 9.55 at $p = 0.7$.

9.60. Determine the sample size needed for the power of the test in Problem 9.55 to be 0.95 at $p = 0.7$.

HYPOTHESIS TESTS FOR VARIANCES

9.61. The number of eggs produced annually by individual chicken hens on Old McDonald's Farm is normally distributed with mean 250 and standard deviation 15. The number of eggs produced by 6 randomly chosen hens given a new feed was: 260, 240, 270, 250, 265, 245. Test the null hypothesis $H_0: \sigma = 15$ against the alternative hypothesis $H_a: \sigma \leq 15$ at the 0.1 significance level. Assume that the mean is still 250 (see Problem 9.36).

9.62. Repeat Problem 9.61 under the assumption that the mean with the new feed may no longer be 250.

9.63. The grades in elementary algebra in a school district are normally distributed with mean 73 and standard deviation 9. A new program, designed to reduce the variation in grades, is introduced at a random selection of schools in the district. In a random selection of 51 students in the new program, the standard deviation was 7.4. Test the hypothesis $H_0: \sigma = 9$ against the alternative hypothesis $H_a: \sigma < 9$ at the 0.05 significance level.

9.64. Repeat Problem 9.63 at the 0.01 significance level.

9.65. Test the hypothesis $H_0: \sigma^2 = 10,000$ against the alternative hypothesis $H_a: \sigma^2 \neq 10,000$ at the 0.05 significance level for the sample SAT scores: 520, 540, 475, 510, 400, 550, 425, 600, 430, 515.

Answers to Supplementary Problems

9.39. $P(\bar{X} \geq 40{,}500) = P\left(\frac{\bar{X} - 39{,}000}{3000/\sqrt{16}} \geq \frac{40{,}500 - 39{,}000}{3000/\sqrt{16}}\right) = P(Z \geq 2) = 0.0228$

9.40. $H_0: \mu = 39{,}000$, $H_a: \mu > 39{,}000$; H_0 is simple, H_a is composite.

9.41. 0.0228 (see answer 9.39).

9.42. Since $0.01 < 0.0228 < 0.05$, H_0 would not be rejected at the 0.01 significance level, but would be rejected at the 0.05 level.

9.43. Critical region: $\frac{\bar{x} - 39{,}000}{3000/\sqrt{16}} \geq 1.65$, or $\bar{x} \geq 40{,}237.5$

(*a*) $K(39{,}000) = 0.05$

(*b*) $K(40{,}500) = P(\bar{X} \geq 40{,}237.5)$, given that $\frac{\bar{X} - 40{,}500}{3000/\sqrt{16}}$ is standard normal,

$$= P\left(\frac{\bar{X} - 40{,}500}{3000/\sqrt{16}} \geq \frac{40{,}237.5 - 40{,}500}{3000/\sqrt{16}}\right) = P(Z \geq -0.35) = 0.64$$

(*c*) $K(42{,}000) = P(\bar{X} \geq 40{,}237.5)$, given that $\frac{\bar{X} - 42{,}000}{3000/\sqrt{16}}$ is standard normal

$$= P\left(\frac{\bar{X} - 42{,}000}{3000/\sqrt{16}} \geq \frac{40{,}237.5 - 42{,}000}{3000/\sqrt{16}}\right) = P(Z \geq -2.35) = 0.99$$

9.44. Critical region: $\frac{\bar{x} - 39{,}000}{3000/\sqrt{36}} \geq 1.65$ or $\bar{x} \geq 39{,}825$.

$K(40{,}500) = P(\bar{X} \geq 39{,}825)$, given that $\frac{\bar{X} - 40{,}500}{3000/\sqrt{36}}$ is standard normal

$$= P\left(\frac{\bar{X} - 40{,}500}{3000/\sqrt{36}} \geq \frac{39{,}825 - 40{,}500}{3000/\sqrt{36}}\right) = P(Z \geq -1.35) = 0.91$$

9.45. When $\bar{x} > \mu_0$, the value z of the test statistic is positive, and the P-value for $H_a: \mu > \mu_0$ is half the P-value for $H_a: \mu \neq \mu_0$. Therefore, H_0 is more likely to be rejected when the alternative hypothesis is $H_a: \mu > \mu_0$.

9.46. P-value $= 0.03 < 0.05$; reject H_0.

9.47. Critical region: $\frac{\bar{x} - 70}{9/\sqrt{50}} \geq 1.65$, or $\bar{x} \geq 1.65 \times 9/\sqrt{50} + 70 \approx 72.1$; test value: $\bar{x} = 72.4$; reject H_0.

9.48. P-value $= 0.03 > 0.01$; do not reject H_0. Critical region: $\frac{\bar{x} - 70}{9/\sqrt{50}} \geq 2.33$, or $\bar{x} \geq 2.33 \times 9/\sqrt{50} + 70 \approx 72.97$; test value: $\bar{x} = 72.4$; do not reject H_0.

9.49. $K(71.5) = P(\bar{X} \geq 72.1)$, given that $\dfrac{\bar{X} - 71.5}{9/\sqrt{50}}$ is standard normal

$$= P\left(\frac{\bar{X} - 71.5}{9/\sqrt{50}} \geq \frac{72.1 - 71.5}{9/\sqrt{50}}\right) = P(Z \geq 0.47) = 0.32$$

9.50. Critical region: $\hat{t} \geq 2.13$; test value: $\hat{t} = 2.76$; reject H_0.

9.51. Critical region: $\hat{t} \geq 3.75$; test value: $\hat{t} = 2.76$; do not reject H_0.

9.52. Critical region at 0.05 significance level: $|\hat{t}| \geq 2.78$; test value: $\hat{t} = 2.76$; do not reject H_0. Critical region at 0.01 significance level: $|\hat{t}| \geq 4.60$; test value: $\hat{t} = 2.76$; do not reject H_0.

9.53. $0.01 < P\text{-value} < 0.05$ (P-value $= 0.025$).

9.54. The test value z satisfies $1.65 \leq z < 2.33$. Therefore, $0 < z < 2.58$, which means H_0 will not be rejected in favor of $H_a: \mu \neq \mu_0$ at the 0.01 significance level. To be rejected in favor of $H_a: \mu \neq \mu_0$ at the 0.05 significance level, z would have to satisfy $z > 1.96$, which cannot be determined from $1.65 \leq z < 2.33$.

9.55. P-value $= 0.036 < 0.05$; reject H_0.

9.56. Critical region: $\dfrac{\hat{p} - 0.5}{\sqrt{0.5 \times 0.5/25}} = \dfrac{\hat{p} - 0.5}{0.1} \geq 1.65$, or $\hat{p} \geq 0.665$; test value: $\hat{p} = 0.68 > 0.665$; reject H_0.

9.57. P-value $= 0.036 > 0.01$; do not reject H_0. Critical region: $\dfrac{\hat{p} - 0.5}{0.1} \geq 2.33$, or $\hat{p} \geq 0.733$; test value: $\hat{p} = 0.68 < 0.733$; do not reject H_0.

9.58. Critical region: $\dfrac{\hat{p} - 0.5}{\sqrt{0.5 \times 0.5/n}} = \dfrac{\hat{p} - 0.5}{0.5/\sqrt{n}} \geq 2.33$; $\dfrac{0.68 - 0.5}{0.5/\sqrt{n}} = 0.36\sqrt{n} \geq 2.33$ for $n \geq 42$.

9.59. Critical region: $\hat{p} \geq 0.665$ (see answer 9.56).

$$K(0.7) = P(\hat{P} \geq 0.665), \text{ given that } \frac{\hat{P} - 0.7}{\sqrt{0.7 \times 0.3/25}} \approx \frac{\hat{P} - 0.7}{0.0917} \text{ is standard normal.}$$

$$= P\left(\frac{\hat{P} - 0.7}{0.0917} \geq \frac{0.665 - 0.7}{0.0917}\right) = P(Z \geq -0.38) = 0.648$$

9.60. Want $P(\hat{P} \geq 0.665) = 0.95$, given that $\dfrac{\hat{P} - 0.7}{\sqrt{0.7 \times 0.3/n}} = \dfrac{\hat{P} - 0.7}{\sqrt{0.21/n}}$ is standard normal.

$$P\left(\frac{\hat{P} - 0.7}{\sqrt{0.21/n}} \geq \frac{0.665 - 0.7}{\sqrt{0.21/n}}\right) = P\left(Z \geq -\frac{0.035\sqrt{n}}{\sqrt{0.21}}\right) = 0.95 \quad \text{for} \quad -\frac{0.035\sqrt{n}}{\sqrt{0.21}} = -1.65$$

$n = 466.7$; round up to 467.

9.61. Critical region: $\hat{\chi}^2 \leq 2.20$; test value:

$$\hat{\chi}^2 = \frac{(260 - 250)^2 + (240 - 250)^2 + (270 - 250)^2 + (250 - 250)^2 + (265 - 250)^2 + (245 - 250)^2}{15^2} = 3.78$$

do not reject H_0.

9.62. Critical region: $\hat{\chi}^2 \leq 1.61$; test values: $\bar{x} = 255$, $s^2 = 140$, $\hat{\chi}^2 = \frac{5 \times 140}{15^2} = 3.11$; do not reject H_0.

9.63. Critical region: $\hat{\chi}^2 \leq 34.8$; test value: $\hat{\chi}^2 = \frac{50 \times (7.4)^2}{9^2} = 33.80$; reject H_0.

9.64. Critical region: $\hat{\chi}^2 \leq 29.7$; test value: $\hat{\chi}^2 = 33.80$; do not reject H_0.

9.65. Critical region: $\hat{\chi}^2 \leq 2.70$ or $\hat{\chi}^2 \geq 19.0$; test values: $\bar{x} = 496.5$, $s^2 = 3983.61$, $\hat{\chi}^2 = \frac{9 \times 3983.61}{10{,}000} = 3.59$; do not reject H_0.

Chapter 10

Inference for Two Populations

10.1 CONFIDENCE INTERVALS FOR THE DIFFERENCE OF MEANS

Let X and Y be independent random variables with means μ_X and μ_Y, and standard deviations σ_X and σ_Y, respectively. The object is to obtain a confidence interval for $\mu_X - \mu_Y$, based on independently chosen random samples of size m and n from the X and Y distributions, respectively. We consider the cases where σ_X and σ_Y are known, and where σ_X and σ_Y are unknown.

Note that $\mu_X - \mu_Y$ is the mean of the random variable $\bar{X} - \bar{Y}$, so we can proceed as in Section 8.3, where confidence intervals for the mean of a single random variable are obtained. Also, since X and Y are independent random variables, so are $\bar{X}$ and $\bar{Y}$, and therefore the variance of $\bar{X} - \bar{Y}$ is the sum of the variances of $\bar{X}$ and $\bar{Y}$:

$$\sigma^2_{\bar{X}-\bar{Y}} = \frac{\sigma^2_X}{m} + \frac{\sigma^2_Y}{n}$$

where σ^2_X is the variance of X and σ^2_Y is the variance of Y.

The confidence intervals prescribed for $\mu_X - \mu_Y$ require that the sample means $\bar{X}$ and $\bar{Y}$ be approximately normally distributed. (Actually, it is only required that $\bar{X} - \bar{Y}$ be approximately normally distributed.) For small X-samples ($m < 30$), $\bar{X}$ will be normally distributed if X itself is, and for small Y-samples ($n < 30$), $\bar{Y}$ will be normally distributed if Y is. For large samples (m and $n \geq 30$), the Central Limit Theorem allows us to assume that $\bar{X}$ and $\bar{Y}$ are approximately normally distributed regardless of the distributions of X and Y.

We arrive at the following prescription.

PRESCRIPTION 10.1 (Confidence interval for $\mu_X - \mu_Y$; σ_X and σ_Y known)

Requirements: X and Y are independent random variables with known standard deviations σ_X and σ_Y, respectively; $\bar{X}$ and $\bar{Y}$ are approximately normally distributed.

Let γ be the specified confidence level, and suppose the values $x_1, x_2, \ldots, x_m$ of X and $y_1, y_2, \ldots, y_n$ of Y are obtained in independently chosen random samples of size m and n, respectively. First compute the sample values $\bar{x} = \dfrac{x_1 + x_2 + \cdots + X_m}{m}$ and $\bar{y} = \dfrac{y_1 + y_2 + \cdots + y_n}{n}$. Then complete the following steps.

(1) *Find Critical Z Value*: Find the value z^* of the standard normal random variable Z for which $P(-z^* \leq Z \leq z^*) = \gamma$.

(2) *Compute Margin of Error*: Compute $E = z^* \sqrt{\dfrac{\sigma^2_X}{m} + \dfrac{\sigma^2_Y}{n}}$.

(3) *Determine Confidence Interval*: An approximate 100γ percent confidence interval for $\mu_X - \mu_Y$ is $[\bar{x} - \bar{y} - E, \bar{x} - \bar{y} + E]$.

EXAMPLE 10.1 The weights of two types of mice in a psychology research lab are normally distributed. Type X mice have mean weight 28 grams and standard deviation $\sigma_X = 3$ grams; Type Y mice have mean weight 28 grams and standard deviation $\sigma_Y = 2$ grams. A new diet is designed to increase the average weight of each type. The gram weights of 8 Type X mice under the new diet are:

29, 28, 30, 31, 26, 32, 25, 34

Ten Type Y mice under the new diet have gram weights:

27, 31, 30, 28, 29, 25, 31, 30, 29, 26

Find a 90 percent confidence interval for $\mu_X - \mu_Y$ corresponding to mice on the new diet. Assume $\sigma_X = 3$ grams and $\sigma_Y = 2$ grams.

We use Prescription 10.1. The sample value of $\bar{X}$ is

$$\bar{x} = \frac{29 + 28 + 30 + 31 + 26 + 32 + 25 + 34}{8} = 29.375 \approx 29.38$$

and the sample value of $\bar{Y}$ is

$$\bar{y} = \frac{26 + 31 + 30 + 28 + 29 + 25 + 31 + 30 + 29 + 26}{10} = 28.6$$

From Table A-1, the critical value z^* of the standard normal random variable Z for which $P(-z^* \leq Z \leq z^*) = 0.9$ is $z^* = 1.65$. Therefore,

$$E = z^* \sqrt{\frac{\sigma_X^2}{m} + \frac{\sigma_Y^2}{n}} = 1.65 \sqrt{\frac{9}{8} + \frac{4}{10}} \approx 2.04$$

and the corresponding approximate 90 percent confidence interval for $\mu_X - \mu_Y$ is

$$[(29.38 - 28.60) - 2.04, (29.38 - 28.60) + 2.04] = [-1.26, 2.82]$$

Since 0 is in the confidence interval, we do not have strong evidence that either mean weight under the new diet is greater than the other.

Confidence Intervals for $\mu_X - \mu_Y$ When σ_X and σ_Y are Unknown but Equal

Suppose σ_X and σ_Y are not known but are presumed to be equal. Let $X_1, X_2, \ldots, X_m$ and $Y_1, Y_2, \ldots, Y_n$ be independent random-variable samples corresponding to X and Y, respectively (see Section 8.1). The statistic

$$S_P = \sqrt{\frac{(m-1)S_X^2 + (n-1)S_Y^2}{m+n-2}}$$

where $S_X^2 = \frac{1}{m-1}\sum(X_i - \bar{X})^2$ and $S_Y^2 = \frac{1}{n-1}\sum(Y_i - \bar{Y})^2$ are the sample variances for X and Y, respectively, is called the *pooled estimator* of the common standard deviation of X and Y. If $\bar{X}$ and $\bar{Y}$ are independent normal random variables, it can be shown that the random variable

$$t = \frac{\bar{X} - \bar{Y} - (\mu_X - \mu_Y)}{S_P\sqrt{\frac{1}{m} + \frac{1}{n}}}$$

has a t distribution with $m + n - 2$ degrees of freedom (see Problem 10.6).

PRESCRIPTION 10.2 (Confidence interval for $\mu_X - \mu_Y; \sigma_X, \sigma_Y$ unknown but equal)

Requirements: X and Y are independent random variables; $\bar{X}$ and $\bar{Y}$ are approximately normally distributed; σ_X and σ_Y are unknown but equal.

Let γ be the specified level of confidence, and suppose $x_1, x_2, \ldots, x_m$ and $y_1, y_2, \ldots, y_n$ are independently chosen random samples corresponding to X and Y, respectively. First compute the sample values $\bar{x} = \frac{x_1 + x_2 + \cdots + x_m}{m}$, $\bar{y} = \frac{y_1 + y_2 + \cdots + y_n}{n}$, $s_X^2 = \frac{1}{m-1}\sum(x_i - \bar{x})^2$, $s_Y^2 = \frac{1}{n-1}\sum(y_i - \bar{y})^2$, and $s_P = \sqrt{\frac{(m-1)s_X^2 + (n-1)s_Y^2}{m+n-2}}$. Now complete the following steps.

(1) *Find Critical t Value*: Using a t table (or computer software), find the value t^* of the t random variable with $m+n-2$ degrees of freedom that satisfies $P(-t^* \le t \le t^*) = \gamma$.

(2) *Compute Margin of Error*: Compute $E = t^* s_P \sqrt{\frac{1}{m} + \frac{1}{n}}$.

(3) *Determine Confidence Interval*: An approximate 100γ percent confidence interval for $\mu_X - \mu_Y$ is $[\bar{x} - \bar{y} - E, \bar{x} - \bar{y} + E]$.

EXAMPLE 10.2 Suppose in Example 10.1 that the random weights X and Y of each type of mice had standard deviation 2.5 oz before the new diet. Use the data given there to construct a 90 percent confidence interval for $\mu_X - \mu_Y$ under the assumption that the new standard deviations are unknown but equal.

We use Prescription 10.2. From Example 10.1, $\bar{x} = 29.375$ and $\bar{y} = 28.6$. The values of the sample variances are $s_X^2 = \frac{1}{7}\sum(x_i - 29.375)^2 \approx 9.125$ and $s_Y^2 = \frac{1}{9}\sum(y_i - 28.6)^2 \approx 4.267$, computed with the help of a calculator. The pooled estimator S_p of the common variance has value $s_P = \sqrt{\frac{(m-1)s_X^2 + (n-1)s_Y^2}{m+n-2}} = \sqrt{\frac{7 \times 9.125 + 9 \times 4.267}{16}} \approx 2.53$. From Table A-2, with 16 degrees of freedom, we find that the critical value t^* satisfying $P(-t^* \le t \le t^*) = 0.9$ is $t^* = 1.75$. Therefore, the margin of error is $E = t^* s_P \sqrt{\frac{1}{m} + \frac{1}{n}} = 1.75 \times 2.53\sqrt{\frac{1}{8} + \frac{1}{10}} \approx 2.10$, and the approximate 90 percent confidence interval for $\mu_X - \mu_Y$ is $[(29.38 - 28.6) - 2.10, (29.38 - 28.6) + 2.10] = [-1.32, 2.88]$.

Confidence Intervals for $\mu_X - \mu_Y$ When σ_X and σ_Y are Unknown and Not Necessarily Equal

Small samples: If it is unreasonable to assume that the unknown standard deviations σ_X and σ_Y are equal, then in place of t, you can use the random variable

$$\tau = \frac{\bar{X} - \bar{Y} - (\mu_X - \mu_Y)}{\sqrt{\frac{S_X^2}{m} + \frac{S_Y^2}{n}}}$$

Although τ does not have a t distribution, when m and n are moderate, say $m \ge 5$ and $n \ge 5$, τ can be approximated as a t random variable provided $\bar{X}$ and $\bar{Y}$ are normally distributed. The number of degrees of freedom, in terms of the sample values s_X^2 and s_Y^2, can be taken as the largest integer, denoted by $[k]$, which is less than or equal to

$$k = \frac{\left(\frac{s_X^2}{m} + \frac{s_Y^2}{n}\right)^2}{\frac{1}{m-1}\left(\frac{s_X^2}{m}\right)^2 + \frac{1}{n-1}\left(\frac{s_Y^2}{n}\right)^2}$$

The corresponding confidence interval for $\mu_X - \mu_Y$ is $[\bar{x} - \bar{y} - E, \bar{x} - \bar{y} + E]$, where $E = t^*\sqrt{\frac{s_X^2}{m} + \frac{s_Y^2}{n}}$, t^* being the value of the t random variable with $[k]$ degrees of freedom satisfying $P(-t^* \le t \le t^*) = \gamma$ (see Problem 10.4).

Large samples: If m and n are each 30 or larger, then the above random variable τ is approximately the standard normal random variable Z, and $E = z^*\sqrt{\frac{s_X^2}{m} + \frac{s_Y^2}{n}}$, z^* being the value of Z satisfying $P(-z^* \le Z \le z^*) = \gamma$ (see Problem 10.3).

10.2 HYPOTHESES TESTS FOR THE DIFFERENCE OF MEANS

In the previous section we gave prescriptions to find confidence intervals for $\mu_X - \mu_Y$, where X and Y are independent random variables. Here we consider hypotheses tests for $\mu_X - \mu_Y$. Prescriptions are given for tests in which the null hypothesis is

$$H_0\colon \mu_X - \mu_Y = 0, \text{ equivalently, } H_0\colon \mu_X = \mu_Y$$

and the alternative hypothesis is one of the following.

$$H_a\colon \mu_X - \mu_Y < 0, \text{ equivalently, } H_a\colon \mu_X < \mu_Y$$
$$H_a\colon \mu_X - \mu_Y > 0, \text{ equivalently, } H_a\colon \mu_X > \mu_Y$$
$$H_a\colon \mu_X - \mu_Y \neq 0, \text{ equivalently, } H_a\colon \mu_X \neq \mu_Y$$

As with confidence intervals, we consider the case where σ_X and σ_Y are known and the case where they are unknown. For each of these, both P-value and critical-region tests are prescribed.

PRESCRIPTION 10.3 (P-value hypotheses tests for $\mu_X - \mu_Y$; σ_X and σ_Y known)

Requirements: X and Y are independent random variables with known standard deviations σ_X and σ_Y, respectively; $\bar{X}$ and $\bar{Y}$ are approximately normally distributed.

Let α be the specified level of significance; and suppose that a value $\bar{x}$ of $\bar{X}$ is obtained in a random sample of size m, and a value $\bar{y}$ of $\bar{Y}$ is obtained in an independently chosen random sample of size n. Complete the following steps.

(1) *State Hypotheses:* State null hypothesis $H_0\colon \mu_X = \mu_Y$ and alternative hypothesis H_a.

(2) *Compute Test Statistic:* The test statistic is $Z = \dfrac{\bar{X} - \bar{Y}}{\sqrt{\dfrac{\sigma_X^2}{m} + \dfrac{\sigma_Y^2}{n}}}$ which, assuming H_0 is true, is (approximately) the standard normal random variable. Compute the test value of Z as $z = \dfrac{\bar{x} - \bar{y}}{\sqrt{\dfrac{\sigma_X^2}{m} + \dfrac{\sigma_Y^2}{n}}}$.

(3) *Determine P-value:* Using a standard normal table (or computer software), find the P-value of the test corresponding to H_a:

For $H_a\colon \mu_X < \mu_Y$, the P-value is $P(Z \leq z)$
For $H_a\colon \mu_X > \mu_Y$, the P-value is $P(Z \geq z)$
For $H_a\colon \mu_X \neq \mu_Y$, the P-value is $P(Z \leq -|z|) + P(Z \geq |z|)$ [equivalently, $2P(Z \geq |z|)$]

(4) *Draw Conclusion:* If P-value $\leq \alpha$, then z and $\bar{x} - \bar{y}$ are statistically significant at level α, and H_0 is rejected. If P-value $> \alpha$, then z and $\bar{x} - \bar{y}$ are not statistically significant at level α, and H_0 is not rejected.

EXAMPLE 10.3 With reference to Example 10.1, test the null hypothesis $H_0\colon \mu_X = \mu_Y$ against the alternative hypothesis $H_a\colon \mu_X \neq \mu_Y$ at the 0.10 level of significance by computing the P-value of the test.

The test statistic is

$$Z = \frac{\bar{X} - \bar{Y}}{\sqrt{\dfrac{\sigma_X^2}{m} + \dfrac{\sigma_Y^2}{n}}} = \frac{\bar{X} - \bar{Y}}{\sqrt{\dfrac{3^2}{8} + \dfrac{2^2}{10}}} = \frac{\bar{X} - \bar{Y}}{1.235}$$

which, if $H_0\colon \mu_X = \mu_Y$ is true, is the standard normal random variable. The test value of Z is

$$z = \frac{\bar{x} - \bar{y}}{1.235} = \frac{29.38 - 28.6}{1.235} \approx 0.63$$

Using Table A-1, we find that the P-value for H_a: $\mu_X \neq \mu_Y$ is $2P(Z \geq z) = 2P(Z \geq 0.63) = 2(0.2643) = 0.5286$. Since $0.5286 > 0.10$, the test is not significant at the 0.10 level, and we do not reject the null hypothesis.

Alternative Version of Prescription 10.3

As an alternative to the P-value test, a critical-region test can be prescribed as follows.

PRESCRIPTION 10.3*a* (Critical-region hypotheses tests for $\mu_X - \mu_Y; \sigma_X$ and σ_Y known)

Requirements: X and Y are independent random variables with known standard deviations σ_X and σ_Y, respectively; $\bar{X}$ and $\bar{Y}$ are approximately normally distributed.

(1) and (2) Same as in Prescription 10.3.

(3) *Determine Critical Region:* Using a standard normal table (or computer software), find the critical region corresponding to H_a and α:

For H_a: $\mu_X < \mu_Y$, the critical region is all z scores $z \leq z^*$, where z^* is the (negative) value satisfying $P(Z \leq z^*) = \alpha$ (Fig. 9-3(*a*)).

For H_a: $\mu_X > \mu_Y$, the critical region is all z scores $z \geq z^*$, where z^* is the (positive) value satisfying $P(Z \geq z^*) = \alpha$ (Fig. 9-3(*b*)).

For H_a: $\mu_X \neq \mu_Y$, the critical region is all z scores for which $z \geq z^*$ or $z \leq -z^*$, where z^* is the (positive) value satisfying $P(Z \leq z^*) + P(Z \geq z^*) = \alpha$ [equivalently, $P(Z \geq z^*) = \alpha/2$] (Fig. 9-3(*c*)).

(4) *Draw Conclusion:* If the sample value z of the test statistic lies in the critical region, then z and $\bar{x} - \bar{y}$ are statistically significant at level α, and H_0 is rejected. If z does not lie in the critical region, then z and $\bar{x} - \bar{y}$ are not statistically significant at level α, and H_0 is not rejected.

EXAMPLE 10.4 With reference to Example 10.1, test the null hypothesis H_0: $\mu_X = \mu_Y$ against the alternative hypothesis H_a: $\mu_X \neq \mu_Y$ at the 0.1 level of significance by determining the critical region.

The critical region consists of all z scores for which $z \geq z^*$ or $z \leq -z^*$, where z^* is the (positive) value satisfying $P(Z \geq z^*) = 0.1/2 = 0.05$. From Table A-1, we find that $P(Z \geq z^*) = 0.05$ for $z^* = 1.65$. Therefore, the critical region consists of all z scores ≤ -1.65 or ≥ 1.65. The test z score is 0.63 (Example 10.3). Since 0.63 is not in the critical region, the null hypothesis is not rejected at the 0.1 significance level.

Hypotheses Tests for $\mu_X - \mu_Y$ When σ_X and σ_Y are Unknown but Equal

As in the case of confidence intervals for $\mu_X - \mu_Y$, when the standard deviations σ_X and σ_Y are unknown but equal, we replace the standard deviation of $\bar{X} - \bar{Y}$ by the pooled estimator

$$S_P = \sqrt{\frac{(m-1)S_X^2 + (n-1)S_Y^2}{m+n-2}}$$

where $S_X^2 = \frac{1}{m-1}\sum(X_i - \bar{X})^2$ and $S_Y^2 = \frac{1}{n-1}\sum(Y_i - \bar{Y})^2$ are the sample variances for X and Y, respectively. We also use the t distribution with $m + n - 2$ degrees of freedom in place of the standard normal distribution. The corresponding prescriptions for performing hypotheses tests are as follows.

PRESCRIPTION 10.4 (*P*-value hypotheses tests for $\mu_X - \mu_Y; \sigma_X, \sigma_Y$ unknown but equal)

Requirements: X and Y independent random variables; $\bar{X}$ and $\bar{Y}$ are approximately normally distributed; σ_X, σ_Y unknown but equal.

Let α be the specified level of significance, and suppose $x_1, x_2, \ldots, x_m$ and $y_1, y_2, \ldots, y_n$ are independently chosen random samples corresponding to X and Y, respectively. First compute the sample values $\bar{x} = \dfrac{x_1 + x_2 + \cdots + x_m}{m}$, $\bar{y} = \dfrac{y_1 + y_2 + \cdots + y_n}{n}$, $s_X^2 = \dfrac{1}{m-1}\sum(x_i - \bar{x})^2$, $s_Y^2 = \dfrac{1}{n-1}\sum(y_i - \bar{y})^2$,

and $s_P = \sqrt{\dfrac{(m-1)s_X^2 + (n-1)s_Y^2}{m+n-2}}$. Then complete the following steps.

(1) *State Hypotheses:* State null hypothesis H_0: $\mu_X = \mu_Y$ and alternative hypothesis H_a.
(2) *Compute Test Statistic:* The test statistic is

$$t = \frac{\bar{X} - \bar{Y}}{S_P\sqrt{\dfrac{1}{m} + \dfrac{1}{n}}}$$

which, assuming H_0 is true, is (approximately) the t random variable with $m + n - 2$ degrees of freedom. Compute the test value of t as

$$\hat{t} = \frac{\bar{x} - \bar{y}}{s_P\sqrt{\dfrac{1}{m} + \dfrac{1}{n}}}$$

(3) *Determine P-value:* Using a t table, if adequate, or computer software for a t random variable with $m + n - 2$ degrees of freedom, find the P-value of the test corresponding to H_a:

For H_a: $\mu_X < \mu_Y$, the P-value is $P(t < \hat{t})$
For H_a: $\mu_X > \mu_Y$, the P-value is $P(t > \hat{t})$
For H_a: $\mu_X \neq \mu_Y$, the P-value is $P(t \leq |\hat{t}|) + P(t > |\hat{t}|)$ [equivalently, $2P(t > |\hat{t}|)$]

(4) *Draw Conclusion:* If P-value $\leq \alpha$, then $\hat{t}$ and $\bar{x} - \bar{y}$ are statistically significant at level α, and H_0 is rejected. If P-value $\geq \alpha$, then $\hat{t}$ and $\bar{x} - \bar{y}$ are not statistically significant at level α, and H_0 is not rejected.

Alternative Version of Prescription 10.4

The alternative version of Prescription 10.4 uses the critical region for H_a and α in place of the P-value of the test. The critical region can be determined for various levels of significance and degrees of freedom from Table A-2 in the Appendix.

PRESCRIPTION 10.4*a* (Critical-region hypotheses tests for $\mu_X - \mu_Y; \sigma_X, \sigma_Y$ unknown but equal)

Requirements: X and Y are independent random variables; $\bar{X}$ and $\bar{Y}$ are approximately normally distributed; σ_X, σ_Y unknown but equal.

(1) and (2) Same as in Prescription 10.4.
(3) *Determine Critical Region:* Using a t table with $m + n - 2$ degrees of freedom (or computer software), find the critical region corresponding to H_a and α:

For H_a: $\mu_X < \mu_Y$, the critical region is all values $\hat{t} \leq t^*$, where t^* is the (negative) value satisfying $P(t \leq t^*) = \alpha$ (Fig. 9-4(*a*)).

For H_a: $\mu_X > \mu_Y$, the critical region is all values $\hat{t} \geq t^*$, where t^* is the (positive) value satisfying $P(t \geq t^*) = \alpha$ (Fig. 9-4(*b*)).

For H_a: $\mu_X \neq \mu_Y$, the critical region is all values $\hat{t}$ for which $\hat{t} \geq t^*$ or $\hat{t} \leq -t^*$, where t^* is the (positive) value satisfying $P(t \leq -t^*) + P(t \geq t^*) = \alpha$ [equivalently, $P(t \geq t^*) = \alpha/2$] (Fig. 9-4(*c*)).

(4) *Draw Conclusion:* If the sample value $\hat{t}$ of the test statistic lies in the critical region, then $\hat{t}$ and $\bar{x} - \bar{y}$ are statistically significant at level α, and H_0 is rejected. If $\hat{t}$ does not lie in the critical region, then $\hat{t}$ and $\bar{x} - \bar{y}$ are not statistically significant at level α, and H_0 is not rejected.

EXAMPLE 10.5 Suppose in Example 10.1 that the random weights X and Y of each type of mice had standard deviation 2.5 oz before the new diet. Use the data given there to test the null hypothesis H_0: $\mu_X = \mu_Y$ against the alternative hypothesis H_a: $\mu_X \neq \mu_Y$ at the 0.1 level of significance. Assume that the new standard deviations are unknown but equal.

We use Prescription 10.4(*a*). From Example 10.2, $\bar{x} = 29.375, \bar{y} = 28.6$, and $s_P \approx 2.53$. The test value of the test statistic

$$t = \frac{\bar{X} - \bar{Y}}{S_P\sqrt{\frac{1}{m} + \frac{1}{n}}}$$

is

$$\hat{t} = \frac{29.375 - 28.6}{2.53\sqrt{\frac{1}{8} + \frac{1}{10}}} \approx 0.65$$

From Table A-2, with 16 degrees of freedom, we find that the critical value t^* satisfying $P(t \geq t^*) = 0.1/2 = 0.05$ is $t^* = 1.75$. Therefore, the critical region is all t scores $\hat{t} \leq -1.75$ or $\hat{t} \geq 1.75$. Hence, 0.65 is not in the critical region, so the null hypothesis is not rejected at the 0.01 level of significance.

Hypotheses Tests for $\mu_X - \mu_Y$ When σ_X and σ_Y are Unknown and Not Necessarily Equal

Small samples: As with confidence intervals, if it is unreasonable to assume that the unknown standard deviations σ_X and σ_Y are equal, then in place of t you can use as a test statistic the random variable

$$\tau = \frac{\bar{X} - \bar{Y}}{\sqrt{\frac{S_X^2}{m} + \frac{S_Y^2}{n}}}$$

which, if $H_0 : \mu_X = \mu_Y$ is true, has an approximate t distribution when m and n are moderate, say $m \geq 5$ and $n \geq 5$, and $\bar{X}$ and $\bar{Y}$ are normally distributed. To obtain the number of degrees of freedom, first find sample values s_X^2 and s_Y^2, and then compute

$$k = \frac{\left(\frac{s_X^2}{m} + \frac{s_Y^2}{n}\right)^2}{\frac{1}{m-1}\left(\frac{s_X^2}{m}\right)^2 + \frac{1}{n-1}\left(\frac{s_Y^2}{n}\right)^2}$$

The largest integer less than or equal to k is the number of degrees of freedom of τ. Now proceed as in Prescription 10.4 or 10.4*a* with the statistic τ in place of t (see Problem 10.12).

Large samples: If m and n are each 30 or larger, then the above random variable τ is approximately the standard normal random variable Z, assuming H_0: $\mu_Z = \mu_Y$ is true, and you can proceed as in Prescription 10.3 or 10.3*a* with τ in place of Z (see Problem 10.13).

10.3 CONFIDENCE INTERVALS FOR DIFFERENCES OF PROPORTIONS

Suppose it is a presidential election year, and that p_1 and p_2 are two states' respective (unknown) proportions of eligible voters that favor the Democratic candidate. Suppose also that some of the candidate's advisors believe that p_1 is greater than p_2, that is, that $p_1 - p_2$ is positive, whereas others believe that p_1 is less than p_2. If the two states have a nearly equal number of electoral votes, and the candidate can spend time campaigning in only one of them, it would be helpful to obtain a confidence interval for $p_1 - p_2$, based on random samples of size n_1 and n_2, respectively.

Sampling in a situation such as this can be modeled by two independent binomial experiments $b(n_1, p_1)$ and $b(n_2, p_2)$, where p_1 is the probability of success in each of the n_1 trials constituting the first experiment, and p_2 is the probability of success in each of the n_2 trials constituting the second experiment. The collection of all possible proportions of successes in the n_1 trials making up the first experiment defines a random variable $\hat{P}_1$ with mean p_1 and variance $p_1(1-p_1)/n_1$. Similarly, all proportions of successes in the n_2 trials making up the second experiment define a random variable $\hat{P}_2$ with mean p_2 and variance $p_2(1-p_2)/n_2$. Therefore, the mean of $\hat{P}_1 - \hat{P}_2$ is

$$\mu_{\hat{P}_1 - \hat{P}_2} = p_1 - p_2$$

and since we are assuming independence, the variance of $\hat{P}_1 - \hat{P}_2$ is

$$\sigma^2_{\hat{P}_1 - \hat{P}_2} = \frac{p_1(1-p_1)}{n_1} + \frac{p_2(1-p_2)}{n_2}$$

Also, if n_1 and n_2 are large, say $n_1 \geq 30$ and $n_2 \geq 30$, then by the Central Limit Theorem, $\hat{P}_1 - \hat{P}_2$ is approximately normally distributed. As in the case of a single proportion, the variance of $\hat{P}_1 - \hat{P}_2$ is estimated by

$$\frac{\hat{p}_1(1-\hat{p}_1)}{n_1} + \frac{\hat{p}_2(1-\hat{p}_2)}{n_2}$$

where $\hat{p}_1$ and $\hat{p}_2$ are sample values of $\hat{P}_1$ and $\hat{P}_2$ obtained in independently drawn large random samples from the respective binomial populations.

We arrive at the following prescription.

PRESCRIPTION 10.5 (Confidence interval for $p_1 - p_2$)

Requirements: The sample sizes n_1 and n_2 are large, say $n_1 \geq 30$ and $n_2 \geq 30$.

Let γ be the specified confidence level; and suppose that a value $\hat{p}_1$ of $\hat{P}_1$ is obtained in a random sample of size $n_1 \geq 30$, and a value $\hat{p}_2$ of $\hat{P}_2$ is obtained in an independently chosen random sample of size $n_2 \geq 30$. Complete the following steps.

(1) *Find Critical Z Value:* Using a standard normal table (or computer software), find the value z^* of the standard normal random variable Z for which $P(-z^* \leq Z \leq z^*) = \gamma$ [equivalently, $P(0 \leq Z \leq z^*) = \gamma/2$].

(2) *Compute Margin of Error:* Compute $E = z^* \sqrt{\dfrac{\hat{p}_1(1-\hat{p}_1)}{n_1} + \dfrac{\hat{p}_2(1-\hat{p}_2)}{n_2}}$.

(3) *Determine Confidence Interval:* An approximate 100γ percent confidence interval for $p_1 - p_2$ is $[\hat{p}_1 - \hat{p}_2 - E, \hat{p}_1 - \hat{p}_2 + E]$.

EXAMPLE 10.6 A presidential candidate needs either Ohio's 21 electoral votes or Pennsylvania's 23 electoral votes to virtually insure a victory. In a poll of 500 eligible voters in Ohio, 260 favored the candidate over his primary opponent; and in a pole of 600 eligible voters in Pennsylvania, 306 voters favored the candidate. The proportions, $\hat{p}_1 = 0.52$ in Ohio and $\hat{p}_2 = 0.51$ in Pennsylvania, seem to give a slight edge to Ohio, and the candidate has time and money to concentrate on only one of the states. Find a 99 percent confidence interval for $p_1 - p_2$, and interpret the result.

Since the sample sizes are each larger than 30, we can apply Prescription 10.5. From the standard normal table, we find that $P(-z^* \leq Z \leq z^*) = 0.99$ for $z^* = 2.58$. The margin of error is $E = z^* \sqrt{\dfrac{\hat{p}_1(1-\hat{p}_1)}{n_1} + \dfrac{\hat{p}_2(1-\hat{p}_2)}{n_2}} = 2.58\sqrt{\dfrac{0.52 \times 0.48}{500} + \dfrac{0.51 \times 0.49}{600}} \approx 0.078$. The corresponding approximate 99 percent confidence interval for $p_1 - p_2$ is $[(0.52 - 0.51) - 0.078, (0.52 - 0.51) + 0.078] = [-0.068, 0.088]$. Since 0 is in the confidence interval, the slight edge in Ohio's favor may simply be due to chance. The results do not indicate that Ohio is the better choice for a campaign effort (see Problem 10.17).

10.4 HYPOTHESES TESTS FOR DIFFERENCES OF PROPORTIONS

As in the case of confidence intervals for differences of proportions, we consider two independent binomial experiments $b_1(n_1, p_1)$ and $b_2(n_2, p_2)$, where p_1 is the probability of success in each of the n_1 trials constituting the first experiment, and p_2 is the probability of success in each of the n_2 trials constituting the second experiment. The collection of all possible proportions of successes in n_1 trials defines a random variable $\hat{P}_1$, and the collection of all possible proportions of successes in n_2 trials defines a random variable $\hat{P}_2$. The random variable $\hat{P}_1 - \hat{P}_2$ has mean $p_1 - p_2$, and since $\hat{P}_1$ and $\hat{P}_2$ are independent, the variance of $\hat{P}_1 - \hat{P}_2$ is $\dfrac{p_1(1-p_1)}{n_1} + \dfrac{p_2(1-p_2)}{n_2}$. For large samples, say $n_1 \geq 30$ and $n_2 \geq 30$, $\hat{P}_1 - \hat{P}_2$ is approximately normal.

So far, all that we have said applies to confidence intervals. However, there is one major difference between confidence intervals and hypotheses tests for differences of proportions. In the case of a confidence interval for $p_1 - p_2$, the variance of $\hat{P}_1 - \hat{P}_2$ is approximated by $\dfrac{\hat{p}_1(1-\hat{p}_1)}{n_1} + \dfrac{\hat{p}_2(1-\hat{p}_2)}{n_2}$, where $\hat{p}_1$ and $\hat{p}_2$ are sample values of $\hat{P}_1$ and $\hat{P}_2$, respectively. In the case of hypotheses tests, where the null hypothesis is

$$H_0\colon p_1 - p_2 = 0, \qquad \text{equivalently,} \qquad H_0\colon p_1 = p_2$$

we combine the sample data to obtain the *pooled sample proportion* $\hat{p} = \dfrac{x_1 + x_2}{n_1 + n_2}$, x_1 being the number of successes in n_1 trials making up the first experiment, and x_2 the number of successes in n_2 trials making up the second experiment. In terms of the sample values $\hat{p}_1$ and $\hat{p}_2$, $\hat{p}$ can be computed as an average, weighted according to the relative values of n_1 and n_2 (see Problem 10.20):

$$\hat{p} = \frac{n_1\hat{p}_1 + n_2\hat{p}_2}{n_1 + n_2}$$

Note that if $n_1 = n_2$, then the above weighted average simplifies to $\dfrac{\hat{p}_1 + \hat{p}_2}{2}$. Replacing p_1 and p_2 by $\hat{p}$ in the above formula for the variance of $\hat{P}_1 - \hat{P}_2$, we get the estimate

$$\hat{p}(1-\hat{p})\left(\frac{1}{n_1} + \frac{1}{n_2}\right)$$

for the variance of $\hat{P}_1 - \hat{P}_2$. The random-variable statistic whose value is $\hat{p}$ on each pair of samples of sizes n_1 and n_2 respectively, is denoted by $\hat{P}$. We then arrive at the following prescriptions for hypotheses tests concerning $p_1 - p_2$.

PRESCRIPTION 10.6 (*P*-value hypotheses tests for $p_1 - p_2$)

Requirement: The sample sizes are large, say $n_1 \geq 30$ and $n_2 \geq 30$.

Let α be the specified level of significance. Suppose that a value $\hat{p}_1$ of $\hat{P}_1$ is obtained in a random sample of size $n_1 \geq 30$, and a value $\hat{p}_2$ of $\hat{P}_2$ is obtained in an independently chosen random sample of size $n_2 \geq 30$. First compute the pooled sample proportion $\hat{p} = \dfrac{n_1\hat{p}_1 + n_2\hat{p}_2}{n_1 + n_2}$, and then complete the following steps.

(1) *State Hypotheses:* State null hypothesis $H_0: p_1 = p_2$ and alternative hypothesis H_a.

(2) *Compute Test Statistic:* The test statistic is

$$Z = \frac{\hat{P}_1 - \hat{P}_2}{\sqrt{\hat{p}(1-\hat{p})\left(\dfrac{1}{n_1} + \dfrac{1}{n_2}\right)}}$$

which, if H_0 is true, is approximately normally distributed with mean 0 and standard deviation 1. Approximate the sample value of Z as

$$z = \frac{\hat{p}_1 - \hat{p}_2}{\sqrt{\hat{p}(1-\hat{p})\left(\frac{1}{n_1} + \frac{1}{n_2}\right)}}$$

(3) *Determine P-value:* Using a standard normal table (or computer software), find the *P*-value of the test corresponding to H_a:

For H_a: $p_1 < p_2$, the *P*-value is $P(Z \le z)$.

For H_a: $p_1 > p_2$, the *P*-value is $P(Z \ge z)$.

For H_a: $p_1 \ne p_2$, the *P*-value is $P(Z \le -|z|) + P(Z \ge |z|)$ [equivalently, $2P(Z \ge |z|)$].

(4) *Draw Conclusion:* If *P*-value $\le \alpha$, then the values z and $\hat{p}_1 - \hat{p}_2$ are statistically significant at level α, and H_0 is rejected. If *P*-value $> \alpha$, then z and $\hat{p}_1 - \hat{p}_2$ are not statistically significant at level α, and H_0 is not rejected.

Comment

Some statisticians estimate the variance of $\hat{P}_1 - \hat{P}_2$ by $\frac{\hat{p}_1(1-\hat{p}_1)}{n_1} + \frac{\hat{p}_2(1-\hat{p}_2)}{n_2}$, as in the case of confidence intervals, but since the null hypothesis states that the proportions p_1 and p_2 are equal, it seems more natural to pool the sample data rather than treat the sample values separately.

EXAMPLE 10.7 With reference to Example 10.6, test the null hypothesis $H_0 : p_1 = p_2$ against the alternative hypothesis H_a: $p_1 > p_2$ at the 0.01 significance level by computing the *P*-value of the sample results.

The sample proportions in Example 10.6 are $\hat{p}_1 = \frac{260}{500} = 0.52$ and $\hat{p}_2 = \frac{306}{600} = 0.51$, so the pooled proportion is $\hat{p} = \frac{500 \times 0.52 + 600 \times 0.51}{500 + 600} = \frac{260 + 306}{1100} \approx 0.5145$. The corresponding estimate of the test statistic is

$$z = \frac{\hat{p}_1 - \hat{p}_2}{\sqrt{\hat{p}(1-\hat{p})\left(\frac{1}{n_1} + \frac{1}{n_2}\right)}} = \frac{0.52 - 0.51}{\sqrt{0.5145 \times 0.4855 \times \left(\frac{1}{500} + \frac{1}{600}\right)}} \approx 0.33$$

The *P*-value of the test is $P(Z \ge 0.33) = 0.3707$. Since 0.3707 is substantially higher than 0.01, we don't even come close to rejecting the null hypothesis at the 0.01 significance level.

Alternative Version of Prescription 10.6

An alternative version of Prescription 10.6, which uses the critical region determined by the alternative hypothesis and the level of significance, is the following.

PRESCRIPTION 10.6*a* **(Critical-region hypotheses tests for $p_1 - p_2$)**

Requirement: The sample sizes are large, say $n_1 \ge 30$ and $n_2 \ge 30$.

(1) and (2) Same as in Prescription 10.6.

(3) *Determine Critical Region:* Using a standard normal table (or computer software), find the critical region corresponding to H_a and α:

For H_a: $p_1 < p_2$, the critical region is all values $z \le z^*$, where z^* is the (negative) value satisfying $P(Z \le z^*) = \alpha$ (Fig. 9-3(*a*)).

For H_a: $p_1 > p_2$, the critical region is all values $z \ge z^*$, where z^* is the (positive) value satisfying $P(Z \ge z^*) = \alpha$ (Fig. 9-3(*b*)).

For H_a: $p_1 \neq p_2$, the critical region is all values z for which $z \geq z^*$ or $z \leq -z^*$, where z^* is the (positive) value satisfying $P(Z \leq z^*) + P(Z \geq z^*) = \alpha$ [equivalently $P(Z \geq z^*) = \alpha/2$] (Fig. 9-3(*c*)).

(4) *Draw Conclusion:* If the sample value z of the test statistic lies in the critical region, then z and $\hat{p}_1 - \hat{p}_2$ are statistically significant at level α, and H_0 is rejected. If z does not lie in the critical region, then z and $\hat{p}_1 - \hat{p}_2$ are not statistically significant at level α, and H_0 is not rejected.

EXAMPLE 10.8 Test the null hypothesis $H_0 : p_1 = p_2$ in Example 10.6 against the alternative hypothesis H_a: $p_1 > p_2$ at the 0.01 significance level by determining the critical region for the test.

The test statistic is the standard normal random variable Z, and from Table A-1, we see that $P(Z \geq z^*) = 0.01$ for $z^* = 2.33$. From Example 10.7 above, the value of the test statistic is $z = 0.33$, which is far from the critical region. Therefore, with some emphasis, we do not reject the null hypothesis at the 0.01 significance level.

10.5 CONFIDENCE INTERVALS FOR RATIOS OF VARIANCES

So far we have found confidence intervals for differences of means and differences of proportions; and it is possible to find confidence intervals for the difference of variances, σ_X^2 and σ_Y^2, corresponding to two independent normal random variables X and Y, respectively. However, the probability distribution of $\sigma_X^2 - \sigma_Y^2$ is more complicated than the probability distribution of σ_X^2/σ_Y^2. Therefore, since either of these two expressions could be used to compare the two variances, we will determine confidence intervals for σ_X^2/σ_Y^2. We consider the case where μ_X and μ_Y are unknown here, and the case where μ_X and μ_Y are known in the exercises. First, however, a new distribution, called the F distribution, must be introduced.

The *F* Distribution

In Section 8.3 we saw how the standard normal and chi-square distributions could be combined to produce the t distribution, which proved useful in constructing confidence intervals and hypotheses tests for means and their differences. Here, two chi-square distributions are combined to produce the F distribution, which is defined as follows.

Definition: Let $\chi^2(m)$ and $\chi^2(n)$ be independent chi-square random variables with degrees of freedom m and n, respectively. Then, the random variable

$$F = \frac{\chi^2(m)/m}{\chi^2(n)/n}$$

is called an *F random variable with m and n degrees of freedom.*

Properties of the *F* Distribution

The random variable F is also denoted by $F(m, n)$ to emphasize its dependence on the parameters m and n. Note that $F(n, m)$ is not the same as $F(m, n)$; in fact,

$$F(n, m) = \frac{\chi^2(n)/n}{\chi^2(m)/m} = \frac{1}{F(m, n)}$$

The first number m in parentheses for $F(m, n)$ always refers to the degrees of freedom of the chi-square random variable in the numerator of the above definition of F. There is a density curve for each pair (m, n), several of which are illustrated in Fig. 10-1. Note that the F random variable assumes only positive values, since it is a ratio of positive random variables. The density curves are skewed to the right, but the skewing becomes less severe as both m and n increase. The mean and standard deviation of $F(m,n)$ are

$$\mu_F = \frac{n}{n-2} \quad \text{for} \quad n \geq 3 \qquad \text{and} \qquad \sigma_F = \frac{n}{n-2}\sqrt{\frac{2(m+n-2)}{m(n-4)}} \quad \text{for} \quad n \geq 5$$

μ_F is not defined for $n = 1$ or 2, and σ_F is not defined for $n = 1, 2, 3$, or 4. The mode of $F(m,n)$ is $\dfrac{n(m-2)}{m(n+2)}$ for $m \geq 3$; for $m = 1$ or 2, there is no mode.

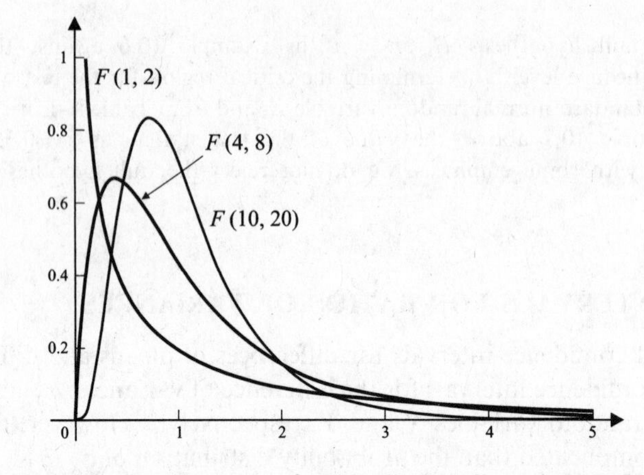

Fig. 10-1 Density curves for $F(1, 2)$, $F(4, 8)$, and $F(10, 20)$.

Confidence Intervals for σ_X^2/σ_Y^2 When μ_X and μ_Y are Unknown

Suppose X and Y are independent normal random variables with unknown variances σ_X^2 and σ_Y^2, respectively. Let $X_1, X_2, \ldots, X_m$ and $Y_1, Y_2, \ldots, Y_n$ be independent random-variable samples corresponding to X and Y, with sample means $\bar{X}$ and $\bar{Y}$, respectively (see Section 8.1). Then

$$\frac{(m-1)S_X^2}{\sigma_X^2} \quad \text{and} \quad \frac{(n-1)S_Y^2}{\sigma_Y^2}$$

where $S_X^2 = \dfrac{1}{m-1}\sum(X_i - \bar{X})^2$ and $S_Y^2 = \dfrac{1}{n-1}\sum(Y_i - \bar{Y})^2$, are independent chi-square random variables with $m-1$ and $n-1$ degrees of freedom, respectively (see Sections 7.4 and 8.5). Dividing the chi-square random variable on the left above by $m-1$ gives S_X^2/σ_X^2, and dividing the one on the right by $n-1$ gives S_Y^2/σ_Y^2. Therefore, by definition, the ratio

$$F(n-1, m-1) = \frac{S_Y^2/\sigma_Y^2}{S_X^2/\sigma_X^2}$$

is an F random variable with $n-1$ degrees of freedom in the numerator and $m-1$ degrees of freedom in the denominator. F can also be written as

$$F(n-1, m-1) = \frac{\sigma_X^2 S_Y^2}{\sigma_Y^2 S_X^2}$$

Let us consider a 98 percent confidence interval for σ_X^2/σ_Y^2. Using an F table, or computer software, we can find constants a and b for which $P(a \leq F(n-1, m-1) \leq b) = 0.98$. For example, we could choose a and b to satisfy

$$P(F \leq a) = 0.01 \qquad \text{and} \qquad P(F \leq b) = 0.99$$

(see Fig. 10-2 and Example 10.9).

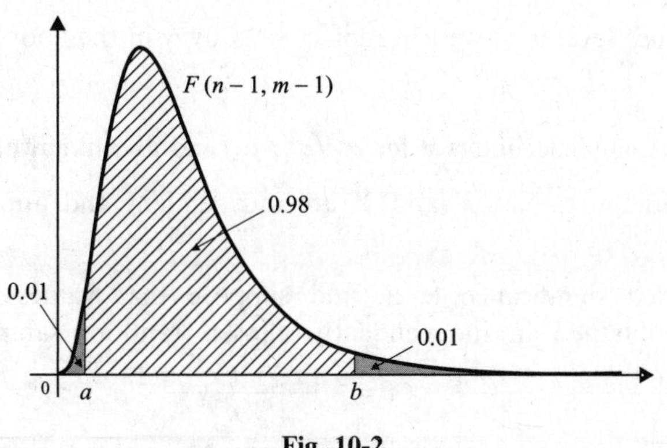

Fig. 10-2

Then

$$P(a \leq F \leq b) = P(F \leq b) - P(F \leq a) = 0.99 - 0.01 = 0.98$$

Substituting $\dfrac{\sigma_X^2 S_Y^2}{\sigma_Y^2 S_X^2}$ for F, we have

$$P\left(a \leq \frac{\sigma_X^2 S_Y^2}{\sigma_Y^2 S_X^2} \leq b\right) = 0.98$$

which is equivalent to

$$P\left(a\frac{S_X^2}{S_Y^2} \leq \frac{\sigma_X^2}{\sigma_Y^2} \leq b\frac{S_X^2}{S_Y^2}\right) = 0.98$$

Therefore,

$$\left[a\frac{S_X^2}{S_Y^2}, b\frac{S_X^2}{S_Y^2}\right]$$

is a random 98 percent confidence interval for σ_X^2/σ_Y^2.

EXAMPLE 10.9 A random sample of size 26, drawn from a normal population X, has sample variance $s_X^2 = 64$; and a random sample of size 16, drawn from a normal population Y, has sample variance $s_Y^2 = 100$. Assuming that X and Y are independent, find a 98 percent confidence interval for σ_X^2/σ_Y^2.

We have $m - 1 = 26 - 1 = 25$, and $n - 1 = 16 - 1 = 15$. We must find a and b for which $P(F(15, 25) \leq a) = 0.01$ and $P(F(15, 25) \leq b) = 0.99$. Tables A-4 to A-7 in the Appendix give values F^* directly for $P(F(15, 25) \leq F^*) = 0.9$ or 0.95 or 0.975 or 0.99. From Table A-7, we find that $b = 2.85$. Table A-7 can also be used indirectly to find the value of a for which $P(F(15, 25) \leq a) = 0.01$ as follows. Since $F(15, 25) = \dfrac{1}{F(25, 15)}$, the following equations are equivalent.

$$P(F(15, 25) \leq a) = 0.01$$

$$P\left(\frac{1}{F(25, 15)} \leq a\right) = 0.01$$

$$P\left(F(25, 15) \geq \frac{1}{a}\right) = 0.01$$

$$P\left(F(25, 15) < \frac{1}{a}\right) = 1 - 0.01 = 0.99$$

From Table A-7, we find that $\dfrac{1}{a} = 3.28$, or $a = \dfrac{1}{3.28}$. Finally, $\dfrac{s_X^2}{s_Y^2} = \dfrac{64}{100} = 0.64$. Therefore, the corresponding 98 percent confidence interval for σ_X^2/σ_Y^2 is $\left[a\dfrac{s_X^2}{s_Y^2}, b\dfrac{s_X^2}{s_Y^2}\right] = \left[\dfrac{1}{3.28} \times 0.64, 2.85 \times 0.64\right] = [0.195, 1.824]$.

In general, if the significance level is γ, we can replace 0.98 by γ in the above example to arrive at the following prescription.

PRESCRIPTION 10.7 (Confidence interval for σ_X^2/σ_X^2; μ_X and μ_Y unknown)

Requirements: The random variables X and Y are independent and approximately normally distributed.

Let γ be the specified significance level, and suppose that values $x_1, x_2, \ldots, x_m$ of X and $y_1, y_2, \ldots, y_n$ of Y are obtained in independently chosen random samples. First compute the sample values $\bar{x} = \dfrac{x_1 + x_2 + \cdots + x_m}{m}$, $\bar{y} = \dfrac{y_1 + y_2 + \cdots + y_n}{n}$, $s_X^2 = \dfrac{1}{m-1}\sum(x_i - \bar{x})^2$, and $s_Y^2 = \dfrac{1}{n-1}\sum(y_i - \bar{y})^2$. Then complete the following steps.

(1) *Find Critical F Values:* Find values F_1^* and F_2^* that satisfy $P(F(m-1, n-1) \le F_1^*) = \dfrac{1+\gamma}{2}$ and $P(F(n-1, m-1) \le F_2^*) = \dfrac{1+\gamma}{2}$ (see Fig. 10.3).

(2) *Determine Confidence Interval:* An approximate 100γ percent confidence interval for σ_X^2/σ_Y^2 is $\left[\dfrac{1}{F_1^*} \times \dfrac{s_X^2}{s_Y^2}, F_2^* \times \dfrac{s_X^2}{s_Y^2}\right]$.

In Example 10.9, $\dfrac{1+\gamma}{2} = \dfrac{1.98}{2} = 0.99$, $m - 1 = 25, n - 1 = 15$, $F_1^* = 3.28$, $F_2^* = 2.85$, and $\dfrac{s_X^2}{s_Y^2} = 0.64$. According to Prescription 10.7, the corresponding 98 percent confidence interval for σ_X^2/σ_Y^2 is $\left[\dfrac{1}{3.28} \times 0.64, 2.85 \times 0.64\right] = [0.195, 1.824]$, as was obtained in Example 10.9.

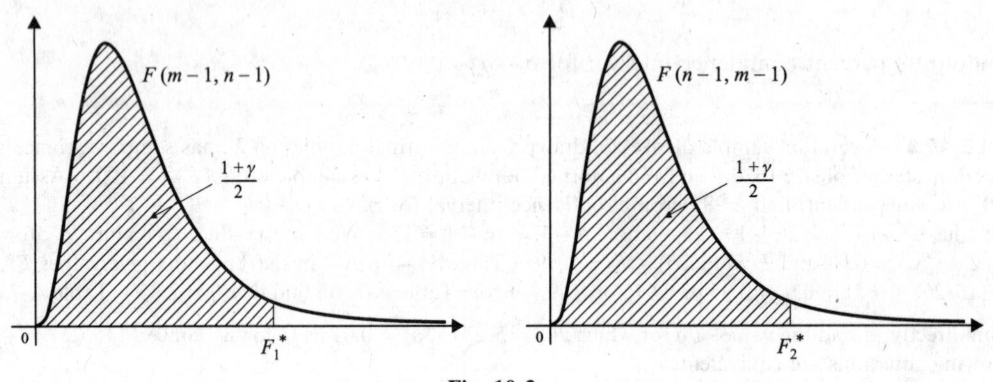

Fig. 10-3

WARNING

Just as in the case for a single variance, the above approximate confidence intervals for σ_X^2/σ_Y^2 are not robust, and can deviate very significantly from the true confidence intervals when X and Y are not normally distributed (see Section 8.5). Hence, the practical use of these confidence intervals is limited.

10.6 HYPOTHESES TESTS FOR RATIOS OF VARIANCES

Suppose X and Y are approximately normally distributed random variables with means μ_X and μ_Y, and unknown variances σ_X^2 and σ_Y^2, respectively. As in the case of confidence intervals, we consider hypotheses tests for σ_X^2/σ_Y^2 when μ_X and μ_Y are unknown. The null hypothesis will be

$$H_0\colon \sigma_X^2/\sigma_Y^2 = 1; \qquad \text{equivalently,} \quad H_0\colon \sigma_X^2 = \sigma_Y^2$$

and the alternative hypothesis will be one of the following:

$$H_a\colon \sigma_X^2/\sigma_Y^2 < 1; \qquad \text{equivalently,} \quad H_a\colon \sigma_X^2 < \sigma_Y^2$$

$$H_a\colon \sigma_X^2/\sigma_Y^2 > 1; \qquad \text{equivalently,} \quad H_a\colon \sigma_X^2 > \sigma_Y^2$$

$$H_a\colon \sigma_X^2/\sigma_Y^2 \neq 1; \qquad \text{equivalently,} \quad H_a\colon \sigma_X^2 \neq \sigma_Y^2$$

Hypotheses Tests for σ_X^2/σ_Y^2 When μ_X and μ_Y are Unknown

In the previous section, confidence intervals for σ_X^2/σ_Y^2 utilized the F random variable

$$F(n-1, m-1) = \frac{\sigma_X^2 S_Y^2}{\sigma_Y^2 S_X^2}$$

where $S_X^2 = \dfrac{1}{m-1}\sum(X_i - \bar{X})^2$ and $S_Y^2 = \dfrac{1}{n-1}\sum(Y_i - \bar{Y})^2$ are the sample variances for X and Y, respectively. If the null hypothesis $H_0\colon \sigma_X^2 = \sigma_Y^2$ is true, then the above random variable becomes

$$F(n-1, m-1) = \frac{S_Y^2}{S_X^2}$$

Also, by the reciprocal property of the F distribution,

$$F(m-1, n-1) = \frac{S_X^2}{S_Y^2}$$

Hence, either S_Y^2/S_X^2 or S_X^2/S_Y^2 can be used as the test statistic. Proceeding as in the case of confidence intervals, we arrive at the following prescription.

PRESCRIPTION 10.8 (*P*-value hypotheses tests for σ_X^2/σ_Y^2; μ_X, μ_Y unknown)

Requirement: X and Y are approximately normally distributed.

Let α be the level of significance for the test. Suppose the values $x_1, x_2, \ldots, x_m$ of X and $y_1, y_2, \ldots, y_n$ of Y are obtained in independently chosen random samples. Compute the sample values

$$\bar{x} = \frac{x_1 + x_2 + \cdots + x_m}{m}, \quad \bar{y} = \frac{y_1 + y_2 + \cdots + y_n}{n}, \quad s_X^2 = \frac{1}{m-1}\sum(x_i - \bar{x})^2, \text{ and } s_Y^2 = \frac{1}{n-1}\sum(y_i - \bar{y})^2.$$

Then complete the following steps.

(1) *State Hypotheses:* State null hypothesis $H_0 : \sigma_X^2 = \sigma_Y^2$ and alternative hypothesis H_a.
(2) *Compute Test Statistic:* Let S_X^2/S_Y^2 be the test statistic, which, assuming H_0 is true, is approximately an $F(m-1, n-1)$ random variable. Compute the test value as s_X^2/s_Y^2.
(3) *Determine P-value:* Using an F table, if adequate, or computer software, find the P-value of the test corresponding to H_a:

For $H_a\colon \sigma_X^2 < \sigma_Y^2$, the P-value is $P(F(m-1, n-1) \leq s_X^2/s_Y^2)$
For $H_a\colon \sigma_X^2 > \sigma_Y^2$, the P-value is $P(F(m-1, n-1) \geq s_X^2/s_Y^2)$

For $H_a\colon \sigma_X^2 \neq \sigma_Y^2$, the P-value is $\begin{cases} 2P(F(m-1, n-1) \leq s_X^2/s_Y^2) \text{ if } s_X^2/s_Y^2 < 1 \\ 2P(F(m-1, n-1) \geq s_X^2/s_Y^2) \text{ if } s_X^2/s_Y^2 > 1 \end{cases}$

(4) *Draw Conclusion:* If P-value $\leq \alpha$, then the test is statistically significant at level α, and H_0 is rejected in favor of H_a. If P-value $> \alpha$, then the test is not statistically significant at level α, and H_0 is not rejected in favor of H_a.

EXAMPLE 10.10 In Example 10.9, a random sample of size $m = 26$, drawn from a normal population X, has sample variance $s_X^2 = 64$; and a random sample of size $n = 16$, drawn from a normal population Y, has sample variance $s_Y^2 = 100$. The value of the test statistic is $\frac{s_X^2}{s_Y^2} = \frac{64}{100} = 0.64$. Let the null hypothesis be H_0: $\sigma_X^2 = \sigma_Y^2$, and let $\alpha = 0.05$ be the significance level. Then, using computer software, we find that for H_a: $\sigma_X^2 < \sigma_Y^2$, the P-value is $P(F(25, 15) \leq 0.64) \approx 0.16$. Since $0.16 > 0.05$, we would not reject the null hypothesis in favor of H_a: $\sigma_X^2 < \sigma_Y^2$ at the 0.05 significance level. Note that for H_a: $\sigma_X^2 \neq \sigma_Y^2$, the P-value is twice the P-value for H_a: $\sigma_X^2 < \sigma_Y^2$, namely $2 \times 0.16 = 0.32$; so we would also not reject the null hypothesis in favor of H_a: $\sigma_X^2 \neq \sigma_Y^2$ at the 0.05 significance level.

Alternative Version of Prescription 10.8

The following alternative version of Prescription 10.8 replaces the P-value by the critical region for the alternative hypothesis at the specified level of significance α. Because of the limitations of the F tables in the Appendix, all critical regions are defined in terms of the right tail of either the $F(n-1, m-1)$ or $F(m-1, n-1)$ distribution.

PRESCRIPTION 10.8*a* **(Critical-region hypotheses tests for σ_X^2/σ_Y^2; μ_X, μ_Y unknown)**

Requirement: X and Y are approximately normally distributed.

(1) and (2) Same as in Prescription 10.8.

(3) *Determine Critical Region:*

For H_a: $\sigma_X^2 < \sigma_Y^2$, the critical region is all sample values $\frac{s_Y^2}{s_X^2} \geq F^*$, where F^* is the F value satisfying $P(F(n-1, m-1) \leq F^*) = 1 - \alpha$ (Fig. 10-4(*a*)).

For H_a: $\sigma_X^2 > \sigma_Y^2$, the critical region is all sample values $\frac{s_X^2}{s_Y^2} \geq F^*$ where F^* is the F value satisfying $P(F(m-1, n-1) \leq F^*) = 1 - \alpha$ (Fig. 10-4(*b*)).

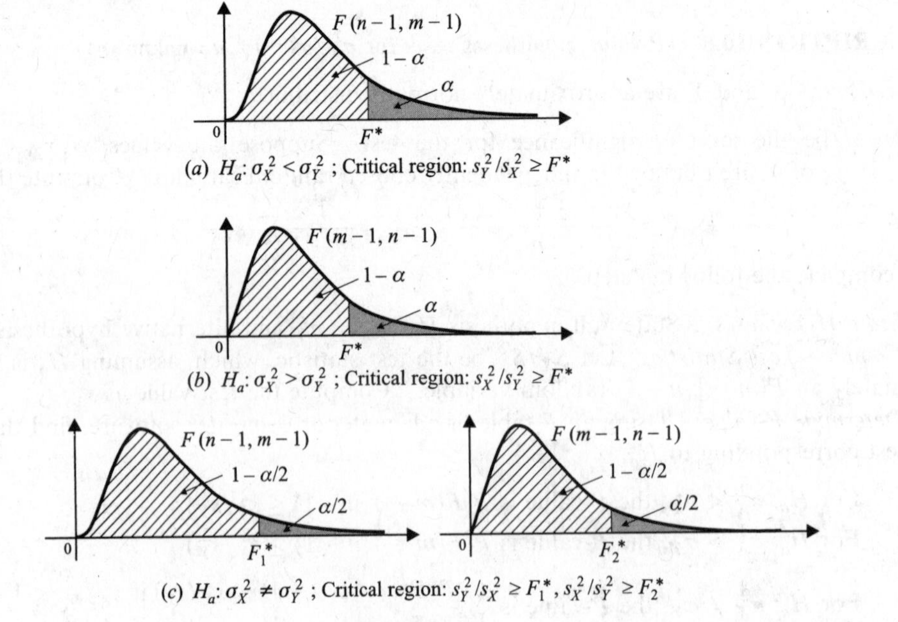

Fig. 10-4

For H_a: $\sigma_X^2 \neq \sigma_Y^2$, the critical region is all values $\frac{s_Y^2}{s_X^2} \geq F_1^*$ or $\frac{s_X^2}{s_Y^2} \geq F_2^*$, where F_1^* is the F value satisfying $P(F(n-1, m-1) \leq F_1^*) = 1 - \alpha/2$, and F_2^* is the F value satisfying $P(F(m-1, n-1) \leq F_2^*) = 1 - \alpha/2$ (Fig. 10-4(*c*)).

(4) *Draw Conclusion:* If the test value s_X^2/s_Y^2 or s_Y^2/s_X^2 lies in its corresponding portion of the critical region, then the test is statistically significant at level α, and H_0 is rejected in favor of H_a. If the test value does not lie in the critical region, then the test is not statistically significant at level α, and H_0 is not rejected in favor of H_a.

EXAMPLE 10.11 In Example 10.10, $s_X^2 = 64$, based on a random sample of size $m = 26$; and $s_Y^2 = 100$, based on a random sample of size $n = 16$. The null hypothesis is H_0: $\sigma_X^2 = \sigma_Y^2$. The critical region at the $\alpha = 0.05$ level of significance for H_a: $\sigma_X^2 < \sigma_Y^2$ is all sample values $\frac{s_Y^2}{s_X^2} \geq F^*$, where F^* is the F value satisfying $P(F(15, 25) \leq F^*) = 1 - \alpha = 0.95$. From Table A-5 in the Appendix, we find that $F^* = 2.09$. The value of the corresponding test statistic is $\frac{s_Y^2}{s_X^2} = \frac{100}{64} \approx 1.56$, which is not in the critical region, so we do not reject H_0: $\sigma_X^2 = \sigma_Y^2$ in favor of H_a: $\sigma_X^2 < \sigma_Y^2$ at the 0.05 significance level.

For the alternative hypothesis H_a: $\sigma_X^2 \neq \sigma_Y^2$, the critical region at the 0.05 significance level contains all values $\frac{s_Y^2}{s_X^2} \geq F_1^*$ or $\frac{s_X^2}{s_Y^2} \geq F_2^*$, where F_1^* is the F value satisfying $P(F(15, 25) \leq F_1^*) = 1 - \alpha/2 = 0.975$, and F_2^* is the F value satisfying $P(F(25, 15) \leq F_2^*) = 1 - \alpha/2 = 0.975$. From Table A-6 in the Appendix, we find that $F_1^* = 2.41$ and $F_2^* = 2.69$. Since $\frac{s_Y^2}{s_X^2} \approx 1.56$ and $\frac{s_X^2}{s_Y^2} = 0.64$, neither statistic is in its corresponding portion of the fundamental region, so the null hypothesis is not rejected in favor of $H_a: \sigma_X^2 \neq \sigma_Y^2$ at the 0.05 significance level.

Solved Problems

CONFIDENCE INTERVALS FOR THE DIFFERENCE OF MEANS

10.1. The scores on a standardized math test in District X are normally distributed with mean 74 and standard deviation 8, while those in District Y are normally distributed with mean 70 and standard deviation 10. A new learning program, which makes extensive use of computers, is introduced in both districts. The mean score under the new system of a random sample of 40 students in District X is $\bar{x} = 75$. In District Y, $\bar{y} = 73$, based on a random sample of 50 students. Find a 95 percent confidence interval for $\mu_X - \mu_Y$ under the new system. Assume $\sigma_X = 8$ and $\sigma_Y = 10$.

We apply Prescription 10.1. From Table A-1, the value z^* of the standard normal random variable Z satisfying $P(-z^* \leq Z \leq z^*) = 0.95$ is $z^* = 1.96$. The margin of error is $E = z^* \sqrt{\frac{\sigma_X^2}{40} + \frac{\sigma_Y^2}{50}} = 1.96 \sqrt{\frac{64}{40} + \frac{100}{50}} \approx 3.72$. Therefore, the corresponding approximate 95 percent confidence interval is $[(75 - 73) - 3.72, (75 - 73) + 3.72] = [-1.72, 5.72]$.

10.2. Suppose the sample standard deviation for X in Problem 10.1, based on the random sample of size 40, is $s_X = 8.8$; and that for Y, based on the sample of 50 students, is $s_Y = 9.2$. Determine an approximate 95 percent confidence interval for $\mu_X - \mu_Y$ under the assumption that σ_X and σ_Y are unknown but equal.

We apply Prescription 10.2. The sample value of the pooled estimator of the common standard deviation is

$$s_P = \sqrt{\frac{(m-1)s_X^2 + (n-1)s_Y^2}{m+n-2}} = \sqrt{\frac{39(8.8)^2 + 49(9.2)^2}{88}} \approx 9.02$$

Table A-2 in the Appendix shows that the value t^* of the t random variable, with 60 degrees of freedom, that satisfies $P(-t^* \le t \le t^*) = 0.95$ is $t^* = 1.98$, and for 120 degrees of freedom, $t^* = 2.00$. Since 88 is approximately midway between 60 and 120, we will use $t^* = 1.99$. (Computer software gives 1.9873.) The margin of error is $E = t^* s_P \sqrt{\frac{1}{m} + \frac{1}{n}} = 1.99 \times 9.02 \sqrt{\frac{1}{40} + \frac{1}{50}} \approx 3.81$. The corresponding approximate 95 percent confidence interval for $\mu_X - \mu_Y$ is $[(75-73) - 3.81, (75-73) + 3.81] = [-1.81, 5.81]$.

10.3. Suppose that the sample standard deviation for the 40 District X students in Problem 10.1 is $s_X = 7.8$, and that for the 50 students in District Y is $s_Y = 9.6$. Compute an approximate 95 percent confidence interval for $\mu_X - \mu_Y$ under the assumption that the random variable

$$T = \frac{\bar{X} - \bar{Y} - (\mu_X - \mu_Y)}{\sqrt{\frac{s_X^2}{40} + \frac{s_Y^2}{50}}}$$

is approximately normally distributed with mean 0 and standard deviation 1.

We follow Prescription 10.1 with $s_X^2 = (7.8)^2 = 60.84$ in place of $\sigma_X^2 = 64$ and $s_Y^2 = (9.6)^2 = 92.16$ in place of $\sigma_Y^2 = 100$. As shown in Problem 10.1, the value z^* of the standard normal random variable Z satisfying $P(-z^* \le Z \le z^*) = 0.95$ is $z^* = 1.96$. The margin of error is $E = z^* \sqrt{\frac{s_X^2}{40} + \frac{s_Y^2}{50}} = 1.96 \sqrt{\frac{60.84}{40} + \frac{92.16}{50}} \approx 3.59$. The corresponding approximate 95 percent confidence interval is $[(75-73) - 3.59, (75-73) + 3.59] = [-1.59, 5.59]$.

10.4. With reference to the mice in Example 10.1, find a 90 percent confidence interval for $\mu_X - \mu_Y$ under the assumption that

$$T = \frac{\bar{X} - \bar{Y} - (\mu_X - \mu_Y)}{\sqrt{\frac{s_X^2}{8} + \frac{s_Y^2}{10}}}$$

is approximately a t random variable whose degrees of freedom are given by the largest integer less than or equal to the expression for k at the end of Section 10.1.

We follow Prescription 10.2 with the margin of error $E = t^* \sqrt{\frac{s_X^2}{m} + \frac{s_Y^2}{n}}$ in place of $E = t^* s_P \sqrt{\frac{1}{m} + \frac{1}{n}}$. From the data in Example 10.1, we find that $s_X^2 \approx 9.125$ and $s_Y^2 \approx 4.276$ (see Example 10.2). Substituting these values, along with $m = 8$ and $n = 10$, into the expression for k, we get

$$k = \frac{\left(\frac{9.125}{8} + \frac{4.276}{10}\right)^2}{\frac{1}{7}\left(\frac{9.125}{8}\right)^2 + \frac{1}{9}\left(\frac{4.276}{10}\right)^2} \approx 11.93$$

We therefore assume that T has 11 degrees of freedom. From Table A-2, we find that the value t^* of the t random variable, with 11 degrees of freedom, that satisfies $P(-t^* \le t \le t^*) = 0.9$ is $t^* = 1.80$. Then

$E = 1.80\sqrt{\frac{9.125}{8} + \frac{4.276}{10}} \approx 2.25$. Using the values $\bar{x} \approx 29.38$ and $\bar{y} = 28.6$ obtained in Example 10.1, we find that the corresponding 90 percent confidence interval for $\mu_X - \mu_Y$ is $[(29.38 - 28.6) - 2.25, (29.38 - 28.6) + 2.25] = [-1.47, 3.03]$.

10.5. What justifies the assumptions made concerning the random variable τ in Problems 10.3 and 10.4?

It can be proven, by methods beyond the level of this text, that the distribution of the random variable τ in Problem 10.3 approaches the standard normal distribution as m and n increase without bound. For our purposes, the assumption that τ is approximately standard normal is justified because the random variables X and Y are normally distributed, and the sample sizes $m = 40$ and $n = 50$ are each larger than 30. Similarly, for our purposes, the assumption in Problem 10.4 that τ is approximately a t random variable with the specified number of degrees of freedom is justified because the random variables X and Y are normally distributed, and the sample sizes $m = 8$ and $n = 10$ are each larger than 5.

10.6. Suppose X and Y are independent random variables, and $\sigma_X^2 = \sigma_Y^2$. Show that the random variable $\frac{\bar{X} - \bar{Y} - (\mu_X - \mu_Y)}{S_P\sqrt{\frac{1}{m} + \frac{1}{n}}}$ defined in Section 10.1 has a t distribution with $m + n - 2$ degrees of freedom.

By definition, if Z and χ^2 are independent, where Z is a standard normal random variable and χ^2 is a chi-square random variable with k degrees of freedom, then $\frac{Z}{\sqrt{\chi^2/k}}$ is a t random variable with k degrees of freedom (see Section 8.3). Let σ^2 be the common value of σ_X^2 and σ_Y^2. The random variable

$$Z = \frac{\bar{X} - \bar{Y} - (\mu_X - \mu_Y)}{\sqrt{\frac{\sigma^2}{m} + \frac{\sigma^2}{n}}} = \frac{\bar{X} - \bar{Y} - (\mu_X - \mu_Y)}{\sigma\sqrt{\frac{1}{m} + \frac{1}{n}}}$$

is standard normal. Also, $\frac{(m-1)S_X^2}{\sigma^2}$ and $\frac{(n-1)S_Y^2}{\sigma^2}$ are independent chi-square random variables with $m - 1$ and $n - 1$ degrees of freedom, respectively (Theorem 7.7). Therefore, $\chi^2 = \frac{(m-1)S_X^2}{\sigma^2} + \frac{(n-1)S_Y^2}{\sigma^2}$ is chi-square with $m - 1 + n - 1 = m + n - 2$ degrees of freedom. When Z is divided by $\sqrt{\chi^2/(m + n - 2)}$, σ cancels, and the resulting quotient is $\frac{\bar{X} - \bar{Y} - (\mu_X - \mu_Y)}{S_P\sqrt{\frac{1}{m} + \frac{1}{n}}}$, where $S_P = \sqrt{\frac{(m-1)S_X^2 + (n-1)S_Y^2}{m + n - 2}}$.

HYPOTHESES TESTS FOR THE DIFFERENCE OF MEANS

10.7. The average GPA for 60 mathematics majors at a particular university is 3.4 with a variance of 0.2; and the GPA for the 50 physics majors at the university is 3.5 with a variance of 0.12. Letting X and Y represent the mathematics and physics GPAs, respectively, and assuming the GPAs are normally distributed, test the null hypothesis H_0: $\mu_X = \mu_Y$ against the alternative hypothesis H_a: $\mu_X < \mu_Y$ at the 0.05 significance level by constructing the P-value for the test.

Following Prescription 10.3, the test statistic is

$$Z = \frac{\bar{X} - \bar{Y}}{\sqrt{\frac{\sigma_X^2}{60} + \frac{\sigma_Y^2}{50}}}$$

which, assuming H_0 is true, is standard normal. The test value of Z is

$$z = \frac{3.4 - 3.5}{\sqrt{\dfrac{0.2}{60} + \dfrac{0.12}{50}}} \approx -1.32$$

The P-value of the test is $P(Z \leq -1.32) \approx 0.09$. Since the P-value is greater than 0.05, the test is not significant at the 0.05 level, and the null hypothesis is not rejected at that level.

10.8. In Problem 10.7, test the null hypothesis H_0: $\mu_X = \mu_Y$ against the alternative hypothesis H_a: $\mu_X < \mu_Y$ at the 0.05 significance level by constructing the critical region for the test.

Following Prescription 10.3*a*, at the 0.05 level, the critical region for H_a: $\mu_X < \mu_Y$ is all values $z \leq z^*$, where $P(Z \leq z^*) = 0.05$. From Table A-1, we find that $z^* = -1.65$. The test value of Z found in Problem 10.7 is $z = -1.32$, which is not in the critical region. Therefore, H_0 is not rejected in favor of H_a at the 0.05 significance level.

10.9. At what significance level would the null hypothesis in Problem 10.7 be rejected in favor of the alternative hypothesis?

H_0 would be rejected at any level greater than or equal to the P-value of the test, which was determined to be 0.09. In particular, H_0 would be rejected in favor of H_a at the 0.1 significance level.

10.10. The annual salaries, in thousands of dollars, of 8 men in middle management at a given company are: 55.5, 64.8, 68.2, 70.2, 52.4, 56.8, 60.6, 72.5, while those for 6 women are: 56.2, 48.8, 58.4, 50.9, 60.2, 54.5. Let X and Y denote the salaries of the men and women, respectively; and assuming normal distributions and equal standard deviations, test H_0: $\mu_X = \mu_Y$ against H_a: $\mu_X > \mu_Y$ at the 0.05 significance level by constructing a critical region for the test.

Following Prescription 10.4*a*, the needed sample values are

$$\bar{x} = \frac{55.5 + 64.8 + 68.2 + 70.2 + 52.4 + 56.8 + 60.6 + 72.5}{8} \approx 62.63$$

$$\bar{y} = \frac{56.2 + 48.8 + 58.4 + 50.9 + 60.2 + 54.5}{6} \approx 54.83$$

$$s_X^2 = \frac{1}{7} \sum (x_i - 62.63)^2 \approx 54.87$$

$$s_Y^2 = \frac{1}{5} \sum (y_i - 54.83)^2 \approx 19.07$$

$$s_P = \sqrt{\frac{7 \times 54.87 + 5 \times 19.07}{12}} \approx 6.32$$

The test statistic is $t = \dfrac{\bar{X} - \bar{Y}}{S_P\sqrt{\dfrac{1}{8} + \dfrac{1}{6}}}$ which, assuming H_0 is true, is a t random variable with 12 degrees of freedom. The test value of t is $\hat{t} = \dfrac{62.63 - 54.83}{6.32\sqrt{\dfrac{1}{8} + \dfrac{1}{6}}} \approx 2.29$. The critical region for H_a: $\mu_X > \mu_Y$ is all values $\hat{t} \geq t^*$, where $P(t \geq t^*) = 0.05$. From Table A-2 in the Appendix, with 12 degrees of freedom, we find that $t^* = 1.78$. Since the test value 2.29 is greater than 1.78, the null hypothesis H_0: $\mu_X = \mu_Y$ is rejected in favor of H_a: $\mu_X > \mu_Y$ at the 0.05 significance level.

10.11. In Problem 10.10, what is the smallest significance level at which the null hypothesis would be rejected in favor of the alternative hypothesis?

The answer is the P-value of the test, which is $P(\bar{X} - \bar{Y} \geq 62.63 - 54.83 = 7.8)$, assuming H_0 is true; equivalently

$$P\left(\frac{\bar{X} - \bar{Y}}{S_p\sqrt{\frac{1}{8} + \frac{1}{6}}} \geq \frac{62.63 - 54.83}{6.32\sqrt{\frac{1}{8} + \frac{1}{6}}} \approx 2.29\right)$$

Using Table 4-2 in the Appendix, with 12 degrees of freedom, we find that the P-value is between 0.01 and 0.025. Using computer software, we find that the P-value is 0.02.

10.12. The sample standard deviations in Problem 10.10 are $s_X = 7.41$ and $s_Y = 4.37$. Such a large difference seems to indicate that the assumption of equality may be unwarranted (see Problem 10.33). Repeat the test using as the test statistic

$$\tau = \frac{\bar{X} - \bar{Y}}{\sqrt{\frac{S_X^2}{m} + \frac{S_Y^2}{n}}}$$

which, assuming H_0 is true, has an approximate t distribution with degrees of freedom equal to the largest integer less than or equal to

$$k = \frac{\left(\frac{s_X^2}{m} + \frac{s_Y^2}{n}\right)^2}{\frac{1}{m-1}\left(\frac{s_X^2}{m}\right)^2 + \frac{1}{n-1}\left(\frac{s_Y^2}{n}\right)^2}$$

We apply Prescription 10.4a, with τ in place of t. The value of the test statistic is

$$\hat{\tau} = \frac{62.63 - 54.83}{\sqrt{\frac{54.87}{8} + \frac{19.07}{6}}} \approx 2.46$$

and the value of k is

$$\frac{\left(\frac{54.87}{8} + \frac{19.07}{6}\right)^2}{\frac{1}{7}\left(\frac{54.87}{8}\right)^2 + \frac{1}{5}\left(\frac{19.07}{6}\right)^2} \approx 11.53$$

The critical region for H_a: $\mu_X > \mu_Y$ is all values $\hat{t} \geq t^*$, where $P(t \geq t^*) = 0.05$. From Table A-2, with 11 degrees of freedom, we find that $t^* = 1.80$. Since the test value 2.46 is greater than 1.80, the null hypothesis H_0: $\mu_X = \mu_Y$ is rejected in favor of H_a at the 0.05 significance level, as in Problem 10.10.

10.13. A random sample of size 100 drawn from a normal population X has sample mean $\bar{x} = 74.8$ sample standard deviation $s_X = 7$; and a random sample of size 150 drawn from a normal population Y has sample mean $\bar{y} = 72$ and sample standard deviation $s_Y = 10$. Test the null hypothesis H_0: $\mu_X = \mu_Y$ against the alternative hypothesis H_a: $\mu_X \neq \mu_Y$ at the 0.01 level of significance. Use τ from Problem 10.12 as the test statistic; for large samples, and assuming H_0 is true, τ is approximately standard normal.

We follow Prescription 10.3a, with τ in place of Z. The value of the test statistic is $\hat{\tau} = \frac{74.8 - 72}{\sqrt{\frac{49}{100} + \frac{100}{150}}} \approx 2.60$. The critical region at the 0.01 significance level for the alternative hypothesis H_a: $\mu_X \neq \mu_Y$ is all values $\hat{\tau} \geq z^*$ or $\hat{\tau} \leq -z^*$, where $P(Z \geq z^*) = 0.01/2 = 0.005$. From Table A-1, we find that $z^* = 2.58$. Since $\hat{\tau} = 2.60$, we reject H_0 in favor of H_a at the 0.01 level.

10.14. Would the null hypothesis H_0: $\mu_X = \mu_Y$ in Problem 10.13 be rejected in favor of H_a: $\mu_X > \mu_Y$ at the 0.01 significance level?

The critical region at the 0.01 significance level for H_a: $\mu_X > \mu_Y$ consists of all values $\hat{\tau} \geq z^*$, where $P(Z \geq z^*) = 0.01$. From Table A-1, we find that $z^* = 2.33$. Since the test value is $\hat{\tau} = 2.60$, H_0 will be rejected in favor of H_a: $\mu_X > \mu_Y$ at the 0.01 significance level. In general, if H_0: $\mu_X = \mu_Y$ is rejected in favor of H_a: $\mu_X \neq \mu_Y$ at any significance level α, then H_0 will be rejected in favor of H_a: $\mu_X > \mu_Y$ at the same level, provided $\bar{x} > \bar{y}$. This is so because the right-tail portion of the critical region for H_a: $\mu_X \neq \mu_Y$ is contained in the critical region for H_a: $\mu_X > \mu_Y$; and therefore any test value of Z that is in the right-tail portion of the critical region for H_a: $\mu_X \neq \mu_Y$ will also be in the critical region for H_a: $\mu_X > \mu_Y$.

CONFIDENCE INTERVALS FOR DIFFERENCES OF PROPORTIONS

10.15. In a random sample of 50 people from eastern states in the U.S.A., 40 said they favored gun control; and 25 out of 48 from western states were in favor of gun control. Find a 95 percent confidence interval for $p_1 - p_2$, where p_1 is the proportion of those in eastern states favoring gun control, and p_2 is the proportion of those in western states favoring gun control, and interpret the result.

We follow Prescription 10.5. The sample proportions are $\hat{p}_1 = \frac{40}{50} = 0.8$ and $\hat{p}_2 = \frac{25}{48} = 0.52$, and the estimated sample standard deviation is $\sqrt{\frac{0.8 \times 0.2}{50} + \frac{0.52 \times 0.48}{48}} = 0.09$. From Table A-1, we find that $P(0 \leq Z \leq z^*) = 0.95/2 = 0.475$ for $z^* = 1.96$. Therefore, the margin of error is $E = 1.96 \times 0.09 = 0.18$, and the 95 percent confidence interval for $p_1 - p_2$ is $[(0.8 - 0.52) - 0.18, (0.8 - 0.52) + 0.18] = [0.10, 0.46]$. Since 0 is not contained in the interval, the sample provides strong evidence that $p_1 > p_2$.

10.16. The sample difference in Problem 10.15 is $\hat{p}_1 - \hat{p}_2 = 0.8 - 0.52 = 0.28$. What is the probability that a difference as large or larger than this would occur if in fact $p_1 = p_2$?

Let $\hat{P}_1$ denote the proportion from eastern states favoring gun control in an arbitrary random sample of size 50, and $\hat{P}_2$ the proportion from western states favoring gun control in an arbitrary random sample of size 48. The random variable $\hat{P}_1 - \hat{P}_2$ is approximately normally distributed with mean $p_1 - p_2$ and variance $\frac{p_1(1-p_1)}{50} + \frac{p_2(1-p_2)}{48}$. We want to compute $P(\hat{P}_1 - \hat{P}_2 \geq 0.28)$, given that $p_1 = p_2$. An estimate $\hat{p}$ of the common value of p_1 and p_2 can be obtained by pooling the data: $\hat{p} = \frac{40 + 25}{50 + 48} \approx 0.66$. Then, $\hat{P}_1 - \hat{P}_2$ is approximately normal with mean 0 and variance $\frac{\hat{p}(1-\hat{p})}{50} + \frac{\hat{p}(1-\hat{p})}{48} = \frac{0.66 \times 0.34}{50} + \frac{0.66 \times 0.34}{48} \approx 0.009$. Therefore, $P(\hat{P}_1 - \hat{P}_2 \geq 0.28) = P\left(\frac{\hat{P}_1 - \hat{P}_2 - 0}{\sqrt{0.009}} \geq \frac{0.28}{\sqrt{0.009}}\right) = P(Z \geq 2.95)$, where Z is the standard normal random variable. From Table A-1, we find that $P(Z \geq 2.95) = 0.0016$. That is, the chances of a difference as large as 0.28 occurring are only 16 in 10,000 if in fact $p_1 = p_2$.

10.17. Repeat Problem 10.16 using the data in Example 10.6.

In Example 10.6, the difference in the sample proportion is $\hat{p}_1 - \hat{p}_2 = 0.52 - 0.51 = 0.01$. The estimate of the common proportion, obtained by pooling the data, is $\hat{p} = \frac{260 + 306}{500 + 600} \approx 0.515$, and the estimated variance of $\hat{P}_1 - \hat{P}_2$ is $\frac{0.515 \times 0.49}{500} + \frac{0.515 \times 0.49}{600} \approx 0.0009$. Then

$$P(\hat{P}_1 - \hat{P}_2 \geq 0.01) = P\left(\frac{\hat{P}_1 - \hat{P}_2}{\sqrt{0.0009}} \geq \frac{0.01}{\sqrt{0.0009}}\right) = P(Z \geq 0.33) = 0.37$$

Hence the chances are 37 in 100 that a difference as large as 0.01 would occur if $p_1 = p_2$. In this case, the data do not provide strong evidence that $p_1 > p_2$.

HYPOTHESES TESTS FOR DIFFERENCES OF PROPORTIONS

10.18. Using the data in Problem 10.15, test the null hypothesis H_0: $p_1 = p_2$ against the alternative hypothesis H_a: $p_1 > p_2$ at the 0.01 significance level by finding the P-value for the test.

We follow Prescription 10.6. The sample proportions are $p_1 = \frac{40}{50} = 0.8$ and $p_2 = \frac{25}{48} = 0.52$, so $p_1 - p_2 = 0.28$. The P-value of the test, computed under the assumption that H_0 is true, is $P(p_1 - p_2 \geq 0.28)$. From Problem 10.16, $P(p_1 - p_2 \geq 0.28) = 0.0016$. Since $0.0016 < 0.01$, the null hypothesis is rejected in favor of H_a: $p_1 > p_2$ at the 0.01 significance level.

10.19. Using the data in Problem 10.15, test the null hypothesis H_0: $p_1 = p_2$ against the alternative hypothesis H_a: $p_1 > p_2$ at the 0.01 significance level by finding the critical region for the test.

We use Prescription 10.6*a*. The sample proportions are $\hat{p}_1 = \frac{40}{50} = 0.8$ and $\hat{p}_2 = \frac{25}{48} = 0.52$; and the pooled proportion is $\hat{p} = \frac{40 + 25}{50 + 48} \approx 0.66$. (Note that $\hat{p}$ can also be computed by the formula given in Prescription 10.6, namely $\hat{p} = \frac{n_1\hat{p}_1 + n_2\hat{p}_2}{n_1 + n_2} = \frac{50 \times 0.8 + 48 \times 0.52}{50 + 48} \approx 0.66$. See Problem 10.20.) The value of the test statistic is

$$z = \frac{\hat{p}_1 - \hat{p}_2}{\sqrt{\hat{p}(1 - \hat{p})\left(\frac{1}{n_1} + \frac{1}{n_2}\right)}} = \frac{8.0 - 0.52}{\sqrt{0.66 \times 0.34\left(\frac{1}{50} + \frac{1}{48}\right)}} \approx 2.93$$

For the alternative hypothesis H_a: $p_1 > p_2$ at the 0.01 significance level, the critical region is all $z \geq z^*$, where $P(Z \geq z^*) = 0.01$, Z being the standard normal random variable. From Table A-1, we find that $z^* = 2.33$. Since $2.93 > 2.33$, the test statistic lies in the critical region, and H_0: $p_1 = p_2$ is rejected in favor of H_a.

10.20. Suppose that x_1 is the number of successes in n_1 trials making up one experiment, x_2 is the number of successes in n_2 trials making up a second experiment. Then $\hat{p}_1 = \frac{x_1}{n_1}$ and $\hat{p}_2 = \frac{x_2}{n_2}$ are the corresponding proportions of successes. Show that the pooled proportion $\hat{p} = \frac{x_1 + x_2}{n_1 + n_2}$ can also be computed by the formula $\hat{p} = \frac{n_1\hat{p}_1 + n_2\hat{p}_2}{n_1 + n_2}$.

From the equation $\hat{p}_1 = \frac{x_1}{n_1}$, we get $x_1 = n_1\hat{p}_1$, and from $\hat{p}_2 = \frac{x_2}{n_2}$, we get $x_2 = n_2\hat{p}_2$. Substituting for x_1 and x_2 in the formula $\hat{p} = \frac{x_1 + x_2}{n_1 + n_2}$, we get $\hat{p} = \frac{n_1\hat{p}_1 + n_2\hat{p}_2}{n_1 + n_2}$, as desired.

CONFIDENCE INTERVALS FOR RATIOS OF VARIANCES

10.21. Find a 98 percent confidence interval for σ_X/σ_Y in Example 10.9.

The 98 percent confidence interval for σ_X^2/σ_Y^2 found in Example 10.9 is [0.195, 1.824]. The corresponding 98 percent confidence interval for σ_X/σ_Y is $[\sqrt{0.195}, \sqrt{1.824}] = [0.442, 1.351]$.

10.22. The sample variance of 11 one-liter bottles of wine bottled in summer was $S_X^2 = 50\ (\text{ml})^2$, and the sample variance of 16 one-liter bottles of wine bottled in winter was $S_Y^2 = 60\ (\text{ml})^2$. Assuming that the volume of liquid in the bottles is normally distributed, find a 90 percent confidence interval for σ_X^2/σ_Y^2.

We follow Prescription 10.7, where $m = 11 - 1 = 10$, $n = 16 - 1 = 15$, and $\frac{\gamma+1}{2} = \frac{1.90}{2} = 0.95$. From Table A-5, we find that $P(F(10,15) \le F_1^*) = 0.95$ for $F_1^* = 2.54$; and $P(F(15,10) \le F_2^*) = 0.95$ for $F_2^* = 2.85$. The corresponding 90 percent confidence interval is $\left[\frac{1}{2.54} \times \frac{50}{60}, 2.85 \times \frac{50}{60}\right] = [0.328, 2.375]$.

10.23. A random sample of six values of a random variable X is:

$$32, \quad 40, \quad 25, \quad 31, \quad 24, \quad 28$$

and independently obtained eight sample values of a random variable Y are:

$$15, \quad 14, \quad 18, \quad 12, \quad 20, \quad 16, \quad 17, \quad 16$$

Assuming that X and Y are normally distributed, find a 90 percent confidence interval for σ_X^2/σ_Y^2 and one for σ_X/σ_Y.

We follow Prescription 10.7. The sample values are $\bar{x} = \frac{32+40+25+31+24+28}{6} = \frac{180}{6} = 30$, $\bar{y} = \frac{15+14+18+12+20+16+17+16}{8} = \frac{128}{8} = 16$, $s_X^2 = \frac{1}{5}\sum(x_i - 30)^2 = 34$, and $s_Y^2 = \frac{1}{7}\sum(y_i - 15)^2 = 6$. Also, $\frac{1+\gamma}{1} = \frac{1.90}{2} = 0.95$. From Table A-5, we find that $P(F(5,7) \le F_1^*) = 0.95$ for $F_1^* = 3.97$, $P(F(7,5) \le F_2^*) = 0.95$ for $F_2^* = 4.88$. The corresponding 90 percent confidence interval for σ_X^2/σ_Y^2 is $\left[\frac{1}{3.97} \times \frac{34}{6}, 4.88 \times \frac{34}{6}\right] = [1.43, 27.65]$. The corresponding 90 percent confidence interval for σ_X/σ_Y is $[\sqrt{1.43}, \sqrt{27.65}] = [1.20, 5.26]$.

10.24. A statistician reports that $[0.250, 1.265]$ is a 98 percent confidence interval for σ_X^2/σ_Y^2 based on a random sample of $m = 41$ values from a normal distribution X and $n = 31$ values from a normal distribution Y, independent of X. What is the sample value of S_X^2/S_Y^2?

We follow Prescription 10.7, where $m - 1 = 40$, $n - 1 = 30$, $\frac{1+\gamma}{2} = \frac{1.98}{2} = 0.99$. The 98 percent confidence interval is $\left[\frac{1}{F_1^*} \times \frac{s_X^2}{s_Y^2}, F_2^* \times \frac{s_X^2}{s_Y^2}\right]$, where F_1^* and F_2^* satisfy $P(F(40,30) \le F_1^*) = 0.99$ and $P(F(30,40) \le F_2^*) = 0.99$. From Table A-7, we find that $F_1^* = 2.30$ and $F_2^* = 2.20$. Therefore, $\frac{s_X^2}{s_Y^2}$ must satisfy the equations $\frac{1}{2.30} \times \frac{s_X^2}{s_Y^2} = 0.250$ and $2.20 \times \frac{s_X^2}{s_Y^2} = 1.265$. From the first equation, we get $\frac{s_X^2}{s_Y^2} = 2.30 \times 0.250 = 0.575$. Checking this value in the second equation, we do get $2.20 \times 0.575 = 1.265$. Hence, the ratio of sample variances is 0.575.

10.25. Suppose X and Y are independent normal random variables with means μ_X and μ_Y, and variances σ_X^2 and σ_Y^2, respectively; and let $X_1, X_2, \ldots, X_m$ and $Y_1, Y_2, \ldots, Y_n$ be independent random-variable samples corresponding to X and Y. Show that

$$\frac{\sigma_X^2 \times \frac{1}{n}\sum(Y_i - \mu_Y)^2}{\sigma_Y^2 \times \frac{1}{m}\sum(X_i - \mu_X)^2}$$

is an $F(n, m)$ random variable.

The random variables $\dfrac{X_i - \mu_X}{\sigma_X}$ and $\dfrac{Y_i - \mu_Y}{\sigma_Y}$ are independent standard normal random variables. By definition, $\sum\left(\dfrac{X_i - \mu_X}{\sigma_X}\right)^2$ is a chi-square random variable with m degrees of freedom, and $\sum\left(\dfrac{Y_i - \mu_Y}{\sigma_Y}\right)^2$ is a chi-square random variable with n degrees of freedom (see Section 7.4). Therefore,

$$\frac{\dfrac{1}{n}\sum\left(\dfrac{Y_i - \mu_Y}{\sigma_Y}\right)^2}{\dfrac{1}{m}\sum\left(\dfrac{X_i - \mu_X}{\sigma_X}\right)^2}$$

is an $F(n, m)$ random variable, as defined in Section 10.5. By multiplying numerator and denominator of this fraction by $\sigma_X^2\sigma_Y^2$, the result is

$$\frac{\sigma_X^2 \times \dfrac{1}{n}\sum(Y_i - \mu_Y)^2}{\sigma_Y^2 \times \dfrac{1}{m}\sum(X_i - \mu_X)^2}$$

10.26. Let X and Y be independent random variables with known means μ_X and μ_Y, respectively. Give a prescription for a 100γ percent confidence interval for σ_X^2/σ_Y^2.

By following the reasoning leading to Prescription 10.7, with $\dfrac{1}{m}\sum(X_i - \mu_X)^2$ in place of S_X^2, and $\dfrac{1}{n}\sum(Y_j - \mu_Y)^2$ in place of S_Y^2, we arrive at the 100γ percent confidence interval

$$\left[\frac{1}{F_1^*} \times \frac{\dfrac{1}{m}\sum(X_i - \mu_X)^2}{\dfrac{1}{n}\sum(Y_i - \mu_Y)^2},\quad F_2^* \times \frac{\dfrac{1}{m}\sum(X_i - \mu_X)^2}{\dfrac{1}{n}\sum(Y_i - \mu_Y)^2}\right]$$

for σ_X^2/σ_Y^2, where F_1^* and F_2^* satisfy $P(F(m, n) \le F_1^*) = \dfrac{1+\gamma}{2}$, and $P(F(n, m) \le F_2^*) = \dfrac{1+\gamma}{2}$.

10.27. Use the result of Problem 10.26 to find a 90 percent confidence interval for σ_X^2/σ_Y^2 in Problem 10.23, assuming that $\mu_X = 30$ and $\mu_Y = 16$.

The sample values are $\dfrac{1}{6}\sum(x_i - 30)^2 \approx 28.33$ and $\dfrac{1}{8}\sum(y_i - 16)^2 = 5.25$. Also, $\dfrac{1+\gamma}{2} = \dfrac{1.90}{2} = 0.95$. From Table A-5, we find that $P(F(6, 8) \le F_1^*) = 0.95$ for $F_1^* = 3.58$, and $P(F(8, 6) \le F_2^*) = 0.95$ for $F_2^* = 4.15$. Using the result of Problem 10.26, the corresponding 90 percent confidence interval for σ_X^2/σ_Y^2 is $\left[\dfrac{1}{3.58} \times \dfrac{28.33}{5.25}, 4.15 \times \dfrac{28.33}{5.25}\right] = [1.51, 22.39]$. Hence, knowing the values of μ_X and μ_Y, which enables us to increase by one the number of degrees of freedom in the numerator and denominator of the F random variable, results in a smaller confidence interval than the one obtained in Problem 10.23.

HYPOTHESES TESTS FOR RATIOS OF VARIANCES

10.28. In Problem 10.22, the sample variance of 11 one-liter bottles of wine bottled in summer was $s_X^2 = 50\,(\text{ml})^2$ and the sample variance of 16 one-liter bottles of wine bottled in winter was $s_Y^2 = 60\,(\text{ml})^2$. Assuming that the volume of liquid in the bottles is normally distributed, test the null hypothesis H_0: $\sigma_X^2 = \sigma_Y^2$ against the alternative hypothesis H_a: $\sigma_X^2 < \sigma_Y^2$ at the 0.05 significance level by finding the critical region for rejecting the null hypothesis.

We follow Prescription 10.8*a*, where $m - 1 = 10$, $n - 1 = 15$, and $1 - \alpha = 0.95$. The critical region is all values $s_Y^2/s_X^2 \geq F^*$, where F^* is the F value satisfying $P(F(15, 10) \leq F^*) = 0.95$. From Table A-5, we find that $F^* = 2.85$. The test value of s_Y^2/s_X^2 is $60/50 = 1.2$. Therefore, we do not reject H_0 at the 0.05 significance level.

10.29. Find the P-value for the test in Problem 10.28.

We follow Prescription 10.8, where $m - 1 = 10$, $n - 1 = 15$, and H_a: $\sigma_X^2 < \sigma_Y^2$. The P-value for the test is the probability that a value of s_X^2/s_Y^2 as small or smaller than $50/60 = 0.83$ would occur if the null hypothesis H_0: $\sigma_X^2 = \sigma_Y^2$ were true; that is, $P(F(10, 15) \leq 0.83)$. Table A-4 in the Appendix can tell us only that this probability is greater than 0.1. Using computer software, we find that $P(F(10, 15) \leq 0.83) = 0.39$.

10.30. For the data in Problem 10.23, test the null hypothesis H_0: $\sigma_X^2 = \sigma_Y^2$ against the alternative hypothesis H_a: $\sigma_X^2 > \sigma_Y^2$ at the 0.05 significance level by finding the critical region for the alternative hypothesis.

We follow Prescription 10.8*a*, where $m - 1 = 5$, $n - 1 = 7$, and $1 - \alpha = 0.95$. The critical region is all values $s_X^2/s_Y^2 \geq F^*$, where F^* is the F value satisfying $P(F(5, 7) \leq F^*) = 0.95$. From Table A-5 in the Appendix, we find that $F^* = 3.97$. The test value of s_X^2/s_Y^2 is $34/6 = 5.67$. Since $5.67 > 3.97$, we reject H_0 at the 0.05 significance level.

10.31. With reference to the previous problem, would H_0: $\sigma_X^2 = \sigma_Y^2$ be rejected in favor of H_a: $\sigma_X^2 > \sigma_Y^2$ at the 0.01 significance level.

The critical region at the 0.01 significance level is all values $s_X^2/s_Y^2 \geq F^*$, where F^* is the F value satisfying $P(F(5, 7) \leq F^*) = 0.99$. From Table A-7, we find that $F^* = 7.46$. Since the test value 5.67 is less than 7.46, we would not reject H_0 in favor of H_a at the 0.01 significance level.

10.32. Determine the P-value for the test in Problem 10.30.

We follow Prescription 10.8, where $m - 1 = 5$, $n - 1 = 7$, and H_a: $\sigma_X^2 > \sigma_Y^2$. The P-value for the test is the probability that a value of s_X^2/s_Y^2 as large or larger than $34/6 = 5.67$ would occur if the null hypothesis H_0: $\sigma_X^2 = \sigma_Y^2$ were true; that is, $P(F(5, 7) \geq 5.67)$. Tables A-6 and A-7 in the Appendix can tell us only that this probability is less than 0.025 and greater than 0.01. Using computer software, we find that $P(F(5, 7) \geq 5.67) = 0.02$.

10.33. In Problem 10.10, we tested H_0: $\mu_X = \mu_Y$ against H_a: $\mu_X > \mu_Y$ under the assumption that $\sigma_X = \sigma_Y$. Test this assumption at the 0.05 significance level. That is, test H_0: $\sigma_X^2 = \sigma_Y^2$ against H_a: $\sigma_X^2 \neq \sigma_Y^2$, using the data of Problem 10.10.

We follow Prescription 10.8*a*, where $m - 1 = 7$, $n - 1 = 5$, and $1 - \alpha/2 = 0.975$. The critical region is all values $s_Y^2/s_X^2 \geq F_1^*$ or $s_X^2/s_Y^2 \geq F_2^*$, where F_1^* is the F value satisfying $P(F(5, 7) \leq F_1^*) = 0.975$, and F_2^* is the F value satisfying $P(F(7, 5) \leq F_2^*) = 0.975$. From Table A-6, we find that $F_1^* = 5.29$ and $F_2^* = 6.85$. The test value of $s_Y^2/s_X^2 = 19.07/54.87 = 0.35$, and the test value of $s_X^2/s_Y^2 = 54.87/19.07 = 2.88$. Since neither test value is in the corresponding portion of the critical region, we do not reject H_0 at the 0.05 significance level.

Supplementary Problems

CONFIDENCE INTERVALS FOR THE DIFFERENCE OF MEANS

10.34. Use the data in Example 10.1 to find a 95 percent confidence interval for $\mu_X - \mu_Y$.

10.35. Use the data in Problem 10.1 to find a 90 percent confidence interval for $\mu_X - \mu_Y$.

10.36. Given that X and Y are independent normally distributed random variables with equal but unknown variances, find a 98 percent confidence interval for $\mu_X - \mu_Y$, based on the independently obtained random samples

$$x_1 = 12.5, \quad x_2 = 14.2, \quad x_3 = 10.8, \quad x_4 = 11.5, \quad x_5 = 10.1, \quad x_6 = 12.9$$

and

$$y_1 = 10.2, \quad y_2 = 10.5, \quad y_3 = 11.4, \quad y_4 = 9.8, \quad y_5 = 12.1$$

10.37. Suppose the random variables X and Y in Problem 10.36 do not necessarily have equal variances. Then the statistic τ defined in Section 10.1 can be used to determine a confidence interval for $\mu_X - \mu_Y$. τ is approximately a t random variable with $[k]$ degrees of freedom, where k is defined along with τ in Section 10.1, and $[k]$ is the largest integer less than or equal to k. Find k and $[k]$ for the data in Problem 10.36.

10.38. Use the result of Problem 10.37 to find a 98 percent confidence interval for $\mu_X - \mu_Y$.

10.39. A random sample of 50 values of a normal random variable X gave sample values $\bar{x} = 114.8$, $s_X^2 = 70.4$; and an independently obtained random sample of 60 values of a normal random variable Y gave sample values $\bar{y} = 110.6$, $s_Y^2 = 48.2$. Assuming X and Y are independent, find a 95 percent confidence interval for $\mu_X - \mu_Y$. Use the statistic τ defined in Section 10.1, and assume that τ is approximately normally distributed.

10.40. Why was the statistic τ defined in Section 10.1 assumed to be an approximate t random variable in Problem 10.37 and an approximate normal random variable in Problem 10.39?

HYPOTHESES TESTS FOR DIFFERENCES OF MEANS

10.41. X and Y are independent random variables with variances $\sigma_X^2 = 125$, $\sigma_Y^2 = 150$. Independently obtained random samples of 35 values of X and 40 values of Y have sample means $\bar{x} = 102.8$, $\bar{y} = 98.1$. Test the null hypothesis H_0: $\mu_X = \mu_Y$ against the alternative hypothesis H_a: $\mu_X \neq \mu_Y$ at the 0.05 significance level by computing the P-value of the test.

10.42. Perform the test in Problem 10.41 by determining the critical region for the test.

10.43. Using the data in Problem 10.41, test the null hypothesis H_0: $\mu_X = \mu_Y$ against the alternative hypothesis H_a: $\mu_X > \mu_Y$ at the 0.05 significance level by computing the P-value of the test.

10.44. Using the data in Problem 10.41, test the null hypothesis H_0: $\mu_X = \mu_Y$ against the alternative hypothesis H_a: $\mu_X > \mu_Y$ at the 0.05 significance level by determining the critical region of the test.

10.45. In Problem 10.36, test the null hypothesis H_0: $\mu_X = \mu_Y$ against the alternative hypothesis H_a: $\mu_X \neq \mu_Y$ at the 0.10 significance level by determining the P-value of the test, assuming computer software is available.

10.46. In Problem 10.36, test the null hypothesis H_0: $\mu_X = \mu_Y$ against the alternative hypothesis H_a: $\mu_X \neq \mu_Y$ at the 0.10 significance level by determining the critical region for the test.

10.47. In Problem 10.36, test the null hypothesis H_0: $\mu_X = \mu_Y$ against the alternative hypothesis H_a: $\mu_X > \mu_Y$ at the 0.10 significance level by determining the P-value of the test, assuming computer software is available.

10.48. In Problem 10.36, test the null hypothesis H_0: $\mu_X = \mu_Y$ against the alternative hypothesis H_a: $\mu_X > \mu_Y$ at the 0.10 significance level by determining the critical region for the test.

10.49. With reference to Problems 10.41 and 10.42, find $K(7)$; that is the power of the test at $\mu_{X-Y} = \mu_X - \mu_Y = 7$.

10.50. With reference to Problems 10.43 and 10.44, find $K(7)$.

CONFIDENCE INTERVALS FOR DIFFERENCES OF PROPORTIONS

10.51. Suppose sample proportions $\hat{p}_1 = 0.58$ and $\hat{p}_2 = 0.52$ are obtained in independent random samples of size 36 and 44, respectively. Find a 98 percent confidence interval for the difference $p_1 - p_2$ of the corresponding population proportions.

10.52. When asked if they believed a woman would be elected president in the next 20 years, 22 out of 40 randomly selected men said yes, and an independent survey, 33 out of 48 randomly selected women said yes. Let p_1 and p_2 denote the proportions of all men and women, respectively, that believe a woman will be elected president in the next 20 years. Find a 95 percent confidence interval for $p_1 - p_2$.

10.53. Use the data in Problem 10.52 to find 90 percent and 98 percent confidence intervals for $p_1 - p_2$, and compare these with the 95 percent confidence interval obtained in Problem 10.52.

10.54. It is desired to obtain a margin of error of at most 0.02 in a confidence interval for the difference $p_1 - p_2$ of population proportions at the 0.95 confidence level, based on two independent random samples, each of size n. How large must n be? (Hint: Use the inequality $\hat{p}_1(1 - \hat{p}_1) + \hat{p}_2(1 - \hat{p}_2) \leq 0.5$.)

10.55. Suppose $[-0.25, 0.25]$ is a confidence interval for $p_1 - p_2$, based on independent random samples, each of size 36. If $\hat{p}_1 = 0.62$, find $\hat{p}_2$ and the confidence level of the interval?

HYPOTHESES TESTS FOR DIFFERENCES OF PROPORTIONS

10.56. Using the data in Problem 10.51, test the null hypothesis H_0: $p_1 = p_2$ against the alternative hypothesis H_a: $p_1 > p_2$ at the 0.1 significance level by computing the P-value of the test.

10.57. Perform the test in Problem 10.56 by determining the critical region for the test.

10.58. Using the data in Problem 10.52, test the null hypothesis H_0: $p_1 = p_2$ against the alternative hypothesis H_a: $p_1 < p_2$ at the 0.1 significance level by computing the P-value of the test.

10.59. Perform the test in Problem 10.58 by determining the critical region for the test.

10.60. Using the data in Problem 10.52, test the null hypothesis H_0: $p_1 = p_2$ against the alternative hypothesis H_a: $p_1 \neq p_2$ at the 0.1 significance level by computing the P-value of the test.

10.61. Perform the test in Problem 10.60 by determining the critical region for the test.

CONFIDENCE INTERVALS FOR RATIOS OF VARIANCES

10.62. Find the mean and standard deviation of the random variable $F(16, 20)$.

10.63. Find a and b for which $P(F(10, 12) \leq a) = 0.05$ and $P(F(10, 12) \leq b) = 0.95$.

10.64. Find positive numbers a and b for which $P(a \leq F(15,8) \leq b) = 0.95$.

10.65. A random sample of size 31, drawn from a normal population X, has sample variance $S_X^2 = 42.25$; and an independently drawn random sample size 41, drawn from a normal population Y, has sample variance $S_Y^2 = 23.04$. Assuming X and Y are independent; find a 95 percent confidence interval for σ_X^2/σ_Y^2.

10.66. A random sample of five values of a normal random variable X is: 10, 12, 18, 27, 13; and a random sample of 6 values of a normal random variable Y is: 23, 24, 31, 26, 28, 30. Find a 90 percent confidence interval for σ_X/σ_Y.

HYPOTHESES TESTS FOR RATIOS OF VARIANCES

10.67. Using the data in Problem 10.65, test the null hypothesis H_0: $\sigma_X^2 = \sigma_Y^2$ against the alternative hypothesis H_a: $\sigma_X^2 > \sigma_Y^2$ at the 0.05 significance level by finding the P-value of the test, assuming computer software is available.

10.68. Perform the test in Problem 10.67 by determining the critical region for the test.

10.69. Using the data in Problem 10.65, test the null hypothesis H_0: $\sigma_X^2 = \sigma_Y^2$ against the alternative hypothesis H_a: $\sigma_X^2 \neq \sigma_Y^2$ at the 0.05 significance level by finding the P-value of the test, assuming computer software is available.

10.70. Perform the test in Problem 10.69 by finding the critical region for the test.

10.71. Using the data in Problem 10.66, test the null hypothesis H_0: $\sigma_X^2 = \sigma_Y^2$ against the alternative hypothesis H_a: $\sigma_X^2 > \sigma_Y^2$ at the 0.1 significance level by finding the P-value of the test, assuming computer software is available.

10.72. Perform the test in Problem 10.71 by finding the critical region for the test.

10.73. Using the data in Problem 10.66, test the null hypothesis H_0: $\sigma_X^2 = \sigma_Y^2$ against the alternative hypothesis H_a: $\sigma_X^2 \neq \sigma_Y^2$ at the 0.1 significance level by finding the P-value of the test, assuming computer software is available.

10.74. Perform the test in Problem 10.73 by finding the critical region for the test.

Answers to Supplementary Problems

10.34. [−1.645, 3.195].

10.35. [−1.256, 2.806].

10.36. [−0.984, 3.384].

10.37. $k = 8.462$, $[k] = 8$.

10.38. [−0.948, 3.348].

10.39. [1.285, 7.115].

10.40. The sample sizes are small (6 and 5) in Problem 10.37, and they are large (50 and 60) in Problem 10.39.

10.41. P-value $= 0.08$; do not reject H_0.

10.42. Critical region: $|z| \geq 1.96$; test value: $z = 1.74$; do not reject H_0.

10.43. P-value $= 0.04$; reject H_0.

10.44. Critical region: $z \geq 1.65$; test value: $z = 1.74$; reject H_0.

10.45. P-value $= 0.16$; do not reject H_0.

10.46. Critical region: $|\hat{t}| \geq 1.83$; test value: $\hat{t} = 1.55$; do not reject H_0.

10.47. P-value $= 0.08$; reject H_0.

10.48. Critical region: $\hat{t} \geq 1.38$; test value: $\hat{t} = 1.55$; reject H_0.

10.49. 0.74.

10.50. 0.83.

10.51. [−0.20, 0.32].

10.52. [−0.340, 0.065].

10.53. 90 percent [−0.307, 0.032]; 98 percent [−0.378, 0.103]; the 90 percent confidence interval is contained in the 95 percent confidence interval (Problem 10.52), which is contained in the 98 percent confidence interval, or, the higher the degree of confidence wanted, the larger the interval must be.

10.54. $n \geq 4802$.

10.55. $\hat{p}_2 = 0.62$; 97.11 percent.

10.56. P-value $= 0.29$; do not reject H_0.

10.57. Critical region: $z \geq 1.28$; test value: $z = 0.54$; do not reject H_0.

10.58. P-value $= 0.09$; reject H_0.

10.59. Critical region: $z \leq -12.8$; test value: $z = -1.33$; reject H_0.

10.60. P-value $= 0.18$; do not reject H_0.

10.61. Critical region: $|z| \geq 1.65$; test value: $z = -1.33$; do not reject H_0.

10.62. $\mu_F \approx 1.11$, $\sigma_F \approx 0.57$.

10.63. $a = 1/2.91$, $b = 2.75$.

10.64. $a = 1/3.20$, $b = 4.10$.

10.65. [0.95, 3.69].

10.66. [0.93, 5.29].

10.67. P-value = 0.0366; reject H_0.

10.68. Critical region: $s_X^2/s_Y^2 \geq 1.74$; test value: $s_X^2/s_Y^2 = 1.83$; reject H_0.

10.69. P-value = 0.0733; do not reject H_0.

10.70. Critical region: $s_X^2/s_Y^2 \geq 1.94$ or $s_Y^2/s_X^2 \geq 2.01$; test values: $s_X^2/s_Y^2 = 1.83$, $s_Y^2/s_X^2 = 0.55$; do not reject H_0.

10.71. P-value = 0.0659; reject H_0.

10.72. Critical region: $s_X^2/s_Y^2 \geq 3.52$; test value: $s_X^2/s_Y^2 = 4.47$; reject H_0.

10.73. P-value = 0.1318; do not reject H_0.

10.74. Critical region: $s_X^2/s_Y^2 \geq 5.19$ or $s_Y^2/s_X^2 \geq 6.26$; test values: $s_X^2/s_Y^2 = 4.47$, $s_Y^2/s_X^2 = 0.22$; do not reject H_0.

Chapter 11

Chi-Square Tests and Analysis of Variance

11.1 CHI-SQUARE GOODNESS-OF-FIT TEST

The chi-square distribution can be used to determine how well experimental data match expected values in a probability model. For example, if we toss a fair coin 10 times, the expected number of heads is 5, as is the expected number of tails. However, we could get more or fewer heads than tails in 10 tosses. It would not be very surprising to get 6 heads and 4 tails (probability ≈ 0.2). It is even possible to get 10 heads with a fair coin, but the probability of 10 straight heads is only about 0.001, so if this actually happened, we might begin to doubt that the coin is fair. We expect some variation in the experimental data due to chance, but a great deal of variation from the expected number of heads and tails would make us suspect that the fair-coin model is not very accurate. Where do we draw the line? The *chi-square test*, which is based on the following theorem, addresses this question. The test provides a technical tool for comparing the expected outcomes of an experiment with the actual outcomes that occur.

Theorem 11.1: Let $a_1, a_2, \ldots, a_k$ be the possible outcomes of an experiment, with corresponding probabilities $p_1, p_2, \ldots, p_k$. For each performance of n independent trials of the experiment, np_j is the expected number of occurrences of a_j; suppose f_j is the actual number of occurrences of a_j, where $f_1 + f_2 + \cdots + f_k = n$. Then for large values of n, say $np_j \geq 5$ for each j, the random variable

$$\chi^2 = \frac{(f_1 - np_1)^2}{np_1} + \frac{(f_2 - np_2)^2}{np_2} + \cdots + \frac{(f_k - np_k)^2}{np_k}$$

is approximately chi-square with $k - 1$ degrees of freedom.

Null Hypothesis and Test Statistic

In applications of Theorem 11.1, the probabilities $p_1, p_2, \ldots, p_k$ are not known, but a probability model of their values is conjectured. The null hypothesis is

$$H_0\colon P(a_1) = p_1, \qquad P(a_2) = p_2, \qquad \ldots, \qquad P(a_k) = p_k$$

Experimental data are then gathered and a value $\hat{\chi}^2$ of the test statistic χ^2 is computed. If $\hat{\chi}^2 = 0$, then the experimental data exactly match the conjectured expected values. In general, the smaller $\hat{\chi}^2$ is, the more support there is for the null hypothesis; the larger $\hat{\chi}^2$ is, the less support there is for the null hypothesis.

Multinomial Random Variable

A more technical statement of the chi-square test involves the notion of a multinomial random variable which is a generalization of a binomial random variable. In the binomial case, n independent trials of an experiment having two possible outcomes, success and failure, are performed. The trials are called independent because the probability of success is the same for each trial. The binomial random variable X is the number of successes in the n trials. Note that the random variable $Y = n - X$ is the number of failures in the n trials. In the multinomial case, n independent trials of an experiment having k possible outcomes, $a_1, a_2, \ldots, a_k$, are performed. The trials are independent because the probability of a_j, $j = 1, 2, \ldots, k$, is the same for each trial. The random variable X_j is the number of times a_j occurs in the n trials; and the k random variables $X_1, X_2, \ldots, X_k$, taken collectively, are called a *multinomial random variable*, denoted simply by X. Note that these random variables are not independent since $X_k = n - (X_1 + X_2 + \cdots + X_{k-1})$. The frequency f_j referred to in Theorem 11.1 is the value of X_j obtained in n trials. In terms of the multinomial random variable X, the null hypothesis H_0 in the chi-square goodness-of-fit test states that the data are a random sample of outcomes for X, while the alternative hypothesis H_a states the data are not a random sample of outcomes for X.

Performing the Test: *P*-value and Critical Region

An experiment consisting of n independent trials is performed, and the frequencies $f_1, f_2, \ldots, f_k$ of outcomes $a_1, a_2, \ldots, a_k$ are determined, where $f_1 + f_2 + \cdots + f_k = n$. Using these frequencies, a test value $\hat{\chi}^2$ of the above χ^2 is computed. Then the P-value of the test is the probability that a test value as large or larger than $\hat{\chi}^2$ would occur if H_0 were true. That is, the P-value is $P(\chi^2 \geq \hat{\chi}^2)$, assuming $k - 1$ degrees of freedom. If a level of significance α is specified, then H_0 is rejected if P-value $\leq \alpha$; H_0 is not rejected if P-value $> \alpha$. Equivalently, the critical region for the test consists of all values of χ^2 that are greater than or equal to χ^*, where χ^* is the critical value satisfying $P(\chi^2 \geq \chi^*) = \alpha$ (see Fig. 11-1); H_0 is rejected if $\hat{\chi}^2$ is in the critical region; H_0 is not rejected if $\hat{\chi}^2$ is not in the critical region.

Note that the alternative hypothesis H_a: $P(a_j) \neq p_j$ is multidirectional in terms of the k probabilities $P(a_j)$, $j = 1, 2, \ldots, k$. However, the test is one-sided in the chi-square random variable since the alternative hypothesis is equivalent to the hypothesis $\chi^2 \geq \chi^*$.

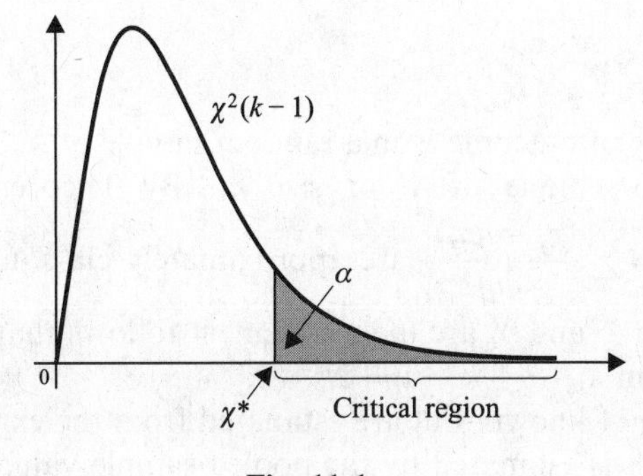

Fig. 11-1

EXAMPLE 11.1 A die is tossed 60 times, and the frequency of each face is as indicated in the chart:

Face (a_j)	1	2	3	4	5	6
Frequency (f_j)	5	7	5	14	13	16

Assume that the die is fair, and apply the chi-square goodness-of-fit test at the 0.05 level of significance.

If the die is fair, then $p_j = \frac{1}{6}$, and $np_j = 60 \cdot \frac{1}{6} = 10$ for $j = 1, 2, \ldots, 6$. The test value is

$$\hat{\chi}^2 = \frac{(5-10)^2}{10} + \frac{(7-10)^2}{10} + \frac{(5-10)^2}{10} + \frac{(14-10)^2}{10} + \frac{(13-10)^2}{10} + \frac{(16-10)^2}{10} = 12$$

There are $6 - 1 = 5$ degrees of freedom. The P-value is $P(\chi^2 \geq 12) = 0.0348$, using computer software. Since 0.0348 is less than 0.05, the hypothesis that the die is fair is rejected at the 0.05 level of significance. If computer software is not available to compute $P(\chi^2 \geq 12)$, Table A-3 in the Appendix can be used to determine that the one-sided critical region for 5 degrees of freedom at the 0.05 level of significance is all values of χ^2 that are greater than

$\chi^* = 11.1$. Since 12 is greater than 11.1, 12 is in the critical region, and the hypothesis that the die is fair is rejected at the 0.05 level of significance.

11.2 CHI-SQUARE TEST FOR EQUAL DISTRIBUTIONS

In the previous section, we used a chi-square random variable to test whether experimental data conformed to a hypothesized probability distribution. The chi-square random variable can also be used to test whether two or more independent multinomial random variables with the same outcomes have the same probability distributions. For example, suppose that a group of subjects is randomly broken up into two categories before the flu season; each person in one category will receive a Type-1 flu shot, and each person in the other category will receive a Type-2 flu shot. The possible outcomes for each category are: no flu, a mild case of flu, and a severe case of flu. The hypothesis that each type of shot has the same effect is tested by constructing a chi-square random variable in terms of the expected frequency of each outcome and the observed frequency of each outcome.

Null Hypothesis

More generally, suppose X and Y are independent multinomial random variables, each with outcomes $a_1, a_2, \ldots, a_k$. Let p_j be the probability of outcome a_j in the distribution of X, and let q_j be the probability of outcome a_j in the distribution of Y, for $j = 1, 2, \ldots, k$. Note that q_j is not necessarily equal to $1 - p_j$. The null hypothesis is

$$H_0\colon p_j = q_j \qquad \text{for} \qquad j = 1, 2, \ldots, k$$

Test Statistic

Suppose f_j is the frequency of outcome a_j in a random sample of X of size m, and g_j is the frequency of outcome a_j in a random sample of Y of size n. By Theorem 11.1, the random variables $\chi_X^2 = \sum \frac{(f_j - mp_j)^2}{mp_j}$ and $\chi_Y^2 = \sum \frac{(g_j - nq_j)^2}{nq_j}$ are approximately chi-square, each with $k - 1$ degrees of freedom. Furthermore, since X and Y are independent, it follows that $\chi_X^2 + \chi_Y^2$ is approximately chi-square with degrees of freedom $k - 1 + k - 1 = 2k - 2$.

In practice, p_j and q_j are not known, but are estimated from the experimental data. Since the null hypothesis is that $p_j = q_j$, each is estimated by the pooled sample value

$$\hat{p}_j = \frac{f_j + g_j}{m + n}$$

for $j = 1, 2, \ldots, k$. Figure 11-2 illustrates the estimated probabilities when $k = 5$.

When p_j and q_j are replaced by $\hat{p}_j$ in the random variable $\chi_X^2 + \chi_Y^2$, $k - 1$ degrees of freedom are lost, so the random variable

$$\chi^2 = \sum \frac{(f_j - m\hat{p}_j)^2}{m\hat{p}_j} + \sum \frac{(g_j - n\hat{p}_j)^2}{n\hat{p}_j}$$

is approximately chi-square with $2k - 2 - (k - 1) = k - 1$ degrees of freedom; χ^2 is the test statistic for the test for equality.

Performing the Test: *P*-value and Critical Region

A random sample of m values of X results in frequencies $f_1, f_2, \ldots, f_k$ of outcomes $a_1, a_2, \ldots, a_k$, where $f_1 + f_2 + \cdots + f_k = m$; and an independently obtained random sample of n values of Y results in frequencies $g_1, g_2, \ldots, g_k$ of $a_1, a_2, \ldots, a_k$, where $g_1 + g_2 + \cdots + g_k = n$. Using these frequencies, the estimated probabilities $\hat{p}_j$ are computed, along with the expected frequencies $m\hat{p}_j$ and $n\hat{p}_j$,

a_j	a_1	a_2	a_3	a_4	a_5	Totals
f_j	f_1	f_2	f_3	f_4	f_5	$\Sigma f_j = m$
g_j	g_1	g_2	g_3	g_4	g_5	$\Sigma g_j = n$
Totals	f_1+g_1	f_2+g_2	f_3+g_3	f_4+g_4	f_5+g_5	$m+n$
$\hat{p}_j$	$\frac{f_1+g_1}{m+n}$	$\frac{f_2+g_2}{m+n}$	$\frac{f_3+g_3}{m+n}$	$\frac{f_4+g_4}{m+n}$	$\frac{f_5+g_5}{m+n}$	$\Sigma \hat{p}_j = 1$

Fig. 11-2 Probabilities $\hat{p}_j$ estimated from frequencies f_j and g_j.

$j = 1, 2, \ldots, k$. Then the corresponding value $\hat{\chi}^2$ of χ^2 is determined. The P-value of the test is the probability that a test value as large or larger than $\hat{\chi}^2$ would occur if H_0 were true. That is, the P-value is $P(\chi^2 \geq \hat{\chi}^2)$, assuming $k-1$ degrees of freedom. If a level of significance α is specified, then H_0 is rejected if P-value $\leq \alpha$; H_0 is not rejected if P-value $> \alpha$. Equivalently, the critical region for the test consists of all values of χ^2 that are greater than or equal to χ^*, where χ^* is the critical value satisfying $P(\chi^2 \geq \chi^*) = \alpha$ (see Fig. 11-1); H_0 is rejected if $\hat{\chi}^2$ is in the critical region; H_0 is not rejected if $\hat{\chi}^2$ is not in the critical region. As in the chi-square goodness-of-fit test in Section 11.1, this test is also one-sided in the chi-square random variable; the alternative hypothesis H_a is equivalent to the hypothesis: $\chi^2 \geq \chi^*$.

EXAMPLE 11.2 The freshman math grades of 250 males and 210 females at a university were distributed as indicated in the following table.

		Grades					
		A	B	C	D	F	Totals
Gender	Male	35	42	85	48	40	250
	Female	28	50	77	35	20	210
	Totals	63	92	162	83	60	460

Use the chi-square random variable to test, at the 0.05 significance level, the hypothesis the grade distributions are the same.

By pooling the $m = 250$ male and $n = 210$ female frequencies in each grade category, we obtain the following estimated probabilities:

$$\hat{p}_A = \frac{63}{460}, \quad \hat{p}_B = \frac{92}{460}, \quad \hat{p}_C = \frac{162}{460}, \quad \hat{p}_D = \frac{83}{460}, \quad \hat{p}_F = \frac{60}{460}$$

The expected frequencies $m\hat{p}_j$ for the males are:

$$m\hat{p}_A = 250 \times \frac{63}{460} = 34.24 \qquad m\hat{p}_B = 250 \times \frac{92}{460} = 50 \qquad m\hat{p}_C = 250 \times \frac{162}{460} = 88.04$$

$$m\hat{p}_D = 250 \times \frac{83}{460} = 45.11 \qquad m\hat{p}_F = 250 \times \frac{60}{460} = 32.61$$

and the expected frequencies $n\hat{p}_j$ for the females are:

$$n\hat{p}_A = 210 \times \frac{63}{460} = 28.76 \qquad n\hat{p}_B = 210 \times \frac{92}{460} = 42 \qquad n\hat{p}_C = 210 \times \frac{162}{460} = 73.96$$

$$n\hat{p}_D = 210 \times \frac{83}{460} = 37.89 \qquad n\hat{p}_F = 210 \times \frac{60}{460} = 27.39$$

The corresponding chi-square test value is

$$\hat{\chi}^2 = \frac{(35-34.24)^2}{34.24} + \frac{(42-50)^2}{50} + \frac{(85-88.04)^2}{88.04} + \frac{(48-45.11)^2}{45.11} + \frac{(40-32.61)^2}{32.61}$$
$$+ \frac{(28-28.76)^2}{28.76} + \frac{(50-42)^2}{42} + \frac{(77-73.96)^2}{73.96} + \frac{(35-37.89)^2}{37.89} + \frac{(20-27.39)^2}{27.39}$$
$$\approx 7.14$$

There are $k = 5$ grades and $k - 1 = 4$ degrees of freedom. The critical chi-square region for 4 degrees of freedom at the 0.05 significance level is all values greater than or equal to 9.49. Since $7.14 < 9.49$, we do not reject the hypothesis that the grade distributions for males is the same as that for females.

Extension to More Than Two Distributions

The chi-square test for equality of two multinomial random variables X and Y can be extended to three or more independent multinomial random variables $X_1, X_2, \ldots, X_r$, each having the same number of outcomes $a_1, a_2, \ldots, a_k$. The null hypothesis is

$$H_0\colon p_{1j} = p_{2j} = \cdots = p_{rj}; \quad j = 1, 2, \ldots, k$$

where p_{ij} is the probability of outcome a_j in the distribution of X_i. Let f_{ij} denote the frequency of outcome a_j in n_i trials corresponding to X_i. For each j ($j = 1, 2, \ldots, k$), the common value of p_{ij} ($i = 1, 2, \ldots, r$) in the null hypothesis is estimated by $\hat{p}_j$, obtained by pooling the frequencies f_{ij} ($i = 1, 2, \ldots, r$):

$$\hat{p}_j = \frac{f_{1j} + f_{2j} + \cdots + f_{rj}}{n_1 + n_2 + \cdots + n_r} = \frac{f_{\cdot j}}{n}$$

where $f_{\cdot j} = f_{1j} + f_{2j} + \cdots + f_{rj}$ is the sum of the frequencies corresponding to outcome a_j for all r multinomial random variables, and $n = n_1 + n_2 + \cdots + n_r$ is the total number of trials (see Fig. 11-3). The number of degrees of freedom before the probability estimates are made is $r(k - 1)$, corresponding to r independent multinomial random variables, each one consisting of $k - 1$ independent frequency counts. Only $k - 1$ degrees of freedom are lost by the estimates since $\hat{p}_k = 1 - (\hat{p}_1 + \hat{p}_2 + \cdots + \hat{p}_{k-1})$. Hence, after the estimates are made, there are $r(k-1) - (k-1) = (r-1)(k-1)$ degrees of freedom. See Problem 11.14 for an example of the chi-square test of equality for three multinomial random variables.

	a_1	a_2	$\cdots$	a_k	Totals
X_1	f_{11}	f_{12}	$\cdots$	f_{1k}	n_1
X_2	f_{21}	f_{22}	$\cdots$	f_{2k}	n_2
$\vdots$	$\vdots$	$\vdots$	$\vdots$	$\vdots$	$\vdots$
X_r	f_{r1}	f_{r2}	$\cdots$	f_{rk}	n_r
Totals	$f_{\cdot 1}$	$f_{\cdot 2}$	$\cdots$	$f_{\cdot k}$	n
$\hat{p}_j$	$\frac{f_{\cdot 1}}{n}$	$\frac{f_{\cdot 2}}{n}$	$\cdots$	$\frac{f_{\cdot k}}{n}$	$\Sigma \hat{p}_j = 1$

Fig. 11-3 Probabilities $\hat{p}_j$ estimated from frequencies f_{ij}.

11.3 CHI-SQUARE TEST FOR INDEPENDENT ATTRIBUTES

The chi-square random variable can also be applied in testing whether attributes are independent. For example, suppose n math students are classified according to stress experienced at final-exam time (Attribute X) and grades received in the final exam (Attribute Y). Stress is classified as L (low),

M (medium), and H (high); while grades are classified as A, B, C, D, and F. Cross-classifying the students according to the three categories of attribute X and the five categories of attribute Y results in Fig. 11-4, which contains $3 \times 5 = 15$ cells within the margins. For example, the cell labeled LA contains all students that experience low stress and get an A in the final exam, while the cell labeled HC contains all students that experience high stress and get a C in the final exam. Such a table of cell counts is called a 3×5 *contingency table*.

		Y (Grades)				
		A	*B*	*C*	*D*	*F*
	L	*LA*	*LB*	*LC*	*LD*	*LF*
X (Stress)	*M*	*MA*	*MB*	*MC*	*MD*	*MF*
	H	*HA*	*HB*	*HC*	*HD*	*HF*

Fig. 11-4 Contingency table.

A different group of n students would most likely result in a different contingency table. When all such tables are considered, a probability can be associated with each cell. For example, $P(LA)$ is the probability that a randomly chosen student will experience low stress and get an A in the final exam, while $P(HC)$ is the probability that a student will experience high stress and get a C in the final exam. Similarly, a probability can be associated with each marginal category. For example, $P(L)$ is the probability that a student will experience low stress, and $P(C)$ is the probability that a student will get a C in the final exam.

Attributes X and Y are, by definition, independent if the probability corresponding to each cell of Fig. 11-4 is equal to the product of the probability in the row margin of the cell with the probability in the column margin of the cell. That is, if

$$P(LA) = P(L)P(A), \qquad P(LB) = P(L)P(B), \qquad \ldots, \qquad P(HF) = P(H)P(F)$$

fifteen equations in all. Usually, the actual values of the probabilities in these equations are not known, but must be estimated from samples. The estimated probabilities most likely will not satisfy the equations exactly, so to test the hypothesis that X and Y are independent, a chi-square random variable χ^2 is constructed in terms of the expected and observed frequencies of the cross categories. The value of χ^2 on a particular random sample of students is then used to test the hypothesis that attributes X and Y are independent.

Contingency Table of Probabilities

To be more specific, and also more general, let X and Y be attributes associated with individuals in a population. Suppose that X can be classified into mutually disjoint categories $A_1, A_2, \ldots, A_r$, and Y can be classified into mutually disjoint categories $B_1, B_2, \ldots, B_c$. The probability $P(A_iB_j)$ that a randomly chosen individual in the population can be classified into both category A_i and category B_j will be denoted by p_{ij}. Figure 11-5 is an $r \times c$ *contingency table of probabilities*, where p_{ij} is in the ith row and jth column, $i = 1, 2, \ldots, r$; $j = 1, 2, \ldots, c$.

The right column margin of Fig. 11-5 contains the probabilities $p_{i\cdot}$, $i = 1, 2, \ldots, r$, where the dot after the subscript i indicates summation through index j. For example, when $i = 1$,

$$p_{1\cdot} = p_{11} + p_{12} + \cdots + p_{1c} = P(A_1)$$

Similarly, the lower row margin contains the probabilities $p_{\cdot j}$, $j = 1, 2, \ldots, c$, where the dot before the j indicates summation through index i. For example, when $j = 2$,

$$p_{\cdot 2} = p_{12} + p_{22} + \cdots + p_{r2} = P(B_2)$$

		Y				
		B_1	B_2	$\cdots$	B_c	
X	A_1	p_{11}	p_{12}	$\cdots$	p_{1c}	$p_{1\cdot}$
	A_2	p_{21}	p_{22}	$\cdots$	p_{2c}	$p_{2\cdot}$
	$\vdots$	$\vdots$	$\vdots$	$\vdots$	$\vdots$	$\vdots$
	A_r	p_{r1}	p_{r2}	$\cdots$	p_{rc}	$p_{r\cdot}$
		$p_{\cdot 1}$	$p_{\cdot 2}$	$\cdots$	$p_{\cdot c}$	1

Fig. 11-5 Contingency table of probabilities.

The sum of all of the probabilities p_{ij} within the margins is 1, as is the sum of the marginal probabilities $p_{i\cdot}$, as is the sum of the marginal probabilities $p_{\cdot j}$.

Contingency Table of Frequencies

A random sample of n individuals in the population results in an $r \times c$ *contingency table of frequencies*, as illustrated in Fig. 11-6. For example, f_{12} denotes the number of individuals in the cross category A_1B_2, while f_{21} denotes the number of individuals in cross category A_2B_1.

		Y				
		B_1	B_2	$\cdots$	B_c	Totals
X	A_1	f_{11}	f_{12}	$\cdots$	f_{1c}	$f_{1\cdot}$
	A_2	f_{21}	f_{22}	$\cdots$	f_{2c}	$f_{2\cdot}$
	$\vdots$	$\vdots$	$\vdots$	$\vdots$	$\vdots$	$\vdots$
	A_r	f_{r1}	f_{r2}	$\cdots$	f_{rc}	$f_{r\cdot}$
	Totals	$f_{\cdot 1}$	$f_{\cdot 2}$	$\cdots$	$f_{\cdot c}$	n

Fig. 11-6 Contingency table of frequencies.

The marginal frequency at the right of each row in Fig. 11-6 is the sum of the c frequencies preceding it, and the marginal frequency at the bottom of each column is the sum of the r frequencies above it. For example,

$$f_{1\cdot} = f_{11} + f_{12} + \cdots + f_{1c} \qquad \text{and} \qquad f_{\cdot 2} = f_{12} + f_{22} + \cdots + f_{r2}$$

The sum of all of the frequencies f_{ij} within the margins is n, as is the sum of the marginal frequencies $f_{i\cdot}$, as is the sum of the marginal frequencies $f_{\cdot j}$. When all possible samples of n individuals in the population are considered, it follows from Theorem 11.1 that the random variable

$$\sum \frac{(f_{1j} - np_{1j})^2}{np_{1j}} + \sum \frac{(f_{2j} - np_{2j})^2}{np_{2j}} + \cdots + \sum \frac{(f_{rj} - np_{rj})^2}{np_{rj}}$$

is approximately chi-square with $rc - 1$ degrees of freedom, assuming n is large. Note that, in each summation, j runs from 1 to c.

Null Hypothesis and Test Statistic

The null hypothesis is that attributes X and Y are independent. Equivalently,

$$H_0\colon p_{ij} = p_{i\cdot} \times p_{\cdot j}$$

for each pair ij. In practice, the values of $p_{i\cdot}$ and $p_{\cdot j}$ are unknown, but are estimated as

$$\hat{p}_{i\cdot} = \frac{f_{i\cdot}}{n} = \frac{f_{i1} + f_{i2} + \cdots + f_{ic}}{n} \quad \text{and} \quad \hat{p}_{\cdot j} = \frac{f_{\cdot j}}{n} = \frac{f_{1j} + f_{2j} + \cdots + f_{rj}}{n}$$

When $p_{i\cdot}$ and $p_{\cdot j}$ are replaced by their estimates $\hat{p}_{i\cdot}$ and $\hat{p}_{\cdot j}$ in the chi-square random variable above, $r - 1 + c - 1$ degrees of freedom are lost; and the resulting random variable

$$\chi^2 = \sum \frac{(f_{1j} - n \times \hat{p}_{1\cdot} \times \hat{p}_{\cdot j})^2}{n \times \hat{p}_{1\cdot} \times \hat{p}_{\cdot j}} + \sum \frac{(f_{2j} - n \times \hat{p}_{2\cdot} \times \hat{p}_{\cdot j})^2}{n \times \hat{p}_{2\cdot} \times \hat{p}_{\cdot j}} + \cdots + \sum \frac{(f_{rj} - n \times \hat{p}_{r\cdot} \times \hat{p}_{\cdot j})^2}{n \times \hat{p}_{r\cdot} \times \hat{p}_{\cdot j}}$$

is approximately chi-square with $rc - 1 - (r - 1 + c - 1) = (r - 1)(c - 1)$ degrees of freedom, assuming the null hypothesis is true. In each summation, j runs from 1 to c. The random variable χ^2 is the test statistic.

Performing the Test: *P*-value and Critical Region

A random sample of n individuals in the population results in the frequencies f_{ij} of cross categories A_iB_j, where $i = 1, 2, \ldots, r$, $j = 1, 2, \ldots, c$, and $\sum f_{ij} = n$. Using these frequencies, the estimated probabilities $\hat{p}_{i\cdot}$ and $\hat{p}_{\cdot j}$ are computed, followed by the expected frequencies $n \times p_{i\cdot} \times p_{\cdot j}$, and then a value $\hat{\chi}^2$ of the test statistic χ^2 is determined. The *P*-value of the test is the probability that a test value as large or larger than $\hat{\chi}^2$ would occur if H_0 were true. That is, the *P*-value is $P(\chi^2 \geq \hat{\chi}^2)$, assuming $(r - 1)(c - 1)$ degrees of freedom. If a level of significance α is specified, then H_0 is rejected if *P*-value $\leq \alpha$; H_0 is not rejected if *P*-value $> \alpha$. Equivalently, the critical region for the test consists of all values of χ^2 that are greater than or equal to χ^*, where χ^* is the critical value satisfying $P(\chi^2 \geq \chi^*) = \alpha$ (see Fig. 11-1); H_0 is rejected if $\hat{\chi}^2$ is in the critical region; H_0 is not rejected if $\hat{\chi}^2$ is not in the critical region.

As in the chi-square tests in Sections 11.1 and 11.2, this test is also one sided in the chi-square random variable; the alternative hypothesis H_a is equivalent to the hypothesis: $\chi^2 \geq \chi^*$.

EXAMPLE 11.3 Let's consider Example 11.2 from the point of view of independence of attributes rather than equality of distributions. That is, 460 freshmen are cross classified according to gender and grades, as indicated in the table.

		Grades					
		A	B	C	D	F	Totals ($f_{i\cdot}$)
Gender	Male	35	42	85	48	40	250
	Female	28	50	77	35	20	210
	Totals ($f_{\cdot j}$)	63	92	162	83	60	460

Use the chi-square random variable to test, at the 0.05 significance level, the hypothesis that the attributes of gender and grades are independent.

Here $r = 2$ and $c = 5$. The subscripts $i = 1, 2$ correspond to male, female, respectively; and the subscripts $j = 1, 2, 3, 4, 5$, correspond to grades A, B, C, D, F, respectively. The probability estimates for gender are

$$\hat{p}_{1\cdot} = \frac{f_{1\cdot}}{n} = \frac{250}{460}, \quad \hat{p}_{2\cdot} = \frac{f_{2\cdot}}{n} = \frac{210}{460}$$

and those for grades are

$$\hat{p}_{\cdot 1} = \frac{f_{\cdot 1}}{n} = \frac{63}{460}, \quad \hat{p}_{\cdot 2} = \frac{f_{\cdot 2}}{n} = \frac{92}{460}, \quad \hat{p}_{\cdot 3} = \frac{f_{\cdot 3}}{n} = \frac{162}{460},$$

$$\hat{p}_{\cdot 4} = \frac{f_{\cdot 4}}{n} = \frac{83}{460}, \quad \hat{p}_{\cdot 5} = \frac{f_{\cdot 5}}{n} = \frac{60}{460}$$

The expected frequency estimate for males getting As is

$$n \times \hat{p}_{1\cdot} \times \hat{p}_{\cdot 1} = 460 \times \frac{250}{460} \times \frac{63}{460} = 34.24$$

Similarly, the other nine cross-classification frequency estimates are

$$n \times \hat{p}_{1\cdot} \times \hat{p}_{\cdot 2} = 50, \qquad n \times \hat{p}_{1\cdot} \times \hat{p}_{\cdot 3} = 88.04, \qquad n \times \hat{p}_{1\cdot} \times \hat{p}_{\cdot 4} = 45.11, \qquad n \times \hat{p}_{1\cdot} \times \hat{p}_{\cdot 5} = 32.61$$

and

$$n \times \hat{p}_{2\cdot} \times \hat{p}_{\cdot 1} = 28.76, \qquad n \times \hat{p}_{2\cdot} \times \hat{p}_{\cdot 2} = 42, \qquad n \times \hat{p}_{2\cdot} \times \hat{p}_{\cdot 3} = 73.96,$$

$$n \times \hat{p}_{2\cdot} \times \hat{p}_{\cdot 4} = 37.89, \qquad n \times \hat{p}_{2\cdot} \times \hat{p}_{\cdot 5} = 27.39$$

The corresponding value of the chi-square test statistic is

$$\begin{aligned}\hat{\chi}^2 &= \frac{(35-34.24)^2}{34.24} + \frac{(42-50)^2}{50} + \frac{(85-88.04)^2}{88.04} + \frac{(48-45.11)^2}{45.11} + \frac{(40-32.61)^2}{32.61} \\ &\quad + \frac{(28-28.76)^2}{28.76} + \frac{(50-42)^2}{42} + \frac{(77-73.96)^2}{73.96} + \frac{(35-37.89)^2}{37.89} + \frac{(20-27.39)^2}{27.39} \\ &\approx 7.14\end{aligned}$$

There are $(r-1)(c-1) = 1 \times 4 = 4$ degrees of freedom. The critical chi-square region for 4 degrees of freedom at the 0.05 significance level is all values greater than or equal to 9.49. Since $7.14 < 9.49$, we do not reject the hypothesis that the attributes of gender and grades are independent.

Comparing the Chi-Square Tests for Equal Distributions and for Independent Attributes

We see that the value of the chi-square statistic in Example 11.3 is the same as that in Example 11.2. In fact, the chi-square test for equality of independent multinomial distributions always gives the same result as that for independence of cross classified attributes. That is, suppose an $r \times c$ frequency table is given. The table can be interpreted as a table of observed frequencies for r independent multinomial random variables $X_1, X_2, \ldots, X_r$, each with the same c outcomes (Fig. 11-3, where $k = c$), or as a table of observed frequencies for a cross-classification of a collection of r attributes with a collection of c attributes (Fig. 11-6). Let a significance level α be specified. Then, on the basis of the chi-square test, the hypothesis that the r multinomial random variables have the same distributions will be rejected at level α if and only if the hypothesis that the r attributes are independent of the c attributes is rejected at level α (see Problems 11.13, 11.14, and 11.15).

Although the results are the same, the collection of the data is different in the two cases. In the case of the r multinomial random variables, r independent random samples of sizes $n_1, n_2, \ldots, n_r$ are collected, one sample for each random variable; in the attribute case, a single random sample of size n is collected, and the data are then cross-classified into an $r \times c$ contingency table. The two cases are comparable, and give equivalent results, when $n = n_1 + n_2 + \cdots + n_r$.

11.4 ONE-WAY ANALYSIS OF VARIANCE

In Section 10.2 we tested the hypothesis that two independent random variables have the same mean. When their variances were unknown, we further assumed that the random variables were approximately normal and that the variances, although unknown, were equal (see Prescriptions 10.4 and 10.4(a)). Here we test the hypothesis that m independent approximately normal random variables with the same unknown variance have the same mean. In the case of two distributions, a t random variable was used as the test statistic. Here it is more convenient to use an F random variable as the test statistic. Both statistics measure in ratio form the variation between the distributions as being separate in relation to the variation within a single distribution obtained by pooling the data from each one. "Analysis of variance" is the technical expression used to describe this measure of variation. The analysis is "one-way" since only the row random variables are compared.

One-Way Random-Samples Table

Suppose that $X_1, X_2, \ldots, X_m$ are m independent normal random variables with unknown means $\mu_1, \mu_2, \ldots, \mu_m$ and unknown but common variance σ^2. We wish to test the null hypothesis

$$H_0\colon \mu_1 = \mu_2 = \cdots = \mu_m$$

As illustrated in Fig. 11-7, let $X_{i1}, X_{i2}, \ldots, X_{in}$ be a random-variable sample of X_i of size n for $i = 1, 2, \ldots, m$.

Random variables	Random samples				Sample means
X_1	X_{11}	X_{12}	...	X_{1n}	$\bar{X}_1$
X_2	X_{21}	X_{22}	...	X_{2n}	$\bar{X}_2$
$\vdots$	$\vdots$	$\vdots$	$\vdots$	$\vdots$	$\vdots$
X_m	X_{m1}	X_{m2}	...	X_{mn}	$\bar{X}_m$
	Grand sample mean				$\bar{X}$

Fig. 11-7 One-way random-samples table.

The right margin in Fig. 11-7 contains the individual sample means $\bar{X}_i$, where

$$\bar{X}_i = \frac{X_{i1} + X_{i2} + \cdots + X_{in}}{n}, \qquad i = 1, 2, \ldots, m$$

The *grand sample mean* in the lower right corner, obtained by pooling the random samples, is

$$\bar{X} = \frac{\bar{X}_1 + \bar{X}_2 + \cdots + \bar{X}_m}{m}$$

Square Variations

The *total square variation*, V_T, of the samples in Figure 11-7 is the sum of the squares of the deviations of the pooled samples from the grand sample mean:

$$V_T = \sum (X_{1j} - \bar{X})^2 + \sum (X_{2j} - \bar{X})^2 + \cdots + \sum (X_{mj} - \bar{X})^2$$

where in each summation, j runs from 1 to n. The grand sample mean $\bar{X}$ is an estimator of the grand parameter mean $\mu = (\mu_1 + \mu_2 + \cdots + \mu_m)/m$, and $\bar{X}_i - \bar{X}$ is an estimator of $\mu_i - \mu$. If the null hypothesis were true, then $\mu_i - \mu = 0$, which means that $\bar{X}_i - \bar{X}$ can be used as a measure of the disagreement of the data and the null hypothesis. With this in mind, the *square deviation between the row samples*, denoted by V_R, is defined as

$$V_R = n \sum (\bar{X}_i - \bar{X})^2$$

The *square variation due to random error*, denoted by V_e, is defined as

$$V_e = V_T - V_R$$

A formula for V_e as a sum of squares is given in Problem 11.22. Some important properties of these square variations can be summarized in the following theorem (see Problem 11.23).

Theorem 11.2: V_e/σ^2 is chi-square with $mn - m$ degrees of freedom regardless of whether the null hypothesis $H_0\colon \mu_1 = \mu_2 = \cdots = \mu_m$ is true or not. If H_0 is true, then V_R/σ^2 is chi-square with $m - 1$ degrees of freedom, and V_T/σ^2 is chi-square with $mn - 1$ degrees of freedom; and all three chi-square random variables are independent.

The Test Statistic

It follows from Theorem 11.2 and the definition of the F distribution (Section 10.5) that if H_0 is true, then

$$F = \frac{V_R/(m-1)}{V_e/(mn-m)}$$

is an F random variable with $m-1$ and $mn-m$ degrees of freedom. F will be the statistic to test H_0. Motivation for this choice of test statistic is provided by the following theorem (see Problem 11.24).

Theorem 11.3: $E(V_e/(mn-m)) = \sigma^2$, and $E(V_R/(m-1)) = \sigma^2 + \frac{n}{m-1}\sum(\mu_i - \mu)^2$, where $\mu = (\mu_1 + \mu_2 + \cdots + \mu_m)/m$.

Theorem 11.3 says that if the null hypothesis is true (meaning $\mu = \mu_i$ for $i = 1, 2, \ldots, m$), then $E(V_R/(m-1))$ will equal σ^2, and therefore sample values of F should be close to 1; the more the means μ_i differ, the larger sample values of F are likely to be.

One-Way Analysis-of-Variance Table

In applications of the F test for equal means, a one-way analysis-of-variance table is usually constructed as illustrated in Fig. 11-8.

Square variation	Degrees of freedom	Mean square	F
Between row samples V_R	$m-1$	$\frac{V_R}{m-1}$	$\frac{V_R/(m-1)}{V_e/(mn-m)}$
Random error V_e	$mn-m$	$\frac{V_e}{mn-m}$	
Total V_T	$mn-1$		

Fig. 11-8 One-way analysis-of-variance table.

Performing the Test: *P*-value and Critical Region

Random collections of n sample values $x_{i1}, x_{i2}, \ldots, x_{in}$ for each random variable X_i, $i = 1, 2, \ldots, m$, are independently obtained. Using these, values $\bar{x}_i$ of the sample means $\bar{X}_i$ are computed, as well as a value $\bar{x}$ of the grand mean $\bar{X}$. The corresponding values

$$v_T = \sum(x_{1j} - \bar{x})^2 + \sum(x_{2j} - \bar{x})^2 + \cdots + \sum(x_{mj} - \bar{x})^2,$$

$$v_R = n\sum(\bar{x}_i - \bar{x})^2, \qquad \text{and} \qquad v_e = v_T - v_R$$

of V_T, V_R, and V_e are then computed, and finally the value

$$\hat{F} = \frac{v_R/(m-1)}{v_e/(mn-m)}$$

of the test statistic F is determined. The P-value of the test is the probability that a test value as large or larger than $\hat{F}$ would occur if H_0: $\mu_1 = \mu_2 = \cdots = \mu_m$ were true. That is, the P-value is $P(F \geq \hat{F})$, assuming $m-1$ and $mn-m$ degrees of freedom. If a level of significance α is specified, then H_0 is rejected if P-value $\leq \alpha$; H_0 is not rejected if P-value $> \alpha$. Equivalently, the critical region for the test consists of all values of F that are greater than or equal to F^*, where F^* is the critical value satisfying $P(F \geq F^*) = \alpha$ (see Fig. 11-9); H_0 is rejected if $\hat{F}$ is in the critical region; H_0 is not rejected if $\hat{F}$ is not in the critical region.

As in the chi-square tests in Sections 11.1, 11.2, and 11.3, this test is also one sided in the test statistic; the alternative hypothesis H_a is equivalent to the hypothesis $F \geq F^*$.

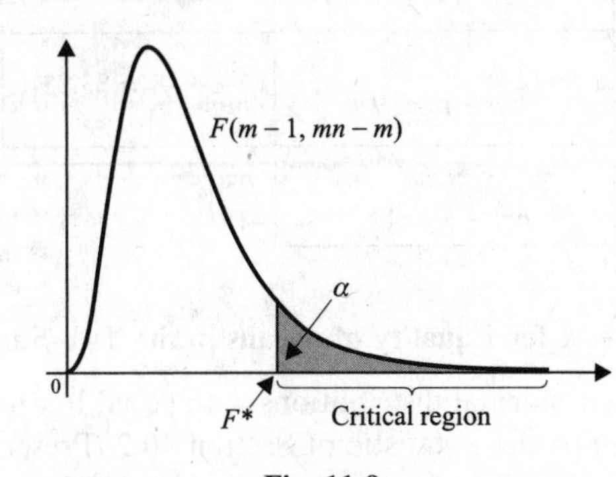

Fig. 11-9

EXAMPLE 11.4 A random sample of size 4 is taken from each of three independent normal random variables, X_1, X_2, X_3, resulting in the following table of sample values.

X_1	13	11	16	22
X_2	16	8	21	11
X_3	15	12	25	10

Assuming that the three random variables have equal variances, test, at the 0.05 significance level, the hypothesis that X_1, X_2, X_3 have the same mean.

The sample means are

$$\bar{x}_1 = \frac{13 + 11 + 16 + 22}{4} = 15.5, \qquad \bar{x}_2 = \frac{16 + 8 + 21 + 11}{4} = 14, \qquad \bar{x}_3 = \frac{16 + 12 + 25 + 10}{4} = 15.5$$

and the grand sample mean is $\bar{x} = \dfrac{15.5 + 14 + 15.5}{3} = 15$. The total square variation is

$$\begin{aligned}\nu_T &= (13-15)^2 + (11-15)^2 + (16-15)^2 + (22-15)^2 + (16-15)^2 + (8-15)^2 + (21-15)^2 \\ &\quad + (11-15)^2 + (15-15)^2 + (12-15)^2 + (25-15)^2 + (10-15)^2 \\ &= 306\end{aligned}$$

The square deviation between row samples is

$$\nu_R = 4[(15.5-15)^2 + (14-15)^2 + (15.5-15)^2] = 6$$

and the square variation due to random error is

$$\nu_e = 306 - 6 = 300$$

To determine the degrees of freedom of the test statistic, we note that $m - 1 = 3 - 1 = 2$, and $mn - m = 3 \times 4 - 3 = 9$. Therefore, the test statistic is

$$F(2, 9) = \frac{6/2}{300/9} = 0.09$$

The critical region at the 0.05 significance level consists of all test values greater than or equal to 4.26. Since $0.09 < 4.26$, the test value is not in the critical region, and we do not reject the null hypothesis that the means of X_1, X_2, X_3 are equal. The corresponding analysis-of-variance table is as follows.

Square variation	Degrees of freedom	Mean square	F
Between row samples $\nu_R = 6$	$m - 1 = 2$	$\frac{\nu_R}{m-1} = 3$	$\frac{\nu_R/(m-1)}{\nu_e/(mn-m)} = 0.09$
Random error $\nu_e = 300$	$mn - m = 9$	$\frac{\nu_e}{mn-m} = 33.33$	
Total $\nu_T = 306$	$mn - 1 = 11$		

Comparing the *t* Test and the *F* test for Equality of Means in the Two-Sample Case

In the case of two independent normal distributions with equal but unknown variances, we can use either the F statistic defined here or the t statistic of Section 10.2 (Prescriptions 4 and 4(a)) to test the hypothesis that the distributions have equal means (see Problem 11.18). The t test and the F test for equal means against H_a: $\mu_1 \neq \mu_2$ will have the same P-value and therefore will give the same result at any significance level. To see that the P-values are equal, we first note that, in the two-sample case, $F = t^2$, where $F = F(1, n-1)$ (see Problem 11.19). The alternative hypothesis H_a: $\mu_1 \neq \mu_2$ is two-sided in the t test, and therefore the P-value is equal to $P(|t| \geq |\hat{t}|)$, where $\hat{t}$ is the sample value of t. The P-value in the F test is $P(F \geq \hat{F})$, where $\hat{F}$ is the test value. Since $|t| = \sqrt{F}$ and $|\hat{t}| = \sqrt{\hat{F}}$, the two P values are equal.

11.5 TWO-WAY ANALYSIS OF VARIANCE

In one-way analysis of variance, only row variables are compared. In two-way analysis of variance, we compare both row and column variables defined by a frequency table of two cross-classified attributes, A and B, where A is classified into r categories $A_1, A_2, \ldots, A_r$, and B is classified into c categories $B_1, B_2, \ldots, B_c$. For example, as illustrated in Fig. 11-10, where $r = 2$ and $c = 3$, the population may consist of college students; attribute A is gender, classified into A_1: Male and A_2: Female, and attribute B is age, classified into B_1: Below 20 years, B_2: 20 to 25 years, and B_3: Over 25 years. Each entry A_iB_j in the table represents the number of students in both category A_i and category B_j.

	Gender		
Age	A_1B_1	A_1B_2	A_1B_3
	A_2B_1	A_2B_2	A_2B_3

Fig. 11-10 Cross classification of attributes.

Assumptions for Two-Way Analysis of Variance

In general, consider a population classified according to two attributes A and B. A sample of size n from the population results in an $r \times c$ cross-classification table such as Fig. 11-11, where the entry in row i and column j is the number of individuals in the population falling into categories A_i and B_j. We assume that the entry in row i and column j is a value of a normal random variable X_{ij} with mean μ_{ij} and standard deviation σ. That is, all rc random variables X_{ij} have the same standard deviation; they are also assumed to be independent. Then, for $i = 1, 2, \ldots, r$, the row sample mean

$$\bar{X}_{i\cdot} = \frac{X_{i1} + X_{i2} + \cdots + X_{ic}}{c}$$

has mean $\mu_{i\cdot} = (\mu_{i1} + \mu_{i2} + \cdots + \mu_{ic})/c$. For $j = 1, 2, \ldots, c$, the column sample mean

$$\bar{X}_{\cdot j} = \frac{X_{1j} + X_{2j} + \cdots + X_{rj}}{r}$$

has mean $\mu_{\cdot j} = (\mu_{1j} + \mu_{2j} + \cdots + \mu_{ry})/r$. The grand sample mean is defined to be

$$\bar{X} = \frac{\bar{X}_{1\cdot} + \bar{X}_{2\cdot} + \cdots + \bar{X}_{r\cdot}}{r}; \qquad \text{equivalently,} \qquad \bar{X} = \frac{\bar{X}_{\cdot 1} + \bar{X}_{\cdot 2} + \cdots + \bar{X}_{\cdot c}}{c}$$

$\bar{X}$ has mean $\mu = (\mu_{1\cdot} + \mu_{2\cdot} + \cdots + \mu_{r\cdot})/r$ which is also equal to $\mu = (\mu_{\cdot 1} + \mu_{\cdot 2} + \cdots + \mu_{\cdot c})/c$.

	Attribute B				Row sample means
Attribute A	X_{11}	X_{12}	...	X_{1c}	$\bar{X}_{1\cdot}$
	X_{21}	X_{22}	...	X_{2c}	$\bar{X}_{2\cdot}$
	$\vdots$	$\vdots$	$\vdots$	$\vdots$	$\vdots$
	X_{r1}	X_{r2}	...	X_{rc}	$\bar{X}_{r\cdot}$
Column sample means	$\bar{X}_{\cdot 1}$	$\bar{X}_{\cdot 2}$	...	$\bar{X}_{\cdot c}$	$\bar{X}$
					Grand sample mean

Fig. 11-11 Two-way cross-classification table.

Null Hypotheses

In a two-way analysis of variance, there are two null hypotheses, one saying that the row means are equal:

$$H_0^{(R)}: \mu_{1j} = \mu_{2j} = \cdots = \mu_{rj} \qquad \text{for} \qquad j = 1, 2, \ldots, c$$

and the other saying that the column means are equal:

$$H_0^{(C)}: \mu_{i1} = \mu_{i2} = \cdots = \mu_{ic} \qquad \text{for} \qquad i = 1, 2, \ldots, r$$

With reference to cross-classification of college students in Fig. 11-10, $H_0^{(R)}$ says that the gender distributions are equal, and $H_0^{(C)}$ says that the age distributions are equal.

Square Variations

Analogous to the one-way case, for two-way analysis, the *total square variation*, denoted by V_T, is the sum of the squares of the deviations of all rc random variables from the grand mean:

$$V_T = \sum (X_{1j} - \bar{X})^2 + \sum (X_{2j} - \bar{X})^2 + \cdots + \sum (X_{rj} - \bar{X})^2$$

where j runs from 1 to c in each summation. The *square deviation between rows*, denoted by V_R, is defined as

$$V_R = c \sum (\bar{X}_{i\cdot} - \bar{X})^2$$

where i runs from 1 to r. The *square deviation between columns*, denoted by V_C, is defined as

$$V_C = r \sum (\bar{X}_{\cdot j} - \bar{X})^2$$

where j runs from 1 to c. The *square variation due to random error*, denoted by V_e, is defined as

$$V_e = V_T - (V_R + V_C)$$

In one-way analysis of variance, we were able to say that the distribution of V_e does not depend on whether the null hypothesis is true or not (Theorem 11.2). To obtain a similar result here, we assume that the means μ_{ij} of the random variables X_{ij} satisfy the equations $\mu_{ij} = \mu + \alpha_i + \beta_j$, where $\sum \alpha_i = 0$ and $\sum \beta_j = 0$ (see Problems 11.27–11.30). Then, as in the one-way case, some important properties of the square variations can be summarized in the following theorem.

Theorem 11.4: V_e/σ^2 is chi-square with $(r-1)(c-1)$ degrees of freedom regardless of whether either null hypothesis is true. If $H_0^{(R)}$ is true, then V_R/σ^2 is chi-square with $r-1$ degrees of freedom; and if $H_0^{(C)}$ is true, then V_C/σ^2 is chi-square with $c-1$ degrees of freedom. If both $H_0^{(R)}$ and $H_0^{(C)}$ are true, then V_T/σ^2 is chi-square with $rc-1$ degrees of freedom; and all four chi-square random variables are independent.

The Test Statistics

It follows from Theorem 11.4 and the definition of the F distribution (Section 10.5) that if $H_0^{(R)}$ is true, then

$$F^{(R)} = \frac{V_R/(r-1)}{V_e/(r-1)(c-1)}$$

is an F random variable with $r-1$ and $(r-1)(c-1)$ degrees of freedom. $F^{(R)}$ is the test statistic used to test $H_0^{(R)}$. Similarly, if the null hypothesis $H_0^{(C)}$ is true, then

$$F^{(C)} = \frac{V_C/(c-1)}{V_e/(r-1)(c-1)}$$

is an F random variable with $c-1$ and $(r-1)(c-1)$ degrees of freedom. $F_0^{(C)}$ is the test statistic used to test $H_0^{(C)}$. Motivation for the choice of test statistics is provided by the following theorem.

Theorem 11.5: $E(V_e/(r-1)(c-1)) = \sigma^2$, $E(V_R/(r-1)) = \sigma^2 + \dfrac{c}{r-1}\sum(\mu_{i\cdot} - \mu)^2$, and $E(V_C/(c-1)) = \sigma^2 + \dfrac{r}{c-1}\sum(\mu_{\cdot j} - \mu)^2$.

Theorem 11.5 says that if $H_0^{(R)}$ is true, then $E(V_R/(r-1))$ will equal σ^2, and sample values of $F^{(R)}$ will tend to be close to 1; the more the row means differ, the larger sample values of $F^{(R)}$ are likely to be. Similarly, if $H_0^{(C)}$ is true, then $E(V_C/(c-1))$ will equal σ^2, and sample values of $F^{(C)}$ will tend to be close to 1; the more the column means differ, the larger sample values of $F^{(C)}$ are likely to be.

Two-Way Analysis-of-Variance Table

In applications of an F test for equal row means and for equal column means, a *two-way analysis-of-variance table* is usually constructed as illustrated in Fig. 11-12.

Performing the Tests: *P*-value and Critical Regions

A total of rc random values x_{ij}, one for each random variable X_{ij}, are independently obtained. Using these, the sample-mean values $\bar{x}_{i\cdot} = (x_{i1} + x_{i2} + \cdots + x_{ic})/c$, $\bar{x}_{\cdot j} = (x_{1j} + x_{2j} + \cdots + x_{rj})/r$,

Square variation	Degrees of freedom	Mean square	F
Between rows V_R	$r-1$	$\dfrac{V_R}{r-1}$	$\dfrac{V_R/(r-1)}{V_e/(r-1(c-1)}$
Between columns V_C	$c-1$	$\dfrac{V_C}{c-1}$	$\dfrac{V_C/(c-1)}{V_e/(r-1)(c-1)}$
Random error V_e	$(r-1)(c-1)$	$\dfrac{V_e}{(r-1)(c-1)}$	
Total V_T	$rc-1$		

Fig. 11-12 Two-way analysis-of-variance table.

and $\bar{x} = (\bar{x}_{1\cdot} + \bar{x}_{2\cdot} + \cdots + \bar{x}_{r\cdot})/r$ (equivalently, $\bar{x} = (\bar{x}_{\cdot 1} + \bar{x}_{\cdot 2} + \cdots + \bar{x}_{\cdot c})/c$) are computed. The corresponding values

$$\nu_T = \sum (x_{1j} - \bar{x})^2 + \sum (x_{2j} - \bar{x})^2 + \cdots + \sum (x_{rj} - \bar{x})^2,$$

$$\nu_R = c \sum (\bar{x}_{i\cdot} - \bar{x})^2, \qquad \nu_C = r \sum (\bar{x}_{\cdot j} - \bar{x})^2, \qquad \text{and} \qquad \nu_e = \nu_T - (\nu_R + \nu_C)$$

are then computed. Finally the values

$$\hat{F}^{(R)} = \frac{\nu_R/(r-1)}{\nu_e/(r-1)(c-1)} \qquad \text{and} \qquad \hat{F}^{(C)} = \frac{\nu_C/(c-1)}{\nu_e/(r-1)(c-1)}$$

of the test statistics are determined.

The P-value of the row test is the probability that a test value as large or larger than $\hat{F}^{(R)}$ would occur if $H_0^{(R)}$ were true. That is, the P-value is $P(F \geq \hat{F}^{(R)})$, assuming $r-1$ and $(r-1)(c-1)$ degrees of freedom. If a level of significance α is specified, then $H_0^{(R)}$ is rejected if P-value $\leq \alpha$; $H_0^{(R)}$ is not rejected if P-value $> \alpha$. Equivalently, the critical region for the test consists of all values of F that are greater than or equal to F^*, where F^* is the critical value satisfying $P(F \geq F^*) = \alpha$, assuming $r-1$ and $(r-1)(c-1)$ degrees of freedom (see Fig. 11-2); $H_0^{(R)}$ is rejected if $\hat{F}^{(R)}$ is in the critical region; $H_0^{(R)}$ is not rejected if $\hat{F}^{(R)}$ is not in the critical region.

The P-value of the column test is the probability that a test value as large or larger than $\hat{F}^{(C)}$ would occur if $H_0^{(C)}$ were true. That is, the P-value is $P(F \geq \hat{F}^{(C)})$, assuming $c-1$ and $(r-1)(c-1)$ degrees of freedom. If a level of significance α is specified, then $H_0^{(C)}$ is rejected if P-value $\leq \alpha$; $H_0^{(C)}$ is not rejected if P-value $> \alpha$. Equivalently, the critical region for the test consists of all values of F that are greater than or equal to F^*, where F^* is the critical value satisfying $P(F \geq F^*) = \alpha$, assuming $c-1$ and $(r-1)(c-1)$ degrees of freedom (see Fig. 11-9); $H_0^{(C)}$ is rejected if $\hat{F}^{(C)}$ is in the critical region; $H_0^{(C)}$ is not rejected if $\hat{F}^{(C)}$ is not in the critical region.

As in the one-way analysis of variance, these tests are also one sided; the alternative hypothesis H_a in each case is equivalent to the hypothesis: $F \geq F^*$.

EXAMPLE 11.5 Three types of indoor lighting: A_1, A_2, and A_3, were tried on three types of flower: B_1, B_2, and B_3, grown from seed. The average heights in cm after 12 weeks of growth are indicated in the table.

(*a*) Test, at the 0.05 significance level, whether there is a significant difference in growth due to lighting.
(*b*) Test, at the 0.05 significance level, whether there is a significant difference in growth due to flower type.

		Flowers			
		B_1	B_2	B_3	Row sample means
	A_1	16	24	19	19.67
Lighting	A_2	15	25	18	19.33
	A_3	21	31	15	22.33
Column sample means		17.33	26.67	17.33	20.44 (Grand sample mean)

From the table, we see that the grand sample mean is $\bar{x} = 20.44$. The total variation is

$$\nu_T = (16 - 20.44)^2 + (24 - 20.44)^2 + (19 - 20.44)^2 + (15 - 20.44)^2 + (25 - 20.44)^2$$
$$+ (18 - 20.44)^2 + (21 - 20.44)^2 + (31 - 20.44)^2 + (15 - 20.44)^2 = 232.22$$

The row sample means are $\bar{x}_{1.} = 19.67$, $\bar{x}_{2.} = 19.33$, and $\bar{x}_{3.} = 22.33$; and the square deviation between rows is

$$\nu_R = 3[(19.67 - 20.44)^2 + (19.33 - 20.44)^2 + (22.33 - 20.44)^2] = 16.19$$

The column sample means are $\bar{x}_{.1} = 17.33$, $\bar{x}_{.2} = 26.67$, and $\bar{x}_{.3} = 17.33$; and the square deviation between columns is

$$\nu_C = 3[(17.33 - 20.44)^2 + (26.67 - 20.44)^2 + (17.33 - 20.44)^2] = 174.47$$

Therefore, the square variation due to random error is

$$\nu_e = 232.22 - (174.47 + 16.19) = 41.56$$

(*a*) The degrees of freedom of the row test statistic are $r - 1 = 2$ and $(r - 1)(c - 1) = 4$; and the value of the row test statistic is

$$\hat{F}^{(R)} = \frac{16.19/2}{41.56/4} = 0.78$$

From Table A-5 in the Appendix, the critical region for the row test at the significance level 0.05 consists of all test values greater than or equal to 6.94. Since $0.78 < 6.94$, the test value is not in the critical region, and we do not reject the null hypothesis that the column means are equal. Equivalently, we conclude that there is not a significant difference in growth due to the type of lighting.

(*b*) The degrees of freedom of the column test statistic are $c - 1 = 2$ and $(r - 1)(c - 1) = 4$; and the value of the column test statistic is

$$\hat{F}^{(C)} = \frac{174.47/2}{41.56/4} = 8.40$$

From Table A-5, the critical region for column test at the significance level 0.05 consists of all test values greater than or equal to 6.94. Since $8.40 > 6.94$, the test value is in the critical region, and we reject the null hypothesis that the column means are equal. Equivalently, we conclude that there is a significant difference in growth due to the type of flower. The corresponding analysis-of-variance table is

Square variation	Degrees of freedom	Mean square	F
Between rows $V_R = 16.19$	$r - 1 = 2$	$\frac{V_R}{r-1} = 8.10$	$\hat{F}^{(R)} = 0.78$
Between columns $V_C = 174.47$	$c - 1 = 2$	$\frac{V_C}{c-1} = 87.24$	$\hat{F}^{(C)} = 8.40$
Random error $V_e = 41.56$	$(r-1)(c-1) = 4$	$\frac{V_e}{(r-1)(c-1)} = 10.39$	
Total $V_T = 232.22$	$rc - 1 = 8$		

Solved Problems

CHI-SQUARE GOODNESS-OF-FIT TEST

11.1. A pair of dice is tossed 360 times, and the frequency of each sum is indicated in the chart.

Sum	2	3	4	5	6	7	8	9	10	11	12
Frequency	8	24	35	37	44	65	51	42	26	14	14

Would you say that the dice are fair on the basis of the chi-square test?

The null hypothesis is

$$H_0\text{: } P(2) = \frac{1}{36} = P(12), \qquad P(3) = \frac{2}{36} = P(11), \qquad P(4) = \frac{3}{36} = P(10),$$

$$P(5) = \frac{4}{36} = P(9), \qquad P(6) = \frac{5}{36} = P(8), \qquad P(7) = \frac{6}{36}$$

The following table lists the 11 expected frequencies np_j, where $n = 360$.

Sum	2	3	4	5	6	7	8	9	10	11	12
Expected frequency	10	20	30	40	50	60	50	40	30	20	10

The chi-square test sum is

$$\hat{\chi}^2 = \frac{(8-10)^2}{10} + \frac{(24-20)^2}{20} + \frac{(35-30)^2}{30} + \frac{(37-40)^2}{40} + \frac{(44-50)^2}{50} + \frac{(65-60)^2}{60} + \frac{(51-50)^2}{50}$$
$$+ \frac{(42-40)^2}{40} + \frac{(26-30)^2}{30} + \frac{(14-20)^2}{20} + \frac{(14-10)^2}{10} \approx 7.45$$

From Table A-3 in the Appendix, with 10 degrees of freedom, we find that the probability of a sum as large or larger than 6.74 is 0.75, and the probability of a sum as large or larger than 9.34 is 0.5. Hence, the probability of getting 7.45 or larger is between 0.5 and 0.75 (using computer software, the probability is 0.68), which is strong evidence that the dice are fair. More precisely, the null hypothesis would not be rejected at any significance level less than 0.68.

11.2. Over the years, the grades in a certain college professor's class are typically as follows: 10 percent As, 20 percent Bs, 50 percent Cs, 15 percent Ds, and 5 percent Fs. The grades for her current class of 100 are 16 As, 28 Bs, 46 Cs, 10 Ds, and 0 Fs. Test the hypothesis that the current class is typical by a chi-square test at the 0.05 significance level.

The null hypothesis, expected frequency, and actual frequency are shown in the following table.

Grade	A	B	C	D	F
H_0: probability =	0.1	0.2	0.5	0.15	0.05
Expected frequency	10	20	50	15	5
Actual frequency	16	28	46	10	0

The chi-square test sum is

$$\hat{\chi}^2 = \frac{(16-10)^2}{10} + \frac{(28-20)^2}{20} + \frac{(46-50)^2}{50} + \frac{(10-15)^2}{15} + \frac{(0-5)^2}{5} \approx 13.79$$

From Table A-3, with $5 - 1 = 4$ degrees of freedom, the critical region consists of all values greater than or equal to 9.49. Since $13.79 > 9.49$, we reject the hypothesis that the class is typical.

11.3. A bag is supposed to contain 20 percent red beans and 80 percent white beans. A random sample of 50 beans from the bag contains 16 red and 34 white. Apply the chi-square test at the 0.05 significance level to either reject or not reject the hypothesis that the contents are as advertised.

If the contents are 20 percent red and 80 percent white, then $P(\text{red}) = p_1 = 0.2$, and $P(\text{white}) = p_2 = 0.8$; $np_1 = 50 \times 0.2 = 10$ and $np_2 = 50 \times 0.8 = 40$. The test chi-square value is

$$\hat{\chi}^2 = \frac{(16-10)^2}{10} + \frac{(34-40)^2}{40} = 4.5$$

From Table A-3, with one degree of freedom, the critical region consists of all values greater than or equal to 3.84. Since $4.5 > 3.84$, we reject the hypothesis that the bag contains 20 percent red and 80 percent white beans.

11.4. A coin is tossed 100 times, resulting in 60 heads (H) and 40 tails (T). Apply the chi-square test at the 0.05 significance level to either reject or not reject the hypothesis that the coin is fair.

The null hypothesis is: $P(H) = p_1 = 0.5$, $P(T) = p_2 = 0.5$. We have $n = 100$, so $np_1 = 100 \times 0.5 = 50 = np_2$. The test chi-square value is

$$\hat{\chi}^2 = \frac{(60-50)^2}{50} + \frac{(40-50)^2}{50} = 4$$

From Table A-3, with one degree of freedom, the critical region consists of all values greater than or equal to 3.84. Since $4 > 3.84$, we reject the hypothesis that the coin is fair.

11.5. Suppose a coin is tossed 100 times, resulting in x heads. For what values of x will the null hypothesis that the coin is fair not be rejected on the basis of the chi-square test at the 0.05 significance level?

If x is the number of heads, then $100 - x$ is the number of tails. For the hypothesis of fairness not to be rejected at the 0.05 level, the test chi-square sum must satisfy (see Problem 11.4)

$$\hat{\chi}^2 = \frac{(x-50)^2}{50} + \frac{(100-x-50)^2}{50} < 3.84$$

which simplifies to

$$\frac{(x-50)^2}{50} + \frac{(50-x)^2}{50} < 3.84 \quad \text{or} \quad \frac{2(x-50)^2}{50} < 3.84$$

Simplifying further,

$$(x-50)^2 < 25 \times 3.84 \quad \text{or} \quad |x-50| < \sqrt{25 \times 3.84} \approx 9.80 < 10$$

Hence, x must satisfy $|x - 50| < 10$ which is equivalent to $-10 < x - 50 < 10$, or $40 < x < 60$. Therefore, if there are more than 40 but fewer than 60 heads in 100 tosses, the hypothesis of fairness will not be rejected at the 0.05 significance level.

CHI-SQUARE TEST FOR EQUAL DISTRIBUTIONS

11.6. At what significance level would the null hypothesis in Example 11.2 be rejected?

The chi-square test value obtained in Example 11.2 is 7.14, and the number of degrees of freedom is 4. From Table A-3, we see that any test value equal to or greater than 5.39 will be in the critical region at significance level 0.25. Hence, the null hypothesis will be rejected at the 0.25 significance level. Using computer software, we find that the P-value of the test, which is defined as $P(\chi^2 \geq 7.14)$ for 4 degrees of freedom, is 0.13. Hence, the null hypothesis will be rejected at any significance level greater than or equal to 0.13.

11.7. A random group of 40 people younger than 50 years was given a flu shot, and a second random group of 60 people 50 years or older was given the same flu shot. Each member of the groups was classified according to whether the member did not get the flu (N), had a mild case of the flu (M), or had a severe case of the flu (S). The frequencies in each group are as indicated in the following table.

		Reaction			
		N	M	S	Totals
Age	Under 50 years	30	6	4	40
	50 years or older	36	12	12	60
	Totals	66	18	16	100

Use a chi-square random variable to test, at the 0.05 significance level, the hypothesis that the reactions to the shot are the same in each group.

By pooling the subjects under 50 years and those 50 years and over in each reaction group, we get the following estimated probabilities:

$$\hat{p}_N = \frac{66}{100} = 0.66, \qquad \hat{p}_M = \frac{18}{100} = 0.18, \qquad \hat{p}_S = \frac{16}{100} = 0.16$$

The expected frequencies $m\hat{p}_j$ for the $m = 40$ subjects under 50 years are

$$m\hat{p}_N = 40 \times 0.66 = 26.4, \qquad m\hat{p}_M = 40 \times 0.18 = 7.2, \qquad m\hat{p}_S = 40 \times 0.16 = 6.4$$

and the expected frequencies $n\hat{p}_j$ for the $n = 60$ subjects 50 years and over are

$$n\hat{p}_N = 60 \times 0.66 = 39.6, \qquad n\hat{p}_M = 60 \times 0.18 = 10.8, \qquad n\hat{p}_S = 60 \times 0.16 = 9.6$$

The corresponding chi-square test value is

$$\hat{\chi}^2 = \frac{(30-26.4)^2}{26.4} + \frac{(6-7.2)^2}{7.2} + \frac{(4-6.4)^2}{6.4} + \frac{(36-39.6)^2}{39.6} + \frac{(12-10.8)^2}{10.8} + \frac{(12-9.6)^2}{9.6}$$
$$\approx 2.65$$

There are $k = 3$ reaction levels and $k - 1 = 2$ degrees of freedom. From Table A-3, the critical chi-square region for 2 degrees of freedom at the 0.05 significance level is all values greater than or equal to 5.99. Since $2.65 < 5.99$, we do not reject the hypothesis that the reactions to the shot are the same in each group.

11.8. At what significance level would the null hypothesis in Problem 11.7 be rejected?

The chi-square test value obtained in Problem 11.7 is 2.65, and the number of degrees of freedom is 2. From Table A-3, we see that any test value equal to or greater than 1.39 will be in the critical region at significance level 0.50. Hence, the null hypothesis will be rejected at the 0.50 significance level. Using computer software, we find that the *P*-value of the test, which is defined as $P(\chi^2 \geq 2.65)$ for 2 degrees of freedom, is 0.27. Hence, the null hypothesis will be rejected at any significance level greater than or equal to 0.27.

11.9. Salaries for 200 males and 300 females at a certain company are as indicated in the following frequency table, where the notation $[a, b)$ means a salary greater than or equal to a but less than b.

Salaries in thousands of dollars

		1	2	3	4	5	
		[20, 30)	[30, 40)	[40, 50)	[50, 60)	[60, –)	Totals
Gender	Male	20	34	46	60	40	200
	Female	45	78	90	62	25	300
	Totals	65	112	136	122	65	500

Use the chi-square random variable to test, at the 0.05 significance level, the hypothesis the salary distributions are the same.

By pooling the male and female frequencies in each salary grade, we obtain the following estimated probabilities:

$$\hat{p}_1 = \frac{65}{500} = 0.13, \qquad \hat{p}_2 = \frac{112}{500} = 0.224, \qquad \hat{p}_3 = \frac{136}{500} = 0.272,$$

$$\hat{p}_4 = \frac{122}{500} = 0.244, \qquad \hat{p}_5 = \frac{65}{500} = 0.13$$

The expected frequencies $m\hat{p}_j$ for the $m = 200$ males are:

$$m\hat{p}_1 = 200 \times 0.13 = 26, \qquad m\hat{p}_2 = 200 \times 0.224 = 44.8, \qquad m\hat{p}_3 = 200 \times 0.272 = 54.4,$$

$$m\hat{p}_4 = 200 \times 0.244 = 48.8, \qquad m\hat{p}_5 = 200 \times 0.13 = 26$$

and the expected frequencies $n\hat{p}_j$ for the $n = 300$ females are:

$$n\hat{p}_1 = 300 \times 0.13 = 39, \qquad n\hat{p}_2 = 300 \times 0.224 = 67.2, \qquad n\hat{p}_3 = 300 \times 0.272 = 81.6,$$

$$n\hat{p}_4 = 300 \times 0.244 = 73.2, \qquad n\hat{p}_5 = 300 \times 0.13 = 39$$

The corresponding chi-square test value is

$$\hat{\chi}^2 = \frac{(20-26)^2}{26} + \frac{(34-44.8)^2}{44.8} + \frac{(46-54.4)^2}{54.4} + \frac{60-48.8)^2}{48.8} + \frac{(40-26)^2}{26}$$
$$+ \frac{(45-39)^2}{39} + \frac{(78-67.2)^2}{67.2} + \frac{(90-81.6)^2}{81.6} + \frac{(62-73.2)^2}{73.2} + \frac{(25-39)^2}{39}$$
$$\approx 25.66$$

There are $k = 5$ salary grades and $k - 1 = 4$ degrees of freedom. The critical chi-square region for 4 degrees of freedom at the 0.05 significance level is all values greater than or equal to 9.49. Since $25.66 > 9.49$, we reject the hypothesis that the salary distribution for males is the same as that for females.

11.10. What is the probability that the chi-square value of 25.66 obtained in Problem 11.9, or higher, would occur if the male and female salaries were equally distributed?

From Table A-3, with 4 degrees of freedom, we see that the probability that a test value equal to or greater than 18.5 is 0.001. Hence, the probability of getting 25.66 or higher is less than 0.001. Using computer software, we find that $P(\chi^2 \geq 25.66)$ for 4 degrees of freedom is 0.000 04. Hence, there are only 4 chances in 100,000, or 1 in 25,000, that a chi-square value this large would occur if the male and female salaries were equally distributed.

CHI-SQUARE TEST FOR INDEPENDENT ATTRIBUTES

11.11. A random group of 800 eligible voters was cross-classified according to annual income and party affiliation, as indicated in the following table. In the table, [20, 40) signifies income of at least $20,000 but less than $40,000; [40, 60) means at least $40,000 but less than $60,000, and [60,000, –) means $60,000 and over. Apply a chi-square test for independence of annual income and party affiliation at the 0.05 significance level.

Annual income

		[20, 40)	[40, 60)	[60, –)	Totals ($f_{i\cdot}$)
	Democratic	125	225	70	420
Party	Republican	60	200	120	380
	Totals ($f_{\cdot j}$)	185	425	190	800

The contingency table has $r = 2$ rows, where Democratic affiliation corresponds to $i = 1$ and Republican affiliation corresponds to $i = 2$; there are $c = 3$ columns, where $j = 1$ corresponds to the salary range [20, 40), $j = 2$ corresponds to [40, 60), and $j = 3$ corresponds to [60, –). The estimated row probabilities are:

$$\hat{p}_{1\cdot} = \frac{420}{800} = 0.525, \qquad \hat{p}_{2\cdot} = \frac{380}{800} = 0.475$$

and the estimated column probabilities are:

$$\hat{p}_{\cdot 1} = \frac{185}{800}, \qquad \hat{p}_{\cdot 2} = \frac{425}{800}, \qquad \hat{p}_{\cdot 3} = \frac{190}{800}$$

The expected frequency estimates are:

$$n \times \hat{p}_{1\cdot} \times \hat{p}_{\cdot 1} = 800 \times 0.525 \times \frac{185}{800} = 97.125, \qquad n \times \hat{p}_{1\cdot} \times \hat{p}_{\cdot 2} = 223.125, \qquad n \times \hat{p}_{1\cdot} \times \hat{p}_{\cdot 3} = 99.75,$$

$$n \times \hat{p}_{2\cdot} \times \hat{p}_{\cdot 1} = 87.875, \qquad n \times \hat{p}_{2\cdot} \times \hat{p}_{\cdot 2} = 201.875, \qquad n \times \hat{p}_{2\cdot} \times \hat{p}_{\cdot 3} = 90.25$$

The test value of the chi-square statistic is:

$$\begin{aligned}\hat{\chi}^2 &= \frac{(125 - 97.125)^2}{97.125} + \frac{(225 - 223.125)^2}{223.125} + \frac{(70 - 99.75)^2}{99.75} \\ &+ \frac{(60 - 87.875)^2}{87.875} + \frac{(200 - 201.875)^2}{201.875} + \frac{(120 - 90.25)^2}{90.25} \\ &\approx 35.56\end{aligned}$$

There are $(r - 1)(c - 1) = 1 \times 2 = 2$ degrees of freedom. From Table A-3, the critical region for 2 degrees of freedom at the 0.05 significance level is all test values greater than or equal to 5.99. Since $35.56 > 5.99$, we reject the hypothesis that annual income and party affiliation are independent.

11.12. Estimate the P-value for the test in Problem 11.12, and interpret the result.

The P-value is the probability that a test value as large or larger than 35.56 would occur, at 2 degrees of freedom, if the attributes of annual income and party affiliation were independent. From Table A-3 in the Appendix, we can conclude only that the P-value is less than 0.001. Using computer software, we find that $P(\chi^2 \geq 35.56) = 0.000\,000\,02$. Hence there are only 2 changes in 100 million, or 1 in 50 million, that such a large test statistic would occur if the attributes of annual income and party affiliation were independent.

11.13. A random group of 300 males was cross-classified according to age and total cholesterol level, as indicated in the table below.

Total cholesterol

		Under 200 Low	200–239 Medium	240 or higher High	Totals ($f_{i\cdot}$)
Age	20–34	66	24	8	98
	35–54	54	48	22	124
	55–74	18	50	10	78
Totals ($f_{\cdot j}$)		138	122	40	300

Use the chi-square random variable to test, at the 0.01 significance level, the hypothesis that the attributes of age and cholesterol level are independent.

The contingency table has $r = 3$ rows, where age bracket 20–34 corresponds to $i = 1$, 35–54 corresponds to $i = 2$, and 55–74 corresponds to $i = 3$. There are $c = 5$ columns, where $j = 1$ corresponds to low cholesterol level, $j = 2$ corresponds to medium, and $j = 3$ corresponds to high. The estimated row probabilities are:

$$\hat{p}_{1\cdot} = \frac{98}{300}, \qquad \hat{p}_{2\cdot} = \frac{124}{300}, \qquad \hat{p}_{3\cdot} = \frac{78}{300}$$

and the estimated column probabilities are:

$$\hat{p}_{\cdot 1} = \frac{138}{300}, \qquad \hat{p}_{\cdot 2} = \frac{122}{300}, \qquad \hat{p}_{\cdot 3} = \frac{40}{300}$$

The expected cross-classification frequency estimates, where $n = 300$, are:

$$n \times \hat{p}_{1\cdot} \times \hat{p}_{\cdot 1} = 45.08, \qquad n \times \hat{p}_{1\cdot} \times \hat{p}_{\cdot 2} = 39.853, \qquad n \times \hat{p}_{1\cdot} \times \hat{p}_{\cdot 3} = 13.067,$$

$$n \times \hat{p}_{2\cdot} \times \hat{p}_{\cdot 1} = 57.04, \qquad n \times \hat{p}_{2\cdot} \times \hat{p}_{\cdot 2} = 50.427, \qquad n \times \hat{p}_{2\cdot} \times \hat{p}_{\cdot 3} = 16.533,$$

$$n \times \hat{p}_{3\cdot} \times \hat{p}_{\cdot 1} = 35.88, \qquad n \times \hat{p}_{3\cdot} \times \hat{p}_{\cdot 2} = 31.72, \qquad n \times \hat{p}_{3\cdot} \times \hat{p}_{\cdot 3} = 10.4$$

The test value of the chi-square statistic is:

$$\hat{\chi}^2 = \frac{(66 - 45.08)^2}{45.08} + \frac{(24 - 39.853)^2}{39.853} + \frac{(8 - 13.067)^2}{13.067} + \frac{(54 - 57.04)^2}{57.04} + \frac{(48 - 50.427)^2}{50.427}$$
$$+ \frac{(22 - 16.533)^2}{16.533} + \frac{(18 - 35.88)^2}{35.88} + \frac{(50 - 31.72)^2}{31.72} + \frac{(10 - 10.4)^2}{10.4}$$
$$\approx 39.53$$

There are $(r - 1)(c - 1) = 2 \times 2 = 4$ degrees of freedom. From Table A-3, the critical region for 4 degrees of freedom at the 0.01 significance level is all test values greater than or equal to 13.3. Since $39.53 > 13.3$, the test value is in the critical region, and we reject the hypothesis that age and total cholesterol level are independent.

11.14. Consider the table in Problem 11.13 as a frequency table for three independent multinomial random variables, X_1, X_2, X_3, where X_i distributes the number of subjects in its corresponding age bracket among the three cholesterol levels. Test, at the 0.01 significance level, the hypothesis that the random variables have the same distribution.

The null hypothesis is H_0: $p_{11} = p_{21} = p_{31}$; $p_{12} = p_{22} = p_{32}$; $p_{13} = p_{23} = p_{33}$, where p_{ij} is the probability of cholesterol level j in age bracket i. We estimate the probability of cholesterol level j by pooling the frequencies in the jth column of the table:

$$\hat{p}_1 = \frac{138}{300}, \qquad \hat{p}_2 = \frac{122}{300}, \qquad \hat{p}_3 = \frac{40}{300}$$

The expected frequencies corresponding to the 98 subjects in age bracket 20–34 are:

$$98 \times \frac{138}{300} = 45.08, \qquad 98 \times \frac{122}{300} = 39.853, \qquad 98 \times \frac{40}{300} = 13.067$$

Those corresponding to the 124 subjects in age bracket 35–54 are:

$$124 \times \frac{138}{300} = 57.04, \qquad 124 \times \frac{122}{300} = 50.427, \qquad 124 \times \frac{40}{300} = 16.533$$

and those corresponding to the 78 subjects in age bracket 55–74 are:

$$78 \times \frac{138}{300} = 35.88, \qquad 78 \times \frac{122}{300} = 31.72, \qquad 78 \times \frac{40}{300} = 10.4$$

Note that these 9 frequencies also occurred as cross-classification frequencies in Problem 11.13. The test value of the chi-square statistic here will also be the same as in Problem 11.13, namely, $\hat{\chi}^2 \approx 39.53$. Finally, since there are also 4 degrees of freedom here, the critical region at the 0.01 significance level is the same as in Problem 11.13, namely all test values greater than 13.3. We therefore reject the hypothesis that the three multinomial random variables have the same distribution.

11.15. Suppose the frequency data in an $r \times c$ contingency table for cross-classified attributes is the same as the frequency data in an $r \times c$ table for r independent multinomial random variables, each with the same c possible outcomes. Show that the test value of the chi-square statistic is the same in each case.

In the cross-classification case, the probability estimates are

$$\hat{p}_{ij} = \hat{p}_{i\cdot} \times \hat{p}_{\cdot j} = \frac{f_{i\cdot}}{n} \times \frac{f_{\cdot j}}{n}$$

(see Fig. 11-6), and the expected frequency estimates are

$$n \times \hat{p}_{i\cdot} \times \hat{p}_{\cdot j} = f_{i\cdot} \times \frac{f_{\cdot j}}{n}$$

In the multinomial random variables case, the probability estimates are

$$\hat{p}_j = \frac{f_{\cdot j}}{n}$$

(see Fig. 11-3), and the expected frequency estimates are

$$n_i \times \hat{p}_j = n_i \times \frac{f_{\cdot j}}{n}$$

Since $n_i = f_{i1} + f_{i2} + \cdots + f_{ic} = f_{i\cdot}$, it follows that the expected frequency estimates are the same, and therefore the test values of the chi-square statistic are the same. Note that the number of degrees of freedom is $(r-1)(c-1)$ in each case.

ONE-WAY ANALYSIS OF VARIANCE

11.16. The average gas mileage, in miles per gallon, of a random sample of compact cars, five from each of three manufactures, is given in the table. Assume that the average gas mileage for each of the three makes of cars is normally distributed, and that the three distributions have the same variance. Test, at the 0.01 significance level, the hypothesis that the three distributions have the same mean.

X_1	32.5	30.2	34.6	31.3	29.8
X_2	28.9	29.6	30.2	30.6	29.1
X_3	34.8	36.2	31.8	33.7	35.3

As indicated in the table, the three distributions are labeled X_1, X_2, X_3, respectively. The respective test values of the sample means for X_1, X_2, X_3 are

$$\bar{x}_1 = \frac{32.5 + 30.2 + 34.6 + 31.3 + 29.8}{5} = 31.68, \qquad \bar{x}_2 = \frac{28.9 + 29.6 + 30.2 + 30.6 + 29.1}{5} = 29.68,$$

$$\bar{x}_3 = \frac{34.8 + 36.2 + 31.8 + 33.7 + 35.3}{5} = 34.36$$

and the grand sample mean is $\bar{x} = \dfrac{31.68 + 29.68 + 34.36}{3} = 31.91$. The total square variation is

$$\begin{aligned} \nu_T = {} & (32.5 - 31.91)^2 + (30.2 - 31.91)^2 + (34.6 - 31.91)^2 + (31.3 - 31.91)^2 + (29.8 - 31.92)^2 \\ & + (28.9 - 31.91)^2 + (29.6 - 31.91)^2 + (30.2 - 31.91)^2 + (30.6 - 31.91)^2 + (29.1 - 31.91)^2 \\ & + (34.8 - 31.91)^2 + (36.2 - 31.91)^2 + (31.8 - 31.91)^2 + (33.7 - 31.91)^2 + (35.3 - 31.91)^2 \\ = {} & 83.73. \end{aligned}$$

The square deviation between row samples is

$$\nu_R = 5[(31.68 - 31.91)^2 + (29.68 - 31.91)^2 + (34.36 - 31.91)^2] = 55.14$$

and the square variation due to random error is

$$\nu_e = 83.73 - 55.14 = 28.59$$

The degrees of freedom are $m - 1 = 3 - 1 = 2$ and $mn - m = 3 \times 5 - 3 = 12$, and the test statistic is

$$\hat{F}(2, 12) = \frac{55.14/2}{28.59/12} = 11.57$$

From Table A-7, the critical region at the 0.01 significance level consists of all test values greater than or equal to 6.93. Since $11.57 > 6.93$, the test value is in the critical region, and we reject the null hypothesis that the means of X_1, X_2, X_3 are equal. The corresponding analysis-of-variance table is as follows.

Square variation	Degrees of freedom	Mean square	F
Between row samples $\nu_R = 55.14$	$m - 1 = 2$	$\dfrac{\nu_R}{m-1} = 27.57$	$\dfrac{V_R/(m-1)}{V_e/(mn-m)} = 11.57$
Random error $\nu_e = 28.59$	$mn - m = 12$	$\dfrac{\nu_e}{mn-m} = 2.38$	
Total $\nu_T = 83.73$	$mn - 1 = 14$		

11.17. Determine the P-value for the test in Problem 11.16, and interpret the result.

The P-value for the test is the probability that a value of the test statistic equal to or greater than 11.57 would occur if the hypothesis that the three distributions have the same mean were true. From Table A-7 in the Appendix, with 2 and 12 degrees of freedom, we can conclude only that the P-value is less than 0.01. Using computer software, we find that the P-value is 0.0016. Hence, there are only 16 chances in 10,000 or 1 in 625 that such a result would occur if each of the three car makes had the same average gas mileage.

11.18. In the case of two independent normal distributions with equal but unknown variances, we can use either the F statistic (Section 11.4) or the t statistic (Section 10.2) to test the hypothesis that the distributions have equal means. Apply both tests to X_1 and X_2 from Problem 11.17 at the 0.05 significance level.

The table of sample values is as shown in the table below.

X_1	32.5	30.2	34.6	31.3	29.8
X_2	28.9	29.6	30.2	30.6	29.1

We first apply analysis of variance using the F statistic. From Problem 11.17, the test values of the sample means for X_1 and X_2 are $\bar{x}_1 = 31.68$ and $\bar{x}_2 = 29.68$. The test value of the grand sample mean is $\bar{x} = \dfrac{31.68 + 29.68}{2} = 30.68$. The total square variation is

$$\begin{aligned}\nu_T &= (32.5 - 30.68)^2 + (30.2 - 30.68)^2 + (34.6 - 30.68)^2 + (31.3 - 30.68)^2 + (29.8 - 30.68)^2 \\ &\quad + (28.9 - 30.68)^2 + (29.6 - 30.68)^2 + (30.2 - 30.68)^2 + (30.6 - 30.68)^2 + (29.1 - 30.68)^2 \\ &= 27.14\end{aligned}$$

The square deviation between row samples is

$$\nu_R = 5[(31.68 - 30.68)^2 + (29.68 - 30.68)^2] = 10$$

and the square variation due to random error is

$$\nu_e = 27.14 - 10 = 17.14$$

The degrees of freedom are $m - 1 = 2 - 1 = 1$ and $mn - m = 2 \times 5 - 2 = 8$, and the test statistic is

$$\hat{F}(1, 8) = \frac{10/1}{17.13/8} = 4.67$$

From Table A-5, the critical region, at the 0.05 significance level, consists of all test values greater than or equal to 5.32. Since $4.67 < 5.32$, the test value is not in the critical region, and we do not reject the null hypothesis that the means of X_1, X_2 are equal.

We now apply the t test from Section 10.2, following Prescription 4(a) from that section, where the random variables X and Y in Prescription 4(a) are represented here by X_1 and X_2, respectively. Also, m and n from Prescription 4(a) are both equal to 5 here. We then compute the following test values:

$$\bar{x}_1 = 31.68, \qquad \bar{x}_2 = 29.68$$

$$s_{X_1}^2 = \frac{1}{4}[(32.5 - 31.68)^2 + (30.2 - 31.68)^2 + (34.6 - 31.68)^2 + (31.3 - 31.68)^2 + (29.8 - 31.68)^2] = 3.77$$

$$s_{X_2}^2 = \frac{1}{4}[(28.9 - 29.68)^2 + (29.6 - 29.68)^2 + (30.2 - 29.68)^2 + (30.6 - 29.68)^2 + (29.1 - 29.68)^2] = 0.52$$

$$s_P = \sqrt{\frac{4 \times 3.77 + 4 \times 0.52}{8}} = 1.46, \qquad \hat{t} = \frac{31.68 - 29.68}{1.46\sqrt{\frac{1}{5} + \frac{1}{5}}} = 2.17$$

From Table A-2, the critical region, at the 0.05 significance level with 8 degrees of freedom, for the alternative hypothesis H_a: $\mu_1 \neq \mu_2$ consists of all values $\hat{t} \geq 2.31$, or $\hat{t} \leq -2.31$. Since 2.17 satisfies neither of these inequalities, 2.17 is not in the critical region, and the null hypothesis H_0: $\mu_1 = \mu_2$ is not rejected in favor of H_a: $\mu_1 \neq \mu_2$, which is the same result obtained by the F test.

11.19. Note that, in Problem 11.18, $\hat{t}^2 = (2.17)^2 = 4.7 = \hat{F}$, allowing for rounding. Show that in general, $F = t^2$ in the two-sample case, where F is the test statistic of Section 11.4 for testing equality of means, and t is the test statistic of Section 10.2, Prescriptions 4 and 4(a), for testing equality of means based on random samples of the same size.

First note that, by definition, $F(m,n) = \dfrac{\chi^2(m)/m}{\chi^2(n)/n}$ and $t^2(n) = \dfrac{Z^2}{\chi^2(n)/n} = \dfrac{\chi^2(1)/1}{\chi^2(n)/n}$. Therefore, $F(1,n) = t^2(n)$. To see how this equality applies here, let the random variables X and Y in the t test of Prescriptions 4 and 4(a) be denoted by X_1 and X_2 with means μ_1 and μ_2, respectively. Also, $m = n$ in the t test since both random samples have size n; and $m = 2$ in the F test since there are only two random variables X_1 and X_2. In both tests, the null hypothesis is H_0: $\mu_1 = \mu_2$, and the alternative hypothesis is H_a: $\mu_1 \neq \mu_2$. Substituting X_1 for X and X_2 for Y, and $m = n$ in the formula for the t statistic in Prescription 4 gives

$$t = \frac{\bar{X}_1 - \bar{X}_2}{S_p\sqrt{\dfrac{1}{n} + \dfrac{1}{n}}} = \frac{\bar{X}_1 - \bar{X}_2}{S_p\sqrt{\dfrac{2}{n}}}$$

where
$$S_P = \sqrt{\frac{(n-1)S_{X_1}^2 + (n-1)S_{X_2}^2}{n+n-2}} = \sqrt{\frac{\sum(X_{1j} - \bar{X}_1)^2 + \sum(X_{2j} - \bar{X}_2)^2}{2(n-1)}}$$

each summation going from $j = 1$ to $j = n$. Then

$$t^2 = \frac{n(\bar{X}_1 - \bar{X}_2)^2}{(\sum(X_{1j} - \bar{X}_1)^2 + \sum(X_{2j} - \bar{X}_2)^2)/(n-1)}$$

On the other hand, substituting $m = 2$ in the formula for the F statistic, we get

$$F = \frac{V_R/(m-1)}{V_e/(mn-m)} = \frac{V_R}{V_e/2(n-1)}, \quad \text{where} \quad V_R = n((\bar{X}_1 - \bar{X})^2 + (\bar{X}_2 - \bar{X})^2)$$

and
$$V_e = \sum(X_{1j} - \bar{X}_1)^2 + \sum(X_{2j} - \bar{X}_2)^2$$

(see Problem 11.22). Substituting $\bar{X} = \dfrac{\bar{X}_1 + \bar{X}_2}{2}$ in the formula for V_R and simplifying, gives $V_R = \dfrac{n}{2}(\bar{X}_1 - \bar{X}_2)^2$. Therefore

$$F = \frac{\dfrac{n}{2}(\bar{X}_1 - \bar{X}_2)^2}{(\sum(X_{1j} - \bar{X}_1)^2 + \sum(X_{2j} - \bar{X}_2)^2)/2(n-1)}$$

Finally, the 2s cancel, resulting in $F = t^2$, as desired.

11.20. Both the F test and the t test in Problem 11.18 were applied under the assumption that X_1 and X_2 have equal variances. Apply the two-sample F test described in Prescription 10.8(a) of Section 10.6 to test the hypothesis that $\sigma_{X_1}^2 = \sigma_{X_2}^2$ against the alternative hypothesis $\sigma_{X_1}^2 \neq \sigma_{X_2}^2$ at the 0.05 significance level.

The random variables X and Y in Prescription 8(a) are represented here by X_1 and X_2, respectively. Also, m and n from Prescription 8(a) are both equal to 5 here. From Problem 11.18, we have the test values $s_{X_1}^2 = 3.77$ and $s_{X_2}^2 = 0.52$. For the alternative hypothesis H_a: $\sigma_{X_1}^2 \neq \sigma_{X_2}^2$, the critical region is all values $s_{X_2}^2/s_{X_1}^2 \geq F^*$ or $s_{X_1}^2/s_{X_2}^2 \geq F^*$, where F^* is the F value satisfying $P(F(4,4) \leq F^*) = 1 - 0.05/2 = 0.975$. From Table A-6, we find that $F^* = 9.6$. We have $s_{X_1}^2/s_{X_2}^2 = 3.77/0.52 = 7.25$, and $s_{X_2}^2/s_{X_1}^2 = 0.52/3.77 = 0.14$. Since neither value is in the critical region, we do not reject the null hypothesis that $\sigma_{X_1}^2 = \sigma_{X_2}^2$.

11.21. Let $X_1, X_2, \ldots, X_m$ be m random variables with sample means $\bar{X}_1, \bar{X}_2, \ldots, \bar{X}_m$, respectively, each based on random samples of size n. Let $\bar{X} = \dfrac{\bar{X}_1 + \bar{X}_2 + \cdots + \bar{X}_m}{m}$. Show that, for each $i = 1, 2, \ldots, m$, $\sum (X_{ij} - \bar{X}_i)(\bar{X}_i - \bar{X}) = 0$, where j runs from 1 to n in the summation.

Let i be any fixed integer from 1 to m. Then

$$\sum (X_{ij} - \bar{X}_i)(\bar{X}_i - \bar{X}) = (\bar{X}_i - \bar{X}) \sum (X_{ij} - \bar{X}_i) = (\bar{X}_i - \bar{X})(n\bar{X}_i - n\bar{X}_i) = 0$$

11.22. With reference to the definitions of V_T, V_R, and V_e in Section 11.4, show that

$$V_e = \sum (X_{1j} - \bar{X}_1)^2 + \sum (X_{2j} - \bar{X}_2)^2 + \cdots + \sum (X_{mj} - \bar{X}_m)^2$$

where in each summation, j runs from 1 to n.

By definition, $V_e = V_T - V_R$, where V_T is a sum of summations $\sum (X_{ij} - \bar{X})^2$; j runs from 1 to n in each summation, and there is one summation for each integer i from 1 to m (see Section 11.4). V_R is a sum of terms $n(\bar{X}_i - \bar{X})^2$; one term for each i from 1 to m. Keeping i fixed and letting j run from 1 to n, we have

$$\begin{aligned}
\sum (X_{ij} - \bar{X})^2 &= \sum (X_{ij} - \bar{X}_i + \bar{X}_i - \bar{X})^2 \\
&= \sum (X_{ij} - \bar{X}_i)^2 + 2\sum (X_{ij} - \bar{X}_i)(\bar{X}_i - \bar{X}) + \sum (\bar{X}_i - \bar{X})^2 \\
&= \sum (X_{ij} - \bar{X}_i)^2 + 0 \text{ (Problem 11.20)} + n(\bar{X}_i - \bar{X})^2 \\
&= \sum (X_{ij} - \bar{X}_i)^2 + n(\bar{X}_i - \bar{X})^2
\end{aligned}$$

Adding the terms on the right side of the equality by letting i run from 1 to m, and then subtracting V_R, we get the desired result.

11.23. Sketch a proof of Theorem 11.2.

From Problem 11.22,

$$V_e/\sigma^2 = \sum \frac{(X_{1j} - \bar{X}_1)^2}{\sigma^2} + \sum \frac{(X_{2j} - \bar{X}_2)^2}{\sigma^2} + \cdots + \sum \frac{(X_{mj} - \bar{X}_m)^2}{\sigma^2}$$

Each summation on the right side is a chi-square random variable with $n - 1$ degrees of freedom (Theorem 7.7). Furthermore, the mn random variables X_{ij} are independent. Therefore, the sum of the summations on the right side is a chi-square random variable with $m(n - 1) = mn - m$ degrees of freedom. Also, each $\bar{X}_i$ is normally distributed with mean μ_i and variance σ^2/n. If H_0 is true, then $\bar{X}$ is the sample mean of $\bar{X}_i$. Therefore, if H_0 is true, then $V_R/\sigma^2 = \sum \dfrac{(\bar{X}_i - \bar{X})^2}{\sigma^2/n}$ is a chi-square random variable with $m - 1$ degrees of freedom (Theorem 7.7). Finally, since $V_T/\sigma^2 = V_R/\sigma^2 + V_e/\sigma^2$, it can be shown that V_T/σ^2 is chi-square with $mn - m + m - 1 = mn - 1$ degrees of freedom, provided H_0 is true; and it can also be shown that all three chi-square random variables are independent.

11.24. Sketch a proof of Theorem 11.3.

The expected value of a chi-square random variable with k degrees of freedom is k. Since V_e/σ^2 is chi-square with $mn - m$ degrees of freedom (Theorem 11.2), it follows that $E(V_e/\sigma^2) = mn - m$, and therefore $E(V_e/(mn - m)) = \sigma^2$. To determine $E(V_R/(m - 1))$, first consider

$$\begin{aligned}
\sum (\bar{X}_i - \bar{X})^2 &= \sum (\bar{X}_i^2 - 2\bar{X}_i\bar{X} + \bar{X}^2) \\
&= \sum \bar{X}_i^2 - 2m\bar{X}^2 + m\bar{X}^2 \\
&= \sum \bar{X}_i^2 - m\bar{X}^2
\end{aligned}$$

where i runs from 1 to m in the summation. Therefore, since $V_R/(m-1) = \frac{n}{m-1}\sum(\bar{X}_i - \bar{X})^2$, it follows that

$$E(V_R/(m-1)) = \frac{n}{m-1}\sum E(\bar{X}_i^2) - \frac{nm}{m-1}E(\bar{X}^2)$$

Now $E(X^2) = \sigma^2 + \mu^2$ for any random variable X with mean μ and variance σ^2. $\bar{X}_i$ has mean μ_i and variance σ^2/n, while $\bar{X}$ has mean $\mu = \frac{\mu_1 + \mu_2 + \cdots + \mu_m}{m}$ and variance σ^2/mn. Therefore,

$$\begin{aligned} E(V_R/(m-1)) &= \frac{n}{m-1}\sum\left(\frac{\sigma^2}{n} + \mu_i^2\right) - \frac{nm}{m-1}\left(\frac{\sigma^2}{mn} + \mu^2\right) \\ &= \frac{n}{m-1}\frac{m}{n}\sigma^2 + \frac{n}{m-1}\sum \mu_i^2 - \frac{1}{m-1}\sigma^2 - \frac{nm}{m-1}\mu^2 \\ &= \sigma^2 + \frac{n}{m-1}\sum(\mu_i^2 - \mu^2) \end{aligned}$$

Now

$$\begin{aligned} \sum(\mu - \mu_i)^2 &= \sum(\mu^2 - 2\mu\mu_i + \mu_i^2) = \sum\mu^2 - 2\mu\sum\mu_i + \sum\mu_i^2 \\ &= m\mu^2 - 2m\mu^2 + \sum\mu_i^2 = \sum\mu_i^2 - m\mu^2 \\ &= \sum(\mu_i^2 - \mu^2) \end{aligned}$$

Therefore, $E(V_R/(m-1)) = \sigma^2 + \frac{n}{m-1}\sum(\mu - \mu_i)^2$, as stated in the theorem.

11.25. Show that for each $i = 1, 2, \ldots, m$, the mean μ_i of X_i is related to the grand mean $\mu = (\mu_1 + \mu_2 + \cdots + \mu_m)/m$ by the equation $\mu_i = \mu + \alpha_i$, where $\sum \alpha_i = 0$.

$$\sum\alpha_i = \sum(\mu - \mu_i) = \sum\mu - \sum\mu_i = m\mu - m\mu = 0$$

TWO-WAY ANALYSIS OF VARIANCE

11.26. Find the P-value for the row test and for the column test in Example 11.5.

The P-value for the row test is the probability that an F-value, with 2 and 4 degrees of freedom, as large as 0.78 would occur if the hypothesis $H_0^{(R)}$: $H_{1j} = H_{2j} = \mu_{3j}$ for $j = 1, 2, 3$ were true. From Table A-4 in the Appendix, we can conclude only that the P-value is greater than 0.1. Using computer software, we find that the P-value is 0.52. The P-value for the column test is the probability that an F-value, with 2 and 4 degrees of freedom, as large as 8.40 would occur if the hypothesis $H_0^{(C)}$: $\mu_{i1} = \mu_{i2} = \mu_{i3}$ for $i = 1, 2, 3$ were true. From Tables A-5 and A-6, we see that the P-value is between 0.025 and 0.05. Using computer software, we find that the P-value is 0.037.

11.27. Suppose that $\mu_{ij} = \mu + \alpha_i + \beta_j$, where $\sum_i \alpha_i = 0$ and $\sum_j \beta_j = 0$ (the notation indicates that i runs from 1 to r in the first summation, and j runs from 1 to c in the second summation). Show that $\alpha_i = \mu_{i\cdot} - \mu$ and $\beta_j = \mu_{\cdot j} - \mu$.

From the given equation $\mu_{ij} = \mu + \alpha_i + \beta_j$, we get

$$c\alpha_i = \sum_j \alpha_i = \sum_j(\mu_{ij} - \mu - \beta_j) = \sum_j \mu_{ij} - \sum_j \mu - \sum_j \beta_j = c\mu_{i\cdot} - c\mu - 0$$

Then, dividing both sides of $c\alpha_i = c\mu_{i\cdot} - c\mu$, we get $\alpha_i = \mu_{i\cdot} - \mu$. The proof that $\beta_j = \mu_{\cdot j} - \mu$ is similar.

11.28. Suppose that $\mu_{ij} = \mu + \alpha_i + \beta_j$, where $\alpha_i = \mu_{i\cdot} - \mu$ and $\beta_j = \mu_{\cdot j} - \mu$. Show that $\sum_i \alpha_i = 0$ and $\sum_j \beta_j = 0$.

From the given equation $\mu_{ij} = \mu + \alpha_i + \beta_j$, we get

$$\sum_i \alpha_i = \sum_i (\mu_{ij} - \mu - \beta_j) = r\mu_{.j} - r\mu - r\beta_j = r(\mu_{.j} - \mu - \beta_j) = 0$$

The proof that $\sum_j \beta_j = 0$ is similar.

11.29. Let μ_{ij} be the entry in the ith row and jth column of the matrix $\begin{bmatrix} 9 & 11 & 8 & 12 \\ 4 & 6 & 3 & 7 \\ 2 & 4 & 1 & 5 \end{bmatrix}$. Show that $\mu_{ij} = \mu + \alpha_i + \beta_j$, where $\sum_i \alpha_i = 0$ and $\sum_j \beta_j = 0$.

By Problem 11.28, it is sufficient to show that $\mu_{ij} = \mu + \alpha_i + \beta_j$, where $\alpha_i = \mu_{i.} - \mu$, $\beta_j = \mu_{.j} - \mu$, and $\mu = \dfrac{\mu_{1.} + \mu_{2.} + \cdots + \mu_{r.}}{r} = \dfrac{\mu_{.1} + \mu_{.2} + \cdots + \mu_{.c}}{c}$ is the grand mean. Averaging the row means gives $\mu_{1.} = 10$, $\mu_{2.} = 5$, $\mu_{3.} = 3$. The grand mean is $\mu = \dfrac{10 + 5 + 3}{3} = 6$. Therefore, $\alpha_1 = 10 - 6 = 4$, $\alpha_2 = 5 - 6 = -1$, and $\alpha_3 = 3 - 6 = -3$. Averaging the column means gives $\mu_{.1} = 5$, $\mu_{.2} = 7$, $\mu_{.3} = 4$, $\mu_{.4} = 8$. Therefore, $\beta_1 = 5 - 6 = -1$, $\beta_2 = 7 - 6 = 1$, $\beta_3 = 4 - 6 = -2$, and $\beta_4 = 8 - 6 = 2$.

We must now verify that $\mu_{ij} = \mu + \alpha_i + \beta_j$ for all 12 means in the given matrix. For example, $\mu + \alpha_1 + \beta_1 = 6 + 4 + (-1) = 9 = \mu_{11}$ and $\mu + \alpha_1 + \beta_2 = 6 + 4 + 1 = 11 = \mu_{12}$. Continuing this way, we will find that $\mu + \alpha_i + \beta_j = \mu_{ij}$ holds in all 12 cases.

11.30. Let μ_{ij} be the entry in the ith row and jth column of the matrix $\begin{bmatrix} 21 & 12 & 12 \\ 15 & 18 & 21 \\ 9 & 6 & 12 \end{bmatrix}$.

(*a*) Show that the property $\mu_{ij} = \mu + \alpha_i + \beta_j$, where $\sum_i \alpha_i = 0$ and $\sum_j \beta_j = 0$, is *not* satisfied for all of the entries in the matrix.

(*b*) Replace the entry in the ith row and jth column of the matrix with $\hat{\mu}_{ij} = \mu_{i.} + \mu_{.j} - \mu$ to obtain a new matrix that does satisfy the property $\hat{\mu}_{ij} = \mu + \alpha_i + \beta_j$, where $\sum_i \alpha_i = 0$ and $\sum_j \beta_j = 0$.

(*a*) If the matrix did satisfy the desired property, then by Problem 11.27, the equation $\mu_{ij} = \mu_{i.} + \mu_{.j} - \mu$ would have to hold for each μ_{ij} in the matrix. Checking μ_{11}, we see that $\mu_{1.} = \dfrac{21 + 12 + 12}{3} = 15$, $\mu_{.1} = \dfrac{21 + 15 + 9}{3} = 15$, and $\mu = \dfrac{21 + 12 + 12 + 15 + 18 + 21 + 9 + 6 + 12}{9} = 14$. Then $\mu_{1.} + \mu_{.1} - \mu = 15 + 15 - 14 = 16$, but $\mu_{11} = 21$. Therefore, the matrix does not have the desired property.

(*b*) To construct a matrix with the desired property, we first compute the remaining two row means and two column means, which are $\mu_{2.} = 18$, $\mu_{3.} = 9$, $\mu_{.2} = 12$, $\mu_{.3} = 15$. We already have $\hat{\mu}_{11} = 16$. We then compute $\hat{\mu}_{12} = \mu_{1.} + \mu_{.2} - \mu = 15 + 12 - 14 = 13$, $\hat{\mu}_{13} = \mu_{1.} + \mu_{.3} - \mu = 15 + 15 - 14 = 16$, and continuing this way, we get $\hat{\mu}_{21} = 19$, $\hat{\mu}_{22} = 16$, $\hat{\mu}_{23} = 19$, $\hat{\mu}_{31} = 10$, $\hat{\mu}_{32} = 7$, and $\hat{\mu}_{33} = 10$. The new matrix is $\begin{bmatrix} 16 & 13 & 16 \\ 19 & 16 & 19 \\ 10 & 7 & 10 \end{bmatrix}$ which, by its construction, has the desired property.

Supplementary Problems

CHI-SQUARE GOODNESS-OF-FIT TEST

11.31. In 150 tosses of a coin, 90 heads and 60 tails were observed. Test the hypothesis that the coin is fair by a chi-square test at the 0.05 significance level.

11.32. Repeat the test in Problem 11.31 at the 0.01 level of significance.

11.33. A random-digit generator on a calculator gave the distribution of digits shown in the table. Test the hypothesis that the digits are random by a chi-square test at the 0.05 significance level.

Digit	0	1	2	3	4	5	6	7	8	9
Frequency	11	11	9	8	8	11	9	11	13	9

11.34. The standard normal random variable Z, with mean 0 and standard deviation 1, has a probability distribution, in terms of class intervals I_1, I_2, I_3, I_4, I_5, as shown in the first two columns of the following table. The third column shows the class frequencies of 100 z scores chosen at random from some population. Apply a chi-square test at the 0.05 significance level to the hypothesis that the z scores are a sample from a standard normal population.

I_j	p_j	f_j
$(-\infty, -1.5)$	0.0668	7
$[-1.5, -0.5)$	0.2417	15
$[-0.5, 0.5)$	0.3830	45
$[0.5, 1.5)$	0.2417	25
$[1.5, \infty)$	0.0668	8

11.35. Use the class frequency distribution of Problem 11.34 to apply a chi-square test at the 0.05 significance level to the hypothesis that the following 50 test scores are approximately normally distributed.

30	66	71	78	88	40	66	72	78	79
42	67	72	80	90	52	67	73	80	90
55	68	74	82	92	60	68	74	83	93
60	68	75	84	93	62	70	76	84	94
64	70	76	85	95	65	70	78	86	97

11.36. It is estimated that the political preference in a certain community is as follows: 50 percent Democrat, 25 percent Republican, 15 percent Independent, 10 percent other. A random sample of 200 people resulted in 90 Democrats, 65 Republicans, 25 Independents, and 20 other. Test the hypothesis that the estimate is correct at the 0.1 significance level.

CHI-SQUARE TEST FOR EQUAL DISTRIBUTIONS

11.37. Independently obtained random samples of two independent multinomial random variables, X and Y, each with outcomes a_1, a_2, a_3, a_4, resulted in the following contingency table of frequencies. Apply a chi-square test at the 0.05 significance level to the hypothesis that X and Y have the same probability distribution.

	a_1	a_2	a_3	a_4	Totals
X	25	45	15	15	100
Y	45	50	35	10	140
Totals	70	95	50	25	240

11.38. Perform the test in Problem 10.37 at the 0.01 significance level.

11.39. Each die of a pair of unbalanced dice, one red and one white, is tossed 200 times, resulting in the following frequency distribution for the faces of the dice. Apply a chi-square test at the 0.01 significance level to the hypothesis that the dice have the same probability distribution.

Side landing face-up

	1	2	3	4	5	6	Totals
Red die	30	20	42	10	41	57	200
White die	44	30	24	20	50	32	200
Totals	74	50	66	30	91	89	400

11.40. Random samples of 125 male graduates and 100 female graduates of a certain college resulted in the following frequency table for the number of semesters in which a natural science was studied. Apply a chi-square test at the 0.05 significance level to the hypothesis that male and female students at the college take the same amount of natural science courses.

Semesters of natural science

	1	2	3	4	Totals
Males	5	6	50	64	125
Females	8	14	34	44	100
Totals	13	20	84	108	225

11.41. Random samples of 200 first-year students and 150 transfer students at a given college resulted in the following frequency table for the number of high-school years in which a foreign language was studied. Apply a chi-square test at the 0.01 significance level to the hypothesis that first-year and transfer students have the same high-school foreign language backgrounds.

Years of foreign-language study in high-school

	0	1	2	3	4	Totals
First-year students	10	11	75	61	43	200
Transfer students	20	18	54	30	28	150
Totals	30	29	129	91	71	350

11.42. Independently obtained random samples of three independent multinomial random variables, X, Y, and Z, each with outcomes a_1, a_2, a_3, resulted in the following contingency table of frequencies. Apply a chi-square test at the 0.05 significance level to the hypothesis that X, Y, and Z have the same probability distribution.

	a_1	a_2	a_3	Totals
X	33	25	12	70
Y	46	20	24	90
Z	50	14	26	90
Totals	129	59	62	250

11.43. Repeat the test in Problem 11.42 at the 0.01 significance level.

CHI-SQUARE TESTS FOR INDEPENDENT ATTRIBUTES

11.44. Seventy-five exercise programs were rated for quality of exercise and motivational value. Each attribute was classified as good, fair, or poor, and the cross-classification frequencies are indicated in the following table. Apply a chi-square test at the 0.05 significance level to the hypothesis that quality of exercise and motivational value are independent.

		Motivational value			
		Good	Fair	Poor	Totals
Exercise value	Good	15	6	4	25
	Fair	7	12	6	25
	Poor	5	8	12	25
	Totals	27	26	22	75

11.45. Use Table A-3 in the Appendix to find an approximate *P*-value for the test in Problem 11.44. (If computer software is available, find the exact *P*-value of the test.)

11.46. Sixty supermarket pizzas were rated for taste (fair, good, very good) and price (high, medium, low). The cross-classification results are indicated in the following frequency contingency table. Apply a chi-square test at the 0.05 significance level to the hypothesis that taste and price are independent.

		Price			
		High	Medium	Low	Totals
Taste	Very good	8	6	4	18
	Good	6	8	8	22
	Fair	4	6	10	20
	Totals	18	20	22	60

11.47. Use Table A-3 to find an approximate *P*-value for the test in Problem 11.46. (If computer software is available, find the exact *P*-value of the test.)

11.48. A random sample of 500 students at a given college was cross-classified according to gender and major subject area of study chosen. The results are listed in the following table. Apply a chi-square test at the 0.05 level of significance to the hypothesis that the attributes of gender and major subject area are independent.

		Major area of study				
		Business	Liberal arts	Nursing	Education	Totals
Gender	Male	105	76	15	48	244
	Female	71	94	31	60	256
	Totals	176	170	46	108	500

11.49. Use Table A-3 to find an approximate *P*-value for the test in Problem 11.48. (If computer software is available, find the exact *P*-value of the test.)

ONE-WAY ANALYSIS OF VARIANCE

11.50. A random sample of size 3 is taken from each of three independent, normally distributed random variables X_1, X_2, X_3 having equal but unknown variances. Test, at the 0.05 level of significance, the hypothesis that X_1, X_2, and X_3 have equal means.

X_1	94	82	84
X_2	102	94	78
X_3	76	68	70

11.51. Use Tables A-4 to A-7 in the Appendix to find an approximate P-value for the test in Problem 11.50. (If computer software is available, find the exact P-value of the test.)

11.52. A home gardener wishes to determine the effect of different fertilizers on the average number of tomatoes produced by her plants. She grows five tomato plants on each of four separate plots, X_1, X_2, X_3, X_4, and uses a different fertilizer treatment on each plot. The number of tomatoes per plant are indicated in the following table. Test, at the 0.05 level of significance, the hypothesis that plots X_1, X_2, X_3, and X_4 have equal average yields.

X_1	14	10	12	16	17
X_2	9	11	12	8	10
X_3	16	15	14	10	18
X_4	10	11	11	13	8

11.53. Repeat the test in Problem 11.52 at the 0.01 significance level.

11.54. Use Tables A-4 to A-7 to find an approximate P-value for the test in Problem 11.52. (If computer software is available, find the exact P-value of the test.)

TWO-WAY ANALYSIS OF VARIANCE

11.55. The table in Problem 11.50 is repeated here, but interpreted as a table obtained by cross-classifying attributes A and B, in which A has three categories A_1, A_2, A_3, and B has three categories B_1, B_2, B_3. Test, at the 0.05 significance level, the hypothesis that the row means are equal.

		Attribute B			
		B_1	B_2	B_3	Row sample means
Attribute A	A_1	94	82	84	86.67
	A_2	102	94	78	91.33
	A_3	76	68	70	71.33
	Column sample means	90.67	81.33	77.33	83.11

Grand sample mean

11.56. Use the data in Problem 11.55 to test, at the 0.05 significance level, whether the column means are equal.

11.57. Using Tables A-4 to A-7, approximate the respective *P*-value of the tests in Problems 11.55 and 11.56. (If computer software is available, find the exact *P*-values of the tests.)

11.58. The table in Problem 11.52 is repeated here, but interpreted as a table obtained by cross classifying four types of fertilizers with 5 types of tomato plants. The entry in the row i and column j of the table represents the yield from type j tomato plant treated with type i fertilizer. Test, at the 0.05 significance level, the hypothesis that the row means are equal.

Plant type

		B_1	B_2	B_3	B_4	B_5	Row sample means
Fertilizer type	A_1	14	10	12	16	17	13.8
	A_2	9	11	12	8	10	10
	A_3	16	15	14	10	18	14.6
	A_4	10	11	11	13	8	10.6
	Column sample means	12.25	11.75	12.25	11.75	13.25	12.25

Grand sample mean

11.59. Use the data in Problem 11.58 to test, at the 0.05 significance level, whether the column means are equal.

11.60. Using Tables A-4 to A-7, approximate the respective *P*-value of the tests in Problems 11.58 and 11.59. (If computer software is available, find the exact *P*-values of the tests.)

Answers to Supplementary Problems

11.31. Critical region: $\hat{\chi}^2 \geq 3.84$; test value: $\hat{\chi}^2 = 6.00$; reject hypothesis that coin is fair (*P*-value $= 0.0143$).

11.32. Critical region $\hat{\chi}^2 \geq 6.63$; test value: $\hat{\chi}^2 = 6.00$; do not reject hypothesis that coin is fair (*P*-value $= 0.0143$).

11.33. Critical region: $\hat{\chi}^2 \geq 16.9$; test value: $\hat{\chi}^2 = 2.4$; do not reject hypothesis that the digits are random (*P*-value $= 0.983$).

11.34. Critical region: $\hat{\chi}^2 \geq 9.49$; test value: $\hat{\chi}^2 = 4.96$; do not reject hypothesis that the z scores are from a normal population (*P*-value $= 0.291$).

11.35. Sample mean: $\bar{x} = 73.84$; sample standard deviation: $s = 14.40$; z score: $z = (x - \bar{x})/s$; class frequencies of z scores:

I_j	p_j	f_j
$(-\infty, -1.5)$	0.0668	4
$[-1.5, -0.5)$	0.2417	8
$[-0.5, 0.5)$	0.3830	22
$[0.5, 1.5)$	0.2417	15
$[1.5, \infty)$	0.0668	1

Critical region: $\hat{\chi}^2 \geq 9.49$; test value: $\hat{\chi}^2 = 4.28$; do not reject hypothesis that the test scores are approximately normally distributed (P-value = 0.369).

11.36. Critical region: $\hat{\chi}^2 \geq 6.25$; test value: $\hat{\chi}^2 = 6.33$; reject hypothesis that the estimate is correct (P-value = 0.097).

11.37. Critical region: $\hat{\chi}^2 \geq 7.81$; test value: $\hat{\chi}^2 = 8.55$; reject hypothesis that X and Y have the same probability distribution (P-value = 0.036).

11.38. Critical region: $\hat{\chi}^2 \geq 11.3$; test value: $\hat{\chi}^2 = 8.55$; do not reject hypothesis that X and Y have the same probability distribution (P-value = 0.036).

11.39. Critical region: $\hat{\chi}^2 \geq 15.1$; test value: $\hat{\chi}^2 = 20.8$; reject hypothesis that the dice have the same probability distribution (P-value = 0.0009).

11.40. Critical region: $\hat{\chi}^2 \geq 7.81$; test value: $\hat{\chi}^2 = 7.96$; reject hypothesis that males and females take the same amount of science courses (P-value = 0.047).

11.41. Critical region: $\hat{\chi}^2 \geq 13.3$; test value: $\hat{\chi}^2 = 15.34$; reject hypothesis that first-year and transfer students have the same high-school foreign language backgrounds (P-value = 0.004).

11.42. Critical region: $\hat{\chi}^2 \geq 9.49$; test value: $\hat{\chi}^2 = 9.83$; reject hypothesis that X, Y, and Z have the same probability distribution (P-value = 0.0434).

11.43. Critical region: $\hat{\chi}^2 \geq 13.3$; test value: $\hat{\chi}^2 = 9.83$; do not reject hypothesis that X, Y, and Z have the same probability distribution (P-value = 0.0434).

11.44. Critical region: $\hat{\chi}^2 \geq 9.49$; test value: $\hat{\chi}^2 = 13.1$; reject hypothesis that quality of exercise and motivational value are independent.

11.45. $0.01 < P\text{-value} < 0.025$ (P-value = 0.011).

11.46. Critical region: $\hat{\chi}^2 \geq 9.49$; test value: $\hat{\chi}^2 = 4.09$; do not reject hypothesis that taste and price are independent.

11.47. $0.25 < P\text{-value} < 0.5$ (P-value = 0.393).

11.48. Critical region: $\hat{\chi}^2 \geq 7.81$; $\hat{\chi}^2 = 15.09$; reject hypothesis that gender and major subject area are independent.

11.49. $0.001 < P\text{-value} < 0.005$ (P-value = 0.002).

11.50.

Square variation	Degrees of freedom	Mean square	F
Between row samples $\nu_R = 656.89$	$m - 1 = 2$	$\dfrac{\nu_R}{m-1} = 328.45$	$\dfrac{V_R/(m-1)}{V_e/(mn-m)} = 4.74$
Random error $\nu_e = 416$	$mn - m = 6$	$\dfrac{\nu_e}{mn-m} = 69.33$	
Total $\nu_T = 1072.89$	$mn - 1 = 8$		

Critical region: $\hat{F}(2, 6) \geq 5.14$; test value: $\hat{F} = 4.74$; do not reject the hypothesis of equal means.

11.51. $0.05 < P\text{-value} < 0.1$ (P-value = 0.0583).

11.52.

Square variation	Degrees of freedom	Mean square	F
Between row samples $\nu_R = 78.55$	$m - 1 = 3$	$\dfrac{\nu_R}{m-1} = 26.18$	$\dfrac{V_R/(m-1)}{V_e/(mn-m)} = 4.59$
Random error $\nu_e = 91.2$	$mn - m = 16$	$\dfrac{\nu_e}{mn-m} = 5.7$	
Total $\nu_T = 169.75$	$mn - 1 = 19$		

Critical region: $\hat{F}(3,16) \geq 3.24$; test value: $\hat{F} = 4.59$; reject the hypothesis of equal average yields.

11.53. Critical region: $\hat{F}(3,16) \geq 5.29$; test value: $\hat{F} = 4.59$; do not reject the hypothesis of equal average yields.

11.54. $0.01 < P\text{-value} < 0.025$ (P-value $= 0.0167$).

11.55.

Square variation	Degrees of freedom	Mean square	F
Between rows $V_R = 656.89$	$r - 1 = 2$	$\dfrac{V_R}{r-1} = 328.45$	$\hat{F}^{(R)} = 9.72$
Between columns $V_C = 280.89$	$c - 1 = 2$	$\dfrac{V_C}{c-1} = 140.45$	$\hat{F}^{(C)} = 4.16$
Random error $V_e = 135.11$	$(r-1)(c-1) = 4$	$\dfrac{V_e}{(r-1)(c-1)} = 33.78$	
Total $V_T = 1072.89$	$rc - 1 = 8$		

Critical region: $\hat{F}^{(R)}(2,4) \geq 6.94$; test value: $\hat{F}^{(R)} = 9.72$; reject hypothesis that row means are equal.

11.56. Critical region: $\hat{F}^{(C)}(2,4) \geq 6.94$; test value: $\hat{F}^{(C)} = 4.16$; do not reject hypothesis that column means are equal.

11.57. Problem 11.55: $0.025 < P\text{-value} < 0.05$ (P-value $= 0.0291$); Problem 11.56: P-value > 0.1 (P-value $= 0.1054$).

11.58.

Square variation	Degrees of freedom	Mean square	F
Between rows $V_R = 78.55$	$r - 1 = 3$	$\dfrac{V_R}{r-1} = 26.18$	$\hat{F}^{(R)} = 3.69$
Between columns $V_C = 6$	$c - 1 = 4$	$\dfrac{V_C}{c-1} = 1.5$	$\hat{F}^{(C)} = 0.21$
Random error $V_e = 85.2$	$(r-1)(c-1) = 12$	$\dfrac{V_e}{(r-1)(c-1)} = 7.1$	
Total $V_T = 169.75$	$rc - 1 = 19$		

Critical region: $\hat{F}^{(R)}(3,12) \geq 3.49$; test value: $\hat{F}^{(R)} = 3.69$; reject hypothesis that row means are equal.

11.59. Critical region: $\hat{F}^{(C)}(4,12) \geq 3.26$; test value: $\hat{F}^{(R)} = 0.21$; do not reject hypothesis that column means are equal.

11.60. Problem 11.58: $0.025 < P\text{-value} < 0.05$ (P-value $= 0.0432$); Problem 11.59: P-value > 0.1 (P-value $= 0.9279$).

Appendix

Table A-1 Standard normal distribution

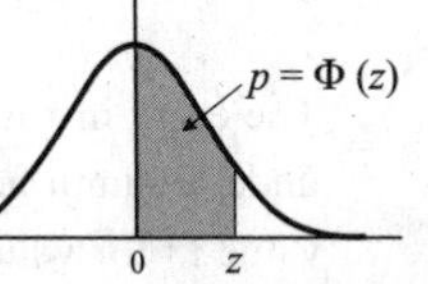

The table entries are the probabilities p for which $P(0 \leq Z \leq z)$, where z ranges from 0.00 to 3.99.

z	0	1	2	3	4	5	6	7	8	9
0.0	.0000	.0040	.0080	.0120	.0160	.0199	.0239	.0279	.0319	.0369
0.1	.0398	.0438	.0478	.0517	.0557	.0596	.0636	.0675	.0714	.0764
0.2	.0793	.0832	.0871	.0910	.0948	.0987	.1026	.1064	.1103	.1141
0.3	.1179	.1217	.1255	.1293	.1331	.1368	.1406	.1443	.1480	.1517
0.4	.1554	.1591	.1628	.1664	.1700	.1736	.1772	.1808	.1844	.1879
0.5	.1915	.1950	.1985	.2019	.2054	.2088	.2123	.2157	.2190	.2224
0.6	.2258	.2291	.2324	.2357	.2389	.2422	.2454	.2486	.2518	.2549
0.7	.2580	.2612	.2642	.2673	.2704	.2734	.2764	.2794	.2823	.2852
0.8	.2881	.2910	.2939	.2967	.2996	.3023	.3051	.3078	.3106	.3133
0.9	.3159	.3186	.3212	.3238	.3264	.3289	.3315	.3340	.3365	.3389
1.0	.3413	.3438	.3461	.3485	.3508	.3531	.3554	.3577	.3599	.3621
1.1	.3643	.3665	.3686	.3708	.3729	.3749	.3770	.3790	.3810	.3830
1.2	.3849	.3869	.3888	.3907	.3925	.3944	.3962	.3980	.3997	.4015
1.3	.4032	.4049	.4066	.4082	.4099	.4115	.4131	.4147	.4162	.4177
1.4	.4192	.4207	.4222	.4236	.4251	.4265	.4279	.4292	.4306	.4319
1.5	.4332	.4345	.4357	.4370	.4382	.4394	.4406	.4418	.4429	.4441
1.6	.4452	.4463	.4474	.4484	.4495	.4505	.4515	.4525	.4535	.4545
1.7	.4554	.4564	.4573	.4582	.4591	.4599	.4608	.4616	.4625	.4633
1.8	.4641	.4649	.4656	.4664	.4671	.4678	.4686	.4693	.4699	.4706
1.9	.4713	.4719	.4726	.4732	.4738	.4744	.4750	.4756	.4761	.4767
2.0	.4772	.4778	.4783	.4788	.4793	.4798	.4803	.4808	.4812	.4817
2.1	.4821	.4826	.4830	.4834	.4838	.4842	.4846	.4850	.4854	.4857
2.2	.4861	.4864	.4868	.4871	.4875	.4878	.4881	.4884	.4887	.4890
2.3	.4893	.4896	.4898	.4901	.4904	.4906	.4909	.4911	.4913	.4916
2.4	.4918	.4920	.4922	.4925	.4927	.4929	.4931	.4932	.4934	.4936
2.5	.4938	.4940	.4941	.4943	.4945	.4946	.4948	.4949	.4951	.4952
2.6	.4953	.4955	.4956	.4957	.4959	.4960	.4961	.4962	.4963	.4964
2.7	.4965	.4966	.4967	.4968	.4969	.4970	.4971	.4972	.4973	.4974
2.8	.4974	.4975	.4976	.4977	.4977	.4978	.4979	.4979	.4980	.4981
2.9	.4981	.4982	.4982	.4983	.4984	.4984	.4985	.4985	.4986	.4986
3.0	.4987	.4987	.4987	.4988	.4988	.4989	.4989	.4989	.4990	.4990
3.1	.4990	.4991	.4991	.4991	.4992	.4992	.4992	.4992	.4993	.4993
3.2	.4993	.4993	.4994	.4994	.4994	.4994	.4994	.4995	.4995	.4995
3.3	.4995	.4995	.4995	.4996	.4996	.4996	.4996	.4996	.4996	.4997
3.4	.4997	.4997	.4997	.4997	.4997	.4997	.4997	.4997	.4997	.4998
3.5	.4998	.4998	.4998	.4998	.4998	.4998	.4998	.4998	.4998	.4998
3.6	.4998	.4998	.4999	.4999	.4999	.4999	.4999	.4999	.4999	.4999
3.7	.4999	.4999	.4999	.4999	.4999	.4999	.4999	.4999	.4999	.4999
3.8	.4999	.4999	.4999	.4999	.4999	.4999	.4999	.4999	.4999	.4999
3.9	.5000	.5000	.5000	.5000	.5000	.5000	.5000	.5000	.5000	.5000

Table A-2 The t distribution

The entry in row k (degrees of freedom) under column heading p (probability) is the value t^* for which $P(0 \leq t \leq t^*) = p$.

k \ p	0.05	0.1	0.2	0.25	0.3	0.4	0.45	0.475	0.49	0.495
1	.158	.325	.727	1.000	1.376	3.08	6.31	12.71	31.82	63.66
2	.142	.289	.617	.816	1.061	1.89	2.92	4.30	6.96	9.92
3	.137	.277	.584	.765	.978	1.64	2.35	3.18	4.54	5.84
4	.134	.271	.569	.741	.941	1.53	2.13	2.78	3.75	4.60
5	.132	.267	.559	.727	.920	1.48	2.02	2.57	3.36	4.03
6	.131	.265	.553	.718	.906	1.44	1.94	2.45	3.14	3.71
7	.130	.263	.549	.711	.896	1.42	1.90	2.36	3.00	3.50
8	.130	.262	.546	.706	.889	1.40	1.86	2.31	2.90	3.36
9	.129	.261	.543	.703	.883	1.38	1.83	2.26	2.82	3.25
10	.129	.260	.542	.700	.879	1.37	1.81	2.23	2.76	3.17
11	.129	.260	.540	.697	.876	1.36	1.80	2.20	2.72	3.11
12	.128	.259	.539	.695	.873	1.36	1.78	2.18	2.68	3.06
13	.128	.259	.538	.694	.870	1.35	1.77	2.16	2.65	3.01
14	.128	.258	.537	.692	.868	1.34	1.76	2.14	2.62	2.98
15	.128	.258	.536	.691	.866	1.34	1.75	2.13	2.60	2.95
16	.128	.258	.535	.690	.865	1.34	1.75	2.12	2.58	2.92
17	.128	.257	.534	.689	.863	1.33	1.74	2.11	2.57	2.90
18	.127	.257	.534	.688	.862	1.33	1.73	2.10	2.55	2.88
19	.127	.257	.533	.688	.861	1.33	1.73	2.09	2.54	2.86
20	.127	.257	.533	.687	.860	1.32	1.72	2.09	2.53	2.84
21	.127	.257	.532	.686	.859	1.32	1.72	2.08	2.52	2.83
22	.127	.256	.532	.686	.858	1.32	1.72	2.07	2.51	2.82
23	.127	.256	.532	.685	.858	1.32	1.71	2.07	2.50	2.81
24	.127	.256	.531	.685	.857	1.32	1.71	2.06	2.49	2.80
25	.127	.256	.531	.684	.856	1.32	1.71	2.06	2.48	2.79
26	.127	.256	.531	.684	.856	1.32	1.71	2.06	2.48	2.78
27	.127	.256	.531	.684	.855	1.31	1.70	2.05	2.47	2.77
28	.127	.256	.530	.683	.855	1.31	1.70	2.05	2.47	2.76
29	.127	.256	.530	.683	.854	1.31	1.70	2.04	2.46	2.76
30	.127	.256	.530	.683	.854	1.31	1.70	2.04	2.46	2.75
40	.126	.255	.529	.681	.851	1.30	1.68	2.02	2.42	2.70
60	.126	.254	.527	.679	.848	1.30	1.67	2.00	2.39	2.66
120	.126	.254	.526	.677	.845	1.29	1.66	1.98	2.36	2.62
∞	.126	.253	.524	.674	.842	1.28	1.645	1.96	2.33	2.58

Source: R. A. Fisher and F. Yates, *Statistical Tables for Biological, Agricultural and Medical Research*, published by Longman Group Ltd., London (previously published by Oliver and Boyd, Edinburgh), and by permission of the authors and publishers.

Table A-3 The chi-square distribution

The entry in row k (degrees of freedom) under column heading p (probability) is the value χ^* for which $P(0 \le \chi^2 \le \chi^*) = p$.

k \ p	0.005	0.01	0.025	0.05	0.10	0.25	0.50	0.75	0.90	0.95	0.975	0.99	0.995	0.999
1	.0000	.0002	.0010	.0039	.0158	.102	.455	1.32	2.71	3.84	5.02	6.63	7.88	10.8
2	.0100	.0201	.0506	.103	.211	.575	1.39	2.77	4.61	5.99	7.38	9.21	10.6	13.8
3	.0717	.115	.216	.352	.584	1.21	2.37	4.11	6.25	7.81	9.35	11.3	12.8	16.3
4	.207	.297	.484	.711	1.06	1.92	3.36	5.39	7.78	9.49	11.1	13.3	14.9	18.5
5	.412	.554	.831	1.15	1.61	2.67	4.35	6.63	9.24	11.1	12.8	15.1	16.7	20.5
6	.676	.872	1.24	1.64	2.20	3.45	5.35	7.84	10.6	12.6	14.4	16.8	18.5	22.5
7	.989	1.24	1.69	2.17	2.83	4.25	6.35	9.04	12.0	14.1	16.0	18.5	20.3	24.3
8	1.34	1.65	2.18	2.73	3.49	5.07	7.34	10.2	13.4	15.5	17.5	20.1	22.0	26.1
9	1.73	2.09	2.70	3.33	4.17	5.90	8.34	11.4	14.7	16.9	19.0	21.7	23.6	27.9
10	2.16	2.56	3.25	3.94	4.87	6.74	9.34	12.5	16.0	18.3	20.5	23.2	25.2	29.6
11	2.60	3.05	3.82	4.57	5.58	7.58	10.3	13.7	17.3	19.7	21.9	24.7	26.8	31.3
12	3.07	3.57	4.40	5.23	6.30	8.44	11.3	14.8	18.5	21.0	23.3	26.2	28.3	32.9
13	3.57	4.11	5.01	5.89	7.04	9.30	12.3	16.0	19.8	22.4	24.7	27.7	29.8	34.5
14	4.07	4.66	5.63	6.57	7.79	10.2	13.3	17.1	21.1	23.7	26.1	29.1	31.3	36.1
15	4.60	5.23	6.26	7.26	8.55	11.0	14.3	18.2	22.3	25.0	27.5	30.6	32.8	37.7
16	5.14	5.81	6.91	7.96	9.31	11.9	15.3	19.4	23.5	26.3	28.8	32.0	34.3	39.3
17	5.70	6.41	7.56	8.67	10.1	12.8	16.3	20.5	24.8	27.6	30.2	33.4	35.7	40.8
18	6.26	7.01	8.23	9.39	10.9	13.7	17.3	21.6	26.0	28.9	31.5	34.8	37.2	42.3
19	6.84	7.63	8.91	10.1	11.7	14.6	18.3	22.7	27.2	30.1	32.9	36.2	38.6	43.8
20	7.43	8.26	9.59	10.9	12.4	15.5	19.3	23.8	28.4	31.4	34.2	37.6	40.0	45.3
21	8.03	8.90	10.3	11.6	13.2	16.3	20.3	24.9	29.6	32.7	35.5	38.9	41.4	46.8
22	8.64	9.54	11.0	12.3	14.0	17.2	21.3	26.0	30.8	33.9	36.8	40.3	42.8	48.3
23	9.26	10.2	11.7	13.1	14.8	18.1	22.3	27.1	32.0	35.2	38.1	41.6	44.2	49.7
24	9.89	10.9	12.4	13.8	15.7	19.0	23.3	28.2	33.2	36.4	39.4	43.0	45.6	51.2
25	10.5	11.5	13.1	14.6	16.5	19.9	24.3	29.3	34.4	37.7	40.6	44.3	46.9	52.6
26	11.2	12.2	13.8	15.4	17.3	20.8	25.3	30.4	35.6	38.9	41.9	45.6	48.3	54.1
27	11.8	12.9	14.6	16.2	18.1	21.7	26.3	31.5	36.7	40.1	43.2	47.0	49.6	55.5
28	12.5	13.6	15.3	16.9	18.9	22.7	27.3	32.6	37.9	41.3	44.5	48.3	51.0	56.9
29	13.1	14.3	16.0	17.7	19.8	23.6	28.3	33.7	39.1	42.6	45.7	49.6	52.3	58.3
30	13.8	15.0	16.8	18.5	20.6	24.5	29.3	34.8	40.3	43.8	47.0	50.9	53.7	59.7
40	20.7	22.2	24.4	26.5	29.1	33.7	39.3	45.6	51.8	55.8	59.3	63.7	66.8	73.4
50	28.0	29.7	32.4	34.8	37.7	42.9	49.3	56.3	63.2	67.5	71.4	76.2	79.5	86.7
60	35.5	37.5	40.5	43.2	46.5	52.3	59.3	67.0	74.4	79.1	83.3	88.4	92.0	99.6
70	43.3	45.4	48.8	51.7	55.3	61.7	69.3	77.6	85.5	90.5	95.0	100	104	112
80	51.2	53.5	57.2	60.4	64.3	71.1	79.3	88.1	96.6	102	107	112	116	125
90	59.2	61.8	65.6	69.1	73.3	80.6	89.3	98.6	108	113	118	124	128	137
100	67.3	70.1	74.2	77.9	82.4	90.1	99.3	109	118	124	130	136	140	149

Source: E. S. Pearson and H. O. Hartley, *Biometrika Tables for Statisticians*, Vol. 1 (1966), Table 8, pages 137 and 138, by permission.

Table A-4 90th percentiles for the F distribution

The entry in column m, row n is the value F^* for which $P(0 \le F(m, n) \le F^*) = 0.90$.
m = degrees of freedom in numerator; n = degrees of freedom in denominator

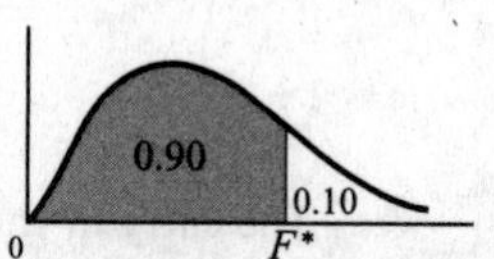

n \ m	1	2	3	4	5	6	7	8	9	10	12	15	20	24	25	30	40	60	120	∞
1	39.9	49.5	53.6	55.8	57.2	58.2	58.9	59.4	59.9	60.2	60.7	61.2	61.7	62.0	62.1	62.3	62.5	62.8	63.1	63.3
2	8.53	9.00	9.16	9.24	9.29	9.33	9.35	9.37	9.38	9.39	9.41	9.42	9.44	9.45	9.45	9.46	9.47	9.47	9.48	9.49
3	5.54	5.46	5.39	5.34	5.31	5.28	5.27	5.25	5.24	5.23	5.22	5.20	5.18	5.18	5.17	5.17	5.16	5.15	5.14	5.13
4	4.54	4.32	4.19	4.11	4.05	4.01	3.98	3.95	3.94	3.92	3.90	3.87	3.84	3.83	3.83	3.82	3.80	3.79	3.78	3.76
5	4.06	3.78	3.62	3.52	3.45	3.40	3.37	3.34	3.32	3.30	3.27	3.24	3.21	3.19	3.19	3.17	3.16	3.14	3.12	3.11
6	3.78	3.46	3.29	3.18	3.11	3.05	3.01	2.98	2.96	2.94	2.90	2.87	2.84	2.82	2.81	2.80	2.78	2.76	2.74	2.72
7	3.59	3.26	3.07	2.96	2.88	2.83	2.78	2.75	2.72	2.70	2.67	2.63	2.59	2.58	2.57	2.56	2.54	2.51	2.49	2.47
8	3.46	3.11	2.92	2.81	2.73	2.67	2.62	2.59	2.56	2.54	2.50	2.46	2.42	2.40	2.40	2.38	2.36	2.34	2.32	2.29
9	3.36	3.01	2.81	2.69	2.61	2.55	2.51	2.47	2.44	2.42	2.38	2.34	2.30	2.28	2.27	2.25	2.23	2.21	2.18	2.16
10	3.29	2.92	2.73	2.61	2.52	2.46	2.41	2.38	2.35	2.32	2.28	2.24	2.20	2.18	2.17	2.16	2.13	2.11	2.08	2.06
11	3.23	2.86	2.66	2.54	2.45	2.39	2.34	2.30	2.27	2.25	2.21	2.17	2.12	2.10	2.10	2.08	2.05	2.03	2.00	1.97
12	3.18	2.81	2.61	2.48	2.39	2.33	2.28	2.24	2.21	2.19	2.15	2.10	2.06	2.04	2.03	2.01	1.99	1.96	1.93	1.90
13	3.14	2.76	2.56	2.43	2.35	2.28	2.23	2.20	2.16	2.14	2.10	2.05	2.01	1.98	1.98	1.96	1.93	1.90	1.88	1.85
14	3.10	2.73	2.52	2.39	2.31	2.24	2.19	2.15	2.12	2.10	2.05	2.01	1.96	1.94	1.93	1.91	1.89	1.86	1.83	1.80
15	3.07	2.70	2.49	2.36	2.27	2.21	2.16	2.12	2.09	2.06	2.02	1.97	1.92	1.90	1.89	1.87	1.85	1.82	1.79	1.76
16	3.05	2.67	2.46	2.33	2.24	2.18	2.13	2.09	2.06	2.03	1.99	1.94	1.89	1.87	1.86	1.84	1.81	1.78	1.75	1.72
17	3.03	2.64	2.44	2.31	2.22	2.15	2.10	2.06	2.03	2.00	1.96	1.91	1.86	1.84	1.83	1.81	1.78	1.75	1.72	1.69
18	3.01	2.62	2.42	2.29	2.20	2.13	2.08	2.04	2.00	1.98	1.93	1.89	1.84	1.81	1.80	1.78	1.75	1.72	1.69	1.66
19	2.99	2.61	2.40	2.27	2.18	2.11	2.06	2.02	1.98	1.96	1.91	1.86	1.81	1.79	1.78	1.76	1.73	1.70	1.67	1.63
20	2.97	2.59	2.38	2.25	2.16	2.09	2.04	2.00	1.96	1.94	1.89	1.84	1.79	1.77	1.76	1.74	1.71	1.68	1.64	1.61
21	2.96	2.57	2.36	2.23	2.14	2.08	2.02	1.98	1.95	1.92	1.87	1.83	1.78	1.75	1.74	1.72	1.69	1.66	1.62	1.59
22	2.95	2.56	2.35	2.22	2.13	2.06	2.01	1.97	1.93	1.90	1.86	1.81	1.76	1.73	1.73	1.70	1.67	1.64	1.60	1.57
23	2.94	2.55	2.34	2.21	2.11	2.05	1.99	1.95	1.92	1.89	1.84	1.80	1.74	1.72	1.71	1.69	1.66	1.62	1.59	1.55
24	2.93	2.54	2.33	2.19	2.10	2.04	1.98	1.94	1.91	1.88	1.83	1.78	1.73	1.70	1.70	1.67	1.64	1.61	1.57	1.53
25	2.92	2.53	2.32	2.18	2.09	2.02	1.97	1.93	1.89	1.87	1.82	1.77	1.72	1.69	1.68	1.66	1.63	1.59	1.56	1.52
26	2.91	2.52	2.31	2.17	2.08	2.01	1.96	1.92	1.88	1.86	1.81	1.76	1.71	1.68	1.67	1.65	1.61	1.58	1.54	1.50
27	2.90	2.51	2.30	2.17	2.07	2.00	1.95	1.91	1.87	1.85	1.80	1.75	1.70	1.67	1.66	1.64	1.60	1.57	1.53	1.49
28	2.89	2.50	2.29	2.16	2.06	2.00	1.94	1.90	1.87	1.84	1.79	1.74	1.69	1.66	1.65	1.63	1.59	1.56	1.52	1.48
29	2.89	2.50	2.28	2.15	2.06	1.99	1.93	1.89	1.86	1.83	1.78	1.73	1.68	1.65	1.64	1.62	1.58	1.55	1.51	1.47
30	2.88	2.49	2.28	2.14	2.05	1.98	1.93	1.88	1.85	1.82	1.77	1.72	1.67	1.64	1.63	1.61	1.57	1.54	1.50	1.46
40	2.84	2.44	2.23	2.09	2.00	1.93	1.87	1.83	1.79	1.76	1.71	1.66	1.61	1.57	1.57	1.54	1.51	1.47	1.42	1.38
60	2.79	2.39	2.18	2.04	1.95	1.87	1.82	1.77	1.74	1.71	1.66	1.60	1.54	1.51	1.50	1.48	1.44	1.40	1.35	1.29
120	2.75	2.35	2.13	1.99	1.90	1.82	1.77	1.72	1.68	1.65	1.60	1.54	1.48	1.45	1.44	1.41	1.37	1.32	1.26	1.19
∞	2.71	2.30	2.08	1.94	1.85	1.77	1.72	1.67	1.63	1.60	1.55	1.49	1.42	1.38	1.38	1.34	1.30	1.24	1.17	1.00

Table A-5 95th percentiles for the F distribution

The entry in column m, row n is the value F^* for which $P(0 \leq F(m,n) \leq F^*) = 0.95$.
$m =$ degrees of freedom in numerator
$n =$ degrees of freedom in denominator

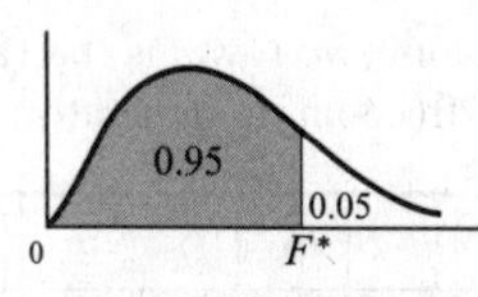

n \ m	1	2	3	4	5	6	7	8	9	10	12	15	20	24	25	30	40	60	120	∞
1	161	200	216	225	230	234	237	239	241	242	244	246	248	249	249	250	251	252	253	254
2	18.5	19.0	19.2	19.2	19.3	19.3	19.4	19.4	19.4	19.4	19.4	19.4	19.4	19.5	19.5	19.5	19.5	19.5	19.5	19.5
3	10.1	9.55	9.28	9.12	9.01	8.94	8.89	8.85	8.81	8.79	8.74	8.70	8.66	8.64	8.63	8.62	8.59	8.57	8.55	8.53
4	7.71	6.94	6.59	6.39	6.26	6.16	6.09	6.04	6.00	5.96	5.91	5.86	5.80	5.77	5.77	5.75	5.72	5.69	5.66	5.63
5	6.61	5.79	5.41	5.19	5.05	4.95	4.88	4.82	4.77	4.74	4.68	4.62	4.56	4.53	4.52	4.50	4.46	4.43	4.40	4.37
6	5.99	5.14	4.76	4.53	4.39	4.28	4.21	4.15	4.10	4.06	4.00	3.94	3.87	3.84	3.83	3.81	3.77	3.74	3.70	3.67
7	5.59	4.74	4.35	4.12	3.97	3.87	3.79	3.73	3.68	3.64	3.57	5.51	3.44	3.41	3.40	3.38	3.34	3.30	3.27	3.23
8	5.32	4.46	4.07	3.84	3.69	3.58	3.50	3.44	3.39	3.35	3.28	3.22	3.15	3.12	3.11	3.08	3.04	3.01	2.97	2.93
9	5.12	4.26	3.86	3.63	3.48	3.37	3.29	3.23	3.18	3.14	3.07	3.01	2.94	2.90	2.89	2.86	2.83	2.79	2.75	2.71
10	4.96	4.10	3.71	3.48	3.33	3.22	3.14	3.07	3.02	2.98	2.91	2.85	2.77	2.74	2.73	2.70	2.66	2.62	2.58	2.54
11	4.84	3.98	3.59	3.36	3.20	3.09	3.01	2.95	2.90	2.85	2.79	2.72	2.65	2.61	2.60	2.57	2.53	2.49	2.45	2.40
12	4.75	3.89	3.49	3.26	3.11	3.00	2.91	2.85	2.80	2.75	2.69	2.62	2.54	2.51	2.50	2.47	2.43	2.38	2.34	2.30
13	4.67	3.81	3.41	3.18	3.03	2.92	2.83	2.77	2.71	2.67	2.60	2.53	2.46	2.42	2.41	2.38	2.34	2.30	2.25	2.21
14	4.60	3.74	3.34	3.11	2.96	2.85	2.76	2.70	2.65	2.60	2.53	2.46	2.39	2.35	2.34	2.31	2.27	2.22	2.18	2.13
15	4.54	3.68	3.29	3.06	2.90	2.79	2.71	2.64	2.59	2.54	2.48	2.40	2.33	2.29	2.28	2.25	2.20	2.16	2.11	2.07
16	4.49	3.63	3.24	3.01	2.85	2.74	2.66	2.59	2.54	2.49	2.42	2.35	2.28	2.24	2.23	2.19	2.15	2.11	2.06	2.01
17	4.45	3.59	3.20	2.96	2.81	2.70	2.61	2.55	2.49	2.45	2.38	2.31	2.23	2.19	2.18	2.15	2.10	2.06	2.01	1.96
18	4.41	3.55	3.16	2.93	2.77	2.66	2.58	2.51	2.46	2.41	2.34	2.27	2.19	2.15	2.14	2.11	2.06	2.02	1.97	1.92
19	4.38	3.52	3.13	2.90	2.74	2.63	2.54	2.48	2.42	2.38	2.31	2.23	2.16	2.11	2.11	2.07	2.03	1.98	1.93	1.88
20	4.35	3.49	3.10	2.87	2.71	2.60	2.51	2.45	2.39	2.35	2.28	2.20	2.12	2.08	2.07	2.04	1.99	1.95	1.90	1.84
21	4.32	3.47	3.07	2.84	2.68	2.57	2.49	2.42	2.37	2.32	2.25	2.18	2.10	2.05	2.05	2.01	1.96	1.92	1.87	1.81
22	4.30	3.44	3.05	2.82	2.66	2.55	2.46	2.40	2.34	2.30	2.23	2.15	2.07	2.03	2.02	1.98	1.94	1.89	1.84	1.78
23	4.28	3.42	3.03	2.80	2.64	2.53	2.44	2.37	2.32	2.27	2.20	2.13	2.05	2.01	2.00	1.96	1.91	1.86	1.81	1.76
24	4.26	3.40	3.01	2.78	2.62	2.51	2.42	2.36	2.30	2.25	2.18	2.11	2.03	1.98	1.97	1.94	1.89	1.84	1.79	1.73
25	4.24	3.39	2.99	2.76	2.60	2.49	2.40	2.34	2.28	2.24	2.16	2.09	2.01	1.96	1.96	1.92	1.87	1.82	1.77	1.71
26	4.23	3.37	2.98	2.74	2.59	2.47	2.39	2.32	2.27	2.22	2.15	2.07	1.99	1.95	1.94	1.90	1.85	1.80	1.75	1.69
27	4.21	3.35	2.96	2.73	2.57	2.46	2.37	2.31	2.25	2.20	2.13	2.06	1.97	1.93	1.92	1.88	1.84	1.79	1.73	1.67
28	4.20	3.34	2.95	2.71	2.56	2.45	2.36	2.29	2.24	2.19	2.12	2.04	1.96	1.91	1.91	1.87	1.82	1.77	1.71	1.65
29	4.18	3.33	2.93	2.70	2.55	2.43	2.35	2.28	2.22	2.18	2.10	2.03	1.94	1.90	1.89	1.85	1.81	1.75	1.70	1.64
30	4.17	3.32	2.92	2.69	2.53	2.42	2.33	2.27	2.21	2.16	2.09	2.01	1.93	1.89	1.88	1.84	1.79	1.74	1.68	1.62
40	4.08	3.23	2.84	2.61	2.45	2.34	2.25	2.18	2.12	2.08	2.00	1.92	1.84	1.79	1.78	1.74	1.69	1.64	1.58	1.51
60	4.00	3.15	2.76	2.53	2.37	2.25	2.17	2.10	2.04	1.99	1.92	1.84	1.75	1.70	1.69	1.65	1.59	1.53	1.47	1.39
120	3.92	3.07	2.68	2.45	2.29	2.18	2.09	2.20	1.96	1.91	1.83	1.75	1.66	1.61	1.60	1.55	1.50	1.43	1.35	1.25
∞	3.84	3.00	2.60	2.37	2.21	2.10	2.01	1.94	1.88	1.83	1.75	1.67	1.57	1.52	1.51	1.46	1.39	1.32	1.22	1.00

Source: E. S. Pearson and H. O. Hartley, *Biometrika Tables for Statisticians*, Vol. 2 (1972), Table 5, page 178, by permission.

Table A-6 97.5 percentiles for the F distribution

The entry in column m, row n is the value F^* for which $P(0 \le F(m,n) \le F^*) = 0.975$.
$m =$ degrees of freedom in numerator; $n =$ degrees of freedom in denominator

n \ m	1	2	3	4	5	6	7	8	9	10	12	15	20	24	25	30	40	60	120	∞
1	647.8	799.5	864.2	899.6	921.9	937.1	948.2	956.7	963.3	968.6	976.7	984.9	993.1	997.3	998.1	1001	1006	1010	1014	1018
2	38.5	39.0	39.2	39.2	39.3	39.3	39.4	39.4	39.4	39.4	39.4	39.4	39.4	39.5	39.5	39.5	39.5	39.5	39.5	39.5
3	17.4	16.0	15.4	15.1	14.9	14.7	14.6	14.5	14.5	14.4	14.3	14.3	14.2	14.1	14.1	14.1	14.0	14.0	13.9	13.9
4	12.2	10.6	9.98	9.60	9.36	9.20	9.07	8.98	8.90	8.84	8.75	8.66	8.56	8.51	8.50	8.46	8.41	8.36	8.31	8.26
5	10.0	8.43	7.76	7.39	7.15	6.98	6.85	6.76	6.68	6.62	6.52	6.43	6.33	6.28	6.27	6.23	6.18	6.12	6.07	6.02
6	8.81	7.26	6.60	6.23	5.99	5.82	5.70	5.60	5.52	5.46	5.37	5.27	5.17	5.12	5.11	5.07	5.01	4.96	4.90	4.85
7	8.07	6.54	5.89	5.52	5.29	5.12	4.99	4.90	4.82	4.76	4.67	4.57	4.47	4.42	4.40	4.36	4.31	4.25	4.20	4.14
8	7.57	6.06	5.42	5.05	4.82	4.65	4.53	4.43	4.36	4.30	4.20	4.10	4.00	3.95	3.94	3.89	3.84	3.78	3.73	3.67
9	7.21	5.71	5.08	4.72	4.48	4.32	4.20	4.10	4.03	3.96	3.87	3.77	3.67	3.61	3.60	3.56	3.51	3.45	3.39	3.33
10	6.94	5.46	4.83	4.47	4.24	4.07	3.95	3.85	3.78	3.72	3.62	3.52	3.42	3.37	3.35	3.31	3.26	3.20	3.14	3.08
11	6.72	5.26	4.63	4.28	4.04	3.88	3.76	3.66	3.59	3.53	3.43	3.33	3.23	3.17	3.16	3.12	3.06	3.00	2.94	2.88
12	6.55	5.10	4.47	4.12	3.89	3.73	3.61	3.51	3.44	3.37	3.28	3.18	3.07	3.02	3.01	2.96	2.91	2.85	2.79	2.72
13	6.41	4.97	4.35	4.00	3.77	3.60	3.48	3.39	3.31	3.25	3.15	3.05	2.95	2.89	2.88	2.84	2.78	2.72	2.66	2.60
14	6.30	4.86	4.24	3.89	3.66	3.50	3.38	3.29	3.21	3.15	3.05	2.95	2.84	2.79	2.78	2.73	2.67	2.61	2.55	2.49
15	6.20	4.77	4.15	3.80	3.58	3.41	3.29	3.20	3.12	3.06	2.96	2.86	2.76	2.70	2.69	2.64	2.59	2.52	2.46	2.40
16	6.12	4.69	4.08	3.73	3.50	3.34	3.22	3.12	3.05	2.99	2.89	2.79	2.68	2.63	2.61	2.57	2.51	2.45	2.38	2.32
17	6.04	4.62	4.01	3.66	3.44	3.28	3.16	3.06	2.98	2.92	2.82	2.72	2.62	2.56	2.55	2.50	2.44	2.38	2.32	2.25
18	5.98	4.56	3.95	3.61	3.38	3.32	3.10	3.01	2.93	2.87	2.77	2.67	2.56	2.50	2.49	2.44	2.38	2.32	2.26	2.19
19	5.92	4.51	3.90	3.56	3.33	3.17	3.05	2.96	2.88	2.82	2.72	2.62	2.51	2.45	2.44	2.39	2.33	2.27	2.20	2.14
20	5.87	4.46	3.86	3.51	3.29	3.13	3.01	2.91	2.84	2.77	2.68	2.57	2.46	2.41	2.40	2.35	2.29	2.22	2.16	2.09
21	5.83	4.42	3.82	3.48	3.25	3.09	2.97	2.87	2.80	2.73	2.64	2.53	2.42	2.37	2.36	2.31	2.25	2.18	2.11	2.04
22	5.79	4.38	3.78	3.44	3.22	3.05	2.93	2.84	2.76	2.70	2.60	2.50	2.39	2.33	2.32	2.27	2.21	2.14	2.08	2.00
23	5.75	4.35	3.75	3.41	3.18	3.02	2.90	2.81	2.73	2.67	2.57	2.47	2.36	2.30	2.29	2.24	2.18	2.11	2.04	1.97
24	5.72	4.32	3.72	3.38	3.15	2.99	2.87	2.78	2.70	2.64	2.54	2.44	2.33	2.27	2.26	2.21	2.15	2.08	2.01	1.94
25	5.69	4.29	3.69	3.35	3.13	2.97	2.85	2.75	2.68	2.61	2.51	2.41	2.30	2.24	2.23	2.18	2.12	2.05	1.98	1.91
26	5.66	4.27	3.67	3.33	3.10	2.94	2.82	2.73	2.65	2.59	2.49	2.39	2.28	2.22	2.21	2.16	2.09	2.03	1.95	1.88
27	5.63	4.24	3.65	3.31	3.08	2.92	2.80	2.71	2.63	2.57	2.47	2.36	2.25	2.19	2.18	2.13	2.07	2.00	1.93	1.85
28	5.61	4.22	3.63	3.29	3.06	2.90	2.78	2.69	2.61	2.55	2.45	2.34	2.23	2.17	2.16	2.11	2.05	1.98	1.91	1.83
29	5.59	4.20	3.61	3.27	3.04	2.88	2.76	2.67	2.59	2.53	2.43	2.32	2.21	2.15	2.14	2.09	2.03	1.96	1.89	1.81
30	5.57	4.18	3.59	3.25	3.03	2.87	2.75	2.65	2.57	2.51	2.41	2.31	2.20	2.14	2.12	2.07	2.01	1.94	1.87	1.79
40	5.42	4.05	3.46	3.13	2.90	2.74	2.62	2.53	2.45	2.39	2.29	2.18	2.07	2.01	1.99	1.94	1.88	1.80	1.72	1.64
60	5.29	3.93	3.34	3.01	2.79	2.63	2.51	2.41	2.33	2.27	2.17	2.06	1.94	1.88	1.87	1.82	1.74	1.67	1.58	1.48
120	5.15	3.80	3.23	2.89	2.67	2.52	2.39	2.30	2.22	2.16	2.05	1.95	1.82	1.76	1.75	1.69	1.61	1.53	1.43	1.31
∞	5.02	3.69	3.12	2.79	2.57	2.41	2.29	2.19	2.11	2.05	1.94	1.83	1.71	1.64	1.63	1.57	1.48	1.39	1.27	1.00

Table A-7 99th percentiles for the F distribution

The entry in column m, row n is the value F^* for which $P(0 \le F(m,n) \le F^*) = 0.99$.
m = degrees of freedom in numerator
n = degrees of freedom in denominator

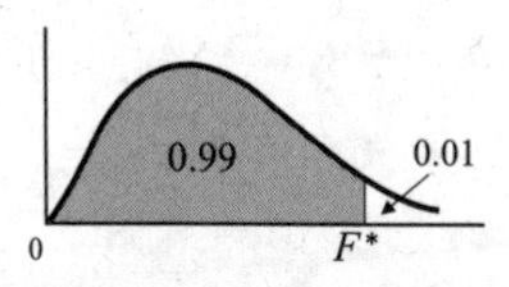

n \ m	1	2	3	4	5	6	7	8	9	10	12	15	20	24	25	30	40	60	120	∞
1	4052	5000	5403	5625	5764	5859	5928	5981	6023	6056	6106	6157	6209	6235	6240	6261	6287	6313	6339	6366
2	98.5	99.0	99.2	99.2	99.3	99.3	99.4	99.4	99.4	99.4	99.4	99.4	99.4	99.5	99.5	99.5	99.5	99.5	99.5	99.5
3	34.1	30.8	29.5	28.7	28.2	27.9	27.7	27.5	27.3	27.2	27.1	26.9	26.7	26.6	26.6	26.5	26.4	26.3	26.2	26.1
4	21.2	18.0	16.7	16.0	15.5	15.2	15.0	14.8	14.7	14.5	14.4	14.2	14.0	13.9	13.9	13.8	13.7	13.7	13.6	13.5
5	16.3	13.3	12.1	11.4	11.0	10.7	10.5	10.3	10.2	10.1	9.89	9.72	9.55	9.47	9.45	9.38	9.29	9.20	9.11	9.02
6	13.7	10.9	9.78	9.15	8.75	8.47	8.26	8.10	7.98	7.87	7.72	7.56	7.40	7.31	7.30	7.23	7.14	7.06	6.97	6.88
7	12.2	9.55	8.45	7.85	7.46	7.19	6.99	6.84	6.72	6.62	6.47	6.31	6.16	6.07	6.06	5.99	5.91	5.82	5.74	5.65
8	11.3	8.65	7.59	7.01	6.63	6.37	6.18	6.03	5.91	5.81	5.67	5.52	5.36	5.28	5.26	5.20	5.12	5.03	4.95	4.86
9	10.6	8.02	6.99	6.42	6.06	5.80	5.61	5.47	5.35	5.26	5.11	4.96	4.81	4.73	4.71	4.65	4.57	4.48	4.40	4.31
10	10.0	7.56	6.55	5.99	5.64	5.39	5.20	5.06	4.94	4.85	4.71	4.56	4.41	4.33	4.31	4.25	4.17	4.08	4.00	3.91
11	9.65	7.21	6.22	5.67	5.32	5.07	4.89	4.74	4.63	4.54	4.40	4.25	4.10	4.02	4.01	3.94	3.86	3.78	3.69	3.60
12	9.33	6.93	5.95	5.41	5.06	4.82	4.64	4.50	4.39	4.30	4.16	4.01	3.86	3.78	3.76	3.70	3.62	3.54	3.45	3.36
13	9.07	6.70	5.74	5.21	4.86	4.62	4.44	4.30	4.19	4.10	3.96	3.82	3.66	3.59	3.57	3.51	3.43	3.34	3.25	3.17
14	8.86	6.51	5.56	5.04	4.70	4.46	4.28	4.14	4.03	3.94	3.80	3.66	3.51	3.43	3.41	3.35	3.27	3.18	3.09	3.00
15	8.68	6.36	5.42	4.89	4.56	4.32	4.14	4.00	3.89	3.80	3.67	3.52	3.37	3.29	3.28	3.21	3.13	3.05	2.96	2.87
16	8.53	6.23	5.29	4.77	4.44	4.20	4.03	3.89	3.78	3.69	3.55	3.41	3.26	3.18	3.16	3.10	3.02	2.93	2.84	2.75
17	8.40	6.11	5.19	4.67	4.34	4.10	3.93	3.79	3.68	3.59	3.46	3.31	3.16	3.08	3.07	3.00	2.92	2.83	2.75	2.65
18	8.29	6.01	5.09	4.58	4.25	4.01	3.84	3.71	3.60	3.51	3.37	3.23	3.08	3.00	2.98	2.92	2.84	2.75	2.66	2.57
19	8.18	5.93	5.01	4.50	4.17	3.94	3.77	3.63	3.52	3.43	3.30	3.15	3.00	2.92	2.91	2.84	2.76	2.67	2.58	2.49
20	8.10	5.85	4.94	4.43	4.10	3.87	3.70	3.56	3.46	3.37	3.23	3.09	2.94	2.86	2.84	2.78	2.69	2.61	2.52	2.42
21	8.02	5.78	4.87	4.37	4.04	3.81	3.64	3.51	3.40	3.31	3.17	3.03	2.88	2.80	2.79	2.72	2.64	2.55	2.46	2.36
22	7.95	5.72	4.82	4.31	3.99	3.76	3.59	3.45	3.35	3.26	3.12	2.98	2.83	2.75	2.73	2.67	2.58	2.50	2.40	2.31
23	7.88	5.66	4.76	4.26	3.94	3.71	3.54	3.41	3.30	3.21	3.07	2.93	2.78	2.70	2.69	2.62	2.54	2.45	2.35	2.26
24	7.82	5.61	4.72	4.22	3.90	3.67	3.50	3.36	3.26	3.17	3.03	2.89	2.74	2.66	2.64	2.58	2.49	2.40	2.31	2.21
25	7.77	5.57	4.68	4.18	3.86	3.63	3.46	3.32	3.22	3.13	2.99	2.85	2.70	2.62	2.60	2.54	2.45	2.36	2.27	2.17
26	7.72	5.53	4.64	4.14	3.82	3.59	3.42	3.29	3.18	3.09	2.96	2.82	2.66	2.58	2.57	2.50	2.42	2.33	2.23	2.13
27	7.68	5.49	4.60	4.11	3.78	3.56	3.39	3.26	3.15	3.06	2.93	2.78	2.63	2.55	2.54	2.47	2.38	2.29	2.20	2.10
28	7.64	5.45	4.57	4.07	3.75	3.53	3.36	3.23	3.12	3.03	2.90	2.75	2.60	2.52	2.51	2.44	2.35	2.26	2.17	2.06
29	7.60	5.42	4.54	4.04	3.73	3.50	3.33	3.20	3.09	3.00	2.87	2.73	2.57	2.49	2.48	2.41	2.33	2.23	2.14	2.03
30	7.56	5.39	4.51	4.02	3.70	3.47	3.30	3.17	3.07	2.98	2.84	2.70	2.55	2.47	2.45	2.39	2.30	2.21	2.11	2.01
40	7.31	5.18	4.31	3.83	3.51	3.29	3.12	2.99	2.89	2.80	2.66	2.52	2.37	2.29	2.27	2.20	2.11	2.02	1.92	1.80
60	7.08	4.98	4.13	3.65	3.34	3.12	2.95	2.82	2.72	2.63	2.50	2.35	2.20	2.12	2.10	2.03	1.94	1.84	1.73	1.60
120	6.85	4.79	3.95	3.48	3.17	2.96	2.79	2.66	2.56	2.47	2.34	2.19	2.03	1.95	1.93	1.86	1.76	1.66	1.53	1.38
∞	6.63	4.61	3.78	3.32	3.02	2.80	2.64	2.51	2.41	2.32	2.18	2.04	1.88	1.79	1.78	1.70	1.59	1.47	1.32	1.00

Source: E. S. Pearson and H. O. Hartley, *Biometrika Tables for Statisticians*, Vol. 2 (1972), Table 5, page 180, by permission.

Index